Brief Integral Table (continued)

Forms Containing $a + bx$

9. $\int (a + bx)^n \, dx = \begin{cases} \dfrac{(a + bx)^{n+1}}{(n + 1)b} + C, & n \neq -1 \\ \dfrac{1}{b} \ln|a + bx| + C, & n = -1 \end{cases}$

10. $\int x(a + bx)^n \, dx = \begin{cases} \dfrac{1}{b^2(n + 2)}(a + bx)^{n+2} - \dfrac{a}{b^2(n + 1)}(a + bx)^{n+1} + C, & n \neq -1, -2 \\ \dfrac{1}{b^2}[a + bx - a \ln|a + bx|] + C, & n = -1 \\ \dfrac{1}{b^2}\left[\ln|a + bx| + \dfrac{a}{a + bx}\right] + C, & n = -2 \end{cases}$

11. $\int x^2(a + bx)^n \, dx = \dfrac{1}{b^3}\left[\dfrac{(a + bx)^{n+3}}{n + 3} - 2a\dfrac{(a + bx)^{n+2}}{n + 2} + a^2\dfrac{(a + bx)^{n+1}}{n + 1}\right] + C, \quad n \neq -1, n \neq -2, n \neq -3$

12. $\int x^m(a + bx)^n \, dx, \quad m > 0, m + n + 1 \neq 0$

    In terms of a lower power of n:

    $\dfrac{x^{m+1}(a + bx)^n}{m + n + 1} + \dfrac{an}{m + n + 1}\int x^m(a + bx)^{n-1} \, dx$

    In terms of a lower power of m:

    $\dfrac{1}{b(m + n + 1)}\left[x^m(a + bx)^{n+1} - ma\int x^{m-1}(a + bx)^n \, dx\right]$

13. $\int x^m \sqrt{a + bx} \, dx = \dfrac{2}{b(2m + 3)}\left[x^m\sqrt{(a + bx)^3} - ma\int x^{m-1}\sqrt{a + bx} \, dx\right]$

    If $m = 1$: $\quad \dfrac{-2(2a - 3bx)\sqrt{(a + bx)^3}}{15b^2} + C$

    If $m = 2$: $\quad \dfrac{2(8a^2 - 12abx + 15b^2x^2)\sqrt{(a + bx)^3}}{105b^3}$

14. $\int \dfrac{\sqrt{a + bx}}{x^m} \, dx = \begin{cases} -\dfrac{1}{(m - 1)a}\left[\dfrac{\sqrt{(a + bx)^3}}{x^{m-1}} + \dfrac{(2m - 5)b}{2}\int \dfrac{\sqrt{a + bx} \, dx}{x^{m-1}}\right], & m \neq 1 \\ 2\sqrt{a + bx} + \sqrt{a} \ln\left|\dfrac{\sqrt{a + bx} - \sqrt{a}}{\sqrt{a + bx} + \sqrt{a}}\right|, & m = 1, a > 0 \end{cases}$

(Table continues inside back cover.)

Calculus
with Applications

THE SMITH BUSINESS SERIES
This book is part of the Smith Business Series of textbooks, which includes:

Finite Mathematics, Third Edition
This is a standard finite mathematics textbook. Published by Brooks/Cole.

Calculus with Applications, Second Edition
This is a business calculus textbook for students in management, life science, and social science. Published by Brooks/Cole.

College Mathematics and Calculus with Applications to Management, Life and Social Sciences, Second Edition
This book combines the material in the finite mathematics and business calculus books. Published by Brooks/Cole.

OTHER BROOKS/COLE TITLES BY KARL J. SMITH

Mathematics: Its Power and Utility, Third Edition
The Nature of Mathematics, Sixth Edition
Essentials of Trigonometry, Second Edition
Trigonometry for College Students, Fifth Edition
Precalculus Mathematics, Fourth Edition
Algebra and Trigonometry

THE SMITH AND BOYLE PRECALCULUS SERIES PUBLISHED BY BROOKS/COLE

Beginning Algebra for College Students, Fourth Edition
Intermediate Algebra for College Students, Fourth Edition
Study Guide for Algebra
College Algebra, Fourth Edition

Consulting Editor: ROBERT J. WISNER

Brooks/Cole Publishing Company
A Division of Wadsworth, Inc.

© 1992 by Wadsworth, Inc., Belmont, California 94002. All rights reserved. No part of this book may be reproduced, stored in a retrieval system, or transcribed, in any form or by any means—electronic, mechanical, photocopying, recording, or otherwise—without the prior written permission of the publisher, Brooks/Cole Publishing Company, Pacific Grove, California 93950, a division of Wadsworth, Inc.

Printed in the United States of America

10 9 8 7 6 5 4 3 2 1

Library of Congress Cataloging-in-Publication Data

Smith, Karl J.
    Calculus with applications/Karl J. Smith.—2nd ed.
      p. cm.
    Includes index.
    ISBN 0-534-16884-1
    1. Calculus. I. Title.
QA303.S63   1992
515—dc20   91-29704
                                                                                              CIP

Sponsoring Editor: PAULA-CHRISTY HEIGHTON
Editorial Assistants: LAINIE GIULIANO AND CAROL ANN BENEDICT
Production Services Coordinator: JOAN MARSH
Production: SUSAN L. REILAND
Manuscript Editor: ADELA C. WHITTEN
Permissions Editor: MARY KAY HANCHARICK
Interior and Cover Design: VERNON T. BOES
Cover Photo: SAN FRANCISCO CONVENTION AND VISITORS BUREAU
Interior Illustration: TECHarts; CARL BROWN
Typesetting: POLYGLOT PTE. LTD, SINGAPORE
Cover Printing: THE LEHIGH PRESS
Printing and Binding: R. R. DONNELLEY & SONS COMPANY

Credits: 1-2-3 is a registered trademark of Lotus Development Corporation. Quattro Pro is a registered trademark of Borland International, Inc. Page 3: Photo courtesy of Sara Hunsaker. Page 53: Photo courtesy of Sara Hunsaker. Page 119: Photo courtesy of Sara Hunsaker. Page 149: Photo courtesy of Sara Hunsaker. Page 191: Photo courtesy of New Balance Athletic Shoe, Inc. Page 281: Photo courtesy of Sara Hunsaker. Page 325: Photo courtesy of Sara Hunsaker.

Karl J. Smith

Calculus
with Applications

2ND EDITION

Brooks/Cole Publishing Company
Pacific Grove, California

This book is dedicated, with love, to my daughter, Melissa A. Smith

Preface

Calculus with Applications can be used in a one-semester or a two-quarter calculus course for students in business, management, or the life or social sciences. Emphasis throughout is to enhance students' understanding of the modeling process and how mathematics is used in real-world applications. The prerequisite for this course is intermediate algebra.

New Edition

It seems as if every new book claims innovation, state-of-the-art production, supplementary materials, readability, abundant problems, and relevant applications. In this new edition I have added over 750 new problems, new figures to enhance the exposition, and have rearranged the material on integration. Specifically, price elasticity of demand is added to Section 3.3, net excess profit to Section 15.3, consumers' surplus and producers' surplus has been expanded, and the material on the chain rule is now introduced with composite functions. The primary integration technique is substitution, which is now presented directly after the antiderivative. The definite integral and the Fundamental Theorem of Calculus are presented in separate sections in this edition, and Riemann sums are introduced in Section 6.5. Probability density functions have been moved to the end of the integration applications chapter, and is now designated as optional.

Why should you use my book? I would like you to look at the content, style, and problems in deciding whether this book will fit your needs.

Content

Every book, of course, must cover the appropriate topics, hopefully in the right order. I believe that it is important to introduce calculus as soon as possible, so Chapter 1 begins with a section called "What Is Calculus?" The rest of the chapter discusses the nature of mathematical modeling while reviewing the important types of functions that students need to know for calculus. This means that the derivative can be introduced in Chapter 2 so students can begin using calculus almost immediately in a variety of real-world applications. Many books begin with a chapter reviewing algebra, but I have decided to put this chapter at the end of the text in an appendix. A class needing to review linear and quadratic equations and inequalities, as well as rational expressions, can begin with the appendix or can review these topics as needed.

Model building is one of the most difficult, yet most important skills we need to teach our students. It is a skill that cannot be learned in a single lesson, or even in a single course. Learning this skill must be a gradual process, and that is how I approach it in this book—in small steps with realistic applications. However, in order to give students experience with true model building for real-life situations, I have included modeling applications at the end of each chapter. These applications are open-ended assignments that require a mathematical model-building approach for their development. A sample essay for the first Modeling Application is given in its entirety in the *Student's Solutions Manual* to illustrate how model building can be developed. These model-building applications, even if not assigned, demonstrate, in a very real way, how the material developed in the rest of the chapter can be used to answer some nontrivial questions (see, for example, the Modeling Application on Cobb–Douglas Production Functions in Chapter 8). They also can help to answer the legitimate question, "Why are we doing this?"

The text is divided into sections of nearly equal size that each take about one class day to develop. Since there are 53 numbered sections (including 14 optional sections), there is ample opportunity to select material that tailors the book to individual classes. There are also 9 additional sections of algebra review in Appendix A.

Style

An author's writing style also distinguishes one textbook from another. My writing style is informal, and I always write with the student in mind. I offer study hints along the way and let the students know what is important. Frequent and abundant examples are provided so that students can understand each step before proceeding to the next. A second color is used to highlight important steps or particular parts of an equation or formula. The chapter reviews list important terms and provide review problems for the material covered in the chapter. Together they emphasize the important ideas in the course.

Problems

The third—and perhaps the most important—factor in deciding on a textbook is the number, quality, and type of problems presented. This is where I have spent a great deal of effort in developing this book. Problems should help to develop students' understanding of the material, not inhibit or thwart it by being obscure. Problems are presented in the problem sets in matched pairs with an answer for one provided in the back. There are about three times more problems than are needed for assignments, so students have the opportunity to practice additional problems, both for the midterm and for the final. Each problem set provides drill problems to develop necessary manipulative skills and a large number of applications to show how the material can be used in business, management, and the life and social sciences.

Supplementary Materials

The final factor often used in selecting a textbook is the type and quality of available supplements. In addition to the answers in the back of the book, several supplements are available:

1. *Student's Solution Manual.* This provides complete solutions to the odd-numbered problems in the book. It also includes a sample essay to illustrate the modeling applications.
2. *Instructor's Manual.* This includes answers to all of the problems in the book. It also has sample essays and additional questions to accompany the modeling applications.
3. *Testing program.* There is a computerized test bank with text-editing capabil-

ities that allows you to create an almost unlimited number of tests or retests of the material. This test bank is also available in printed form for instructors without access to a computer.

4. *The Math Lab: Interactive Applied Calculus.* This is an IBM computer supplement prepared by Chris Avery and Charles Barker. Appendix D lists the programs available in this supplement.

As the author, I am also available to help you create any other set of supplementary materials that you think are necessary or worthwhile for your course.

Acknowledgments

I wish to thank the many reviewers of this manuscript:

John Alberghini
   Manchester Community College
Patricia Bannantine
   Marquette University
Frank Cheek
   University of Wisconsin
Joe S. Evans
   Middle Tennessee State University
J. Ellen Fordham
   Fullerton College
Kevin Hastings
   University of Delaware
Vuryl Klassen
   California State University, Fullerton
David Kurtz
   Rollins College
David Legg
   Indiana-Purdue University at Fort Wayne

Ann Barber Megaw
   University of Texas, Austin
William Ramaley
   Fort Lewis College
Robert Sharpton
   Miami-Dade Community College
Dwayne Snider
   Tarleton State University
Harriette J. Stephens
   State University of New York, Canton
Mary T. Teegarden
   Mesa Junior College
Walter Turner
   Western Michigan University
Joan Wyzkoski Weiss
   Fairfield University
Jan Wynn
   Brigham Young University

I have put considerable effort into making sure that the problems and answers are correct; in addition to myself, others have worked all the problems and examples to ensure their accuracy, and I would like to thank Gary Gislason (University of Alaska, Fairbanks) and Nancy Angle (Cerritos College) for their careful working of all the problems and examples. I continue to thank those who checked the accuracy of the previous edition: Terry Shell, Donna Szott, Pat Bannantine, and Michael Anderson. Their meticulous checking, as well as their numerous suggestions, is greatly appreciated.

A textbook is produced through the collaborative efforts of many people, and I appreciate the extraordinary effort and help of Joan Marsh and Susan Reiland. I would especially like to thank Mary A. McBerty for typing the student's solution manual and the instructor's manual.

Finally, I thank my wife, Linda, and my children, Missy and Shannon, for their help, love, and support.

Karl. J. Smith
Sebastopol, California

Contents

1 Functions 2

MODELING APPLICATION 1
Gaining a Competitive Edge in Business 3

- 1.1 What Is Calculus? 4
- 1.2 Functions and Graphs 16
- 1.3 Linear Functions 21
- 1.4 Quadratic and Polynomial Functions 32
- 1.5 Rational Functions 41
- *1.6 Review 48

2 The Derivative 52

MODELING APPLICATION 2
Instantaneous Acceleration: A Case Study of the Mazda 626 53

- 2.1 Limits 54
- 2.2 Continuity 68
- 2.3 Rates of Change 77
- 2.4 Definition of Derivative 88
- 2.5 Differentiation Techniques, Part I 94
- 2.6 Differentiation Techniques, Part II 100
- 2.7 The Chain Rule 106
- *2.8 Review 114

3 Additional Derivative Topics 118

MODELING APPLICATION 3
Publishing: An Economic Model 119

- 3.1 Implicit Differentiation 120
- 3.2 Differentials 125
- 3.3 Business Models Using Differentiation 131
- *3.4 Related Rates 137
- *3.5 Review 144

* Optional sections.

4 Applications and Differentiation 148

MODELING APPLICATION 4
Health Care Pricing 149

- 4.1 First Derivatives and Graphs 150
- 4.2 Second Derivatives and Graphs 159
- 4.3 Curve Sketching—Relative Maximums and Minimums 169
- 4.4 Absolute Maximum and Minimum 175
- *4.5 Review 186

Cumulative Review for Chapters 2–4 189

5 Exponential and Logarithmic Functions 190

MODELING APPLICATION 5
World Running Records 191

- 5.1 Exponential Functions 192
- 5.2 Logarithmic Functions 198
- 5.3 Logarithmic and Exponential Equations 204
- 5.4 Derivatives of Logarithmic and Exponential Functions 211
- *5.5 Review 219

6 The Integral 222

MODELING APPLICATION 6
Computers in Mathematics 223

- 6.1 The Antiderivative 224
- 6.2 Integration by Substitution 230
- 6.3 The Definite Integral 237
- 6.4 Area Between Curves 244
- 6.5 The Fundamental Theorem of Calculus 254
- *6.6 Numerical Integration 266
- *6.7 Review 275

7 Applications and Integration 280

MODELING APPLICATION 7
Modeling the Nervous System 281

- 7.1 Business Models Using Integration 282
- 7.2 Integration by Parts 292
- 7.3 Using Tables of Integrals 297
- 7.4 Improper Integrals 302

* Optional sections.

*7.5 Probability Density Functions 305
*7.6 Review 319
   Cumulative Review for Chapters 5–7 322

8 *Functions of Several Variables* 324

MODELING APPLICATION 8
The Cobb–Douglas Production Function 325

8.1 Three-Dimensional Coordinate System 326
8.2 Partial Derivatives 333
8.3 Maximum–Minimum Applications 339
8.4 Lagrange Multipliers 346
*8.5 Multiple Integrals 352
*8.6 Correlation and Least Squares Applications 363
*8.7 Review 371

9 *Differential Equations* 374

MODELING APPLICATION 9
The Battle of Trafalgar 375

9.1 First-Order Differential Equations 376
9.2 Applications—Growth Models 381
9.3 Special Second-Order Differential Equations 390
*9.4 Review 395
   Cumulative Review 398

Appendixes 399

A Review of Algebra 399
   1 Algebra Pretests 400
   2 Real Numbers 402
   3 Algebraic Expressions 406
   4 Factoring 412
   5 Linear Equations and Inequalities 418
   6 Quadratic Equations and Inequalities 423
   7 Rational Expressions 428
   8 Review 432
B Computers 434
C Answers to Odd-Numbered Problems 436
   Index 465

* Optional sections.

Calculus

with Applications

2ND EDITION

1 Functions

- 1.1 What Is Calculus?
- 1.2 Functions and Graphs
- 1.3 Linear Functions
- 1.4 Quadratic and Polynomial Functions
- 1.5 Rational Functions
- 1.6 Chapter 1 Review
  Chapter Objectives
  Sample Test

CHAPTER OVERVIEW

In Chapter 1 you are introduced to the building blocks for this course. We begin by asking the question, "What is calculus?" and then introduce the two main threads used for developing the material in this book: the concepts of mathematical models and of graphs. For an algebra diagnostic test and a review of algebra, see Appendix A.

PREVIEW

We begin by reviewing the concept of a function, evaluating functions, and functional notation. Then we discuss graphing functions, and in the remainder of the chapter focus on three very important types of functions: linear, quadratic, and rational. You should notice that each chapter opening presents a modeling application. These are open-ended questions that ask you to write a paper using the given information. You have probably had experience in writing papers in other subjects, but not in mathematics. One of the best ways to learn about the process of mathematical modeling is to start with a little information (the given situation), to state your assumptions, to build a mathematical model and apply the appropriate mathematics, and then to interpret your answer in terms of the given situation. This requires more work and effort than answering a typical homework problem. You are given nine modeling applications in this textbook.

PERSPECTIVE

There are two main ideas in calculus: the notion of a derivative and that of an integral. Each of these requires a thorough understanding not only of the function concept, but also of functional notation. The last chapter of the book considers functions of several variables, which brings us full circle back to the ideas developed in this chapter. In order to focus your attention on the concepts at hand as you progress through this book, you will need to have mastered the material on functions in this first chapter.

MODELING APPLICATION 1

Gaining a Competitive Edge in Business

Solartex manufactures solar collector panels. During the first year of operation, rent, insurance, utilities, and other fixed costs averaged $8,500 per month. Each panel sold for $320 and cost the company $95 in materials and $55 in labor. Since company resources are limited, Solartex cannot spend more than $20,000 in any one month. Sunenergy, another company selling panels, competes directly with Solartex. Last month, Sunenergy manufactured 85 panels at a total cost of $20,475, but the previous month produced only 60 panels at a total cost of $17,100.

After you have finished this chapter, write a paper based on this modeling application. This Modeling Application is continued on page 51.

APPLICATIONS

Management (*Business, Economics, Finance, and Investments*)
Purchasing power of the dollar
  (1.1, Problems 45–54)
Sales graph (1.2, Problem 44)
Supply and demand equations
  (1.3, Problems 60–61; 1.6, Problem 55)
Trend line (1.3, Problems 66–69)
1990 Tax Rate Schedule
  (1.3, Problems 70–75)
Demand, revenue, and break-even points
  (1.4, Problems 51–52; 1.6, Problems 54, 57)
Maximum profit
  (1.4, Problems 53–55; 1.6, Problem 58)
Cost and revenue functions
  (1.4, Problems 56–57)
Equilibrium point for supply and demand
  (1.5, Problem 40; 1.6, Problem 56)
Rate of change (1.6, Problem 53)

Life Sciences (*Biology, Ecology, Health, and Medicine*)
Atmospheric pressure (1.2, Problem 45)

Life Sciences (*continued*)
Response of a nerve (1.2, Problem 47)
Cost of removing a pollutant
  (1.2, Problem 46)
Cost-benefit model for removing pollutants
  (1.5, Problems 28–35; 1.6,
  Test Problems 19–20; 1.6, Problem 59)
Radiology
  (1.5, Problems 36–37; 1.6, Problem 60)
Pressure-volume relationship
  (1.5, Problems 38–39)

Social Sciences (*Demography, Political science, Population, Psychology, Society, and Sociology*)
U.S. population (1.1, Problems 55–58)
Marriage rate (1.1, Problem 59)
Divorce rate (1.1, Problem 60)
Population as predicted by an equation
  (1.3, Problems 62–63, 68)

General Interest
Admission price for a performance
  (1.2, Problem 42)
Taxable income (1.2, Problem 43)
Cost analysis for a car rental
  (1.3, Problems 64–65)
Maintenance costs (1.3, Problem 67)
Cigarette use in high school (1.3, Problem 69)
Highest bridge (1.4, Problem 58)
Skycycle ride across Snake River Canyon
  (1.4, Problem 59)
Stopping distance for a car (1.4, Problem 60)

Modeling Application—
Gaining a Competitive Edge in Business

1.1 What Is Calculus?

You are about to begin a study of calculus specifically oriented for students in management, life sciences, or social sciences.

Before beginning, let us take a few moments to consider the nature of calculus. We begin by comparing the types of problems encountered in elementary mathematics with those encountered in calculus (see Table 1.1).

The ideas of calculus were first considered toward the end of the sixteenth century, but the theory was not developed until the second half of the seventeenth

TABLE 1.1 Concepts and Applications in Elementary Mathematics and Calculus

Elementary mathematics	Calculus
Slope of a line	Slope of a curve
Tangent line to a circle	Tangent line to a general curve
Area of a region bounded by line segments	Area of a region bounded by curves
Average change Average velocity Average acceleration Average of a finite collection of numbers	Instantaneous change Instantaneous velocity Instantaneous acceleration Average value of a function on an interval

century when Gottfried Leibniz (1646–1716) and Isaac Newton (1642–1727) simultaneously, but independently, invented calculus as we know it today.

Up to now, your study of mathematics has focused on the mechanics of mathematics (such as solving equations, drawing graphs, and manipulating symbols). In this course you will not only reinforce such skills but also apply them to particular real-life situations.

Mathematical Models

A real-life situation does not easily lend itself to mathematical analysis because the real world is far too complicated to be precisely and mathematically defined. It is therefore necessary to develop what is known as a **mathematical model**. The model is based on certain assumptions about the real world and is modified by experimentation and accumulation of data. It is then used to predict some future occurrence in the real world. A mathematical model is not static or unchanging but is continually being revised and modified as additional relevant information becomes known.

Some mathematical models are quite accurate, particularly in the physical sciences. For example, the path of a projectile, the distance that an object falls in a vacuum, or the time of sunrise tomorrow all have mathematical models that provide very accurate predictions about future occurrences. On the other hand, in the fields of social science, psychology, and management, models provide much less accurate predictions because they must deal with situations that are often random in character. It is therefore necessary to consider two types of models:

Types of Models

> A **deterministic model** predicts the exact outcome of a situation because it is based on certain known laws.
>
> A **probabilistic model** deals with situations that are random in character and can predict the outcome within a certain stated or known degree of accuracy.

How do we construct a model? We need to observe a real-world problem and make assumptions about the influencing factors. This is called *abstraction*.

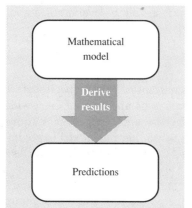

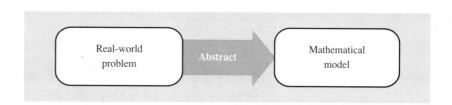

You must know enough about the mechanics of mathematics to *derive results* from the model.

The next step is to gather data. Does the prediction given by the model fit all the known data? If not, we use the data to *modify* the assumptions used to create the model. This is an ongoing process.

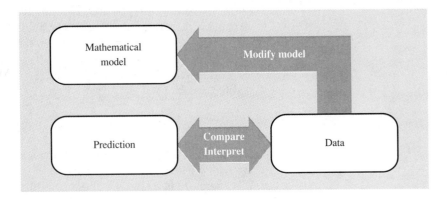

We begin with a rather artificial example of modeling (because we have not as yet developed any content in this course).

A Model for Studying in This Course

How do you plan for success in this course? Suppose we assume that the real-world problem is to obtain a grade of C or better in this course. How do you expect to reach this goal? What are your past experiences with college-level courses in general and with mathematics courses in general? Perhaps your model for studying in this course is quite simple, as shown in Figure 1.1.

Is the model shown in Figure 1.1 a good model? What do we mean by the word "good"? Is the model "true" in some absolute sense, or do we mean that it is "valid" in the sense that it has been checked successfully with a wide range of students and college-level courses?

A procedure that is often used in modeling says that if a question is difficult to answer, start by asking some easier questions. We might rephrase our question about this being a "good" model by asking some easier questions:

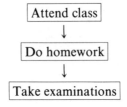

FIGURE 1.1

1. Has this procedure worked with success *for me* in previous college-level mathematics courses?
2. Is there a relationship between class attendance and final grade?
3. Is there a relationship between doing homework and final grade?
4. Will this course be typical of my past experiences in college-level mathematics courses?

Models Should Be Predictive

A good model should be able to predict real-life occurrences. Should the model we build for obtaining a C or better grade in this course be appropriate for a wide variety of students, instructors, methods of instruction, and types of institutions of higher learning? How can we go about making sure that our model is predictive? You are probably interested *only* in making sure that it is predictive for *you* at your school with your instructor.

Different people will build different models. Perhaps you study best under pressure and do not like to plan ahead. If a paper is due on the 15th you will begin on the 14th. Perhaps you are a highly organized person who has already scheduled in advance how much time you have allowed for study and homework in this course.

Perhaps we should try to build a mathematical model. Suppose you count the number of sections you will study in this course. Let this number be n. Also suppose you decide that your grade is determined by the amount of time, t, that you will spend in each section:

$$G = nt$$

Is *this* a good model for predicting your grade? Not yet. The variables are not well defined, nor are the units for measuring each of the variables. Let t be the time in minutes and let G be the grade as a percent, so $0 \leq G \leq 100$. *Furthermore, suppose that to obtain a C or better, G must be at least 70.* Now is this a good model? Not yet. For example, if there are 20 sections and you study each for 10 minutes, then

$$G = (20)(10) = 200$$

does not make sense. Clearly, we need to refine this mathematical model.

Criticize Your Model

Notice that building a model requires several steps and comparisons between the mathematics and the real world. Do not expect to come up with a working model on the first try. One of the most difficult aspects about teaching and learning how to model mathematics is to deal with impatience. Do not expect to "get it right" on your first attempt. For this model we might consider a scaling factor, k:

$$G = knt$$

Next we could do some research and compare whether the time spent on each section by a large number of students and the grades obtained by those students is related by a line (we consider how to do this in Section 8.7). Based on this research we replace this formula by a more predictive formula, say

$$G = \frac{1}{3}(t - 30) + 50$$

Is this a good model? Not yet. We have not yet built into this model the way we prepare for examinations and other course-related requirements. Perhaps you are beginning to say, "When will I finish this model-building process?" This is a reasonable question, but part of the impatience lies in wanting to "get an answer." The modeling process is never complete. We can simply build better and better models.

In the physical sciences, it is possible to build very accurate models, but in the management, life, and social sciences there are often so many variables to consider that the modeling process is rather complex and will require many simplifying assumptions in order to come up with a workable model.

You can also see that we need to build some mathematical skills in order to consider some mathematical models. This is the topic of this book.

Modeling in This Textbook

It is not possible to build mathematical modeling into every task in a textbook such as this because we must all deal with the realities of time and content in a particular course. The best we can do is build a framework *in which you can practice building models*. We have tried to do this with the modeling applications that appear in each chapter.

Mathematical Functions

We begin by reviewing an idea from algebra that is fundamental for the study of calculus—namely, the idea of a **function**.

Definition of a Function

Function

> A **function** of a variable x is a rule f that assigns to each value of x a unique* number $f(x)$, called the *value of the function at x*. A function can be defined by a verbal rule, a table, a graph, or an algebraic formula.

Remember:

1. $f(x)$ is pronounced "f of x."
2. f is a *function* while $f(x)$ is a *number*.
3. The set of replacements for x is called the **domain** of the function.
4. The set of all values $f(x)$ is called the **range** of f.

EXAMPLE 1 *A function defined as a verbal rule.*
Let x be a year from 1990 to 1995, inclusive. We define the function p by the rule that $p(x)$ is the closing price of Xerox stock on January 4 of year x. The domain is the set $\{1990, 1991, 1992, 1993, 1994, 1995\}$, and the range is the set of possible prices for Xerox stock.† For example, $p(1994) = 49\frac{1}{4}$ (or whatever the closing price of Xerox stock was on January 4, 1994). ∎

EXAMPLE 2 *A function defined by a table.*
Let g be a function defined so that $g(x)$ is the average price of gasoline in the year x as given by the following table:

Year	Average price per gallon of gasoline on January 4
1944	$ .21
1954	$ .29
1964	$ .30
1974	$ .53
1984	$1.24

The domain is $\{1944, 1954, 1964, 1974, 1984\}$.

The range is $\{\$.21, \$.29, \$.30, \$.53, \$1.24\}$.

For example, $g(1984) = \$1.24$. ∎

---

* By a unique number, we mean exactly one value.
† If you are familiar with the stock market, you know that Xerox stock prices are quoted in eighths of a dollar. The actual prices of Xerox stock over the years 1990–1995 provide the elements of the range.

EXAMPLE 3   *A function defined by a graph.*

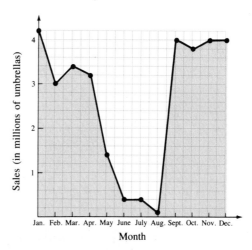

Let s be a function such that $s(x)$ is the sales of umbrellas in millions of units for x, a month in 1992. The domain is {Jan., Feb., ..., Nov., Dec.}, and the range is $.1 \le y \le 4.2$. (Remember that the units on the y-axis are in millions, so this means $100{,}000 \le y \le 4{,}200{,}000$.) For example, $s(\text{Jan.}) \approx 4.2$, $s(\text{Aug.}) \approx .1$. ∎

EXAMPLE 4   *A function defined by an algebraic formula.*
Let f be a function defined by

$$f(x) = 2x - 5$$

This rule says take a number x from the domain, multiply it by 2, and then subtract 5. For example,

$$f(3) = 2(3) - 5 = 1$$ ∎

Implied Domain of a Function

In this book, unless otherwise specified, the *domain is the set of real numbers for which the given function is meaningful.* For example,

$$f(x) = 2x - 5$$

has a domain consisting of all real numbers, while

$$g(x) = \frac{3}{5x - 5}$$

has a domain that excludes $x = 1$ because division by zero is not defined. For the function g, $5x - 5 = 0$ if $x = 1$. We say $g(1)$ does not exist or we say $g(x)$ is not defined at $x = 1$. Finally, if

$$F(x) = \sqrt{x}$$

then $x \geq 0$ is implied since the square root of a number x is real if and only if $x \geq 0$, so $F(x)$ is defined only if $x \geq 0$.

EXAMPLE 5 Give the domain of each of the following functions.

a. $f(x) = \dfrac{x-8}{3}$ **b.** $f(x) = \dfrac{3}{x-8}$ **c.** $f(x) = \sqrt{x+2}$ **d.** $f(x) = \dfrac{1}{\sqrt{x+2}}$

Solution **a.** The domain is all real numbers.

b. The domain is all real numbers except where $x - 8 = 0$ or $x = 8$. We state this domain by simply writing $x \neq 8$.

c. The domain is all real numbers for which

$$x + 2 \geq 0$$
$$x \geq -2$$

This can be stated in **set-builder notation** as $\{x \mid x \geq -2\}$, which is read

"The set of all x, such that $x \geq -2$"
" $\{$ x $\mid$ $x \geq -2\}$ "

However, set-builder notation is too formal for our work in this course, so we will just write $x \geq -2$ for the domain.

d. This is similar to part **c**, but division by zero must also be excluded, so we see that the domain is $x \geq -2$. ■

Functional Notation — Evaluating a Function

Example 4 illustrates a very important process called **evaluating a function**. To find $f(3)$ you substitute 3 for every occurrence of x in the formula for $f(x)$. This process is illustrated in Examples 6 and 7.

EXAMPLE 6 Given f and g defined by $f(x) = 2x - 5$ and $g(x) = x^2 + 4x + 3$, find the indicated values:

a. $f(1)$ **b.** $g(2)$ **c.** $g(-3)$ **d.** $f(-3)$

Solution **a.** The number $f(1)$ is found by replacing x by 1 in the expression

$$f(x) = 2x - 5$$
$$f(1) = 2(1) - 5$$
$$= -3$$

WARNING *This is a very important example.*

b. $g(2)$: $g(x) = x^2 + 4x + 3$
$$g(2) = (2)^2 + 4(2) + 3$$
$$= 4 + 8 + 3$$
$$= 15$$

c. $g(-3)$: $\quad g(-3) = (-3)^2 + 4(-3) + 3$
$= 9 - 12 + 3$
$= 0$

d. $f(-3) = 2(-3) - 5$
$= -11$

A function may be evaluated by using a variable.

EXAMPLE 7 Let F and G be defined by $F(x) = x^2 + 1$ and $G(x) = (x + 1)^2$. Then:
a. $F(w) = w^2 + 1$
b. $G(t) = (t + 1)^2$
$= t^2 + 2t + 1$

c. $F(-a) = (-a)^2 + 1$
$= a^2 + 1$
d. $G(-a) = (-a + 1)^2$
$= a^2 - 2a + 1$

e. $F(w + h) = (w + h)^2 + 1$
$= w^2 + 2wh + h^2 + 1$
f. $F(w^3) = (w^3)^2 + 1$
$= w^6 + 1$

g. $[G(a)]^2 = [(a + 1)^2]^2\quad$ since $G(a) = (a + 1)^2$
$= (a + 1)^4$

h. $F[G(x)] = F[(x + 1)^2]$
$= [(x + 1)^2]^2 + 1$
$= (x + 1)^4 + 1$

i. $F(\sqrt{t}) = (\sqrt{t})^2 + 1 = t + 1$

j. $G(x + h) = [(x + h) + 1]^2$
$= (x + h)^2 + 2(x + h) + 1$
$= x^2 + 2xh + h^2 + 2x + 2h + 1$

In calculus, functional notation is used to carry out manipulations such as those shown in Example 8.

WARNING *Because of the extensive later use of this manipulation, be sure you thoroughly understand Example 8.*

EXAMPLE 8 Find $\dfrac{f(x + h) - f(x)}{h}$ for each function given below.

a. $f(x) = x^2$, where $x = 5$. We substitute 5 for every occurrence of x in the formula, but do not substitute for h.

$$\frac{f(5 + h) - f(5)}{h} = \frac{(5 + h)^2 - 5^2}{h} \qquad \text{Since } f(x) = x^2$$

$$= \frac{25 + 10h + h^2 - 25}{h}$$

$$= \frac{(10 + h)h}{h}$$

$$= 10 + h$$

b. $f(x) = 2x^2 + 1$, where $x = 1$:

$$\frac{f(1+h) - f(1)}{h} = \frac{[2(1+h)^2 + 1] - [2(1)^2 + 1]}{h} \quad \text{Since } f(x) = 2x^2 + 1$$

$$= \frac{[2(1 + 2h + h^2) + 1] - (2 + 1)}{h}$$

$$= \frac{2h^2 + 4h + 3 - 3}{h}$$

$$= 2h + 4$$

c. $f(x) = x^2 + 3x - 2$:

$$\frac{f(x+h) - f(x)}{h} = \frac{[(x+h)^2 + 3(x+h) - 2] - (x^2 + 3x - 2)}{h}$$

$$= \frac{x^2 + 2xh + h^2 + 3x + 3h - 2 - x^2 - 3x + 2}{h}$$

$$= \frac{2xh + h^2 + 3h}{h}$$

$$= 2x + 3 + h \qquad \blacksquare$$

Functional notation can be used to work a wide variety of applied problems, as shown in Examples 9–12.

EXAMPLE 9 What is the average change per year in the price per gallon of gasoline from 1974 ($.53) to 1984 ($1.24)?

Solution The price of gasoline changed by $.71 per gallon during this 10-year period ($1.24 − $.53 = $.71). Thus the average per-year change in the price is

$$\text{Change in the price of gas} = \frac{\$1.24 - \$.53}{1984 - 1974} = \frac{\$.71}{10} \quad \longleftarrow \text{10 years}$$

In functional notation, if g is the average price of gasoline, then $g(x)$ is the average price of gas in year x. Thus,

$$\text{Change in the price of gas} = \frac{g(1984) - g(1974)}{1984 - 1974} = \$.071 \qquad \blacksquare$$

EXAMPLE 10 What is the average change per year in the price per gallon of gasoline from 1974 to a year h years later?

Solution Notice that h years after 1974 is $1974 + h$, so the average is

$$\frac{g(1974 + h) - g(1974)}{h} \qquad \blacksquare$$

EXAMPLE 11   In 1990 the U.S. population was 249.6 million. If we assume a growth rate of 2%, the population (in millions) t years after 1990 can be approximated by the formula

$$P(t) = 249.6(1.02)^t$$

What is the expected population for 1999?

Solution   Since $1999 - 1990 = 9$, $t = 9$. Thus:

$P(9) = 249.6(1.02)^9$ BY CALCULATOR: $\boxed{1.02}$ $\boxed{y^x}$ $\boxed{9}$ $\boxed{\times}$ $\boxed{249.6}$ $\boxed{=}$

≈ 298.3 DISPLAY: 298.2951051

The 1999 U.S. population will be about 298.3 million.

Composition of Functions

Suppose a farmer sells eggs to Safeway. If the farmer's price for a dozen eggs is x dollars, then there is a function f that can be used to describe $f(x)$, the total cost of those eggs to Safeway. Note that x and $f(x)$ are not the same because Safeway must pay for ordering, shipping, and distributing the eggs. Moreover, in order to determine the price to the consumer, Safeway must add an appropriate markup. Suppose this markup function is called g. Then the price of the eggs to the consumer is

$$g[f(x)] \quad \text{and not} \quad g(x)$$

since the markup must be on Safeway's *total* cost of the eggs and not just on the price the farmer charges for the eggs. This process of evaluating a function of a function illustrates the idea of composition of functions.

Consider two functions f and g such that f is a function from a set X to a set Y and g is a function from set Y to set Z as illustrated by Figure 1.2.

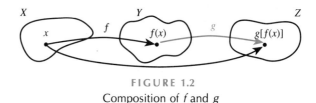

FIGURE 1.2
Composition of f and g

The value in the set Z is the number $g[f(x)]$ and defines a function from X to Z called the **composition of functions f and g**.

Composite Functions

> Let X, Y, and Z be sets of real numbers. Let f be a function from X to Y and g be a function from Y to Z. Then the **composite function** is the function from X to Z defined by
>
> $$g[f(x)]$$

EXAMPLE 12 If $f(x) = x^2$ and $g(x) = x + 4$, find:
a. $g[f(x)]$
b. $f[g(x)]$
c. $g[f(-1)]$
d. $f[g(5)]$

Solution
a. $g[f(x)] = g[x^2]$
$= x^2 + 4$

b. $f[g(x)] = f[x + 4]$
$= (x + 4)^2$
$= x^2 + 8x + 16$

c. $g[f(-1)] = (-1)^2 + 4$
$= 5$

d. $f[g(5)] = f[9]$
$= 9^2$
$= 81$

Notice from Examples 12a and 12b that $g[f(x)] \neq f[g(x)]$.

1.1

Problem Set

1. What is a mathematical model?
2. Why are mathematical models necessary or useful?
3. What is a function? (Use your own words.)
4. If $y = f(x)$, what is the difference between the symbols f and $f(x)$, if any?
5. One of the following examples expresses y as a function of x and the other does not. Which is a function? Explain your answer.
   a. y is the closing price of IBM stock on March 3 of year x.
   b. x is the closing price of Tandy stock on July 1 of year y.
6. One of the following examples is a function of x and the other is not. Which is a function? Explain your answer.
   a. $y = x^2$
   b. $x = y^2$
7. Use the table in Example 2 on page 8 to find and interpret each of the following expressions.
   a. $g(1954)$
   b. $g(1974)$
8. Use the graph in Example 3 on page 9 to find and interpret the following expressions.
   a. $s(\text{March})$
   b. $s(\text{June})$
9. Use $f(x) = 2x - 5$ (see Example 4) to find
   a. $f(8)$
   b. $f(-3)$
10. Use $f(x) = 2x - 5$ (see Example 4) to find
    a. $f(-5)$
    b. $f(7)$

Evaluate the functions in Problems 11–16.

11. Let $f(x) = 5x - 3$. Find
    a. $f(2)$
    b. $f(10)$
    c. $f(-15)$
    d. $f(100)$
12. Let $g(x) = 4x - 10$. Find
    a. $g(-2)$
    b. $g(5)$
    c. $g(-10)$
    d. $g(50)$
13. Let $h(x) = 6 - 4x$. Find
    a. $h(0)$
    b. $h(8)$
    c. $h(-7)$
    d. $h(100)$
14. Let $k(x) = 10 - 3x$. Find
    a. $k(-4)$
    b. $k(-5)$
    c. $k(6)$
    d. $k(10)$
15. Let $m(x) = x^2 - 3x + 1$. Find
    a. $m(0)$
    b. $m(1)$
    c. $m(2)$
    d. $m(3)$
16. Let $n(x) = x^2 + x - 3$. Find
    a. $n(0)$
    b. $n(1)$
    c. $n(2)$
    d. $n(3)$

State the domain of each function in Problems 17–20.

17. a. $f(x) = 5x^2 - 3x + 2$
    b. $f(x) = 6x^2 + 5x - \sqrt{17}$
18. a. $f(x) = \dfrac{2x - 5}{3}$
    b. $f(x) = \dfrac{3}{2x - 5}$
19. a. $f(x) = \dfrac{3x + 2}{x + 5}$
    b. $f(x) = \dfrac{\sqrt{2x - 1}}{2x + 1}$
20. a. $f(x) = \sqrt{2x + 6}$
    b. $f(x) = \dfrac{1}{\sqrt{2x + 6}}$

In Problems 21–36, let $f(x) = 5x - 2$ and $g(x) = 2x^2 - 4x - 5$. Evaluate and simplify:

21. a. $f(t)$
    b. $f(w)$
22. a. $g(s)$
    b. $g(t)$
23. a. $f(t + h)$
    b. $f(s + t)$
24. a. $g(t + h)$
    b. $g(s + t)$
25. $f(t + h + 8)$
26. $g(t - h - 3)$

27. $g(3 + h)$
28. $g(t - 2)$
29. $f(2x^2)$
30. $f(2x^2 - 4x)$
31. $g[f(x)]$
32. $f[g(x)]$
33. $f[g(5x)]$
34. $g[f(2x)]$
35. $g[g(x)]$
36. $f[f(x)]$

Find $[f(x + h) - f(x)]/h$ for each function in Problems 37–44.
37. $f(x) = 2x$
38. $f(x) = 5x$
39. $f(x) = 2x^2$
40. $f(x) = 5x^2$
41. $f(x) = 2x^2 - 3$
42. $f(x) = 5x^2 - 3x$
43. $f(x) = x^2 - 2x + 1$
44. $f(x) = 3x^2 - 2x + 4$

APPLICATIONS

For Problems 45–54, use the following table, which reflects the purchasing power of the dollar from October 1944 to October 1984 (Source: U.S. Bureau of Labor Statistics, Consumer Division). Let x represent the year; let the domain be the set $\{1944, 1954, 1964, 1974, 1984\}$.

Year	Round steak (1 lb)	Sugar (5 lb)	Bread (loaf)	Coffee (1 lb)	Eggs (1 doz)	Milk ($\frac{1}{2}$ gal)	Gasoline (1 gal)
1944	$ .45	$ .34	$ .09	$ .30	$ .64	$ .29	$ .21
1954	.92	.52	.17	1.10	.60	.45	.29
1964	1.07	.59	.21	.82	.57	.48	.30
1974	1.78	1.08	.36	1.31	.84	.78	.53
1984	2.15	1.48	1.29	2.69	1.15	1.08	1.52

Let $r(x)$ = price of 1 lb of round steak $g(x)$ = price of 1 gal of gasoline $c(x)$ = price of 1 lb of coffee
$b(x)$ = price of a loaf of bread $s(x)$ = price of 5 lb of sugar $m(x)$ = price of $\frac{1}{2}$ gal of milk
$e(x)$ = price of a dozen eggs

45. Find: **a.** $r(1954)$ **b.** $m(1954)$
46. Find: **a.** $g(1944)$ **b.** $c(1984)$
47. Find $s(1984) - s(1944)$.
48. Find $b(1984) - b(1944)$.
49. **a.** Find the change in the price of eggs from 1944 to 1984.
    b. Use functional notation to write the change in the price of eggs.
50. **a.** Find the change in the price of round steak from 1944 to 1984.
    b. Use functional notation to write the change in the price of round steak.
51. **a.** Find $\dfrac{g(1944 + 40) - g(1944)}{40}$.
    b. What does the expression in part **a** mean?
52. **a.** Find $\dfrac{m(1944 + 40) - m(1944)}{40}$.
    b. What does the expression in part **a** mean?
53. **a.** What is the average increase in the price of sugar per year from 1944 to 1954? Use functional notation.
    b. What is the average increase in the price of sugar per year from 1944 to 1964? Use functional notation.
    c. What is the average increase in the price of sugar per year from 1944 to 1974? Use functional notation.
    d. What is the average increase in the price of sugar per year from 1944 to 1984? Use functional notation.
    e. What is the average increase in the price of sugar per year from 1944 to 1944 + h, where h is an unspecified number of years? Use functional notation.
54. Repeat Problem 53 for coffee instead of sugar.
55. Use $P(t) = 249.6(1.02)^t$ from Example 11 to estimate the U.S. population in the year 2008.
56. Use $P(t) = 249.6(1.02)^t$ from Example 11 to estimate the U.S. population at the turn of the next century.
57. Use $P(t) = 249.6(1.02)^t$ from Example 11 to estimate the population in 1981. How well does this formula model the population if the actual population was 230 million?
58. Use $P(t) = 249.6(1.02)^t$ from Example 11 to estimate the population in 1900.
59. According to the U.S. Public Health Service, the number of marriages in the United States in 1987 was about 2,421,000 and in 1982 it was about 2,495,000. If $M(x)$

represents the number of marriages in year x,

a. Find $\dfrac{M(1987) - M(1982)}{5}$.

b. Verbally describe $\dfrac{M(1982 + h) - M(1982)}{h}$.

60. According to the U.S Public Health Service, the number of divorces in the United States in 1987 was about 1,157,000 and in 1982 it was about 1,180,000. If $D(x)$ represents the number of divorces in year x,

a. Find $\dfrac{D(1987) - D(1982)}{5}$.

b. Verbally describe $\dfrac{D(1982 + h) - D(1982)}{h}$.

1.2
Functions and Graphs

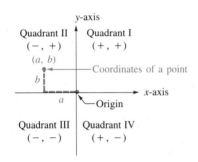

FIGURE 1.3
Cartesian coordinate system

A **two-dimensional coordinate system** consists of two perpendicular coordinate lines in a plane. Usually one of the coordinate lines is horizontal with the positive direction to the right; the other is vertical with the positive direction upward. These coordinate lines are called **coordinate axes**, and the point of intersection is called the **origin**. Note in Figure 1.3 that the axes divide the plane into four parts called the **first, second, third,** and **fourth quadrants**. This two-dimensional coordinate system is also called a **Cartesian coordinate system** in honor of René Descartes, who was the first to describe it in mathematical detail.

Points in a plane are denoted by ordered pairs. The term *ordered pair* refers to two real numbers represented by (a, b), where a is the **first component** and b is the **second component**. The order in which the components are listed is important since $(a, b) \neq (b, a)$ if $a \neq b$.

The horizontal number line is called the **x-axis** (or the *axis of abscissas*), and x represents the first component of the ordered pair. The vertical number line is called the **y-axis** (or the *axis of ordinates*), and y represents the second component of the ordered pair. The plane determined by the x- and y-axes is called the *coordinate plane*, *Cartesian plane*, or *xy-plane*. When we refer to a point (x, y), we mean a point in the coordinate plane whose abscissa is x and whose ordinate is y. To **plot a point (x, y)** means to locate the point with coordinates (x, y) in the plane and represent its location by a dot. (In this book, if variables other than x and y are used, you will be told which represents the first and which represents the second component.

To **graph a function** means to draw a picture of the ordered pairs that *satisfy* the equation in a one-to-one fashion. The set of x-values permitted for a function is called the **domain** of the function and the set of y-values for the function is called the **range**. We will illustrate graphing functions for three different types of models.

EXAMPLE 1 Graph $f(x) = 2x - 5$.

Solution Let $y = f(x)$ so that we can speak about the ordered pair (x, y) instead of the more notationally cumbersome form $(x, f(x))$. One method for graphing a function is to plot ordered pairs (x, y). That is, *you* choose values for x (the first component) and calculate, or find, the corresponding values for $y = f(x)$ (the second component): Let

$x_1 = 0$:	$f(0) = y_1 = 2 \cdot 0 - 5 = -5$	The ordered pair is $(0, -5)$.
$x_2 = 2$:	$f(2) = y_2 = 2 \cdot 2 - 5 = -1$	The ordered pair is $(2, -1)$.
$x_3 = 4$:	$f(4) = y_3 = 2 \cdot 4 - 5 = 3$	The ordered pair is $(4, 3)$.
$x_4 = -2$:	$f(-2) = y_4 = 2 \cdot (-2) - 5 = -9$	The ordered pair is $(-2, -9)$.

Plot the points as shown in Figure 1.4. After you have plotted enough points to see the curve's general shape, connect the points to obtain the curve.

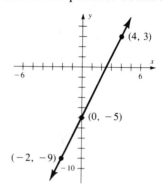

FIGURE 1.4
Graph of $f(x) = 2x - 5$. These points seem to lie on a line. In Section 1.3, we will graph lines much more efficiently than we did here.

EXAMPLE 2 Graph $g(x) = x^2 + 4x + 3$.

Solution Let $y = g(x)$ and

$x_1 = 0$:	$g(0) = y_1 = 0^2 + 4 \cdot 0 + 3 = 3$	Ordered pair $(0, 3)$
$x_2 = 1$:	$g(1) = y_2 = 1^2 + 4 \cdot 1 + 3 = 8$	$(1, 8)$
$x_3 = 2$:	$g(2) = y_3 = 2^2 + 4 \cdot 2 + 3 = 15$	$(2, 15)$
$x_4 = -1$:	$g(-1) = y_4 = (-1)^2 + 4(-1) + 3 = 0$	$(-1, 0)$
$x_5 = -2$:	$g(-2) = y_5 = (-2)^2 + 4(-2) + 3 = -1$	$(-2, -1)$
$x_6 = -3$:	$g(-3) = y_6 = (-3)^2 + 4(-3) + 3 = 0$	$(-3, 0)$
$x_7 = -4$:	$g(-4) = y_7 = (-4)^2 + 4(-4) + 3 = 3$	$(-4, 3)$

Plot the points and draw the curve as shown in Figure 1.5.

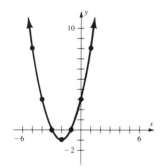

FIGURE 1.5
Graph of $g(x) = x^2 + 4x + 3$. These points do not lie on a line. If the points are connected they form a curve called a **parabola**. We will learn how to efficiently graph parabolas in Section 1.4.

EXAMPLE 3 Sketch the graph of $h(x) = 1/x$.

Solution   Let $y = h(x)$ and

$x_1 = 0$: $y_1 = \frac{1}{0}$ But $y_1 = \frac{1}{0}$ is not defined, so $x = 0$ is not in the domain—*there is no point on the graph corresponding to $x = 0$.*

$x_2 = 1$: $y_2 = \frac{1}{1} = 1$ The ordered pair is $(1, 1)$.
$x_3 = 2$: $y_3 = \frac{1}{2}$ $(2, \frac{1}{2})$
$x_4 = 3$: $y_4 = \frac{1}{3}$ $(3, \frac{1}{3})$
$x_5 = -1$: $y_5 = 1/(-1) = -1$ $(-1, -1)$
$x_6 = -2$: $y_6 = 1/(-2) = -\frac{1}{2}$ $(-2, -\frac{1}{2})$
$x_7 = -3$: $y_7 = 1/(-3) = -\frac{1}{3}$ $(-3, -\frac{1}{3})$

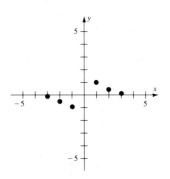

FIGURE 1.6a

If you plot the ordered pairs as shown in Figure 1.6a, you see that they cannot be connected with either a line or a parabola. In fact, you cannot connect all of these points with any smooth curve because no point corresponds to $x = 0$. Consider some additional points close to zero:

$$x = \frac{1}{2}: \quad y = \frac{1}{\frac{1}{2}} = 2 \quad \left(\frac{1}{2}, 2\right)$$

$$x = \frac{1}{3}: \quad y = \frac{1}{\frac{1}{3}} = 3 \quad \left(\frac{1}{3}, 3\right)$$

$$\vdots \qquad\qquad \vdots$$

We can now draw a smooth curve as shown in Figure 1.6b.   ∎

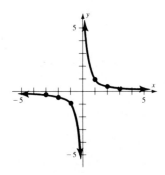

FIGURE 1.6b
Graph of $h(x) = \frac{1}{x}$. This curve is a type of curve that is the graph of what is called a **rational function**. These curves are discussed in Section 1.5.

Graphs of Functions

To summarize, the **graph of a function** f is the set of all points $(x, f(x))$ in a coordinate plane, where x is in the domain of f. That is, the graph of f can be described as the set of all points with coordinates (x, y) such that $y = f(x)$. The graph of a typical function f is shown in Figure 1.7.

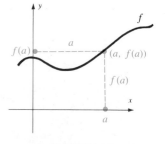

FIGURE 1.7
Graph of a function

Vertical line test   Note that the graph of a function is such that for each a in the domain there is only *one point* $(a, f(a))$ on the graph. This means that every vertical line passes through the graph of a function in at most one point. This is the so-called **vertical line test** for the graphs of functions as illustrated in Example 4.

EXAMPLE 4 Which of the following graphs are functions?

Solution

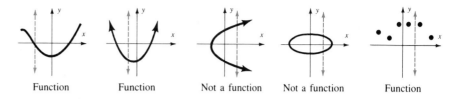

Function   Function   Not a function   Not a function   Function

Imagine a vertical line (shown in color) moving from left to right across the plane—if it passes through more than one point of the graph at one time, then the graph does not represent a function. ∎

Intercepts   There are some points on a graph that are of particular importance to us—the **intercepts**.

Intercepts

If the number zero is in the domain of f, then $f(0)$ is called the **y-intercept** of the graph of f and is the point $(0, f(0))$. This is the point where the graph intersects the y-axis.

If a is a real number in the domain such that $f(a) = 0$, then a is called an **x-intercept** and is the point $(a, 0)$. This is the point where the graph intersects the x-axis. Any number x such that $f(x) = 0$ is called a **zero of the function**.

EXAMPLE 5 Find the domain, range, and intercepts for f defined by the graph below.

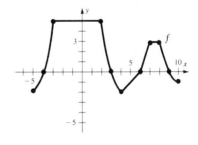

Solution Domain: $-5 \leq x \leq 10$ y-intercept: $(0, 5)$; we usually simply say that the y-intercept is 5. A function will not have more than one y-intercept.
Range: $-2 \leq y \leq 5$
The zeros of the function are $-4, 3, 6,$ and 9.
x-intercepts: $(-4, 0), (3, 0), (6, 0),$ and $(9, 0)$. ∎

1.2
Problem Set

1. Define the graph of a function f.
2. What does it mean when we say the ordered pair (a, b) satisfies the equation $y = f(x)$?
3. What are the x- and y-intercepts of the graph of a function f?
4. What is a zero of a function f?

Use the figures below to find the coordinates of the points in Problems 5–13. For example, point D has coordinates $(-1, f(-1))$.

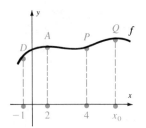
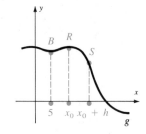

5. Point A
6. Point P
7. Point Q
8. Point B
9. Point R
10. Point S
11. Point C
12. Point T
13. Point U

Find the domain, range, and intercepts of the relations defined by the graphs in Problems 14–19. Also state whether the graph defines a function.

14.
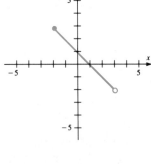

15.
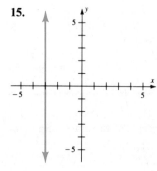

16.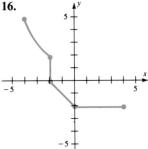

17.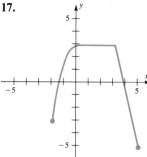

18.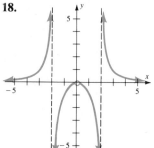

19.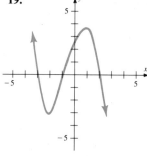

Graph the functions in Problems 20–41 by plotting points.

20. $f(x) = 3x + 1$
21. $f(x) = 2x + 3$
22. $f(x) = x - 4$
23. $f(x) = 6 - 2x$
24. $f(x) = 1 - x$
25. $f(x) = -3x - 1$
26. $g(x) = 2x^2$
27. $g(x) = \frac{1}{2}x^2$
28. $g(x) = \frac{1}{10}x^2$
29. $g(x) = x^2 + 4x + 4$
30. $g(x) = x^2 + 6x + 9$
31. $g(x) = x^2 - 6x + 9$
32. $g(x) = x^2 + 2x - 3$
33. $g(x) = 2x^2 - 4x + 5$
34. $g(x) = 2x^2 - 4x + 4$
35. $h(x) = -1/x$
36. $h(x) = 3/x$
37. $h(x) = -2/x$
38. $h(x) = 1/(x - 1)$
39. $h(x) = 2/(x - 2)$
40. $c(x) = x^3 + 2x - 4$
41. $c(x) = x^3 - 3x + 2$

APPLICATIONS

42. A theater has a capacity of 1,500 seats. The formula relating the number of adult tickets sold, a, and the number of children's tickets sold, c, is $a + c = 1,500$. Graph (a, c) satisfying this relationship. Assume that both a and c are positive.

43. A formula from the Tax Rate Schedule X on the 1990 federal income tax for a single person is

$$T = .28x - 3,250 \qquad 19{,}451 < x \leq 47{,}050$$

where x is the amount on line 37 of Form 1040. Graph (x, T) for $19{,}451 < x \leq 47{,}050$.

44. A supply company finds that the number of computer disks sold in year x is given by $s(x) = 5{,}000 + x^2$, where $x = 0$ corresponds to 1980. Graph the sales for the years 1980 to 1990 (inclusive).

45. The pressure, P, in centimeters of mercury, is given as a function of the depth in meters, d, under water, by using the formula $P = .4d + 7.6$. Graph (d, P) for $0 \le d \le 10$.

46. The cost C (in thousands of dollars) of removing p percent of a certain pollutant is given by the formula $C = 20p/(105 - p)$. Graph (p, C) for $0 \le p \le 100$.

47. The number of responses n (per milliseconds) of a nerve is a function of the length of time t (in milliseconds) since the nerve was stimulated. For a certain nerve, this relationship is $n = 150 - (t - 10)^2$. Graph (t, n) for $0 \le t \le 22$.

1.3
Linear Functions

This section reviews material from previous courses and also provides the notation and concepts required in the remainder of this book. Recall that a first-degree equation with two variables is called a *linear equation*.

Linear Function

A function f is a **linear function** if it can be written in the form

$$f(x) = mx + b \quad \text{or} \quad y = mx + b$$

where m and b are real numbers.

If $m = 0$, then $f(x) = b$, and the graph of $f(x) = b$ is a **horizontal line** as shown in Figure 1.8a. Functions whose graphs are horizontal lines are called **constant functions**. Now let (x_1, y_1) and (x_2, y_2) be any points on a line so that $x_1 = x_2$. Then this line is parallel to the y-axis and is called a **vertical line**, as shown in Figure 1.8b. Note that vertical lines are not the graphs of functions. Vertical lines always have the form $x = c$, where c is some constant ($x = 3$, for example).

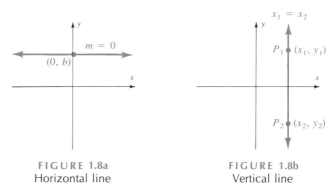

FIGURE 1.8a
Horizontal line

FIGURE 1.8b
Vertical line

A line is determined by two points, so if we know *any* two points on a line we can draw the line by using a straightedge and the given points. In the last section we evaluated formulas to find those points. However, two points that are usually easy to find are the intercepts:

To find the y-intercept: Let $x = 0$ and solve for y.
To find the x-intercept: Let $y = 0$ and solve for x.

EXAMPLE 1 Graph $f(x) = -\frac{3}{2}x + 3$ by plotting the x- and y-intercepts.

Solution If $f(x) = -\frac{3}{2}x + 3$, write $y = -\frac{3}{2}x + 3$.

Let $x = 0$: $y = -\frac{3}{2}(0) + 3$
$y = 3$ The point $(0, 3)$ is the y-intercept.

Let $y = 0$: $0 = -\frac{3}{2}x + 3$
$\frac{3}{2}x = 3$
$x = 2$ The point $(2, 0)$ is the x-intercept.

Draw the line passing through the plotted intercepts as shown in Figure 1.9. ■

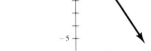

FIGURE 1.9
Graph of $f(x) = -\frac{3}{2}x + 3$

The constants b and m in the form $f(x) = mx + b$ gives us important information about the line we wish to graph. For example, if we let $y = f(x)$, we can write $y = mx + b$. Now we can find the y-intercept. Let $x = 0$:

$$y = m \cdot x + b$$
$$y = b$$

This means that the y-intercept is $(0, b)$. We generally shorten the notation and simply say the y-intercept is b to mean the line passes through the y-axis at the point $(0, b)$.

The second constant m, which is the coefficient of x when the equation is solved for y, tells us the steepness or **slope** of a line. But, first, we need to define what we mean by the slope of a line. This definition requires that we know what is meant by the vertical change (*rise*) relative to a horizontal change (*run*). These ideas are illustrated in Figure 1.10.

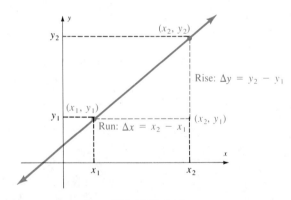

FIGURE 1.10
Slope of a line

Let Δx represent the horizontal change and Δy represent the vertical change.*
Note that $\Delta x = x_2 - x_1$ and $\Delta y = y_2 - y_1$.

---

* Δx is one symbol (not the multiplication of two variables) and is pronounced "delta x" (Δy is pronounced "delta y").

Slope

Let (x_1, y_1) and (x_2, y_2) be points on a line. Then

$$\text{Slope} = \frac{\text{Vertical change}}{\text{Horizontal change}} = \frac{\Delta y}{\Delta x} \quad \text{or} \quad \frac{\text{Rise}}{\text{Run}} = \frac{y_2 - y_1}{x_2 - x_1} = \frac{\Delta y}{\Delta x}$$

If $\Delta x = 0$, then the line is vertical and has *no* slope ($\Delta y/0$ is undefined).
If $\Delta y = 0$, then the line is horizontal and has *zero* slope ($0/\Delta x = 0$).

To show that m in the equation $y = mx + b$ is the slope, consider the line specified by the equation $y = mx + b$, which passes through (x_1, y_1) and (x_2, y_2), with $x_1 \neq x_2$. This means that $y_1 = mx_1 + b$ and $y_2 = mx_2 + b$, so that

$$\begin{aligned}
\text{Slope} = \frac{\Delta y}{\Delta x} &= \frac{y_2 - y_1}{x_2 - y_1} \\
&= \frac{(mx_2 + b) - (mx_1 + b)}{x_2 - x_1} \quad \text{Substitution} \\
&= \frac{mx_2 - mx_1}{x_2 - x_1} \\
&= \frac{m(x_2 - x_1)}{x_2 - x_1} = m
\end{aligned}$$

This discussion tells us that we can find the *y*-intercept and slope of linear equations of the form $y = mx + b$ by inspection, as shown by Example 2.

EXAMPLE 2 Find the slope and *y*-intercept.

		Slope	*y*-intercept	
a.	$y = \frac{1}{2}x + 3$	$m = \frac{1}{2}$	$b = 3; (0, 3)$	By inspection
b.	$y = x - 3$	$m = 1$	$b = -3; (0, -3)$	By inspection
c.	$y = -\frac{2}{3}x + \frac{5}{2}$	$m = -\frac{2}{3}$	$b = \frac{5}{2}; \left(0, \frac{5}{2}\right)$	By inspection

d. $3x + 4y + 8 = 0$
Solve for *y*: $4y = -3x - 8$
$y = -\frac{3}{4}x - \frac{8}{4}$

Thus the slope $m = -\frac{3}{4}$, and the *y*-intercept is $(0, -2)$ since $b = -\frac{8}{4} = -2$. ∎

Since the slope and *y*-intercept are easy to find after the linear equation is solved for *y*, we give a special name to this form of the equation.

Slope–Intercept Form of the Equation of a Line

The **slope–intercept** form of a linear equation is

$$y = mx + b$$

and the graph of this equation is the line having slope m and *y*-intercept $(0, b)$.

This form of the equation of a line can be used for graphing certain lines when it is not convenient to plot points. The procedure is summarized in Figure 1.11. Carefully study this procedure.

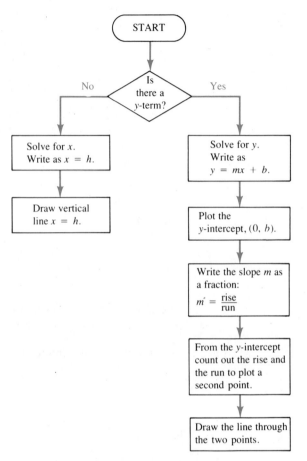

FIGURE 1.11
Procedure for graphing a line by the slope–intercept method

EXAMPLE 3 Graph $y = \frac{1}{2}x + 3$.

Solution   By inspection, $b = 3$ and the slope is $\frac{1}{2}$; the line is graphed by first plotting the y-intercept $(0, 3)$ and then finding a second point by counting out the slope: over 2 and up 1.

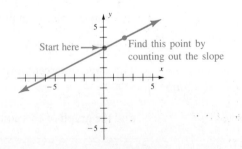

EXAMPLE 4 Graph $3x - 2y - 6 = 0$.

Solution Solve for y:

$$2y = 3x - 6$$
$$y = \tfrac{3}{2}x - 3$$

$b = -3$ and the slope is $\tfrac{3}{2}$; the line is graphed by first plotting the y-intercept $(0, -3)$ and then finding a second point by counting out the slope (over 2, up 3).

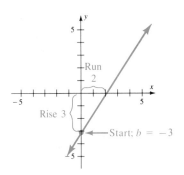

EXAMPLE 5 Graph $4x + 2y - 5 = 0$ for $-1 \leq x \leq 3$.

Solution Solve for y:

$$2y = -4x + 5$$
$$y = -2x + \tfrac{5}{2}$$

Because of the restriction on the domain, we do not want to graph the entire line, just that part with first components as specified by the restriction, $-1 \leq x \leq 3$. The usual convention is to show the entire line as a dashed line and the answer as a solid line. For this example $b = \tfrac{5}{2}$ and the slope is -2; the line is shown as a dashed line. Because of the restriction of the domain, the part of the line with x values between -1 and 3 (inclusive) is shown as a solid line segment.

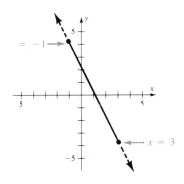

Sometimes a function will have a different equation for different parts of its domain, as shown in Example 6. If each of those equations is linear, then the function is called a **piecewise linear function**.

EXAMPLE 6 Graph $y = x$ if $x \geq 0$
$\phantom{\text{Graph }}y = -x$ if $x < 0$

Solution Graph each of the line segments as shown in the figure.
Remember from algebra the definition of absolute value:

$$|x| = x \quad \text{if } x \geq 0$$

and

$$|x| = -x \quad \text{if } x < 0$$

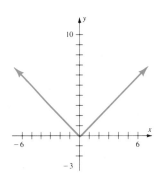

Thus the function in Example 6, the **absolute value function**, can be written $y = |x|$.

EXAMPLE 7 Graph $y = |6 - 2x|$.

Solution We can write this function without absolute value symbols by considering the two parts of the definition of absolute value.

$$y = 6 - 2x \quad \text{if } 6 - 2x \geq 0$$
$$-2x \geq -6$$
$$x \leq 3$$

See Section 1.3 for a review of solving linear inequalities.

And

$$y = -(6 - 2x) \quad \text{if } 6 - 2x < 0$$
$$x > 3$$

Note that $y = -(6 - 2x)$ is the same as $y = 2x - 6$.

This graph is shown at the left.

Two relationships make it easy for us to recognize parallel and perpendicular lines.

Parallel Lines
Perpendicular Lines

Two lines ℓ_1 and ℓ_2 with slopes m_1 and m_2 are

parallel if and only if $m_1 = m_2$, and are
perpendicular if and only if $m_1 m_2 = -1$.

EXAMPLE 8 Find the slope and sketch the indicated lines.

a. Passing through $(2, -3)$ and $(-1, 2)$

b. The line passing through the origin which is parallel to the line passing through $(-4, -1)$ and $(1, 3)$

c. The line passing through $(3, 2)$ perpendicular to the line passing through $(-3, 4)$ and $(5, 6)$

d. The line passing through $(-3, 2)$ and $(-3, 4)$

Solution **a.** $m = \dfrac{2 - (-3)}{-1 - 2} = \dfrac{5}{-3} = -\dfrac{5}{3}$ **b.** $m = \dfrac{3 - (-1)}{1 - (-4)} = \dfrac{4}{5}$

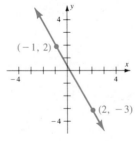

Negative slope

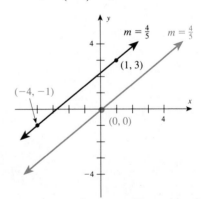

Positive slope

c. $m = \dfrac{6-4}{5-(-3)} = \dfrac{2}{8} = \dfrac{1}{4}$

Slope of perpendicular line is -4

d. $m = \dfrac{4-2}{-3+3}$ ← This is zero, so the fraction is an undefined number.

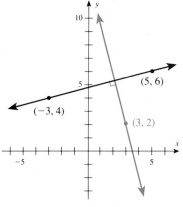

Negative slope

Undefined slope; vertical line

In constructing mathematical models, it is often necessary to write a linear equation using available or given information about the line. Example 9 shows how to do this if you know the slope and y-intercept.

EXAMPLE 9 Find the equation of the line with y-intercept $(0, 5)$ and slope $-\tfrac{2}{3}$.

Solution   Use the equation $y = mx + b$, where $b = 5$ and $m = -\tfrac{2}{3}$.

$$y = -\dfrac{2}{3}x + 5$$

More often than not, unfortunately, when you need to find the equation of a line you will not know the y-intercept. You will, however, know a point and the slope or two points. In these cases it is easier to use another form of the equation of a line called the **point–slope form**. It is easy to derive this equation if we remember that

$$m = \dfrac{y_2 - y_1}{x_2 - x_1}$$

Since (x, y) is any point on the line passing through (x_1, y_1), we have

$$m = \dfrac{y - y_1}{x - x_1}$$

Thus, by multiplying both sides by $x - x_1$, we obtain

$$m(x - x_1) = y - y_1$$

28 CHAPTER ONE FUNCTIONS

Point–Slope Form of the Equation of a Line

> A nonvertical line having slope m and passing through (x_1, y_1) has the equation
>
> $$y - y_1 = m(x - x_1)$$

EXAMPLE 10 Find an equation of the line with slope 3 passing through $(-2, -5)$.

Solution Use the equation $y - y_1 = m(x - x_1)$, where $m = 3$, $x_1 = -2$, and $y_1 = -5$:

$$y - (-5) = 3[x - (-2)]$$
$$y + 5 = 3(x + 2) \quad \blacksquare$$

If you know two points and want the equation, first find the slope and *then* use the point–slope form.

EXAMPLE 11 Find the equation of the line passing through $(-2, 3)$ and $(4, -1)$.

Solution First find the slope:

$$m = \frac{\Delta y}{\Delta x} = \frac{-1 - 3}{4 - (-2)} = \frac{-4}{6} = -\frac{2}{3}$$

Now use the point–slope form (you can use *either* of the given points):

$$(-2, 3): \quad y - 3 = -\frac{2}{3}(x + 2) \quad \text{or} \quad (4, -1): \quad y + 1 = -\frac{2}{3}(x - 4) \quad \blacksquare$$

It is not easy to see that the equations in Example 11 are the same. For this reason, we are often asked to algebraically manipulate the answers into the same form. This form is called the **standard form** of the equation of a line.

Standard Form of the Equation of a Line

> The **standard form** of the equation of a line is
>
> $$Ax + By + C = 0$$
>
> where (x, y) is any point on the line, and A, B, and C are constants (A and B not both zero).

EXAMPLE 12 Change the point–slope forms given in Example 11 to standard form.

Solution

$y - 3 = -\dfrac{2}{3}(x + 2)$ Eliminate fractions (multiply by 3). $\qquad y + 1 = -\dfrac{2}{3}(x + 4)$

$3(y - 3) = -2(x + 2)$ Obtain a 0 on the right. $\qquad 3(y + 1) = -2(x - 4)$

$3y - 9 = -2x - 4$ Eliminate parentheses. $\qquad 3y + 3 = -2x + 8$

$2x + 3y - 5 = 0 \qquad\qquad\qquad\qquad\qquad\qquad\qquad 2x + 3y - 5 = 0 \quad \blacksquare$

Note that both equations given in Example 11 are the same in standard form.

EXAMPLE 13 Find the equation of a line passing through $(7, -2)$ with no slope.

Solution Since the line is vertical, the equation has the form $x = h$ when it passes through (h, k). Thus $x = 7$ is the equation. In standard form, $x - 7 = 0$. ■

WARNING *Do not confuse "no slope" (vertical line) with "zero slope" (horizontal line).*

A mathematical model is often constructed by assuming that the relationship between two variables is linear and then writing an equation using two known data points, as illustrated in Example 14. When relating two values in an applied setting, one value will often depend on the other, so it is customary to designate the dependent one as the **dependent variable**, and the other as the **independent variable**. When x and y are used to represent the variables, x is the independent variable and y the dependent variable.

EXAMPLE 14 A sales executive plotted sales (in millions of dollars) versus the amount spent on advertising (in thousands of dollars) and observed the points as shown in the figure below. It is easy to see that the points lie approximately on a line. By using the points for advertising at $40,000 and $80,000, find the equation of this **trend line**. What sales figure can be predicted for an expenditure of $100,000 on advertising?

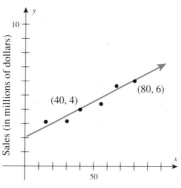

Solution The information gives us points $(40, 4)$ and $(80, 6)$. First find the slope:

$$m = \frac{6 - 4}{80 - 40} = \frac{2}{40} = \frac{1}{20}$$

Next, use the point–slope form using either point:

$$y - y_1 = m(x - x_1)$$

$$y - 4 = \frac{1}{20}(x - 40)$$

$$y = \frac{1}{20}x + 2$$

For advertising expenditures of $100,000, $x = 100$ so

$$y = \frac{1}{20}(100) + 2$$
$$= 7$$

This means the expected sales are $7,000,000.

We conclude by summarizing the various forms of linear equations.

Forms of a Linear Equation

STANDARD FORM:		$Ax + By + C = 0$	(x, y) is any point on the line; A, B, and C are constants; A and B are not both zero
SLOPE–INTERCEPT FORM:		$y = mx + b$	m is the slope; $(0, b)$ is the y-intercept
POINT–SLOPE FORM:		$y - y_1 = m(x - x_1)$	m is the slope; (x_1, y_1) is the known point
HORIZONTAL LINE:		$y = k$	(h, k) is a point on the line
VERTICAL LINE:		$x = h$	(h, k) is a point on the line

1.3
Problem Set

Find the x- and y-intercepts for the lines whose equations are given in Problems 1–8.

1. $y = 2x + 4$
2. $y = 5x - 10$
3. $4x + 3y + 4 = 0$
4. $3x + 2y - 9 = 0$
5. $100x - 250y + 500 = 0$
6. $2x - 5y - 1{,}200 = 0$
7. $y + 2 = 0$
8. $x - 2 = 0$

Find the slope of the line passing through the points in Problems 9–14.

9. $(2, 3)$ and $(5, 4)$
10. $(4, -1)$ and $(-2, 3)$
11. $(5, 2)$ and $(-2, -3)$
12. $(-2, -3)$ and $(4, 5)$
13. $(-2, -3)$ and $(-1, -2)$
14. $(-3, -1)$ and $(-7, -10)$

Find the slope and y-intercept in Problems 15–23.

15. $y = 2x + 4$
16. $y = 5x - 3$
17. $y = 9x + 1$
18. $4x + 3y + 4 = 0$
19. $2x - 3y + 5 = 0$
20. $5x - 2y - 5 = 0$
21. $y - 5 = 0$
22. $y + 9 = 0$
23. $x - 3 = 0$

Graph the lines of the equations in Problems 24–35 by finding the slope and y-intercept.

24. $y = 3x + 4$
25. $y = 2x - 5$
26. $y = -3x + 1$
27. $y = -\frac{1}{4}x + 2$
28. $y = -\frac{2}{3}x - 4$
29. $y = \frac{3}{5}x + \frac{2}{5}$
30. $3x - y + 2 = 0$
31. $x + 3y - 9 = 0$
32. $2x - 3y + 15 = 0$
33. $x = \frac{2}{3}y$
34. $y - 3 = 0$
35. $2x + 5 = 0$

Graph the line segments or piecewise functions given in Problems 36–47.

36. $y = 4x - 2, \quad -3 \le x \le 4$
37. $y = -3x + 2, \quad -4 \le x \le 3$
38. $2x + 5y + 10 = 0, \quad -1 \le x \le 5$
39. $3x - 2y + 8 = 0, \quad -3 \le x \le 4$
40. $2x + y + 5 = 0 \quad \text{if } -3 \le x \le 0$
    $y + 5 = 0 \quad \text{if } 0 < x < 3$
    $x - y - 8 = 0 \quad \text{if } 3 \le x \le 10$

41.
$$y = 3x + 2 \quad \text{if } 0 \leq x \leq 2$$
$$y - 8 = 0 \quad \text{if } 2 < x < 5$$
$$x - 2y + 11 = 0 \quad \text{if } x \geq 5$$
42. $y = 2|x|$
43. $y = |3x - 6|$
44. $y = |2x + 4|$
45. $y = -3|x|$
46. $y = |5 - x|$
47. $y = |4 - 3x|$

Find the standard form of the equation of the line satisfying the conditions given in Problems 48–59.

48. y-intercept 5; slope 6
49. y-intercept -3; slope -2
50. y-intercept 0; slope 0
51. y-intercept 4; slope 0
52. slope 2; passing through $(4, 3)$
53. slope -1; passing through $(-3, 5)$
54. slope $\frac{1}{2}$; passing through $(5, 3)$
55. slope $\frac{3}{5}$; passing through $(4, -3)$
56. passing through $(-3, -1)$ and $(3, 2)$
57. passing through $(5, 6)$ and $(1, -2)$
58. passing through $(4, -2)$ and $(4, 5)$
59. passing through $(5, 6)$ and $(7, 6)$

APPLICATIONS

Problems 60–65 provide some real-world examples of line graphs. One way of finding the equation of the line is to write two data points from the given information (as shown in Example 14) and then to use those points to write the equation. Use the given information to write an equation in standard form of the line described by the problem.

60. The demand for a certain product is related to the price of the item. Suppose a new line of stationery is tested at two stores. At store A, 25 boxes are sold within a month at $5 each, and, at store B, 15 boxes priced at $10 each are sold during the same time. Let x be the price and y be the number of boxes sold. How many boxes would be sold if the price is $2.00?

61. An important factor related to the demand for a product is its supply. The amount of stationery in Problem 60 that can be supplied is also related to the price. At $5 each, 10 boxes can be supplied, and, at $10, 20 boxes can be supplied. Let x be the price and y be the number of boxes supplied. How many boxes could be supplied at $2.00?

62. The population of Florida in 1980 was roughly 9.7 million, and in 1990 it was 12.6 million. Let x be the year (let 1960 be the base year; that is, $x = 0$ represents 1960, so $x = 10$ represents 1970) and y be the population. Use this equation to predict the population in 2000.

63. The population of Texas in 1980 was roughly 14.2 million, and in 1990 it was 17.2 million. Let x be the year (let 1960 be the base year; that is, $x = 0$ represents 1960, so $x = 10$ represents 1970) and y be the population. Use this equation to predict the population in 2000.

64. It costs $90 to rent a car if it is driven 100 miles and $140 if it is driven 200 miles.

65. It costs $60 to rent a car if it is driven 50 miles and $60 if it is driven 260 miles.

Many real-world examples have data points that can be approximated by a line. Consider the given data points to find the equation of a trend line, and then answer the question asked in Problems 66–69.

66. If the sales (in thousands of dollars) of a particular item are plotted for the first five years, it can be noticed that these points lie approximately along a straight line.

Year	1	2	3	4	5
Sales	3	4	5	6	6

Using the points $(2, 4)$ and $(5, 6)$, find the equation of a trend line. What sales figure can be predicted for the 8th year?

67. Suppose the cost for maintenance and repairs is plotted as a function of the number of miles (in thousands) the vehicle has been driven.

Miles	10	20	30	40	50
Cost	100	189	309	400	508

Using the points $(10, 100)$ and $(40, 400)$, find the equation of a trend line. What is the expected cost for 53,500 miles?

68. The population of California (in millions) is shown by the following table:

Year	1950	1960	1970	1980	1990
Population	10.6	17.7	19.9	23.7	28.7

Using the points $(0, 10.6)$ and $(30, 23.7)$, find the equation of a trend line that uses 1950 as a base year $(1950 = 0)$. What is the expected population of California in the year 2000?

69. The use of cigarettes among high school seniors is shown by the following table:

Class of	1983	1984	1985	1986	1987
Percentage	70.6	69.7	68.8	67.6	67.2

Using the points $(3, 70.6)$ and $(7, 67.2)$, find the equation of a trend line that uses 1980 as a base year (1980 = 0). What is the expected percentage of cigarette usage for the class of 1992?

Use the following 1990 U.S. Tax Rate Schedule to answer the questions in Problems 70–75.

70. Graph the tax for income from $0 to $30,000.
71. Write the equation for the tax, T, for the domain $0 < T \le 32{,}450$.
72. Write the equation for the tax, T, for the domain $78{,}401 < T \le 185{,}730$.
73. Graph the tax for income from $80,000 to $180,000.
74. Graph the tax for income from $0 to $180,000.
75. Write the piecewise linear function for the tax T for amounts up to $185,730.

Schedule Y-1—Use if your filing status is Married filing jointly or Qualifying widow(er)

If the amount on Form 1040, line 37, is: Over—	But not over—	Enter on Form 1040, line 38		of the amount over—
$0	$32,450	.... 15%		$0
32,450	78,400	$4,867.50 + 28%		32,450
78,400	185,730	17,733.50 + 33%		78,400
185,730	....	Use **Worksheet** below to figure your tax.		

1.4
Quadratic and Polynomial Functions

Standard Position Parabola

While many real-world situations can be described by the linear models discussed, many others cannot. It is therefore important to be able to build many *different* types of models. In this section, we discuss a model that requires a second-degree equation for its description.

Quadratic Function

A function f is a **quadratic function** if

$$f(x) = ax^2 + bx + c$$

where a, b, and c are real numbers and $a \ne 0$.

Note that if we write $f(x)$ as y, the quadratic function has one first-degree variable and one second-degree variable. If $b = c = 0$, however, the quadratic function has the form

$$y = ax^2$$

and has a graph called a **standard position parabola**.

1.4 QUADRATIC AND POLYNOMIAL FUNCTIONS   33

EXAMPLE 1   Sketch the graph of $y = 2x^2$.

Solution   Begin by finding some ordered pairs that satisfy the equation. Let

$x_1 = 0$:   then $y_1 = 2(0)^2 = 0$;   the point is $(0, 0)$
$x_2 = 1$:   then $y_2 = 2(1)^2 = 2$;   the point is $(1, 2)$
$x_3 = -1$:   then $y_3 = 2(-1)^2 = 2$;   the point is $(-1, 2)$
$x_4 = 2$:   then $y_4 = 2(2)^2 = 8$;   the point is $(2, 8)$
$x_5 = -2$:   then $y_5 = 2(-2)^2 = 8$;   the point is $(-2, 8)$
$\vdots$

These points are plotted and a smooth curve is drawn through them, as shown in Figure 1.12. ∎

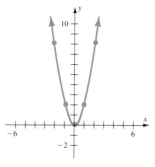

FIGURE 1.12
Graph of $y = 2x^2$

EXAMPLE 2   Sketch the graph of $y = -\frac{1}{2}x^2$.

Solution   Plot the points satisfying this equation. If

$x_1 = 0$:   then $y_1 = 0$;   the point is $(0, 0)$
$x_2 = 1$:   then $y_2 = -\frac{1}{2}$;   the point is $\left(1, -\frac{1}{2}\right)$
$x_3 = -1$:   then $y_3 = -\frac{1}{2}$;   the point is $\left(-1, -\frac{1}{2}\right)$
$x_4 = 2$:   then $y_4 = -2$;   the point is $(2, -2)$
$x_5 = -2$:   then $y_5 = -2$;   the point is $(-2, -2)$
$x_6 = 3$:   then $y_6 = -\frac{9}{2}$;   the point is $\left(3, -\frac{9}{2}\right)$
$x_7 = 4$:   then $y_7 = -8$;   the point is $(4, -8)$

The plotted points yield the curve shown in Figure 1.13.

CALCULATOR COMMENT

There are several brands of calculators in the $80–$100 price range that graph functions. For example, to graph $y = -\frac{1}{2}x^2$ on a TI-81 press the following keys:

| Y= | (−) | .5 | X|T |
| x^2 | GRAPH |

The coordinates on the curve can be see by pressing TRACE.

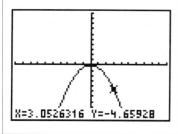

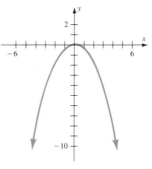

FIGURE 1.13
Graph of $y = -\frac{1}{2}x^2$ ∎

We can now make some general observations based on the special case $y = ax^2$:

1. The graph has a characteristic shape called a **parabola**.
2. If $a > 0$, the parabola open upward; we say it is **concave upward**. If $a < 0$, the parabola opens downward and is **concave downward**.
3. The point $(0, 0)$ is the lowest point if the parabola opens upward $(a > 0)$; $(0, 0)$ is the highest point if the parabola opens downward $(a < 0)$. This highest or lowest point is called the **vertex**.
4. A parabola is **symmetric** with respect to the vertical line passing through the vertex. This means we can calculate points to the right of the vertex and use symmetry to plot the corresponding points to the left of the vertex.
5. Relative to a fixed scale, the magnitude of a determines the "width" of the parabola; small values of $|a|$ yield "wide" parabolas; large values of $|a|$ yield "narrow" parabolas.

Vertex of a Parabola

For graphs of parabolas of the form

$$y - k = a(x - h)^2$$

the vertex is the point (h, k), and we use this point as the starting point for graphing the parabola. This means that we count out units from the point (h, k) instead of from the origin in order to find new points of the parabola, as illustrated in Example 3.

EXAMPLE 3 Sketch the graph of $y + 3 = 2(x - 2)^2$.

Solution By inspection, $(h, k) = (2, -3)$ is the vertex of the parabola. (Compare this equation with the equation graphed in Example 1.) If the only difference between two equations is the (h, k) point, the graphs will be the same except that they will have different vertices, as shown in Figure 1.14a.

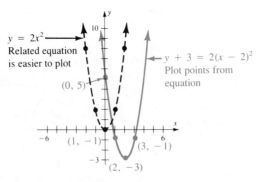

FIGURE 1.14a
Graphs of $y + 3 = 2(x - 2)^2$ and $y = 2x^2$

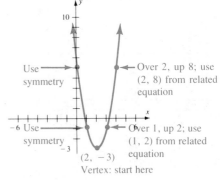

FIGURE 1.14b
Graph of $y = 2x^2$ with points plotted from the vertex $(2, -3)$

Equation: $y + 3 = 2(x - 2)^2$ Related equation: $y = 2x^2$

Let $x_1 = 0$:	$(0, 5)$	$(0, 0)$ vertex
$x_2 = 1$:	$(1, -1)$	$(1, 2)$
$x_3 = -1$:	$(-1, 15)$	$(-1, 2)$
$x_4 = 2$:	$(2, -3)$ vertex	$(2, 8)$
$x_5 = -2$:	$(-2, 29)$	$(-2, 8)$
$x_6 = 3$:	$(3, -1)$	$(3, 18)$

Instead of graphing both of these as in Figure 1.14a, suppose we count out the points of the related equation *from the point (h, k) instead of from the origin*, as shown in Figure 1.14b. Note that this graph *is the same as the graph of the more complicated equation*. This leads us to the conclusion that we can graph the simpler related equation if we remember to count out the points from the vertex rather than from the origin. ∎

EXAMPLE 4   Sketch the graph of $y - \frac{1}{2} = -2(x + \frac{3}{4})^2$.

Solution   We find and plot the vertex $(-\frac{3}{4}, \frac{1}{2})$; we next plot points on the related equation $y = -2x^2$: If

$x_1 = 0,$	then $y_1 = 0$	Vertex
$x_2 = -1,$	then $y_2 = -2$	Count over 1 and down 2 *from the vertex*
$x_3 = -2,$	then $y_3 = -8$	See Figure 1.15

∎

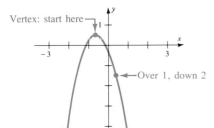

FIGURE 1.15
Graph of $y - \frac{1}{2} = -2(x + \frac{3}{4})^2$

Completing the Square

A general quadratic function

$$y = ax^2 + bx + c$$

can be rewritten in the form

$$y - k = a(x - h)^2$$

by an algebraic procedure called *completing the square*. In order to understand this procedure you need to understand the algebraic form called a *perfect square*:

$$(x + 3)^2 = x^2 + 6x + 9$$
$$(x - 2)^2 = x^2 - 4x + 4$$
$$(x + a)^2 = x^2 + 2ax + a^2$$

A perfect square is a binomial squared. When multiplied out, it is a trinomial in which the first and last terms are the squares of the first and second terms of the binomial. In addition, the middle term is twice the product of the first and second terms of the binomials. This observation gives us a procedure for completing the square for a parabola:

Given: $y = ax^2 + bx + c$ where $a \neq 0$

Step 1: Subtract the constant term from both sides:

$$y - c = ax^2 + bx$$

Step 2: Factor out the coefficient of the squared term:

$$y - c = a\left(x^2 + \frac{b}{a}x\right)$$

Step 3: Find $\frac{1}{2}$ the coefficient of the x-term, square it, and add it to both sides. Notice that since you are adding it inside the parentheses on the right, you must **add the same number**, which is $a\left(\frac{b}{2a}\right)^2 = \frac{b^2}{4a}$, to the left side:

$$y - c + \frac{b^2}{4a} = a\left[x^2 + \frac{b}{a}x + \left(\frac{b}{2a}\right)^2\right]$$

$$y + \frac{-4ac + b^2}{4a} = a\left(x + \frac{b}{2a}\right)^2$$

This is now in the form $y - y_1 = a(x - x_1)^2$. The algebra here looks terrible, but when you are working with numbers, the process is not so difficult, as shown in Example 5.

EXAMPLE 5 Complete the square for the parabola $y = 2x^2 - 8x + 5$.

Solution
$y - 5 = 2(x^2 - 4x)$ Subtract 5 from both sides and factor out the coefficient of x^2.
$y - 5 + 8 = 2(x^2 - 4x + 4)$ Add 2(4) to both sides.
$y + 3 = 2(x - 2)^2$ Factor.

The graph of this parabola is shown in Example 3. ∎

Maximum and Minimum Values

Many applications involve finding the maximum or minimum value of a function f. If f is a quadratic model, then the maximum or minimum value is found by looking at the vertex, (h, k). That is, an equation of the form

$$y - h = a(x - h)^2$$

has vertex (h, k) and can be rewritten in functional form as

$$f(x) = a(x - h)^2 + k$$

If $a > 0$, the parabola opens upward. If $x = h$, then you can see that the minimum value of f is k. You can see that this is true by looking at the graph or noting that $(x - h)^2$ is nonnegative for all x and zero if and only if $x = h$.

If $a < 0$, the parabola opens downward, and the maximum value of f is k.

1.4 QUADRATIC AND POLYNOMIAL FUNCTIONS

EXAMPLE 6 A small manufacturer of custom necklaces determines that profit is related to the number of items produced. If x items are produced per day, and the maximum number that can be produced is ten items, the total profit, in dollars, is determined to be

$$P(x) = 360x - 30x^2 - 600$$

How many necklaces should be produced in order to maximize profit?

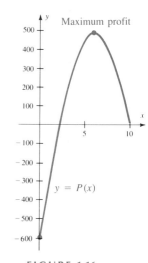

FIGURE 1.16
Graph of $P(x) = -30(x - 6)^2 + 480$

Solution The profit function has as its graph a parabola that opens downward. This can be seen by looking at the negative coefficient on the squared term. Next, to find the maximum profit we need to complete the square:

$$P(x) + 600 = -30(x^2 - 12x)$$
$$P(x) + 600 - 1{,}080 = -30(x^2 - 12x + 6^2) \quad \text{Note: } 6^2(-30) = -1{,}080$$
$$P(x) - 480 = -30(x - 6)^2$$

The vertex is (6, 480) so we see that the maximum value of P is 480, which occurs when $x = 6$. Note from the graph shown in Figure 1.16 that if there were a strike and no necklaces were produced, the daily profit would be -600 (this is a \$600-per-day loss). ∎

CALCULATOR COMMENT

PRESS: [Y=] [360] [X|T]
[−] [30] [X|T] [x^2]
[−] [600] [GRAPH]

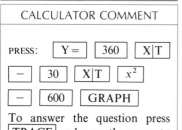

To answer the question press [TRACE] and move the cursor to the highest point on the curve:

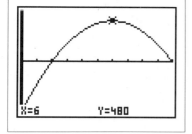

In Example 6 you might ask how it might be possible to come up with the profit function. The **profit function** is found according to the relationship*

PROFIT = REVENUE − COST

Suppose the cost of producing the necklaces in Example 6 is $5x^2 + 40x + 600$. This function,

$$C(x) = 5x^2 + 40x + 600$$

is called a **cost function** (or, to be precise, **total** cost function) because it gives the cost of producing x items. Most cost functions are made up of two parts, a variable that depends on the number of items produced and a fixed part, which does not. The value of $C(0)$ is the **fixed cost**. For this example, $C(0) = 5(0)^2 + 40(0) + 600 = 600$. Suppose also that the price of each necklace is set at $400 - 25x$ dollars. If this function represents the highest price per unit that would sell all x units, it is called the **demand function**. The demand function indicates that the price is determined by the number of necklaces sold. The **revenue function** (i.e., **total** revenue function),

---

* This is what economists refer to as the **total profit function**; see, for example, Samuelson's *Economics*, 13th ed. (1989, New York: McGraw-Hill, p. 425). In this book, we will simply refer to this as the profit function.

$R(x)$, is defined to be the product of the number of items sold and the price:

REVENUE = (NUMBER OF ITEMS)(PRICE PER ITEM)

For this example, $R(x) = x(400 - 25x)$.   ↑ This is also called the demand.

EXAMPLE 7   Draw the graphs of the cost and revenue functions on the same axis and compare with the graph of the profit function in Example 6.

Solution   We have:

REVENUE FUNCTION: $R(x) = x(400 - 25x) = -25x^2 + 400x$

Complete the square: $R(x) = -25(x^2 - 16x)$
$$R(x) - 1{,}600 = -25(x^2 - 16x + 64)$$
$$R(x) - 1{,}600 = -25(x - 8)^2$$

This parabola opens down with vertex at $(8, 1600)$. The domain is $[0, 10]$, and is shown in Figure 1.17.

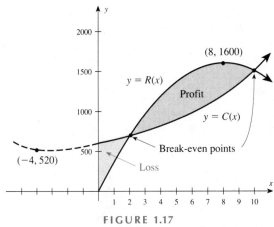

FIGURE 1.17
Graphs of $y = 400x - 25x^2$ and $y = 5x^2 + 40x + 600$

COST FUNCTION: $C(x) = 5x^2 + 40x + 600$

Complete the square: $C(x) - 600 = 5(x^2 + 8x)$
$$C(x) - 600 + 80 = 5(x^2 + 8x + 16)$$
$$C(x) - 520 = 5(x + 4)^2$$

This parabola opens up with vertex at $(-4, 520)$. The domain is $[0, 10]$, and is shown in Figure 1.17.

PROFIT FUNCTION: $P(x) = R(x) - C(x)$
$$= -25x^2 + 400x - (5x^2 + 40x + 600)$$
$$= -30x^2 + 360x - 600$$

This function was graphed in Example 6. Notice that positive profit is shown as the gray section of Figure 1.17 and negative profit (loss) is shown as the colored section of Figure 1.17. ■

Break-Even Analysis

A company will *break even* (that is, profit will balance loss) if the cost and the revenue are equal. The point (or points) at which this occurs is called a **break-even point**.

EXAMPLE 8 Find the break-even point(s) for Example 7.

Solution   Figure 1.17 labels the break-even points. Geometrically, they are the points where the revenue and cost curves intersect. To find these points algebraically, we need to find the x values for which $R(x) = C(x)$.

$$R(x) = C(x)$$
$$400x - 25x^2 = 5x^2 + 40x + 600$$
$$30x^2 - 360x + 600 = 0$$
$$x^2 - 12x + 20 = 0$$
$$(x - 10)(x - 2) = 0$$
$$x = 10, 2$$

The break-even points are at (2, 700) and (10, 1500). ■

Polynomial Functions

Linear and quadratic functions are special types of more general mathematical functions called **polynomial functions**.

Polynomial Function

> A function f is a **polynomial function** in x of degree n if
> $$f(x) = a_n x^n + a_{n-1} x^{n-1} + \cdots + a_1 x + a_0$$
> where $a_0, a_1, a_2, \ldots, a_n$ are real numbers, n is a whole number, and $a_n \neq 0$.

Examples of polynomial functions are

$$f(x) = 5x^3 + 2x^2 - 3x + 5 \qquad g(x) = x^2 - 5x + 1 \qquad h(x) = 6$$

Note that linear and quadratic functions are special types of polynomial functions of degree 1 and 2, respectively. (For a review of the terminology and algebraic simplification of polynomials, see Appendix A.)

EXAMPLE 9
a. $f(x) = x$ is a polynomial.
b. $f(x) = \frac{1}{x}$ is not a polynomial since $\frac{1}{x} = x^{-1}$ and -1 is not a whole number.
c. $f(x) = \frac{1}{6}x$ is a polynomial ($n = 1$, $a_n = \frac{1}{6}$).
d. $f(x) = \sqrt{x} + 3x^2$ is not a polynomial since $\sqrt{x} = x^{1/2}$ and $\frac{1}{2}$ is not a whole number.
e. $f(x) = x^2 + 3x + 4x^{-2}$ is not a polynomial since the exponent of $4x^{-2}$ is not a whole number. ■

1.4 Problem Set

Sketch the graph of each equation in Problems 1–38.

1. $y = x^2$
2. $y = -x^2$
3. $y = -2x^2$
4. $y = 3x^2$
5. $y = -5x^2$
6. $y = 5x^2$
7. $y = \frac{1}{3}x^2$
8. $y = -\frac{1}{3}x^2$
9. $y = \frac{1}{10}x^2$
10. $y = -\frac{1}{10}x^2$
11. $y = \frac{2}{3}x^2$
12. $y = -\frac{2}{3}x^2$
13. $y = (x - 1)^2$
14. $y = -(x + 2)^2$
15. $y = (x + 3)^2$
16. $y = -2(x - 1)^2$
17. $y = \frac{1}{4}(x - 1)^2$
18. $y = -\frac{1}{2}(x + 1)^2$
19. $y - 2 = (x - 1)^2$
20. $y - 2 = 3(x + 2)^2$
21. $y - 2 = -\frac{3}{5}(x - 1)^2$
22. $y + 3 = \frac{2}{3}(x + 2)^2$
23. $y + \frac{2}{3} = (x + \frac{1}{3})^2$
24. $y + \frac{2}{5} = -(x - \frac{3}{5})^2$
25. $y + \frac{2}{5} = (x - \frac{3}{5})^2$
26. $y - .1 = (x + .2)^2$
27. $y = x^2 - 4x + 4$
28. $y = x^2 - 6x + 9$
29. $y = -2x^2 - 2x - 2$
30. $y = 3x^2 + 6x + 3$
31. $y = x^2 + 4x + 5$
32. $y = x^2 - 6x + 11$
33. $y = 2x^2 + 4x$
34. $y = 3x^2 - 6x$
35. $y = -2x^2 - 4x + 1$
36. $y = -2x^2 - 4x + 3$
37. $y = -3x^2 + 12x - 16$
38. $y = 2x^2 - 2x + 5$

Find the maximum or minimum value of y for each of the functions in Problems 39–50.

39. $y = -4(x + 1)^2 + 3$
40. $y = -5(x - 4)^2 + 2$
41. $y = -10(x - 450)^2 + 1,250$
42. $y = 12(x + 30)^2 - 140$
43. $y = 25(x - 560)^2 - 1,400$
44. $y = -150(x + 2,300)^2 + 12,000$
45. $y = -3x^2 - 18x - 41$
46. $y = -4x^2 - 40x - 60$
47. $2x^2 + 12x - y + 31 = 0$
48. $3x^2 - 12x - y + 22 = 0$
49. $6x^2 - 12x + 3y + 18 = 0$
50. $5y - 30x^2 - 180x - 370 = 0$

APPLICATIONS

51. After extensive market research, a consulting firm has determined that the demand for a certain item is $1,040 - 10x$ dollars, where x is the number of items produced. Since the company has fixed costs of 6,650 dollars, the cost function is found to be $C(x) = 6,650 + 500x$.
    a. Find the revenue function.
    b. Find the break-even point(s).

52. The demand for a certain ratchet flange is $50 - x$ dollars, where x is the number of flanges produced. The cost function is found to be $C(x) = 200 + 20x$.
    a. Find the revenue function.
    b. Find the break-even point(s).

53. A manufacturer produces quality boats. The profit, in dollars, is determined to be
$$P(x) = -10(x - 375)^2 + 1,156,250$$
where x is the number of boats.
    a. How many boats should be produced in order to produce a maximum profit?
    b. What is the profit (or loss) if no boats are produced?
    c. What is the maximum profit?

54. Find the maximum profit for Problem 51.
55. Find the maximum profit for Problem 52.
56. Graph the cost and revenue functions of Problem 51 and shade the region representing the company's profit.
57. Graph the cost and revenue functions of Problem 52 and shade the region representing the company's profit.
58. The highest bridge in the world is the bridge over the Royal Gorge of the Arkansas River in Colorado. It is 1,053 ft above the water. If a rock is thrown vertically upward from this bridge with an initial velocity of 64 feet per second, the height h of the rock above the river at time t is described by the function
$$h(x) = -16t^2 + 64t + 1,053$$

What is the maximum height possible for a rock thrown vertically upward from the bridge with an initial velocity of 64 feet per second? After how many seconds will it reach that height?

59. In 1974 Evel Knievel attempted a skycycle ride across the Snake River. Suppose the path of the skycycle is given by the equation

$$d(x) = -.0005(x - 2{,}390)^2 + 3{,}456$$

where $d(x)$ is the height above the canyon floor for a horizontal distance of x feet from the launching ramp. What was Knievel's maximum height?

60. **a.** In most states, drivers are required to know approximately how long it takes to stop their cars at various speeds. Suppose you estimate three car lengths for 30 mph and six car lengths for 60 mph. One car length per 10 mph assumes a linear relationship between speed and distance covered. Write a linear equation where x is the speed of the car and y is the distance traveled by the car in feet. (Assume that one car length = 15 ft.)

b. The scheme for the stopping distance of a car given in part **a** is convenient but not accurate. The stopping distance is more accurately approximated by the quadratic equation

$$y = .071x^2$$

This says that a car requires four times as many feet to stop at 60 miles per hour than at 30 mph. That is, doubling the speed quadruples the braking distance. Graph this quadratic equation and the linear equation from part **a** on the same coordinate axes.

c. Comment on the results from part **b**. At about what speed are the two measures the same?

1.5
Rational Functions

The third type of function used for mathematical models is called a **rational function**.

Rational Function

A function f is a **rational function** if

$$f(x) = \frac{P(x)}{Q(x)}$$

where P is any polynomial function and Q is a polynomial function whose domain excludes values for which $Q(x) = 0$.

Functions such as

$$f(x) = \frac{1}{x} \qquad y = \frac{1}{x-2} \qquad g(x) = \frac{x^2 + 3x - 1}{x - 1}$$

are examples of rational functions. The values $x = 0$, $x = 2$, and $x = 1$, respectively, are excluded from the domains of these three functions according to the definition of a rational function.

In Section 1.2, we graphed $h(x) = \frac{1}{x}$ (Example 3) by plotting points. It is important to remember the general shape of the graph of this function so we repeat it here for easy reference.

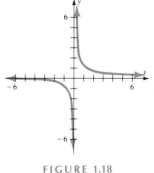

FIGURE 1.18
Graph of $y = \frac{1}{x}$

Note that the domain for the graph in Figure 1.18 consists of all real numbers except $x = 0$. That is, the graph of the function $y = \frac{1}{x}$ does not cross the vertical line $x = 0$. Now consider the graph of the rational function $y = \frac{1}{x-1}$. The definition of rational functions *excludes* from the domain all values for which $x - 1 = 0$, namely, $x = 1$. You can, however, view the equation $x = 1$ as the equation of a vertical line. This line is called a **vertical asymptote** for the curve $y = 1/(x - 1)$ and is illustrated in Example 1a. We also say that this function is *unbounded*. In general, a function is **unbounded** if for any number M there is a value of the function whose numerical value is larger than M or smaller than $-M$.

EXAMPLE 1   Graph the following equations by plotting points:

a. $y = \dfrac{1}{x - 1}$ b. $y = \dfrac{1}{x + 2}$ c. $y = \dfrac{1}{2x - 5}$

Solution   The vertical asymptotes are lines for which the denominators are equal to zero. To find the asymptotes, set the denominator equal to zero and solve:

a. $x - 1 = 0$ b. $x + 2 = 0$ c. $2x - 5 = 0$
$x = 1$ $x = -2$ $x = \frac{5}{2}$

Plot additional points as necessary as shown in Figure 1.19.

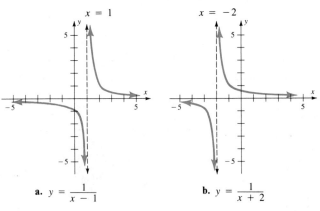

a. $y = \dfrac{1}{x - 1}$ b. $y = \dfrac{1}{x + 2}$

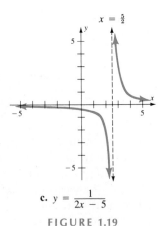

c. $y = \dfrac{1}{2x - 5}$

FIGURE 1.19
Vertical asymptotes

A line is an asymptote for a curve if the distance between the line and the curve becomes smaller and smaller within some suitable domain. The graph of a function cannot cross its vertical asymptotes.

EXAMPLE 2   Graph

$$y = \frac{1}{3x^2 - 5x - 2}$$

Solution   Find the vertical asymptotes and compare them with those in Figure 1.19:

$$3x^2 - 5x - 2 = 0$$
$$(x - 2)(3x + 1) = 0$$
$$x = 2, -\frac{1}{3}$$

Draw the vertical asymptotes and plot additional points.

x	y
0	$-\frac{1}{2}$
1	$-\frac{1}{4}$
3	$\frac{1}{10}$
-1	$\frac{1}{6}$
-2	$\frac{1}{20}$

See Appendix B for an Introduction to Spreadsheets.

```
SPREADSHEET PROGRAM
         A          B
  1      x          y=1/(3x^2-5x-2)
  2      -3         1/(3*A2^2-5*A2-2)
  3      +A2+0.25   replicate
  4      replicate
```

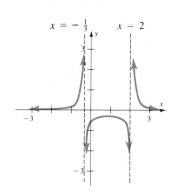

```
   x          y=1/(3x^2-5x-2)
  -3          0.025
  -2.75       0.029038113
  -2.5        0.034188034
  -2.25       0.040920716
  -2          0.05
  -1.75       0.062745098
  -1.5        0.081632653
  -1.25       0.111888112
  -1          0.166666667
  -0.75       0.290909091
  -0.5        0.8
  -0.25      -1.77777778
   0         -0.5
   0.25      -0.32653061
   0.5       -0.26666667
   0.75      -0.24615385
   1         -0.25
   1.25      -0.28070175
   1.5       -0.36363636
   1.75      -0.64
   2          ERR
   2.25       0.516129032
   2.5        0.235294118
   2.75       0.144144144
   3          0.1
```

It is difficult to find all asymptotes with a spreadsheet program.

← The error here is that we are requesting division by zero. This indicates a vertical asymptote.

It would be convenient if we could say that if $f(x) = P(x)/Q(x)$, then any value for which $Q(x) = 0$ would be a vertical asymptote. Unfortunately, such is not the case, as we see in Example 3.

EXAMPLE 3 Graph

$$y = \frac{x}{x^2 - x}$$

Solution Note that if $x^2 - x = 0$ then

$$x(x - 1) = 0$$
$$x = 0, 1$$

There are two values that cause division by zero. However, if we plot points, we see that $x = 1$ is a vertical asymptote and that $x = 0$ is simply a point deleted from the domain. The graph is shown in Figure 1.20.

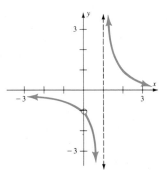

FIGURE 1.20

Graph of $y = \dfrac{x}{x^2 - x}$

CALCULATOR COMMENT

If you are using a graphing calculator, you must pay particular attention to Example 3. Note here, if we press

Y=	X	T	÷	(
X|T	x^2	−	X|T
)	GRAPH

we get the graph shown below; the asymptote is shown, but the deleted point at $x = 0$ is not shown.

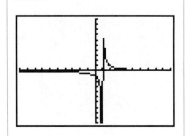

There is a way to decide if a value causing division by zero indicates a deleted point or a vertical asymptote. All we need to do is reduce the fraction and set the resulting denominator equal to zero. Thus

$$\frac{x}{x^2 - x} = \frac{x}{x(x - 1)} = \frac{1}{x - 1}$$

$$x \neq 0, \, x \neq 1$$

Values that cancel indicate deleted points (that is, $x = 0$).

$x = 1$ is an asymptote
Values that do not cancel indicate asymptotes

EXAMPLE 4 Given the following equations, find the vertical asymptotes.

a. $y = \dfrac{x + 4}{x^2 - 16}$ **b.** $y = \dfrac{x + 4}{x^2 - 15}$ **c.** $y = \dfrac{3x^2 - 5x - 2}{x^2 - 5x + 6}$

Solution **a.** $y = \dfrac{x + 4}{x^2 - 16} = \dfrac{x + 4}{(x - 4)(x + 4)} = \dfrac{1}{x - 4}$

$x \neq 4, \, x \neq -4$ $x = 4$ is a vertical asymptote.

$x = -4$ is a deleted point on the graph.

b. $y = \dfrac{x + 4}{x^2 - 15}$ is reduced. The vertical asymptotes are found by solving $x^2 - 15 = 0$. The vertical asymptotes have equations $x = \sqrt{15}, \, x = -\sqrt{15}$.

c. $y = \dfrac{3x^2 - 5x - 2}{x^2 - 5x + 6} = \dfrac{(3x + 1)(x - 2)}{(x - 3)(x - 2)} = \dfrac{3x + 1}{x - 3}$

$x = 3$ is a vertical asymptote.

Note that the graphs in Figures 1.18 and 1.19 and in Examples 2 and 3 not only have vertical asymptotes but also have a **horizontal asymptote**, $y = 0$. As $|x|$ gets larger, each of the named curves gets closer to the line $y = 0$. There will be a horizontal asymptote $y = 0$ whenever the degree of the numerator is less than the degree of the denominator. Study the variations of asymptotes shown in Figure 1.21**a**–**d**.

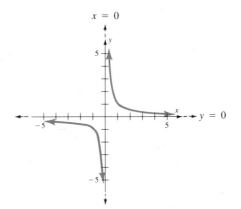

FIGURE 1.21a
Graph of $y = \dfrac{1}{x}$.
Vertical asymptote, $x = 0$; horizontal asymptote, $y = 0$.

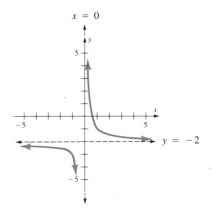

FIGURE 1.21b
Graph of $y = \dfrac{1}{x} - 2$.
Vertical asymptote, $x = 0$; horizontal asymptote, $y = -2$.

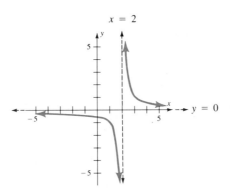

FIGURE 1.21c
Graph of $y = \dfrac{1}{x - 2}$.
Vertical asymptote, $x = 2$; horizontal asymptote, $y = 0$.

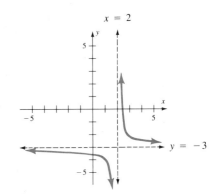

FIGURE 1.21d
Graph of $y = \dfrac{1}{x - 2} - 3$.
Vertical asymptote, $x = 2$; horizontal asymptote, $y = -3$.

Apparently, horizontal asymptotes depend on the relative degrees of the numerator and denominator polynomials. The following summary is useful when graphing rational functions.

Asymptotes

> If $f(x) = P(x)/D(x)$, where $P(x)$ and $D(x)$ are polynomial functions with no common factors, then:
>
> the line $x = r$ is a *vertical asymptote* if $D(r) = 0$;
> the line $y = 0$ is a *horizontal asymptote* if the degree of P is less than the degree of D;
> the line $y = \dfrac{a_n}{b_n}$ is a *horizontal asymptote* if the degree of P is the same as the degree of D and
> $$P(x) = a_n x^n + \cdots + a_0 \text{ and}$$
> $$D(x) = b_n x^n + \cdots + b_0$$

EXAMPLE 5 Graph $y = \dfrac{3x - 1}{2x + 5}$.

Solution First find the vertical asymptotes. Set the denominator equal to 0 and solve.

$$2x + 5 = 0$$
$$2x = -5$$
$$x = -\dfrac{5}{2}$$

Second, find the horizontal asymptotes. The degree of $3x - 1$ and $2x + 5$ is the same, so there is one horizontal asymptote: $y = \tfrac{3}{2}$. The remaining points are found by calculation.

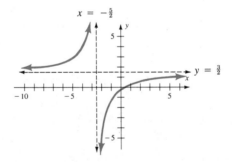

x	y
0	$-\tfrac{1}{5}$
1	$\tfrac{2}{7}$
-1	$-\tfrac{4}{3}$
-2	-7
-3	10
-4	$\tfrac{13}{3}$
-5	$\tfrac{16}{5}$

Cost-Benefit Models

Many real-life situations involve some desired result, or benefit. However, to achieve this result involves a certain cost and it is often necessary to compare the benefit with the cost. An equation that compares costs and benefits is called a **cost-benefit model**. Suppose that the cost, C, of installing five-field electrostatic precipitators for the removal of p percent of pollutants released into the atmosphere from the emissions of a cement factory is given by the formula

$$C(p) = \dfrac{200{,}000p}{105 - p}$$

Example 6 demonstrates the use of this cost-benefit model.

EXAMPLE 6 Use the cost-benefit function given above to find:

a. How much would it cost to remove 95% of the pollutants (as required by the Environmental Protection Agency)?
b. How much does it cost to remove 80% of the pollutants (which the company is presently doing)?
c. How much would it cost to remove 100% of the pollutants?
d. Graph the cost-benefit relationship for $0 \leq p \leq 100$.

Solution a. For $p = 95$, $C = \dfrac{200{,}000(95)}{105 - 95} = 1{,}900{,}000$

b. For $p = 80$, $C = \dfrac{200{,}000(80)}{105 - 80} = 640{,}000$

c. For $p = 100$, $C = \dfrac{200{,}000(100)}{105 - 100} = 4{,}000{,}000$

d. Use the points found above (along with additional ones as needed) to graph the function for $0 \leq p \leq 100$. Note that there is both a horizontal and vertical asymptote.

$$C = \dfrac{200{,}000p}{105 - p} \quad \longleftarrow \text{Vertical asymptote: } 105 - p = 0$$
$$p = 105$$

Horizontal asymptote since the degree of the numerator and denominator is the same; It Is

$$C = \dfrac{200{,}000}{-1}$$
$$= -200{,}000$$

From the coefficients of p

Note: You will not see the graph approach this asymptote because C approaches $-200{,}000$ as $|p|$ gets larger, but we are asked to graph this curve only for $0 \leq p \leq 100$. ∎

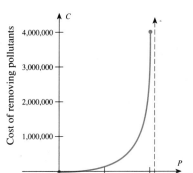

Points:

x	y
0	0
50	188,510
60	268,860
70	391,111
80	636,478
90	1,230,107

1.5
Problem Set

1. What is a rational function?
2. What is a vertical asymptote, and how do you find it?
3. What is a horizontal asymptote, and how do you find it?

Name the horizontal and vertical asymptotes for the curves in Problems 4–15. Also name any deleted points.

4. $f(x) = \dfrac{1}{x + 3}$

5. $f(x) = \dfrac{5}{2x - 3}$

6. $f(x) = \dfrac{2x - 2}{3x^2 - x - 2}$

7. $g(x) = \dfrac{x + 3}{x^2 + 2x - 3}$

8. $g(x) = \dfrac{6}{2x^2 + 3x - 2}$

9. $g(x) = \dfrac{5}{6x^2 + 5x - 6}$

10. $h(x) = \dfrac{2x + 5}{3x - 4}$

11. $h(x) = \dfrac{5x^2 + 3x - 2}{2x^2 + 4x + 1}$

12. $h(x) = \dfrac{6x^2 - 3x + 1}{2x^2 - 5}$

13. $y = \dfrac{3x + 2}{x^2 + x + 2}$

14. $y = \dfrac{x - 5}{x^2 - 3x + 5}$

15. $y = \dfrac{2x + 1}{x^2 + x - 2}$

Graph the functions in Problems 16–27.

16. $y = \dfrac{2}{x}$

17. $y = -\dfrac{1}{x}$

18. $y = \dfrac{1}{x+3}$

19. $y = \dfrac{1}{x-4}$

20. $y = \dfrac{1}{x} + 4$

21. $y = \dfrac{1}{x} + 3$

22. $y = \dfrac{2x^2 + 5x + 3}{x^2 - x - 2}$

23. $y = \dfrac{2x^2 - 3x + 1}{x^2 + 2x - 3}$

24. $y = \dfrac{1}{x^2 - 4}$

25. $y = \dfrac{1}{x^2 - 9}$

26. $y = \dfrac{1}{3x^2 - x - 2}$

27. $y = \dfrac{1}{x^2 + 2x - 3}$

APPLICATIONS

In Problems 28–31, use the cost-benefit model

$$C = \dfrac{40{,}000p}{110 - p}$$

where C is the cost (to the nearest dollar) of removing p% of the pollutants.

28. What is the cost of removing 95% of the pollutants?
29. What is the cost of removing 80% of the pollutants?
30. What is the cost of removing 100% of the pollutants?
31. Graph these relationships for $0 \le p \le 100$.

In Problems 32–35, use the cost-benefit model

$$C = \dfrac{18{,}000p}{100 - p}$$

where C is the cost of removing p% of the pollutants.

32. What is the cost of removing 95% of the pollutants?
33. What is the cost of removing 80% of the pollutants?
34. Is it possible to remove 100% of the pollutants?
35. Graph these relationships for $0 \le p < 100$.

Radiologists must deal with three quantities each time an X-ray is taken:

t = time in seconds that the X-ray machine is on
mA = the current, measured in milliamps
FFD = distance from the X-ray machine to the film

Use this information in Problems 36 and 37.

36. If FFD is held constant, the relationship between mA and t is given by

$$mA = \dfrac{400}{t}$$

Graph this relationship for $0 < t \le 10$.

37. If time is held constant, the relationship between mA and FFD is given by

$$mA = \dfrac{256}{(FFD)^2}$$

Graph this relationship for $0 < FFD \le 16$.

The pressure-volume relationship is

$$\dfrac{\text{original pressure}}{\text{new pressure}} = \dfrac{\text{new volume}}{\text{original volume}}$$

Use this proportion in Problems 38 and 39.

38. If the new pressure is 840 millimeters (mm) of mercury and the new volume is 100 milliliters (ml), write a rational function relating the original pressure and the original volume. Graph this relationship using the pressure as the independent variable.

39. If 400 ml of oxygen is under a pressure of 2,800 mm of mercury, write a rational function relating the new pressure and the new volume. Graph this relationship using the pressure as the independent variable.

40. Suppose that the supply and demand functions for a product are

$$S(p) = p - 20 \quad \text{and} \quad D(p) = \dfrac{800}{p}$$

The point(s) where $S(p) = D(p)$ are called the **equilibrium point(s)**.
 a. Graph S and D on the same coordinate axes and estimate the equilibrium point(s).
 b. Algebraically solve $S(p) = D(p)$.
 c. Where does the supply function cross the p-axis? What is the economic significance of this point?

*1.6

Review

The material of this chapter is reviewed in the following list of objectives. After each objective there are some practice questions.

* Optional section.

For a sample test select the first question of each set and check your answers. The second question for each objective has no answer given. If you are having trouble with a particular type of problem, look back at the indicated section in the text. When you are finished reviewing these objectives, a sample examination is given at the end of this section.

[1.1]
Objective 1.1: Evaluate a function. Let $f(x) = 5x - 3$ and $g(x) = 2x^2 - 3x + 1$ and find the requested values.
1. $f(10)$
2. $f(-3)$
3. $g(-2)$
4. $g(4)$

Objective 1.2: Evaluate a function and simplify. Let $f(x) = 3x^2$ and $g(x) = x^2 - 5$.
5. $f(s + 3)$
6. $g(t - h)$
7. $f(x + h) - f(x)$
8. $\dfrac{g(x + h) - g(x)}{h}$

[1.2]
Objective 1.3: Given a graph, find coordinates of indicated points, the domain, range, and intercepts.

9.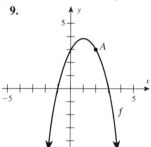
10.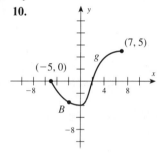
11.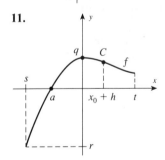
12.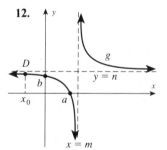

Objective 1.4: Graph a given function by plotting points.
13. $f(x) = 9 - 5x$
14. $g(x) = 25 - 5x - x^2$
15. $h(x) = \dfrac{1}{x - 3}$
16. $t(x) = \dfrac{x^2 - 3}{x}$

Objective 1.5: Find the x- and y-intercepts for a line whose equation is given.
17. $y = 5x - 4$
18. $3x + 2y - 6 = 0$
19. $3x - 4y + 12 = 0$
20. $50x - 250y = 1,000$

Objective 1.6: Find the slope of the line passing through two given points.
21. $(4, 1), (-3, -2)$
22. $(-3, -1), (2, 6)$
23. $(-3, 5), (-1, -5)$
24. $(-1, -2), (-6, -8)$

[1.3]
Objective 1.7: Graph a line by finding the slope and y-intercept.
25. $y = -\tfrac{2}{5}x + 2$
26. $y = \tfrac{2}{3}x - \tfrac{1}{3}$
27. $x + 2y - 8 = 0$
28. $3x - 5y - 10 = 0$

Objective 1.8: Graph a line segment or piecewise linear function.
29. $2x - 3y - 6 = 0; \quad -6 \le x \le 3$
30. $y = -5x + 3; \quad -4 \le x \le 3$
31. $\begin{cases} x - y + 5 = 0 & \text{if } -5 \le x \le 0 \\ x + y + 5 = 0 & \text{if } 0 < x < 5 \end{cases}$
32. $y = 2|x - 1|$

Objective 1.9: Find the standard form of the equation of a line satisfying given conditions.
33. Passing through $(2, 10)$ and $(5, 25)$
34. Slope $-\tfrac{2}{3}$; passing through $(20, -100)$
35. y-intercept -4; slope $\tfrac{1}{5}$
36. no y-intercept and no slope, passing through $(1, 5)$

[1.4]
Objective 1.10: Sketch the graph of a parabola.
37. $y = 2(x + 3)^2$
38. $y + 1 = \tfrac{1}{2}(x - 2)^2$
39. $y - 4 = -\tfrac{2}{3}(x + 2)^2$
40. $y = 3x^2 - 6x - 1$

Objective 1.11: Find the maximum or minimum value of y for a quadratic function.
41. $y + 250 = -\tfrac{1}{3}(x - 1{,}300)^2$
42. $y = -5(x - 300)^2 + 1{,}100$
43. $y = 2x^2 - 12x + 268$
44. $2y - 3(x + 40)^2 - 100 = 0$

[1.5]
Objective 1.12: Name the asymptotes for a rational function.
45. $f(x) = \dfrac{1}{x - 5}$
46. $f(x) = \dfrac{3x + 2}{2x - 5}$
47. $g(x) = \dfrac{4}{x^2 - x - 6}$
48. $g(x) = \dfrac{3x^2 + 2x + 1}{2x^2 - 7x - 4}$

Objective 1.13: Graph rational functions.
49. $f(x) = \dfrac{4}{x}$
50. $f(x) = \dfrac{2}{x + 2}$
51. $g(x) = \dfrac{4}{x} + 3$
52. $g(x) = \dfrac{2x + 2}{x^2 + 3x + 2}$

Objective 1.14: Solve applied problems based on the preceding objectives.
53. *Rate of change.* If $Z(x)$ = the price of Xerox stock on the first trading day in year x, write an expression for the average rate of change in price of Xerox stock from 1985 to 1990.
54. *Revenue.* After market research it was found that the demand equation for a certain product is $4{,}000 - 50x$ dollars,

where x is the number of items produced. The cost function for this product is $C(x) = 25,000 + 1,000x$. Find the revenue function.

55. *Supply.* If a distributor can supply 2,000 items when the cost is $200 and 8,000 if the cost is $400, write the supply equation if you know that supply is a linear function. Assume x = cost is the independent variable.

56. *Equilibrium point.* The demand for the item described in Problem 55 is 7,000 items if the price is $100 and only 1,000 items if the price is $700. What is the equilibrium point for which the supply and demand are the same provided the demand is linear?

57. *Break-even point.* What is the break-even point for the information described in Problem 54?

58. *Maximum profit.* A manufacturer can produce no more than 200 items and it is determined that the profit (in dollars) is given by the following function:

$$P(x) = -10(x - 170)^2 + 14,320$$

What is the maximum profit, and how many items should be manufactured in order to achieve this maximum profit?

59. *Cost-benefit model.* Use the cost-benefit model

$$C(p) = \frac{20,000p}{100 - p}$$

to find the cost of removing 95% of the pollutants.

60. *Radiology technology.* Graph the relationship

$$mA = \frac{200}{t}$$

which relates the time that an X-ray machine is on and the current used in the X-ray. You need graph this function only for $0 < t \leq 2$.

SAMPLE TEST

The following sample test (45 minutes) is intended to review the main ideas of this chapter.

In Problems 1–8, let $f(x) = 2x^2 - 3x + 5$ and $g(x) = 1 - 4x$. Find the values requested in Problems 1–6.

1. $f(3)$
2. $g(10)$
3. $g(t + 1)$
4. $f(t - h)$
5. $\dfrac{g(x + h) - g(x)}{h}$
6. $\dfrac{f(x + h) - f(x)}{h}$

7. Graph g.
8. What are the x- and y-intercepts for f?
9. Find the slope of the line passing through $(-5, 9)$ and $(-3, -7)$.
10. Graph $2x + 3y - 12 = 0$ by finding the slope and y-intercept.
11. Graph $y = |x + 3|$.
12. Graph

$$\begin{cases} x + y + 1 = 0 & \text{for } -3 \leq x \leq 1 \\ y + 2 = 0 & \text{for } 1 < x < 5 \end{cases}$$

13. Find the standard form of the equation of a line passing through $(30, 20)$ with slope -5.
14. Find the equation of a line with zero slope and y-intercept -3.
15. Find the equation of a line with no slope passing through $(-3, -2)$.
16. Sketch $y + 3 = \frac{1}{2}(x - 2)^2$.
17. What is the maximum value of the function $y - 550 = -\frac{2}{3}(x - 40)^2$?
18. What are the asymptotes for the function

$$F(x) = \frac{2x + 1}{x^2 + x - 6}$$

19. Use the cost-benefit model

$$C(p) = \frac{500p}{100 - p}$$

to find the cost of removing 90% of the pollutants.

20. Graph C from Problem 19 for $0 \leq p < 100$.

MODELING
APPLICATION 1

Gaining a Competitive Edge in Business

Solartex manufactures solar collector panels. During the first year of operation, rent, insurance, utilities, and other fixed costs averaged $8,500 per month. Each panel sold for $320 and cost the company $95 in materials and $55 in labor. Since company resources are limited, Solartex cannot spend more than $20,000 in any one month. Sunenergy, another company selling similar panels, competes directly with Solartex. Last month, Sunenergy manufactured 85 panels at a total cost of $20,475, but the previous month produced only 60 panels at a total cost of $17,100.

Mathematical modeling involves creating mathematical equations and procedures to make predictions about the real world. Typical textbook problems focus on limited, specific skills, but when confronted with a real-life example you are often faced with a myriad of "facts" without specific clues on how to fit them together to make predictions. In this book, you will be given a modeling application and asked to write a paper using the given information. You will need to do some research to have adequate data. There are no "right answers" for these papers. In the *Student's Solution Manual*, a paper is presented for this modeling application, but your paper could certainly take a quite different direction.

For this first application, the following questions might help you get started: What is the cost equation for Solartex? How many panels can be manufactured by Solartex given the available capital? What is the minimum number of panels that should be produced? In order to make a profit, revenue must exceed cost; what is Solartex's break-even point? How would you compare Solartex and Sunenergy?

Feel free to supply information that is not given above. For example, a study of market demand could be useful. Suppose you find that at $75 per panel, Solartex (or a real company you have studied) could sell 200 panels per month, but at $450, it could sell only 20 panels per month. On the other hand, if Solartex had to sell the panels at $225 each, it could afford material that would limit it to only 30 panels per month, but at $450 each, it could afford sufficient material to supply 100 panels per month. What is the equilibrium point for this information?

Write a paper based on this modeling application.

2 The Derivative

- 2.1 Limits
- 2.2 Continuity
- 2.3 Rates of Change
- 2.4 Definition of Derivative
- 2.5 Differentiation Techniques, Part I
- 2.6 Differentiation Techniques, Part II
- 2.7 The Chain Rule
- 2.8 Chapter 2 Review
  Chapter Objectives
  Sample Test

CHAPTER OVERVIEW
We focus here on the concept of a derivative and introduce efficient ways of finding the derivative. The derivative is one of the fundamental ideas in all of calculus and is the cornerstone of more advanced mathematics.

PREVIEW
We first introduce the derivative as a rate of change, but soon show that it has many additional useful applications. After just a glimpse of these applications (to be continued in the next chapter), we concentrate on finding derivatives.

PERSPECTIVE
The skills learned in this chapter are used in the next two chapters to develop applications of the derivative. Then the integral is introduced as an "antiderivative"—which will again use the knowledge of the derivatives introduced in this chapter. Two of the single most revolutionary concepts in all of mathematics are the ideas of limits and derivatives. Remember, it took some of the greatest minds in the history of mathematics many years to formulate these ideas, so do not despair if you have trouble understanding them in one evening, or even in one course. Hard work and perseverance will pay off.

MODELING
APPLICATION 2

Instantaneous Acceleration: A Case Study of the Mazda 626

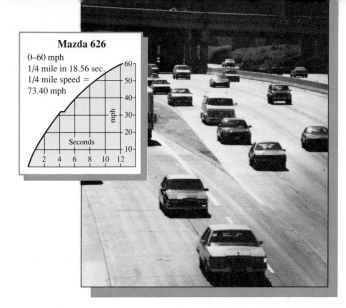

Mazda 626
0–60 mph
1/4 mile in 18.56 sec.
1/4 mile speed = 73.40 mph

Consider the performance of the Mazda 626 Sport Coupe. The graph shown here, taken from an advertisement, shows the car's acceleration.*

After you have finished this chapter, write a paper discussing the average rate of travel, the distance traveled, the velocity, and the acceleration for this car.**

For general guidelines about writing this essay, See Modeling Application 1 on page 51.

*From *Time*, January 4, 1982.
**This modeling application is adapted from Peter A. Lindstrom, "A Beginning Calculus Project," UMAP, Vol. 5, No. 3, pp. 271–276.

APPLICATIONS

Management (*Business, Economics, Finance, and Investments*)
Cost of manufacturing
  (2.1, Problem 69; 2.5, Problem 55)
Rental charges for a fleet of trucks
  (2.2, Problem 46)
Output as a function of the number of workers (2.3, Problems 5–8)
Rate of change of the Gross National Product (2.3, Problems 13–16)
Cost of a per unit increase in production
  (2.3, Problems 47–49; 2.8, Sample Test, Problem 19)
Marginal profit
  (2.3, Problem 50; 2.5, Problem 56; 2.8, Sample Test, Problem 20)
Average rate of change of cost
  (2.3, Problems 51–55)
Marginal cost
  (2.3, Problems 58–59; 2.4, Problems 28–37; 2.5, Problem 55)
Change in the rate of earnings in a corporation (2.5, Problem 59)
Consumer Price Index (2.5, Problem 65)

Management (*continued*)
Demand for a commodity in a free market
  (2.6, Problems 41–44)
Advertising to influence purchasing
  (2.6, Problem 45)
Rate of change of prices in a free market
  (2.6, Problems 46–48)
Enrollment projections (2.7, Problem 59)

Life Sciences (*Biology, Ecology, Health, and Medicine*)
Temperatures at Death Valley
  (2.2, Problem 48)
Rate at which the number of bacteria in a culture change
  (2.4, Problems 38–40; 2.7, Problem 53)
Relationship between current and time on an X-ray machine (2.5, Problems 57–58)
The relationship between a population of foxes and rabbits (2.5, Problems 62–63)
Rate of a liquid flowing into a reservoir
  (2.5, Problem 64)
Effect of a drug in the bloodstream
  (2.6, Problems 49–50)

Social Sciences (*Demography, Political science, Population, Psychology, Society, and Sociology*)
The number of animals available for a psychology experiment (2.1, Problem 70)
Learning theory
  (2.1, Problem 71; 2.5, Problems 60–61; 2.6, Problems 51–52; 2.7, Problems 54, 60)
SAT scores of first-year college students
  (2.3, Problems 9–12)

General Interest
Postal charges (2.2, Problem 47)
Height of a projectile after t seconds
  (2.3, Problems 1–4)
Average speeds in a daily commute
  (2.3, Problems 17–20)
Velocity of an object moving in a straight line (2.4, Problems 41–44)
Interest rate changes (2.7, Problems 55–57)

Modeling Application—
Instantaneous Acceleration: A Case Study of the Mazda 626

2.1 Limits

Mathematical analysis can be divided into two broad categories: *continuous* and *discrete*. Let us consider a few examples to clarify the distinction. The counter on a turnstile may look very similar to the odometer on a car, but the turnstile is a discrete counting device while the odometer is a continuous device. The set of integers is a discrete set, whereas the set of real numbers is not. Calculus was being invented as mathematicians all over the world realized they needed to deal with new notions concerning the transition from discrete to continuous. Calculus is based on a continuous model, and the central key to understanding calculus is the notion of a limit.

Calculus was originally developed intuitively. Over time, every concept was subjected to the most meticulous scrutiny. It was felt that all mathematical thinking should eventually lead to the ideas of calculus and be continuous in nature. Then, in 1956, a landmark text by Kemeny, Snell, and Thompson called *Introduction to Finite Mathematics* was published. It dealt with discrete ideas not contained in calculus. For over 20 years this course has been offered at various colleges and universities but it was never meant to be a replacement for calculus. However, as computers became more and more a part of mathematics and mathematical development, the necessity of treating the continuous ideas of calculus from a discrete standpoint became increasingly important. Many mathematicians are now suggesting that continuous and discrete be accepted as fundamental classifications in mathematics and the mathematics curriculum and that courses in calculus be offered alongside courses in discrete mathematics.

Intuitive Notion of a Limit

We will now turn to the notion of a limit. Imagine a child throwing a ball at a brick wall:

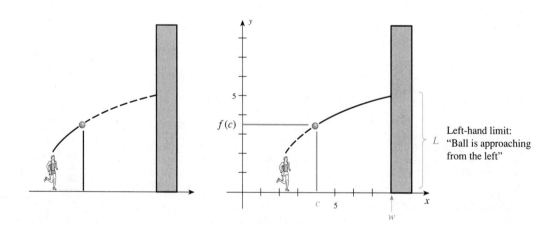

We might say that the ball is getting closer and closer to the wall, and that the wall is a limiting position for the ball.

Can we develop a mathematical model for this situation? In mathematical notation, we can describe the path of the ball by some function, say f, which gives the position of the ball. We see that the point $(c, f(c))$ is a point on the ball's path when x has the value of c. Suppose also that the wall is located at some point $(w, 0)$ on the x-axis. If we write $x \to w^-$ we mean that x approaches the wall from the left (the negative means it is approaching from the left side). If we write $x \to w^+$ we means that x is approaching the wall from the right (the positive means it is approaching from the right side).

The next question that we ask is how high is the ball when it hits the wall? If we mark the location on the wall using the coordinate system, we see the first component must be w. Thus, if L represents the height, then the ball strikes the wall at (w, L).

The mathematical model for this situation uses a notation called *limit notation* as follows:

$$\lim_{x \to w^-} f(x) = L$$

This is read as "the limit of f as x approaches w from the left equals L." In this case, L is called the "left-hand limit." This means that we predict that as x approaches w from the left along the x-axis, the path described by the function f will hit the wall at a height L.

If the projectile approaches from the right, we write

$$\lim_{x \to w^+} f(x) = R$$

The number R is called the "right-hand limit."

If we throw a projectile at a wall from both the left and the right, there is no reason to expect that it will hit at exactly the same height as shown by the following figure. Notice that our wall is now drawn as a line with no thickness:

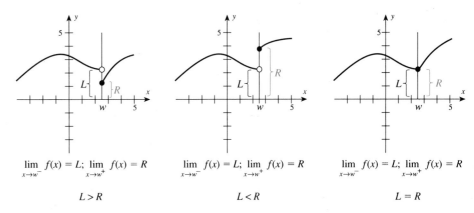

$\lim_{x \to w^-} f(x) = L$; $\lim_{x \to w^+} f(x) = R$ $\lim_{x \to w^-} f(x) = L$; $\lim_{x \to w^+} f(x) = R$ $\lim_{x \to w^-} f(x) = L$; $\lim_{x \to w^+} f(x) = R$

$L > R$ $L < R$ $L = R$

If the left- and right-hand limits are the same, then we write

$$\lim_{x \to w} f(x) = L$$

Notice that this notation does not specify a right- or left-hand limit. That is, there is no plus or minus associated with the number w. If the left- and right-hand limits are not the same, we see that *the limit does not exist at w.*

Limit Notation

> The notation $\lim_{x \to c} f(x) = L$ is read
>
> "the limit of f as x approaches c is L"
>
> and means that for all values of x in the domain of f, the values of $f(x)$ get closer and closer to the number L as x gets closer and closer to (but remains different from) c.

This intuitive definition of limit will suffice for this course. In a more formal course we would define what we mean by "closer and closer."* Notice that the limit as $x \to c$ does not require that the number $f(c)$ exists. Functions with the property that

$$\lim_{x \to c} f(x) = f(c)$$

are said to be **continuous at $x = c$**. We will consider this in the next section.

Limits by Graphing

Figure 2.1 shows the graph of a function f and the number $c = 3$. The arrowheads show possible sequences of values of x approaching $c = 3$ from both the left and the right. As x approaches $c = 3$, the $f(x)$ values get closer and closer to 5. We write this as $\lim_{x \to 3} f(x) = 5$.

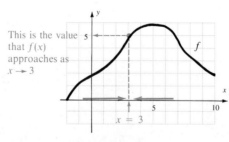

FIGURE 2.1
Graph of f and $\lim_{x \to c} f(x)$

---

* To make this intuitive notion precise, we must define what is meant by "closer and closer." For each $c > 0$ there exists a $d > 0$ such that $|f(x) - L| < c$ whenever $0 < |x - c| < d$. Notice also that we say the limit **is equal to L**. The statement that $0 < |x - c| < d$ implies that $x \neq c$. This exclusion ($x \neq c$) will be particularly important when we evaluate certain limits of rational functions.

EXAMPLE 1 Given the function defined by the graph below, find the following limits:

a. $\lim\limits_{x \to 3^-} f(x)$ **b.** $\lim\limits_{x \to -2^+} f(x)$ **c.** $\lim\limits_{x \to 0} f(x)$ **d.** $\lim\limits_{x \to 5} f(x)$

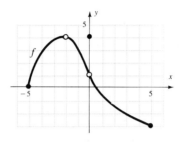

Solution Remember that $f(c)$ does not need to be defined in order to consider a limit. Also remember that an open circle on a graph indicates an excluded point. The following limits are found by inspection:

a. $\lim\limits_{x \to 3^-} f(x) = -2$ **b.** $\lim\limits_{x \to -2^+} f(x) = 4$

c. $\lim\limits_{x \to 0^-} f(x) = 1$ and $\lim\limits_{x \to 0^+} f(x) = 1$, so $\lim\limits_{x \to 0} f(x) = 1$

d. $\lim\limits_{x \to 5^-} f(x) = -3$, so $\lim\limits_{x \to 5} f(x) = -3$. Because the domain of f is the closed interval $[-5, 5]$, we do not need to consider the right-hand limit, $\lim\limits_{x \to 5^+} f(x)$, which implies $x > 5$ (which is not in the domain) ■

EXAMPLE 2 Find the requested limits on $[-6, 5)$.

a. $\lim\limits_{x \to 3^-} f(x)$ **b.** $\lim\limits_{x \to -2^-} f(x)$

c. $\lim\limits_{x \to -2^+} f(x)$ **d.** $\lim\limits_{x \to 5} f(x)$

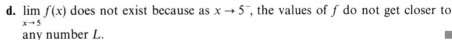

Solution **a.** $\lim\limits_{x \to 3^-} f(x) = 6$ **b.** $\lim\limits_{x \to -2^-} f(x) = 2$

c. $\lim\limits_{x \to -2^+} f(x) = 4$

d. $\lim\limits_{x \to 5} f(x)$ does not exist because as $x \to 5^-$, the values of f do not get closer to any number L. ■

EXAMPLE 3 Find $\lim\limits_{x \to 2} 5$.

Solution Look at the graph of $f(x) = 5$. It is easy to see that regardless of what x is approaching, the value of $f(x)$ is 5. This means that $\lim\limits_{x \to 2} 5 = 5$.

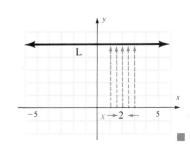

■

| Limit of a Constant | If $f(x) = k$ is a constant, then $$\lim_{x \to c} f(x) = k$$ |

Limits by Table

It is not always convenient (or even possible) to first draw a graph in order to find limits. You can also use a calculator or a computer to construct a table of values for f as $x \to c$.

EXAMPLE 4 Find $\lim_{x \to 4} x^2$.

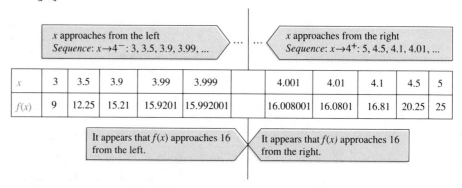

Thus: $\lim_{x \to 4} x^2 = 16$. ∎

After Example 4 you might be saying to yourself, why not just substitute $x = 4$ in x^2 to obtain the limit 16? Well, it is not that easy, as Example 5 illustrates.

EXAMPLE 5 Find $\lim_{x \to 2} \dfrac{x^2 - x - 6}{x - 2}$.

Notice that $f(2)$ is not defined: $f(2) = \dfrac{2^2 + (2) - 6}{2 - 2}$ ← Division by 0

Solution Construct a table of values.

				$x \to 2^-$	Left-hand limit →		← Right-hand limit	$x \to 2^+$			
x	1	1.5	1.9	1.99	1.999	1.9999	2.0001	2.001	2.01	2.1	2.5	3
$f(x)$	4	4.5	4.9	4.99	4.999	4.9999	5.0001	5.001	5.01	5.1	5.5	6

Since the left- and right-hand limits appear to be 5, we say the limit of $f(x)$ as x approaches 2 is 5.

A spreadsheet can be a very valuable and easy-to-use tool when evaluating limits by table (see Appendix B). The following illustration shows this example using a spreadsheet.

SPREADSHEET PROGRAM				
A	B	C	D	E
1 What is c?	[input c here]			
2				
3 n	left approach	f(x)	right approach	f(x)
4				
5 1	+B1-1	(B5^2+B5-6)/(B5-2)	+B1+1	(B5^2+B5-6)/(B5-2)
6 +A5+1	+$B5+(1-(0.1)^A6)	replicate	+$B5+(1-(0.1)^A6)	replicate
7 replicate	replicate		replicate	

OUTPUT:				
What is c?		2		
n	left approach	f(x)	right approach	f(x)
1	1	4	3	6
2	1.9	4.9	2.1	5.1
3	1.99	4.99	2.01	5.01
4	1.999	4.999	2.001	5.001
5	1.9999	4.9999	2.0001	5.0001
6	1.99999	4.99999	2.00001	5.00001
7	1.999999	4.999999	2.000001	5.000001
8	1.9999999	4.9999998999975	2.0000001	5.0000000999981
9	1.99999999	4.999999989982	2.00000001	5.0000000099747

Limits by Using Algebra

In Example 5 we saw that substitution of x for 2 leads to undefined values for the function. This occurs because the domain of f in Example 5 excludes the limiting value for x. Instead of substitution, suppose we try algebraic simplification.

EXAMPLE 6 Find $\lim_{x \to 2} \dfrac{x^2 + x - 6}{x - 2}$.

Solution Instead of finding a table of values (as we did in Example 5 and in the spreadsheet following Example 5), let us simplify the given expression:

$$\lim_{x \to 2} \frac{x^2 + x - 6}{x - 2} = \lim_{x \to 2} \frac{(x + 3)(x - 2)}{x - 2} = \lim_{x \to 2}(x + 3)$$

The above simplification is valid only if $x \neq 2$; but as $x \to 2$ we know that $x \neq 2$, so the simplification is valid. What about the limit of $x + 3$ as x approaches 2? You can complete this by graphing or by a table of values to see that when x is close to 2, $x + 3$ is close to 5. Thus,

$$\lim_{x \to 2} \frac{x^2 + x - 6}{x - 2} = 5$$

Clearly, we need a theorem to help us evaluate limits such as the one in Example 6. As long as f is a polynomial function, the limit can be found by direct evaluation. On the other hand, if f is not a polynomial function, then substitution *may* cause undefined values for f. The procedure is to algebraically simplify the given expression, and then *if the simplified expression is a polynomial,* evaluate it by direct substitution. This is summarized in the following box.

Limit of a Polynomial

> If f is any polynomial function, then
> $$\lim_{x \to c} f(x) = f(c) \quad \text{for any real number } c$$

EXAMPLE 7   Find:

a. $\lim\limits_{x \to 1} (4x^3 - 2x^2 + x - 1)$

b. $\lim\limits_{x \to 3} (x^2 - x)$

c. $\lim\limits_{x \to 3} \dfrac{x-3}{x-3}$

d. $\lim\limits_{x \to 1} \dfrac{x^2 - 1}{x - 1}$

Solution

a. $\lim\limits_{x \to 1} (4x^3 - 2x^2 + x - 1) = 4(1)^3 - 2(1)^2 + 1 - 1 = 2$
since $4x^3 - 2x^2 + x - 1$ is a polynomial

b. $\lim\limits_{x \to 3} (x^2 - x) = 3^2 - 3 = 6 \quad$ since $x^2 - x$ is a polynomial

c. $\lim\limits_{x \to 3} \dfrac{x-3}{x-3} = \lim\limits_{x \to 3} 1 = 1 \quad x \ne 3 \quad$ Limit of a constant

d. $\lim\limits_{x \to 1} \dfrac{x^2 - 1}{x - 1} = \lim\limits_{x \to 1} \dfrac{(x-1)(x+1)}{x - 1} = \lim\limits_{x \to 1} (x + 1) = 2 \quad x \ne 1$ ∎

EXAMPLE 8   Find $\lim\limits_{x \to 0} f(x)$ where $f(x) = \begin{cases} x + 5 & \text{if } x > 0 \\ x & \text{if } x < 0 \end{cases}$.

Solution   Notice that $f(0)$ is not defined, but since the limit is as x approaches 0, it is not necessary that $f(0)$ be defined. It is necessary, however, to consider left- and right-hand limits for this problem.

$\lim\limits_{x \to 0^-} f(x) = \lim\limits_{x \to 0^-} x \qquad$ As x approaches from the left of 0, it must be the case that $x < 0$, so $f(x) = x$ when $x < 0$.

$= 0$

$\lim\limits_{x \to 0^+} f(x) = \lim\limits_{x \to 0^+} (x + 5) \qquad$ As x approaches from the right of 0, $x > 0$.

$= 5$

Since the left- and right-hand limits are not the same, we say $\lim\limits_{x \to 0} f(x)$ does not exist. ∎

It is instructive to compare the results of Example 8 with the graph of the function, as shown in Figure 2.2. After it is graphed, it is easy to see that the left- and right-hand limits are not the same. Compare with the graph of g in the following example.

2.1 LIMITS 61

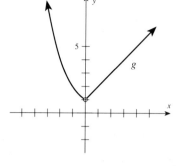

Graph of $f(x) = \begin{cases} x + 5 & \text{if } x > 0 \\ x & \text{if } x < 0 \end{cases}$ Graph of $g(x) = \begin{cases} x + 1 & \text{if } x > 0 \\ x^2 + 1 & \text{if } x < 0 \end{cases}$

FIGURE 2.2
Graphs of the functions in Examples 8 and 9

EXAMPLE 9 Find $\lim_{x \to 0} g(x)$ where $g(x) = \begin{cases} x + 1 & \text{if } x > 0 \\ x^2 + 1 & \text{if } x < 0 \end{cases}$.

Solution
$\lim_{x \to 0^-} g(x) = \lim_{x \to 0^-} (x^2 + 1) = 1$

$\lim_{x \to 0^+} g(x) = \lim_{x \to 0^+} (x + 1) = 1$

Since the left- and right-hand limits are equal, $\lim_{x \to 0} g(x) = 1$. ∎

Sometimes the function is a rational function that cannot be simplified to a polynomial. Sometimes such an expression will have a limit (Example 10) and sometimes it will not (Example 11). In order to evaluate such limits we need a result called the Limit of a Quotient Theorem.

Limit of a Quotient Theorem

Let f and g be two functions whose limits as $x \to c$ exist and $\lim_{x \to c} g(x) \neq 0$. Then

$$\lim_{x \to c} \frac{f(x)}{g(x)} = \frac{\lim_{x \to c} f(x)}{\lim_{x \to c} g(x)}$$

The theorem says that if the limit of the denominator function is not zero, the limit of the ratio of the two functions is the ratio of their limits. In particular, if f and g are polynomial functions, then you can evaluate a rational function by substitution.

EXAMPLE 10 Find $\lim_{x \to 2} \frac{x - 2}{x + 2}$.

Solution $\lim_{x \to 2} \frac{x - 2}{x + 2} = \frac{\lim_{x \to 2}(x - 2)}{\lim_{x \to 2}(x + 2)} = \frac{0}{4} = 0$ ∎

62 CHAPTER TWO THE DERIVATIVE

EXAMPLE 11 Find $\lim_{x \to -2} \dfrac{x-2}{x+2}$.

Solution $\lim_{x \to -2} \dfrac{x-2}{x+2}$ does not exist. You cannot apply the Limit of a Quotient Theorem because the limit of $x + 2$ (the denominator) is 0 as $x \to -2$. In order to see that this limit does not exist, you can use a table of values to see that the quotient increases without limit as $x \to -2^-$ and decreases as $x \to -2^+$, or you can draw the graph as shown in the margin. ∎

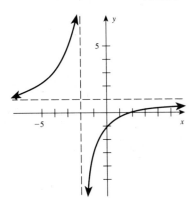

COMPUTER COMMENT

See the program *Limit of f(x)* on the computer disk accompanying this book. It sets up a table of values (either left- or right-hand limits). You can also draw the graph. From the graph and the table of values, you will be able to see if the limit exists, and in most cases will also be able to find the limit.

Several other limit theorems that we will need to use occasionally are stated here for completeness. We have also included the other limit theorems in this list.

Limit Theorems

Let f and g be two functions whose limits as $x \to c$ exist.

Limit of a constant: $\lim_{x \to c} k = k$ for any constant k

Limit of a polynomial: $\lim_{x \to c} f(x) = f(c)$ for any polynomial function f

Limit of a sum: $\lim_{x \to c} [f(x) + g(x)] = \lim_{x \to c} f(x) + \lim_{x \to c} g(x)$

Limit of a difference: $\lim_{x \to c} [f(x) - g(x)] = \lim_{x \to c} f(x) - \lim_{x \to c} g(x)$

Limit of a product: $\lim_{x \to c} [f(x) \cdot g(x)] = \left[\lim_{x \to c} f(x)\right]\left[\lim_{x \to c} f(x)\right]$

Limit of a quotient: $\lim_{x \to c} \dfrac{f(x)}{g(x)} = \dfrac{\lim_{x \to c} f(x)}{\lim_{x \to c} g(x)}$ if $\lim_{x \to c} g(x) \neq 0$

Limit of a power: $\lim_{x \to c} [f(x)]^n = \left[\lim_{x \to c} f(x)\right]^n$ where n is a positive integer

Limit of a root: $\lim_{x \to c} \sqrt[n]{f(x)} = \sqrt[n]{\lim_{x \to c} f(x)}$ where $n \geq 2$ is a positive integer, and both roots are defined.

The expression 0/0 is not a real number; in calculus we call this an **indeterminate form**. As long as an indeterminate form is not obtained, you may evaluate the limit

EXAMPLE 12 Find $\lim_{x \to 1} \frac{\sqrt{x} - 1}{x - 1}$.

Solution Try substitution: $\lim_{x \to 1} \frac{\sqrt{x} - 1}{x - 1} = \frac{\sqrt{1} - 1}{1 - 1} = \frac{1 - 1}{1 - 1} = \frac{0}{0}$ Indeterminate form

WARNING *Do not assume when you obtain an indeterminate form that the limit does not exist.*

You could proceed with a table of values for this example but, instead, we will **rationalize the numerator**. Remember, from algebra, the process of *rationalizing the denominator*. The process here is the same—namely, multiply both numerator and denominator by $\sqrt{x} + 1$:

$$\frac{\sqrt{x} - 1}{x - 1} \cdot \frac{\sqrt{x} + 1}{\sqrt{x} + 1} = \frac{\sqrt{x} \cdot \sqrt{x} - \sqrt{x} + \sqrt{x} - 1}{(x - 1)(\sqrt{x} + 1)}$$

$$= \frac{x - 1}{(x - 1)(\sqrt{x} + 1)}$$

$$= \frac{1}{\sqrt{x} + 1}$$

Thus

$$\lim_{x \to 1} \frac{\sqrt{x} - 1}{x - 1} = \lim_{x \to 1} \frac{1}{\sqrt{x} + 1} = \frac{1}{1 + 1} = \frac{1}{2}$$ ∎

Limits at Infinity

In Section 1.5 we considered asymptotes as an aid for graphing rational functions. The idea of a horizontal asymptote seeks to find the limiting value of a function as x becomes very large or very small. Consider Examples 13 and 14.

EXAMPLE 13 Find the value of $\frac{1}{x}$, if it exists, as x increases without bound. This is called a *limit at infinity*.

Solution The limit theorems are not stated for limits at infinity so we cannot assume they apply. If the limit theorems do not apply, we must rely on a graph or a table. For clarity we will do both.

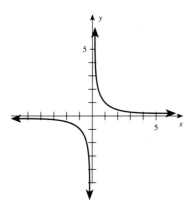

x	1	2	10	100	1,000	10,000
$f(x)$	1	.5	.1	.01	.001	.0001

We say that $\frac{1}{x}$ approaches 0 as x increases without bound, and we symbolize this by $\lim_{x \to \infty} \frac{1}{x} = 0$. ∎

EXAMPLE 14 Find the value of $\frac{1}{x}$ as x decreases without bound.

x	-1	-2	-10	-100	$-1{,}000$	$-10{,}000$
$f(x)$	-1	$-.5$	$-.1$	$-.01$	$-.001$	$-.0001$

We say that $\frac{1}{x}$ approaches 0 as x decreases without bound, and we symbolize this by $\lim\limits_{x \to -\infty} \frac{1}{x} = 0$. ∎

Limits at Infinity

The statement
$$\lim_{x \to \infty} f(x) = L$$
means that $f(x)$ is close to L for very large positive values of x. Also,
$$\lim_{x \to -\infty} f(x) = L$$
means that $f(x)$ is close to L for negative values of x with very large absolute values. Finally,
$$\lim_{|x| \to \infty} f(x) = L$$
means that $f(x)$ is close to L for both positive and negative values of x with very large absolute values.

In Examples 13 and 14 we see that $\lim\limits_{|x| \to \infty} \frac{1}{x} = 0$. It is easy to show that the limit is also 0 for any constant k divided by x, and for $\frac{1}{x^n}$ where n is a positive real number. This is summarized by the following limit theorem.

Limit to Infinity

$$\lim_{|x| \to \infty} \frac{1}{x^n} = 0 \quad \text{and} \quad \lim_{|x| \to \infty} \frac{k}{x^n} = 0 \quad \text{for any constant } k \text{ and } n > 0.$$

EXAMPLE 15 Find $\lim\limits_{x \to \infty} \dfrac{x}{2x + 1}$.

Solution We could construct a table of values. Instead, suppose we multiply the rational expression by 1, written as $\dfrac{\frac{1}{x}}{\frac{1}{x}}$:

$$\frac{x}{2x+1} \cdot \frac{\frac{1}{x}}{\frac{1}{x}} = \frac{1}{2 + \frac{1}{x}}$$

Now, since $\lim\limits_{x \to \infty} \frac{1}{x} = 0$, we see that

$$\lim_{x \to \infty} \frac{1}{2 + \frac{1}{x}} = \frac{1}{2 + 0} = \frac{1}{2}$$ ∎

EXAMPLE 16 Find $\lim_{x \to \infty} \dfrac{3x^2 - 7x + 2}{7x^2 + 2x + 5}$.

Solution   Note that the largest power of x in the expression is x^2, so we multiply the numerator and denominator by $\frac{1}{x^2}$:

$$\lim_{x \to \infty} \frac{3x^2 - 7x + 2}{7x^2 + 2x + 5} \cdot \frac{\frac{1}{x^2}}{\frac{1}{x^2}}$$

Now, since $\dfrac{k}{x}$ and $\dfrac{k}{x^2}$ both approach 0 as x increases without bound, we have

$$= \lim_{x \to \infty} \frac{3 - \frac{7}{x} + \frac{2}{x^2}}{7 + \frac{2}{x} + \frac{5}{x^2}} = \frac{3 - 0 + 0}{7 + 0 + 0}$$

Thus $\lim_{x \to \infty} f(x) = \frac{3}{7}$. ∎

Asymptotes

We can now substantiate the result about horizontal asymptotes stated in Section 1.5. If

$$f(x) = \frac{P(x)}{D(x)}$$

where $P(x)$ and $D(x)$ are polynomial functions with no common factors, then the line $y = 0$ is a *horizontal asymptote* if the degree of $P(x) = a_n x^n + \cdots + a_0$ is less than the degree of $D(x) = b_m x^m + \cdots + b_0$. If $y = f(x)$, then

$$y = \lim_{x \to \infty} \frac{P(x)}{D(x)} = \lim_{x \to \infty} \frac{a_n x^n + \cdots + a_0}{b_m x^m + \cdots + b_0} \qquad \text{where } n < m$$

$$= \lim_{x \to \infty} \frac{a_n x^n + \cdots + a_0}{b_m x^m + \cdots + b_0} \cdot \frac{\frac{1}{x^m}}{\frac{1}{x^m}} \qquad \text{where } n < m$$

$$= \lim_{x \to \infty} \frac{\frac{a_n}{x^{m-n}} + \cdots + \frac{a_0}{x^m}}{b_m + \frac{b_{m-1}}{x} + \cdots + \frac{b_0}{x^m}}$$

$$= \frac{0 + 0 + \cdots + 0}{b_m + 0 + 0 + \cdots + 0} = 0$$

You can similarly show that if the degree of P is equal to the degree of D then the horizontal asymptote is

$$y = \frac{a_n}{b_n}$$

This is left as a problem.

EXAMPLE 17 Graph $y = \dfrac{x^2}{x^2 + 1}$.

Solution There are no vertical asymptotes (no values of x that cause division by zero). The horizontal asymptote(s) are found by taking the limit at infinity:

$$y = \lim_{|x| \to \infty} \frac{x^2}{x^2 + 1} = \lim_{|x| \to \infty} \frac{1}{1 + \dfrac{1}{x^2}} = 1$$

The graph is shown in Figure 2.3.

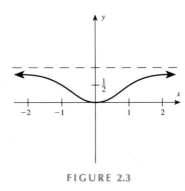

FIGURE 2.3

2.1
Problem Set

Given the functions defined by the graphs in Figure 2.4, find the limits in Problems 1–12.

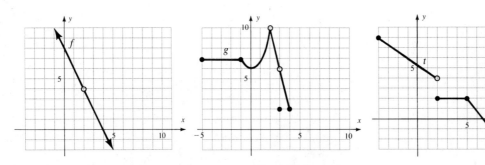

FIGURE 2.4
Functions f, g, and t

1. $\lim\limits_{x \to 4} f(x)$
2. $\lim\limits_{x \to 2} f(x)$
7. $\lim\limits_{x \to 3^+} g(x)$
8. $\lim\limits_{x \to 2^-} t(x)$
3. $\lim\limits_{x \to 0} f(x)$
4. $\lim\limits_{x \to -3} g(x)$
9. $\lim\limits_{x \to 2^+} t(x)$
10. $\lim\limits_{x \to 2} t(x)$
5. $\lim\limits_{x \to -1^-} g(x)$
6. $\lim\limits_{x \to 2^+} g(x)$
11. $\lim\limits_{x \to 4} t(x)$
12. $\lim\limits_{x \to -4} t(x)$

Find the limits by filling in the appropriate values in the tables in Problems 13–16.

13. $\lim\limits_{x \to 5^-} (4x - 5)$

x	2	3	4	4.5	4.9	4.99
$f(x)$	3					

14. $\lim\limits_{x \to 2} \dfrac{x^3 - 8}{x^2 + 2x + 4}$

x	1	1.5	1.9	1.99	1.999	2.5	2.01	2.001
$f(x)$								

15. $\lim\limits_{x \to 2} \dfrac{x^2 + 2x + 4}{x^3 - 8}$

x	1	1.5	1.9	1.99	1.999	2.5	2.01	2.001
$f(x)$								

16. $\lim\limits_{x \to \infty} \dfrac{x^2 + 6x + 9}{x + 3}$

x	1	10	100	1,000	10,000	100,000	1,000,000
$f(x)$								

Find the limits in Problems 17–62.

17. $\lim\limits_{x \to 0} x^8$

18. $\lim\limits_{x \to 2} (x^2 - 4)$

19. $\lim\limits_{x \to 3} (x^2 - 4)$

20. $\lim\limits_{x \to 1} \dfrac{1}{x - 3}$

21. $\lim\limits_{x \to -3} \dfrac{1}{x - 3}$

22. $\lim\limits_{x \to 3} \dfrac{1}{x - 3}$

23. $\lim\limits_{x \to 0} \dfrac{1}{x^2 + 1}$

24. $\lim\limits_{x \to 1} \dfrac{3x + 2}{x - 2}$

25. $\lim\limits_{x \to \infty} \dfrac{1}{x^2 + 1}$

26. $\lim\limits_{x \to \infty} 2x$

27. $\lim\limits_{x \to -1} \dfrac{1}{x^2 + 1}$

28. $\lim\limits_{x \to 2} \dfrac{1}{x^2 - 4}$

29. $\lim\limits_{x \to 2} \dfrac{x^2 - 4}{x - 2}$

30. $\lim\limits_{x \to \infty} (3x - 4)$

31. $\lim\limits_{x \to 3} \dfrac{x^2 + 3x - 10}{x - 2}$

32. $\lim\limits_{x \to 3} \dfrac{x^2 - 8x + 15}{x - 3}$

33. $\lim\limits_{x \to 4} \dfrac{\sqrt{x} - 4}{x - 16}$

34. $\lim\limits_{x \to -5} \dfrac{x^2 + 3x - 10}{x + 5}$

35. $\lim\limits_{x \to 9} \dfrac{\sqrt{x} - 3}{x - 3}$

36. $\lim\limits_{x \to 9} \dfrac{\sqrt{x} - 3}{x - 9}$

37. $\lim\limits_{x \to 2} \dfrac{x^2 - 1}{x - 2}$

38. $\lim\limits_{x \to 4} \dfrac{\sqrt{x} - 2}{x - 2}$

39. $\lim\limits_{x \to 2} \dfrac{x^3 - 8}{x^2 + 2x + 4}$

40. $\lim\limits_{x \to 4} \dfrac{2x^2 - 5x - 12}{x - 4}$

41. $\lim\limits_{x \to 2} \dfrac{x + 2}{x^3 + 8}$

42. $\lim\limits_{x \to 2} \dfrac{x^2 + 2x + 4}{x^3 - 8}$

43. $\lim\limits_{|x| \to \infty} \dfrac{2x^2 - 5x - 3}{x^2 - 9}$

44. $\lim\limits_{x \to 2} \dfrac{6 - x}{2x - 15}$

45. $\lim\limits_{|x| \to \infty} \dfrac{3x - 1}{2x + 3}$

46. $\lim\limits_{|x| \to \infty} \dfrac{x^2 + 6x + 9}{x + 3}$

47. $\lim\limits_{x \to -\infty} \dfrac{5x + 10{,}000}{x - 1}$

48. $\lim\limits_{|x| \to \infty} \dfrac{6x^2 - 5x + 2}{2x^2 + 5x + 1}$

49. $\lim\limits_{x \to \infty} \left(x + 2 + \dfrac{3}{x - 1} \right)$

50. $\lim\limits_{x \to -\infty} \dfrac{4x + 10^6}{x + 1}$

51. $\lim\limits_{x \to 0} \dfrac{1 - \dfrac{1}{x+1}}{x}$

52. $\lim\limits_{x \to 1} \dfrac{1 - \dfrac{1}{x}}{x - 1}$

53. $\lim\limits_{x \to -\infty} \dfrac{3x^2 - 5x + 15}{x + 3}$

54. $\lim\limits_{x \to -\infty} \left(2x - 3 + \dfrac{4}{x + 2} \right)$

55. $\lim\limits_{x \to -\infty} \dfrac{4x^4 - 3x^3 + 2x + 1}{3x^4 - 9}$

56. $\lim\limits_{x \to 1} \dfrac{x^2 + x + 1}{x^3 - 1}$

57. $\lim\limits_{x \to 0} f(x) = \begin{cases} 2x + 3 & \text{if } x > 0 \\ x^2 + 3 & \text{if } x < 0 \end{cases}$

58. $\lim\limits_{x \to 0} f(x) = \begin{cases} x^2 - 1 & \text{if } x > 0 \\ 2x^2 + 1 & \text{if } x < 0 \end{cases}$

59. $\lim\limits_{x \to 1} f(x) = \begin{cases} 2x + 3 & \text{if } x > 0 \\ x^2 + 3 & \text{if } x < 0 \end{cases}$

60. $\lim\limits_{x \to -3} f(x) = \begin{cases} x^2 - 1 & \text{if } x > 0 \\ 2x^2 + 1 & \text{if } x < 0 \end{cases}$

61. $\lim\limits_{x \to 3} f(x) = \begin{cases} x + 7 & \text{if } x > 3 \\ x^2 + 1 & \text{if } x \le 3 \end{cases}$

62. $\lim\limits_{x \to -2} f(x) = \begin{cases} 5x & \text{if } x \ge -2 \\ x^2 + 1 & \text{if } x < -2 \end{cases}$

APPLICATIONS

Use limits to find the horizontal asymptotes for the curves whose equations are given in Problems 63–68.

63. $y = \dfrac{3x^2}{x^2 + 2}$

64. $y = \dfrac{-2x^2}{x^2 + 1}$

65. $y = \dfrac{2x^2 + 5}{3x^2 + 2}$

66. $y = \dfrac{2x^2 + 1}{5x^2}$

67. $y = \dfrac{2x^2}{x^2 - 4}$ **68.** $y = \dfrac{x^2}{x^2 - 1}$

69. The cost of manufacturing a specialized machine tool is a function of the number of items manufactured. This cost is graphed below. Note that there is a jump in the cost after 10,000 items because at that point it is necessary to add a second shift.

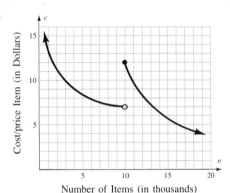

Number of Items (in thousands)

Find the following limits.
a. $\lim\limits_{n \to 10^-} c(n)$ **b.** $\lim\limits_{n \to 10^+} c(n)$
c. $\lim\limits_{n \to 13} c(n)$ **d.** $\lim\limits_{n \to 10} c(n)$

70. The number of live animals during a psychology experiment is shown in the graph. The experiment begins with 2 animals, and after time t_4 there are 14 animals. Find the following limits.

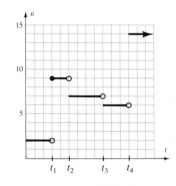

a. $\lim\limits_{t \to t_1^-} n(t)$ **b.** $\lim\limits_{t \to t_1^+} n(t)$
c. $\lim\limits_{t \to t_1} n(t)$ **d.** $\lim\limits_{t \to t_4} n(t)$

71. Learning theory measures the percentage of mastery of a subject as a function of time. The learning curve for a particular learning task is shown below. Note that at time t_1 there is a jump in mastery. Find the following limits.

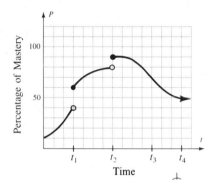

Time

a. $\lim\limits_{t \to t_1} P(t)$ **b.** $\lim\limits_{t \to t_2} P(t)$
c. $\lim\limits_{t \to t_3} P(t)$ **d.** $\lim\limits_{t \to \infty} P(t)$

72. If $f(x) = \dfrac{P(x)}{D(x)}$ where $P(x)$ and $D(x)$ are polynomial functions with no common factors, and degree n, show that the horizontal asymptote is

$$y = \dfrac{a_n}{b_n}$$

73. If $f(x) = \sqrt{x}$, find $\lim\limits_{h \to 0} \dfrac{f(1+h) - f(1)}{h}$.

74. If $f(x) = x^2 + 1$, find $\lim\limits_{h \to 0} \dfrac{f(2+h) - f(2)}{h}$.

75. If $f(x) = \dfrac{1}{x}$, find $\lim\limits_{h \to 0} \dfrac{f(3+h) - f(3)}{h}$.

76. If $f(x) = x^3$, find $\lim\limits_{h \to 0} \dfrac{f(-1+h) - f(-1)}{h}$.

2.2
Continuity

Introduction

You may remember the puzzle in the margin at the top of page 69 from elementary school: the challenge is to draw the figure without lifting your pencil from the paper or retracing any of the lines. In calculus we are concerned with figures that

can be drawn without lifting a pencil from the paper, but we focus our attention on functions. The idea of *continuity* evolved from the notion of a curve "without breaks or jumps" to a rigorous definition given by Karl Weierstrass (1815–1897). Galileo and Leibniz had thought of continuity in terms of the density of points on a curve, but they were in error since the rational numbers have this property of denseness, yet do not form a continuous curve. Another mathematician, J. W. R. Dedekind (1831–1916), took an entirely different approach and concluded that continuity is due to the division of a curve into two parts so that there is one and only one point that makes this division. As Dedekind wrote, "By this commonplace remark, the secret of continuity is to be revealed."* We begin with a discussion of *continuity at a point*. It may seem strange to talk about continuity *at a point*, but it should seem natural to talk about a curve being "discontinuous at a point," as illustrated by Example 1.

EXAMPLE 1   Which of the following curves appear to have a discontinuity at the point $x = 1$?

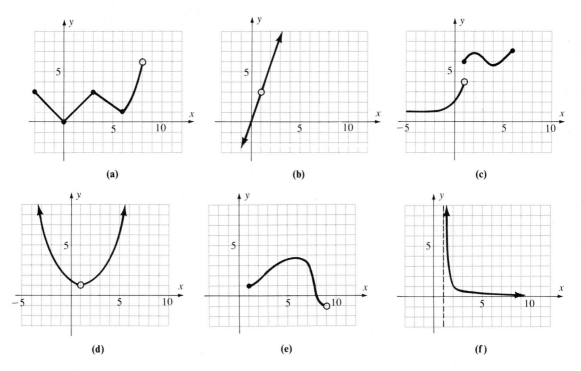

Solution   The functions whose graphs are shown in parts **b, c, d,** and **f** are obviously discontinuous at $x = 1$. If, however, $x = 1$ happens to be the endpoint of an interval (part **e**) we do not say the curve is discontinuous at $x = 1$.   ■

The solution to Example 1 illustrates the idea of continuity from an intuitive standpoint, but we need to define and formalize this idea more completely. There are two essential conditions for a function f to be continuous at a point c. First, $f(c)$ must be defined. For example, the curve in part **e** of Example 1 is not continuous at

* From Carl Boyer, *A History of Mathematics* (New York: Wiley, 1968), p. 607.

its right endpoint because it is not defined at $x = 9$ (the open dot indicates an excluded point).

EXAMPLE 2 Which, if any, of the given functions is continuous at $x = 0$?

a. $f(x) = \dfrac{3x^2 - x + 2}{x}$

b. $g(x) = \dfrac{x^2 - 5x}{x}$

c. $s(x) = \begin{cases} x + 3 & \text{if } x > 0 \\ x^2 & \text{if } x < 0 \end{cases}$

Solution None of these functions is defined at $x = 0$, so none is continuous at $x = 0$. ∎

Definition of Continuity at a Point

A second condition for continuity at a point $x = c$ is that the function makes no jumps at this point. This means that if "x is close to c," then "$f(x)$ must be close to $f(c)$." Looking at Example 1, we see that the graphs in parts **b**, **c**, and **d** jump at the point $x = 1$. We recognize this as the concept of limit and now define the concept of continuity at a point:

Definition of Continuity at a Point

A function f is continuous at a point $x = c$ if

1. $f(c)$ exists and
2. $\lim\limits_{x \to c} f(x) = f(c)$

The conditions of this definition are summarized in Table 2.1.

Test the continuity of each function in Examples 3–6 at the point $x = 1$. If it is not continuous at $x = 1$, tell why.

EXAMPLE 3 $f(x) = \dfrac{x^2 + 2x - 3}{x - 1}$

Solution Not continuous at $x = 1$ (hole), because f is not defined at this point. ∎

EXAMPLE 4 $g(x) = \dfrac{x^2 + 2x - 3}{x - 1}$ if $x \neq 1$ and $g(x) = 6$ if $x = 1$

Solution Note that g is very similar to f in Example 3 except that g is defined at $x = 1$. Now, to test the second condition of continuity, $g(1) = 6$ and

$$\lim_{x \to 1} g(x) = \lim_{x \to 1} \dfrac{x^2 + 2x - 3}{x - 1}$$

$$= \lim_{x \to 1} \dfrac{(x - 1)(x + 3)}{x - 1}$$

$$= \lim_{x \to 1} (x + 3)$$

$$= 4$$

Since $\lim\limits_{x \to 1} g(x) \neq g(1)$, we see that g is not continuous at $x = 1$ (jump). ∎

TABLE 2.1
Holes, Poles, Jumps, and Continuity

$f(c)$ is defined	$f(c)$ is not defined
Continuous at $x = c$ $\lim_{x \to c} f(x)$ exists and is equal to $f(c)$ 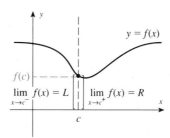	
Not continuous at $x = c$: Hole $\lim_{x \to c} f(x)$ exists and is not equal to $f(c)$ 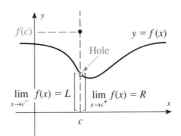	**Not continuous at $x = c$: Hole** $\lim_{x \to c} f(x)$ exists and $f(c)$ is not defined 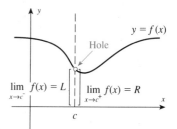
Not continuous at $x = c$: Pole $\lim_{x \to c} f(x)$ does not exist; $f(c)$ is defined 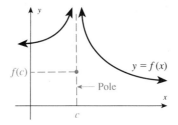	**Not continuous at $x = c$: Pole** $\lim_{x \to c} f(x)$ does not exist; $f(c)$ is not defined 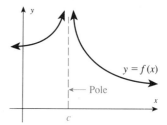
Not continuous at $x = c$: Jump $\lim_{x \to c} f(x)$ does not exist; $f(c)$ is defined 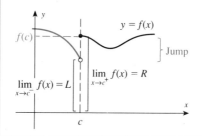	**Not continuous at $x = c$: Jump** $\lim_{x \to c} f(x)$ does not exist; $f(c)$ is not defined 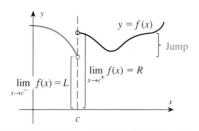

EXAMPLE 5 $h(x) = \dfrac{x^2 + 2x - 3}{x - 1}$ if $x \neq 1$ and $h(x) = 4$ if $x = 1$

Solution Compare h with g of Example 4. We see that both conditions of continuity are satisfied, which means that h is continuous at $x = 1$. ∎

EXAMPLE 6 $m(x) = 3x^3 + 5x^2 - 4x + 1$

Solution $m(1) = 3(1)^3 + 5(1)^2 - 4(1) + 1 = 5$, so m is defined at $x = 1$. Also, $\lim\limits_{x \to 1} m(x) = m(1)$, so the function is continuous at $x = 1$. ∎

Continuity Theorems

Note that Example 6 is a polynomial function, and from the previous section we know that if f is any polynomial function, then

$$\lim_{x \to c} f(x) = f(c)$$

for any real number c. Thus we immediately have the following result:

Continuity of a Polynomial Every polynomial function is continuous at every point in its domain.

If a function is continuous for every point in a given open interval, then we say that the function is continuous over that interval. A function that is not continuous is said to be a **discontinuous function**. For example, polynomials are continuous for all real numbers. On the other hand,

$$f(x) = \frac{x + 1}{x - 1}$$

is continuous for $-1 \leq x \leq 0$ but not for $0 \leq x \leq 2$ since f is undefined at $x = 1$. The function h from Example 5,

$$h(x) = \begin{cases} \dfrac{x^2 + 2x - 3}{x - 1} & \text{if } x \neq 1 \\ 4 & \text{if } x = 1 \end{cases}$$

is continuous for all real numbers. As you can see from the above examples, we are usually concerned with finding points of discontinuity. The following result summarizes the properties of continuity:

Continuity Theorem Let f and g be continuous functions at $x = c$. Then the following functions are also continuous at $x = c$:

  1. $f + g$ 2. $f - g$ 3. fg 4. f/g $(g(c) \neq 0)$

Since most of the functions we will discuss in this book are continuous over certain intervals, our task will be to look for *points of discontinuity*. These points may be values for which the definition of the function changes or values that cause division by zero. We call such points **suspicious points**. We then check for the continuity at the suspicious points and use the continuity theorem for all other points in the interval.

EXAMPLE 7 Let $f(x) = \dfrac{x^2 + 3x - 10}{x - 2}$. Check continuity.

Solution The suspicious point is $x = 2$ since this is a value for which the function $(x - 2)$ is equal to zero. The function f is not defined at this point, so it is *discontinuous at* $x = 2$. By the continuity theorem (part 4), it is continuous at all other points. To graph f, notice that

$$f(x) = \frac{x^2 + 3x - 10}{x - 2} = \frac{(x + 5)(x - 2)}{x - 2} = x + 5 \qquad x \neq 2$$

The graph is shown in Figure 2.5.

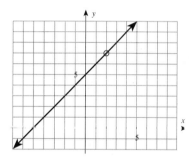

FIGURE 2.5
Graph of $f(x) = \dfrac{x^2 + 3x - 10}{x - 2}$

EXAMPLE 8 Let $g(x) = \begin{cases} \dfrac{x^2 + 3x - 10}{x - 2} & \text{if } x \neq 2 \\ 6 & \text{if } x = 2 \end{cases}$. Check continuity.

Solution The suspicious point is $x = 2$ since it is a value for which the definition of the function changes. Check this value:

Step 1: $g(2) = 6$, so the function is defined at $x = 2$.

Step 2: $\displaystyle\lim_{x \to 2} g(x) = \lim_{x \to 2} \frac{x^2 + 3x - 10}{x - 2}$

$\qquad\qquad\qquad = \displaystyle\lim_{x \to 2} \frac{(x + 5)(x - 2)}{x - 2}$

$\qquad\qquad\qquad = \displaystyle\lim_{x \to 2} (x + 5)$

$\qquad\qquad\qquad = 7$

But $g(2) = 6$ and thus $\displaystyle\lim_{x \to 2} g(x) \neq g(2)$, so the function is *discontinuous at* $x = 2$. The graph of g is the same as f in Figure 2.5 except the point $(2, 6)$ is included.

EXAMPLE 9 Let $G(x) = \begin{cases} \dfrac{x^2 + 3x - 10}{x - 2} & \text{if } x \neq 2 \\ 7 & \text{if } x = 2 \end{cases}$. Check continuity.

Solution Note that we have simply redefined the functional value of g of Example 8 at a single point. Thus

Step 1: $G(2) = 7$ and

Step 2: $\lim_{x \to 2} G(x) = 7$ (from Example 8) and $\lim_{x \to 2} G(x) = G(2)$. Thus the function is continuous for all real numbers.

The graph of G is the same as f in Figure 2.5 except the point $(2, 7)$ is included. Compare this with Example 8. In this example, the point "plugs the hole" to force continuity. ∎

Note in Example 9 that you do not need to check the continuity at other points in the domain. The other points are not suspicious points and you only need to apply the continuity theorem for all of these points.

EXAMPLE 10 Let $f(x) = \begin{cases} 3 - x & \text{if } -5 \leq x < 2 \\ x - 2 & \text{if } 2 \leq x \leq 5 \end{cases}$. Check continuity.

Solution The suspicious point is $x = 2$ since it is a value for which the definition of the function changes. Check continuity by applying the definition: that is, f is defined at $x = 2$ since $f(2) = 0$. In order to find the limit as $x \to 2$ you need to check both the left- and right-hand limits (since the function is defined according to a different rule depending on from which direction x approaches 2).

$$\lim_{x \to 2^-} f(x) = \lim_{x \to 2^-} (3 - x) = 1$$
$$\lim_{x \to 2^+} f(x) = \lim_{x \to 2^+} (x - 2) = 0$$

Since the left- and right-hand limits are different, $\lim_{x \to 2} f(x)$ does not exist, so the function is discontinuous at $x = 2$.

If you draw the graph of f you can see the discontinuity at $x = 2$. Note that f is continuous over $-5 \leq x < 2$ and over $2 \leq x \leq 5$ but not over $-5 \leq x \leq 5$.

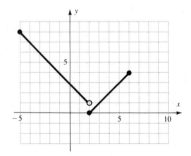

∎

EXAMPLE 11 Let $g(x) = \begin{cases} 2 - x & \text{if } -5 \leq x < 2 \\ x - 2 & \text{if } 2 \leq x < 5 \end{cases}$. Check continuity.

Solution   The suspicious point is $x = 2$. g is defined at $x = 2$ since $g(2) = 0$.

$$\lim_{x \to 2^-} g(x) = \lim_{x \to 2^-} (2 - x) = 0$$
$$\lim_{x \to 2^+} g(x) = \lim_{x \to 2^+} (x - 2) = 0$$

By looking at the graph of g you can see that it is continuous over $-5 \leq x \leq 5$. Note that $g(x) = |x - 2|$. Since $\lim_{x \to 2^+} g(x) = \lim_{x \to 2^-} g(x) = 0$, we can write

$$\lim_{x \to 2} g(x) = 0$$

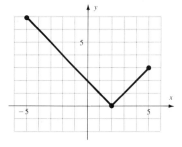

The function is continuous at the suspicious point, so we conclude that g is continuous for all real numbers over $-5 \leq x \leq 5$. ■

You will note that the graphs of the continuous functions in the last few examples can be sketched without lifting pencil from paper. The graphs of the discontinuous functions cannot be so sketched.

2.2
Problem Set

In Problems 1–12 find all suspicious points and tell which of those are points of discontinuity.

1.

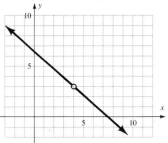

2.

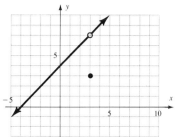

3.

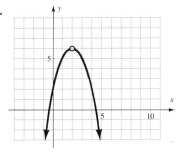

4.

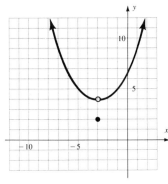

5.

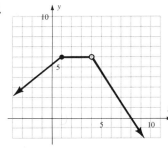

6.

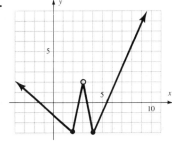

7.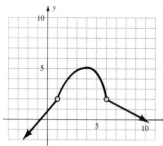

8.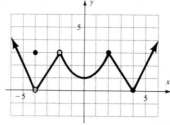

9.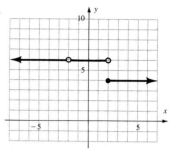

10.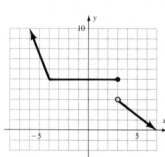

11.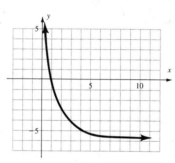

12.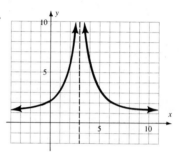

13. Is the graph in Problem 11 continuous at $x = 1$? Explain your answer.

14. Is the graph in Problem 12 continuous at $x = 2.5$? Explain your answer.

15. Is the graph in Problem 11 continuous at $x = .001$? Explain your answer.

16. Is the graph in Problem 12 continuous at $x = 2.9999$? Explain your answer.

Which of the functions described in Problems 17–22 represent continuous functions? State the domain, if possible, for each example.

17. The humidity on a specific day at a given location considered as a function of time

18. The temperature on a specific day at a given location considered as a function of time

19. The selling price of IBM stock on a specific day considered as a function of time

20. The number of unemployed people in the United States during December 1989 considered as a function of time

21. The charges for a telephone call from Los Angeles to New York considered as a function of time

22. The charges for a taxi ride across town considered as a function of mileage

In Problems 23–45 a function along with an interval is given. State whether the function is continuous at all points in this interval. Give the points of discontinuity, if any.

23. $f(x) = \dfrac{1}{x^2 + 5}$, $-5 \le x \le 5$

24. $f(x) = \dfrac{1}{x^2 - 5}$, $-5 \le x \le 5$

25. $f(x) = \dfrac{1}{x^2 - 9}$, $5 \le x \le 10$

26. $f(x) = \dfrac{1}{x^2 - 9}$, $-5 \le x \le 5$

27. $f(x) = \dfrac{x - 1}{x^2 - 1}$, $-5 \le x \le 5$

28. $f(x) = \dfrac{3x}{x^3 - x}$, $-5 \le x \le 5$

29. $f(x) = \dfrac{x + 2}{x^2 - 6x - 16}$, $-5 \le x \le 5$

30. $f(x) = \dfrac{x + 2}{x^2 - 6x - 16}$, $-10 \le x \le 10$

31. $f(x) = \dfrac{1}{x - 3}$, $0 \le x \le 2$

32. $f(x) = \dfrac{1}{x - 3}$, $0 \le x \le 5$

33. $f(x) = \dfrac{x}{(x - 8)^2}$, $0 \le x \le 5$

34. $f(x) = \dfrac{x}{(x - 3)^2}$, $0 \le x \le 5$

35. $f(x) = \begin{cases} \dfrac{1}{x - 3} & -5 \le x \le 5,\ x \ne 3 \\ 4 & x = 3 \end{cases}$

36. $f(x) = \dfrac{x^2 - x - 6}{x + 2}, \quad -5 \le x \le 5$

37. $f(x) = \begin{cases} \dfrac{x^2 - x - 6}{x + 2} & -5 \le x \le 5, x \ne -2 \\ -4 & x = -2 \end{cases}$

38. $f(x) = \begin{cases} \dfrac{x^2 - x - 6}{x + 2} & -5 \le x \le 5, x \ne -2 \\ -5 & x = -2 \end{cases}$

39. $f(x) = \dfrac{x^2 - 3x - 10}{x + 2}, \quad 0 \le x \le 5$

40. $f(x) = \dfrac{x^2 - 3x - 10}{x + 2}, \quad -5 \le x \le 5$

41. $f(x) = \begin{cases} \dfrac{x^2 - 3x - 10}{x + 2} & -5 \le x \le 5, x \ne -2 \\ -3 & x = -2 \end{cases}$

42. $f(x) = \begin{cases} \dfrac{x^2 + x + 1}{x^3 - 1} & 0 \le x \le 5, x \ne 1 \\ 1 & x = 1 \end{cases}$

43. $f(x) = |x|, \quad -5 \le x \le 5$
44. $f(x) = |x - 2|, \quad -5 \le x \le 5$
45. $f(x) = |x + 3|, \quad -5 \le x \le 5$

APPLICATIONS

46. A rental agency will lease trucks at a cost C of $.55 per mile if the annual mileage is under 10,000 miles, but the rate is lowered to $.40 per mile for mileage over 10,000 or more miles.

    a. Is C a continuous function? What is the domain?
    b. If m is the number of miles, we can write $C(m)$ to represent the mileage charge. What is $C(9,000)$? What is $C(11,000)$?
    c. Graph the function C.

47. Postal charges are $.29 for the first ounce and $.23 for each additional ounce or fraction thereof. Let C be the cost function for mailing a letter weighing w ounces.

    a. Is C a continuous function? What is the domain?
    b. What is $C(1.9)$? $C(2.01)$? $C(2.89)$?
    c. Graph the function C for $[0, 10]$.

48. On July 10, 1913, the following temperatures were recorded at Death Valley, California:

Time	8	10	12	1	3	5	7
°F	90	115	123	134	130	128	105

    Let T represent the temperature in degrees Fahrenheit and assume that T is a function of time (measured on a 24-hour clock) so that $T(8) = 90, \ldots, T(12) = 123, \ldots, T(19) = 105$.

    a. Is T a continuous function? What is the domain?
    b. On the day in question, what is the minimum number of times during the day that the temperature was 100°F? 110°F? 80°F?

49. Give an example of a function defined on $1 \le x \le 3$ that is discontinuous at $x = 2$ and continuous elsewhere on the interval.

50. Give an example of a function defined on $3 \le x \le 7$ that is discontinuous at $x = 4$ and $x = 5$ and continuous elsewhere on the interval.

2.3
Rates of Change

Average Rate of Change

Elementary mathematics focuses on formulas and relationships among variables but does not include the analysis of quantities that are in a state of constant change. For example, the following problem might be found in elementary mathematics. If you drive at 55 miles per hour (mph) for a total of 3 hours, how far did you travel? This example uses a formula, $d = rt$ (distance = rate · time), and the answer is $d = 55(3) = 165$. However, this model does not adequately describe the situation in the real world. You could drive for 3 hours at an *average rate* of 55 mph, but you probably could not drive for 3 hours at a *constant rate* of 55 mph—in reality, the rate would be in a state of constant change. Other applications in which rates may not remain constant quickly come to mind:

   Profits changing with sales
   Population changing with the growth rate

TABLE 2.2 Distance and Time for a Commuter Car

Time	Distance from home
6:09 A.M.	0 miles
6:25	16
6:30	21
6:34	25
6:36	26.7
6:37	27.7
6:39	28.5
7:01	34
7:03	35
7:09	38
7:15	43
7:28	50

Property taxes changing with the tax rate
Tumor sizes changing with chemotherapy
The speed of falling objects changing over time

The rate at which one quantity changes relative to another is mathematically described by using a concept called a **derivative**. We will see that the rate of change of one quantity relative to another is mathematically determined by finding the slope of a line drawn tangent to a curve.

We begin with a simple example. Consider the speed of a moving object, say, a car. By *speed* we mean the rate at which the distance traveled varies with time. It has magnitude, but no direction. We can measure speed in two ways: *average speed* and *instantaneous speed*. We begin with the average speed of a commuter driving from home to the office.

To find the average speed we use the formula $d = rt$ or $r = \frac{d}{t}$; that is, we divide the distance traveled by the elapsed time:

$$\text{Average speed} = \frac{\text{distance traveled}}{\text{elapsed time}}$$

EXAMPLE 1 A commuter left home one morning and set the odometer to zero. The times and distances were noted as shown in Table 2.2. Find the average speed for the requested time intervals.

a. 6:09 A.M. to 6:39 A.M. (0–30 min)

b. 6:39 A.M. to 7:09 A.M. (30–60 min)

c. 7:01 A.M. to 7:28 A.M. (52–79 min)

Solution **a.** 0–30 min: Average speed $= \dfrac{28.5 - 0}{\frac{1}{2} - 0} = 57$ mph

b. 30–60 min: Average speed $= \dfrac{38 - 28.5}{1 - \frac{1}{2}} = 19$ mph

c. 52–79 min: Average speed $= \dfrac{50 - 34}{\frac{79}{60} - \frac{52}{60}} \approx 35.6$ mph ∎

We can generalize the work done in Example 1 by writing the following formula:

$$\text{Average speed} = \frac{d_2 - d_1}{t_2 - t_1} \text{ mph}$$

where the car travels d_1 miles in t_1 hours and d_2 miles in t_2 hours. The relationship is shown graphically in Figure 2.6, where the average speed is the slope of the line joining the points (d_1, t_1) and (d_2, t_2). This line is sometimes called the **secant line**.

We see that the average rate of change is $\dfrac{f(t_2) - f(t_1)}{t_2 - t_1}$.

It is worthwhile to state this formula in terms of the usual variable x. Let $x = t_1$ and let $h = t_2 - t_1$, which is the length of the time interval over which we are finding an average. From these substitutions it follows that

$$h = t_2 - x$$
$$h + x = t_2$$

This leads to a general statement of the average rate of change.

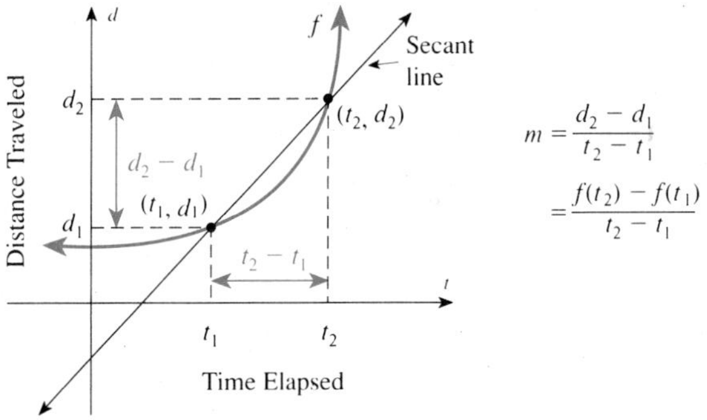

2.3 RATES OF CHANGE 79

FIGURE 2.6
Geometrical interpretation of average speed

Average Rate of Change

> The **average rate of change** of a function f with respect to x over an interval $[x, x + h]$ is given by the formula
>
> $$\frac{f(x + h) - f(x)}{h}$$

EXAMPLE 2 Find the average rate of change of f with respect to x between $x = 3$ and $x = 5$ if $f(x) = x^2 - 4x + 7$.

Solution Given $x = 3$ and $h = 5 - 3 = 2$:

$$f(x + h) = f(3 + 2)$$
$$= f(5)$$
$$= 5^2 - 4(5) + 7$$
$$= 12$$
$$f(x) = f(3)$$
$$= 3^2 - 4(3) + 7$$
$$= 4$$

Average rate of change $= \dfrac{f(x + h) - f(x)}{h}$

$$= \frac{12 - 4}{2}$$
$$= 4$$

The slope of the secant line is 4, as shown in Figure 2.7.

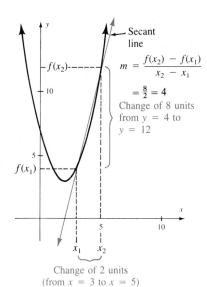

FIGURE 2.7
Average rate of change of f, where $f(x) = x^2 - 4x + 7$ between $x = 3$ and $x = 5$ ■

EXAMPLE 3 Table 2.3 shows the number of divorces for selected years from 1970 to 1987. Find the average divorce rate for the following periods of time:

a. 1970–1980   **b.** 1970–1987   **c.** 1980–1987   **d.** 1985–1987

Solution Average rate of change for f defined by Table 2.3:

$$\frac{f(x+h)-f(x)}{h}$$

TABLE 2.3 Number of U.S. Divorces

Year	Number (in millions)
1970	.708
1975	1.036
1980	1.182
1985	1.187
1987	1.157

a. $x = 1970$ and $h = 10$: $\dfrac{f(1980) - f(1970)}{h} = \dfrac{1.182 - .708}{10}$

$= .0474$

The average divorce rate (rate of change of divorces) between the years 1970 and 1980.

b. $x = 1970$ and $h = 17$: $\dfrac{f(1987) - f(1970)}{17} = \dfrac{1.157 - .708}{17}$

$\approx .0264$

c. $x = 1980$ and $h = 7$: $\dfrac{f(1987) - f(1980)}{7} = \dfrac{1.157 - 1.182}{7}$

$\approx -.0036$

A negative rate of change indicates a decrease in the rate at which divorces are occurring between 1980 and 1987.

d. $x = 1985$ and $h = 2$: $\dfrac{f(1987) - f(1985)}{2} = \dfrac{1.157 - 1.187}{2}$

$= -.015$ ∎

Instantaneous Rate of Change

Note in Examples 2 and 3 that the slopes of the secant lines we found when calculating the average rate of change can vary greatly and really do not tell us much about the rate *at a particular time*. Imagine that a curve called f is defined by a roller coaster track. Consider a fixed location on this track, call it $(x, f(x))$. Now imagine the front car of the roller coaster at some point on the track a horizontal distance of h from x. The slope of the secant line between these points is the average rate of change of the function f between the two points.

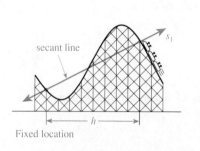

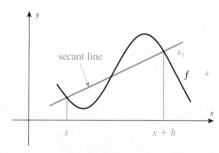

Also imagine the roller coaster car moving along the track toward the fixed location. That is, let $h \to 0$. As the car moves along the track we obtain a sequence of secant lines. The roller coaster is shown at the left and the curve with secant lines is shown at the right.

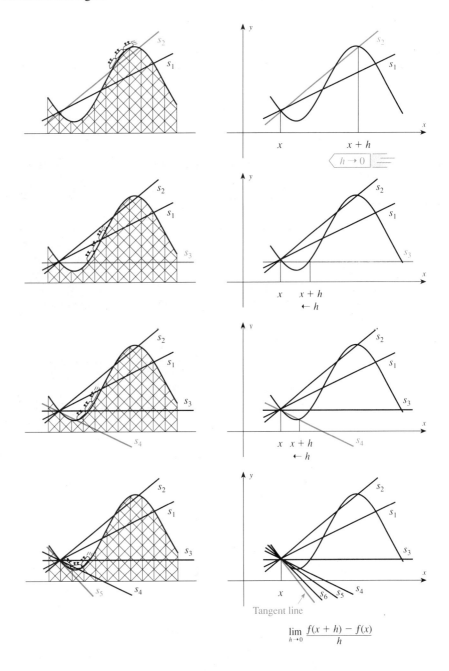

$$\lim_{h \to 0} \frac{f(x+h) - f(x)}{h}$$

As the car gets close to the fixed location, the secant lines approach a limiting line (shown in color). This limiting line for which $h \to 0$ is called the **tangent line**.

Instantaneous Rate of Change

Given a function f and the graph of $y = f(x)$, the *tangent line* at the point $(x, f(x))$ is the line that passes through this point with slope

$$\lim_{h \to 0} \frac{f(x+h) - f(x)}{h}$$

if this limit exists. The slope of the tangent line is also referred to as the *instantaneous rate of change* of the function f with respect to x.

To summarize:

Velocity interpretation	Average rate of change	Instantaneous rate of change
Algebraic interpretation	$\frac{f(x+h) - f(x)}{h}$	$\lim_{h \to 0} \frac{f(x+h) - f(x)}{h}$
Geometric interpretation	Slope of the secant line through the points $[x, f(x)]$ and $[x+h, f(x+h)]$	Slope of the tangent line at the point $[x, f(x)]$

EXAMPLE 4 Find the instantaneous rate of change of the function $f(x) = 5x^2$ with respect to x.

Solution Find $\lim_{h \to 0} \frac{f(x+h) - f(x)}{h}$.

We evaluate this expression in small steps:

1. $f(x)$ is given.
2. $f(x+h) = 5(x+h)^2 = 5x^2 + 10xh + 5h^2$
3. $f(x+h) - f(x) = (5x^2 + 10xh + 5h^2) - 5x^2 = 10xh + 5h^2$
4. $\dfrac{f(x+h) - f(x)}{h} = \dfrac{10xh + 5h^2}{h} = 10x + 5h \quad (h \neq 0)$
5. $\lim_{h \to 0} \dfrac{f(x+h) - f(x)}{h} = \lim_{h \to 0}(10x + 5h) = 10x$ ∎

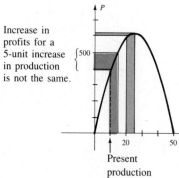

Increase in profits for a 5-unit increase in production is not the same.

FIGURE 2.8
Profit function

Marginal Profit

Now we will use the concept of instantaneous rate of change in a business application. Suppose the profit, P, for a manufacturer is a function of the number of units produced and behaves according to the model

$$P(x) = 50x - x^2$$

The graph of this profit function is shown in Figure 2.8. Now suppose that present production is 10 units. What is the profit?

$$P(10) = 50(10) - 10^2 = 400$$

What is the per unit increase in profit if production is increased from 10 to 20 units?

$$P(20) = 50(20) - 20^2 = 600$$

Increased profit: $P(20) - P(10) = 600 - 400 = 200$

Per unit increase in profit: $\dfrac{P(20) - P(10)}{20 - 10} = \dfrac{200}{10} = 20$

EXAMPLE 5 What is the per unit increase in profit if production is increased from 10 units to:

a. 15 units? **b.** 11 units?

Solution **a.** $\dfrac{P(15) - P(10)}{15 - 10} = \dfrac{525 - 400}{5} = \dfrac{125}{5} = 25$ This is the average rate of change of profit as x increases from 10 to 15.

b. $\dfrac{P(11) - P(10)}{11 - 10} = \dfrac{429 - 400}{1} = \dfrac{29}{1} = 29$ ∎

From Example 5 we see that the per unit increase in profit is different if we increase from 10 to 15 units than if we increase from 10 to 11 units. We would therefore talk about the average per unit increase in profit from 10 to 15 or from 10 to 11 units.

Let us consider this situation in general. An increase in production from x units to $x + h$ units would produce an average per unit increase in profit for a function P as follows:

$$\frac{P(x + h) - P(x)}{x + h - x} = \frac{P(x + h) - P(x)}{h}$$

For the instantaneous rate of change we can let $h \to 0$. *This represents the per unit increase in profit at a production level of x units.* In business this per unit increase is called the **marginal profit** and is defined by

$$\lim_{h \to 0} \frac{P(x + h) - P(x)}{h}$$

if this limit exists.

EXAMPLE 6 Find the marginal profit for the above model; that is, find the marginal profit for

$$P(x) = 50x - x^2$$

Solution **1.** $P(x) = 50x - x^2$ is given.

2. $P(x + h) = 50(x + h) - (x + h)^2 = 50x + 50h - x^2 - 2xh - h^2$

3. $P(x + h) - P(x) = (50x + 50h - x^2 - 2xh - h^2) - (50x - x^2)$
$= 50h - 2xh - h^2$

4. $\dfrac{P(x + h) - P(x)}{h} = \dfrac{(50 - 2x - h)h}{h} = 50 - 2x - h \quad (h \neq 0)$

5. $\lim\limits_{h \to 0} \dfrac{P(x + h) - P(x)}{h} = \lim\limits_{h \to 0} (50 - 2x - h) = 50 - 2x \quad (h \neq 0)$ ∎

We check the results of Example 6 with our previous results:

1. Production level of 10 units where $x = 10$

   $50 - 2(10) - h = 30 - h$

2. Increases: From 10 to 20 ($h = 10$): $30 - h = 30 - 10 = 20$
   From 10 to 15 ($h = 5$): $30 - h = 30 - 5 = 25$
   From 10 to 11 ($h = 1$): $30 - h = 30 - 1 = 29$

3. Marginal profit ($h \neq 0$): $\lim\limits_{h \to 0}(30 - h) = 30$

Velocity

The final application of the concept of an instantaneous rate of change is **velocity**.

EXAMPLE 7 At an amusement park there is a "free fall" ride called "The Edge." The ride involves falling 100 feet in 2.5 seconds. As you are falling, you pass the 16-foot mark at one second and the 64-foot mark at two seconds. What is the velocity at the *instant* the ride passes the 100-foot mark (measured from the top)?

Solution Let $s(t)$ be the distance in feet from the top t seconds after release. Also assume that $s(t) = 16t^2$.*

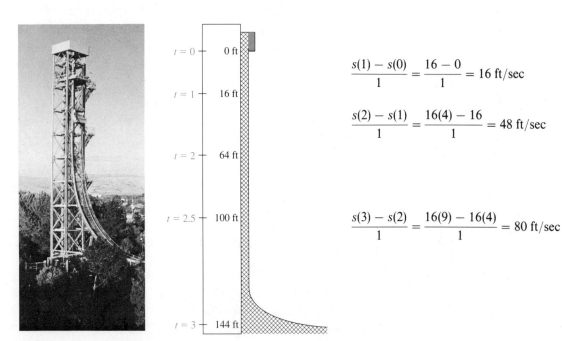

$$\frac{s(1) - s(0)}{1} = \frac{16 - 0}{1} = 16 \text{ ft/sec}$$

$$\frac{s(2) - s(1)}{1} = \frac{16(4) - 16}{1} = 48 \text{ ft/sec}$$

$$\frac{s(3) - s(2)}{1} = \frac{16(9) - 16(4)}{1} = 80 \text{ ft/sec}$$

Courtesy of Great America Theme Park, Santa Clara, CA

* This is the formula for free fall in a vacuum; it is sufficiently accurate for our purposes in this problem.

In general, the average velocity from time $t = t_1$ to $t = t_1 + h$ is given by the following formula (where $h \neq 0$).

$$\text{Average velocity} = \frac{\text{Change in position}}{\text{Change in time}}$$

$$= \frac{s(t_1 + h) - s(t_1)}{h}$$

$$= \frac{16(t_1 + h)^2 - 16t_1^2}{h}$$

$$= \frac{16t_1^2 + 32t_1 h + 16h^2 - 16t_1^2}{h}$$

$$= \frac{32t_1 h + 16h^2}{h}$$

$$= 32t_1 + 16h$$

Now, to find the velocity at a particular instant in time, we simply need to consider the limit as $h \to 0$:*

$$\text{Instantaneous velocity} = \lim_{h \to 0} \frac{s(t_1 + h) - s(t_1)}{h}$$

$$= \lim_{h \to 0} (32t_1 + 16h)$$

$$= 32t_1$$

We can now use these formulas to find the average and instantaneous velocities:

Time interval	Average velocity $32t + 16h$	Instantaneous velocity $32t$
At $t = 0$		$32(0) = 0$
From $t = 0$ to $t = 1$ ($h = 1$)	$32(0) + 16(1) = 16$	
At $t = 1$		$32(1) = 32$
From $t = 1$ to $t = 2$ ($h = 1$)	$32(1) + 16(1) = 48$	
From $t = 1$ to $t = 3$ ($h = 2$)	$32(1) + 16(2) = 64$	
At $t = 2$		$32(2) = 64$
⋮	⋮	⋮

---

* Note that as $h \to 0$ it must be true that $h \neq 0$.

We are given that the ride passes 100 feet at 2.5 seconds, so the instantaneous velocity at that instant is

$$32\left(2\frac{1}{2}\right) = 80 \text{ ft/sec}$$

(By the way, 80 ft/sec is approximately 54 mph.) ∎

2.3
Problem Set

APPLICATIONS

The graph in Figure 2.9 shows the height h of a projectile after t seconds. Find the average rate of change of height (in ft) with respect to the changes in time t (in sec) in Problems 1–4.

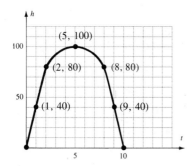

FIGURE 2.9
Height of a projectile in feet t seconds after fired

1. 1 to 7
2. 1 to 5
3. 1 to 2
4. 2 to 9

The graph in Figure 2.10 shows company output as a function of the number of workers. Find the average rate of change of output for the changes in the number of workers in Problems 5–8.

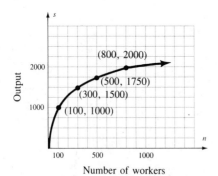

FIGURE 2.10
Output in production relative to the number of employees at Kampbell Construction

5. 100 to 800
6. 300 to 800
7. 500 to 800
8. 300 to 500

The SAT scores of entering first-year college students are shown in Figure 2.11. Find the average yearly rate of change of the scores for the periods of time in Problems 9–12.

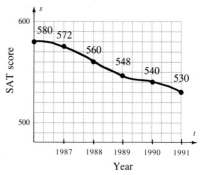

FIGURE 2.11
SAT scores at Riveria College

9. 1986 to 1991
10. 1987 to 1991
11. 1989 to 1991
12. 1990 to 1991

Table 2.4 shows the Gross National Product (GNP) in trillions of dollars for the years 1960–1987. Find the average yearly rate of change of the GNP for the years given in Problems 13–16.

TABLE 2.4 Gross National Product

Year	Dollars (in trillions)
1960	.5153
1970	1.0155
1975	1.5984
1980	2.7320
1986	4.2403
1987	4.5267

13. 1960 to 1987
14. 1970 to 1987
15. 1980 to 1987
16. 1986 to 1987

Table 2.2 gives some distances and commute times for a typical daily commute. Find the average speed for the requested time intervals.

17. 6:09 A.M. to 6:36 A.M.
18. 6:36 A.M. to 7:03 A.M.
19. 7:03 A.M. to 7:28 A.M.
20. 6:09 A.M. to 7:28 A.M.

Trace the curves in Problems 21–26 onto your own paper and draw the secant line passing through P and Q. Next, imagine $h \to 0$ and draw the tangent line at P assuming that Q moves along the curve to the point P. Finally, estimate the slope of the curve at P using the slope of the tangent line you have drawn.

21.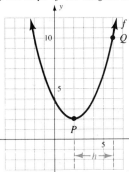

22.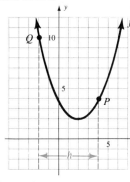

23.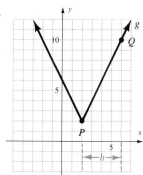

24.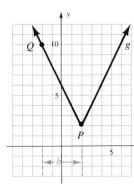

25.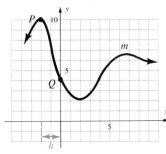

26.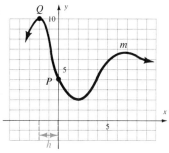

Find the average rate of change for the functions in Problems 27–36.

27. $f(x) = 4 - 3x$ for $x = -3$ to $x = 2$
28. $f(x) = 5x - 1$ for $x = -5$ to $x = -1$
29. $f(x) = 5$ for $x = -3$ to $x = 3$
30. $f(x) = -2$ for $x = 0$ to $x = 5$
31. $f(x) = 3x^2$ for $x = 1$ to $x = 3$
32. $f(x) = x^2 - 3x$ for $x = 0$ to $x = 3$
33. $y = -2x^2 + x + 4$ for $x = 1$ to $x = 4$
34. $y = \sqrt{x}$ for $x = 4$ to $x = 9$
35. $y = \dfrac{-2}{x+1}$ for $x = 1$ to $x = 5$
36. $y = 2x^3 + 7x$ for $x = 1$ to $x = 3$

Find the instantaneous rate of change for the functions in Problems 37–46. These are the same functions as those given in Problems 27–36.

37. $f(x) = 4 - 3x$ for $x = -3$
38. $f(x) = 5x - 1$ for $x = -5$
39. $f(x) = 5$ for $x = -3$
40. $f(x) = -2$ for $x = 0$
41. $f(x) = 3x^2$ for $x = 1$
42. $f(x) = x^2 - 3x$ for $x = 0$
43. $y = -2x^2 + x + 4$ for $x = 1$
44. $y = \sqrt{x}$ for $x = 4$
45. $y = \dfrac{-2}{x+1}$ for $x = 0$
46. $y = 2x^3 + 7x$ for $x = 1$

APPLICATIONS

Suppose the profit, P, for a manufacturer is a function of the number of units produced and behaves according to the model

$$P(x) = 50x - x^2$$

Also suppose that the present production is 20 units. Use this model for Problems 47–50.

47. What is the per unit increase in profit if production is increased from 20 to 30 units?

48. Repeat Problem 47 for an increase from 20 to 25 units.
49. Repeat Problem 47 for an increase from 20 to 21 units.
50. What is the marginal profit for P at $x = 20$?

The cost, C, in dollars for producing x items is given by

$$C(x) = 30x^2 - 100x$$

Use this model for Problems 51–57.

51. Find the average rate of change of cost as x increases from 100 to 200 items.
52. Repeat Problem 51 for an increase from 100 to 110 items.
53. Repeat Problem 51 for an increase from 100 to 101 items.
54. Repeat Problem 51 for an increase from 100 to $(100 + h)$ items.
55. Repeat Problem 51 for an increase from x to $(x + h)$ items.
56. Find $\lim_{h \to 0} \dfrac{C(100 + h) - C(100)}{h}$.
57. Find $\lim_{h \to 0} \dfrac{C(x + h) - C(x)}{h}$.
58. Attach a possible meaning for the result of Problem 56. In business the result of Problem 56 is called the *marginal cost* for the production level of 100. We will discuss this concept later in the text.
59. Attach a possible meaning for the result of Problem 57. In business the result of Problem 57 is called the *marginal cost* for the production level of x. We will discuss this concept later in the text.

2.4
Definition of Derivative

Derivative

The concept of the derivative is a very powerful mathematical idea, and the variety of applications is almost unlimited. In the last section we investigated the instantaneous rate of change of a function f per unit change in x (marginal profit, instantaneous rate of change, velocity, and the slope of a line tangent to a curve at a particular point). All of these ideas can be summarized by a single concept, called the **derivative**:

Definition of Derivative

For a given function f, we define the *derivative of f at x*, denoted by $f'(x)$, to be

$$f'(x) = \lim_{h \to 0} \frac{f(x + h) - f(x)}{h}$$

provided this limit exists. If the limit exists we say f is a *differentiable function at x*.

If the limit does not exist, then we say that f is *not differentiable* at x. We have now developed all of the techniques necessary to apply the definition of derivative to a variety of functions. Also note that as $h \to 0$, $h \neq 0$ so we will assume $h \neq 0$ without stating it when using the five-step process for finding the derivative.

EXAMPLE 1 Use the definition to find the derivative of $y = x^2$.

Solution We carry out the five-step process of the last section:

1. $f(x) = x^2$ is given
2. $f(x + h) = (x + h)^2$ Evaluate f at $x + h$.
   $= x^2 + 2xh + h^2$

3. $f(x + h) - f(x) = (x^2 + 2xh + h^2) - x^2$
$= 2xh + h^2$

4. $\dfrac{f(x + h) - f(x)}{h} = \dfrac{2xh + h^2}{h}$
$= \dfrac{h(2x + h)}{h}$
$= 2x + h$

5. $\lim\limits_{h \to 0} \dfrac{f(x + h) - f(x)}{h} = \lim\limits_{h \to 0}(2x + h)$
$= 2x$

The derivative of $y = x^2$ is $2x$. Sometimes we write $y' = 2x$ or $f'(x) = 2x$. ∎

Tangent Line

EXAMPLE 2 Find the equation of the line tangent to the curve $y = x^2$ at $x = 3$.

Solution From Example 1, $y' = 2x$, so the slope of the tangent line at any point is $2x$. When $x = 3$, then $y = 3^2$, so we are looking for the tangent line passing through $(3, 9)$ with slope equal to the derivative. Remember, the slope of a curve at a point is the value of the derivative at that point. Thus, at $x = 3$, $f'(3) = 2(3) = 6$. Now we can use the point–slope form:

$$y - 9 = 6(x - 3)$$
$$y = 6x - 9 \quad \text{or, in standard form,} \quad 6x - y - 9 = 0 \quad \blacksquare$$

WARNING y' or f' means derivative and does not represent an exponent.

EXAMPLE 3 Find the instantaneous rate of change of profit $P(x) = 80x - 25x^2$, if the present level of production is 50 units.

Solution Evaluate the derivative at $x = 50$ (because that is the present production level). First, find $P'(x)$:

1. $P(x) = 80x - 25x^2$
2. $P(x + h) = 80(x + h) - 25(x + h)^2$
$= 80x + 80h - 25x^2 - 50xh - 25h^2$
3. $P(x + h) - P(x) = (80x + 80h - 25x^2 - 50xh - 25h^2) - (80x - 25x^2)$
$= 80h - 50xh - 25h^2$
4. $\dfrac{P(x + h) - P(x)}{h} = \dfrac{80h - 50xh - 25h^2}{h}$
$= 80 - 50x - 25h$
5. $\lim\limits_{h \to 0} \dfrac{P(x + h) - P(x)}{h} = \lim\limits_{h \to 0}(80 - 50x - 25h)$
$= 80 - 50x$

Thus $P'(x) = 80 - 50x$ and $P'(50) = -2{,}420$. ∎

We use tangent lines to sketch functions in the next chapter. It is worth noting here, though, that if the derivative of a function is positive, then the curve is rising at that point (since the slope of the tangent line is also positive); and if the derivative is negative at a point, then the curve is falling at that point. If the derivative is 0, then the tangent is horizontal. In Example 2, we found the equation of the tangent line of the function $y = x^2$ at $(2, 4)$. This tangent line and others are shown in Figure 2.12.

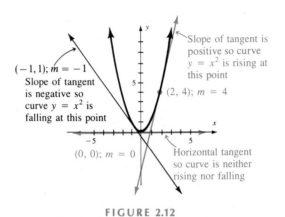

FIGURE 2.12
Graph of $y = x^2$ with tangent lines

Functions That Are Not Differentiable

A derivative may not exist at a particular point. If the limit does not exist at $x = a$, then we say that the function is not differentiable at $x = a$. The concept of differentiability is more easily understood if you relate it to tangent lines. Geometrically, a tangent line will not exist when the graph of the function "has a sharp point," as shown in Figure 2.13a.

In Figure 2.13a, the tangent lines as $h \to 0^-$ and $h \to 0^+$ are not the same, so limit does not exist.

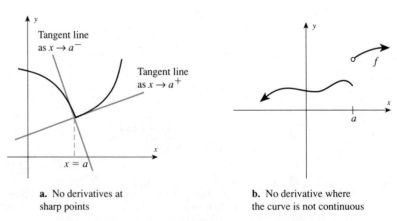

a. No derivatives at sharp points

b. No derivative where the curve is not continuous

FIGURE 2.13
Points on a curve for which the derivative is not defined

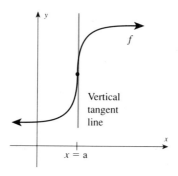

c. No derivative when tangent line is vertical

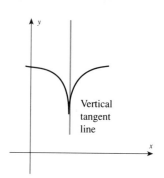

d. No derivative when tangent line is vertical

FIGURE 2.13 *(continued)*

The slope of a curve does not exist at a point for which the derivative is not defined. This means that there might be a vertical tangent line, and when this occurs we also say there is no slope (see Figures 2.13**c** and **d**). A point for which a function is not continuous cannot have a derivative at that point (see Figure 2.13**b**). However, be careful with this statement because a **function may be continuous at $x = a$ but still not be differentiable there** (see Figures 2.13**a, c,** and **d**).

EXAMPLE 4 Let $y = |x|$. Show that y is continuous at $x = 0$ but not differentiable at $x = 0$.

Solution First, $f(x) = |x|$ is continuous at $x = 0$ since $f(0) = |0| = 0$ is defined and $\lim_{x \to 0} |x| = 0$ (see Figure 2.14). Next, find $f'(x)$:

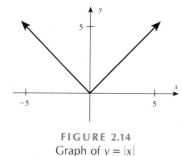

FIGURE 2.14
Graph of $y = |x|$

1. $f(x) = |x|$
2. $f(x + h) = |x + h|$
3. $f(x + h) - f(x) = |x + h| - |x|$
4. $\dfrac{f(x + h) - f(x)}{h} = \dfrac{|x + h| - |x|}{h}$
5. $\lim_{h \to 0} \dfrac{f(x + h) - f(x)}{h} = \lim_{h \to 0} \dfrac{|x + h| - |x|}{h} = \lim_{h \to 0} \dfrac{|0 + h| - |0|}{h}$ at $x = 0$

$= \lim_{h \to 0} \dfrac{|h|}{h}$

Now consider the left- and right-hand limits:

$\lim_{h \to 0^+} \dfrac{|h|}{h} = \lim_{h \to 0^+} \dfrac{h}{h} = 1$ If $h \to 0^+$, then h is positive, so $|h| = h$.

$\lim_{h \to 0^-} \dfrac{|h|}{h} = \lim_{h \to 0^-} \dfrac{-h}{h} = -1$ If $h \to 0^-$, then h is negative, so $|h| = -h$.

Since the left- and right-hand limits are not the same, we see that this limit does not exist. Therefore the derivative does not exist. Thus $f(x) = |x|$ is continuous at $x = 0$ but is not differentiable at $x = 0$. ∎

Example 4 illustrates an important relationship between derivatives and continuity:

> If a function f has a derivative at $x = c$, then it is continuous at $x = c$.
> If a function f is continuous at $x = c$, then it does *not necessarily* have a derivative at $x = c$.

To show that **differentiability implies continuity** it is necessary to establish (1) $f(c)$ is defined, (2) $\lim_{x \to c} f(x)$ exists, and (3) $\lim_{x \to c} f(x) = f(c)$. These conditions all follow from the definition of derivative (we will not show the details here). Example 4 shows that **continuity does not imply differentiability**.

Marginal Cost

In the previous section *marginal profit* was defined. In business and economics the adjective *marginal* means rate of change. Mathematically this means that it is a derivative. Another example of this usage is **marginal cost**, which means the rate of change in cost per unit change in production at an output level of x units. This idea is illustrated with the following example.

EXAMPLE 5 Suppose the total cost in thousands of dollars for manufacturing x thousand items is described by the following equation:

$$C(x) = 2x^2 + 4x + 2{,}500$$

Also suppose that current production is 50,000 items ($x = 50$). Derive the formula for marginal cost and then find the marginal cost for $x = 50$. Also find the actual cost for producing one additional item at this production level.

Solution The formula for the marginal cost is

$$C'(x) = \lim_{h \to 0} \frac{C(x + h) - C(x)}{h}$$

Carry out the five-step procedure.

1. $C(x) = 2x^2 + 4x + 2{,}500$
2. $C(x + h) = 2(x + h)^2 + 4(x + h) + 2{,}500$
3. $C(x + h) - C(x) = [2(x + h)^2 + 4(x + h) + 2{,}500] - [2x^2 + 4x + 2{,}500]$
   $= 2x^2 + 4xh + 2h^2 + 4x + 4h + 2{,}500 - 2x^2 - 4x - 2{,}500$
   $= 4xh + 2h^2 + 4h$
4. $\dfrac{C(x + h) - C(x)}{h} = \dfrac{4xh + 2h^2 + 4h}{h} = 4x + 2h + 4$
5. $\lim_{h \to 0} \dfrac{C(x + h) - C(x)}{h} = \lim_{h \to 0}(4x + 2h + 4) = 4x + 4$

The marginal cost for $x = 50$ (current production is 50,000) is

$$C'(50) = 4(50) + 4 = 204$$

We also want to find the *actual cost* of producing the 51st unit. We need to find

$$\frac{C(51) - C(50)}{1} = 4(50) + 2(1) + 4 = 206$$

This is the number we calculated in step 4 where $x = 50$ and $h = 1$. ∎

The actual cost of producing one unit is very close to the marginal cost. It is usually easier to find the derivative (especially after studying the next two sections) than it is to calculate the actual cost of producing one additional unit. For this reason economists often use marginal cost to approximate the actual cost of producing one additional unit.

2.4
Problem Set

Find the x-value for all points where the functions in Problems 1–6 do not have derivatives.

1.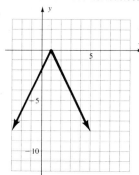
2.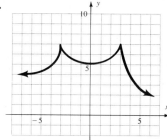
3.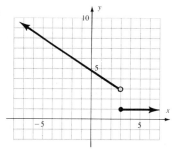
4.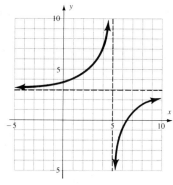
5.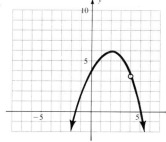
6.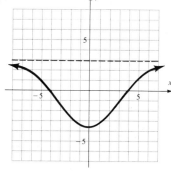

Find the derivative, $f'(x)$, of each of the functions in Problems 7–22 by using the derivative definition.

7. $f(x) = 2x^2$
8. $f(x) = 5x^2$
9. $f(x) = -3x^2$
10. $y = 2x + 1$
11. $y = 4 - 5x$
12. $y = 25 - 250x$
13. $y = 3x^2 + 4x$
14. $y = 3 + 2x - 3x^2$
15. $y = -3x^2 - 50x + 125$
16. $y = x^3$
17. $f(x) = 2x^3$
18. $g(x) = x^3 + x^2$
19. $f(x) = -\dfrac{3}{x}$
20. $f(x) = -\dfrac{2}{x+1}$
21. $y = 2\sqrt{x}$
22. $y = \sqrt{3x}$

Find the equation of the line tangent to the curves in Problems 23–27 at the given point.

23. $y = 5x^2$ at $x = -3$
24. $y = 2x^2$ at $x = 4$
25. $y = 4 - 5x$ at $x = -2$
26. $y = 3x^2 + 4x$ at $x = 0$
27. $y = 3 + 2x - 3x^2$ at $x = -1$

APPLICATIONS

In Problems 28–32 let $C(x) = x^2 - 40x + 2{,}000$ be the total cost function for producing x items where $20 \le x \le 40$.

28. Derive the formula for the marginal cost of this function.

94 CHAPTER TWO THE DERIVATIVE

29. Find the marginal cost for a production level of 20 items.
30. Find the marginal cost for a production level of 30 items.
31. Find the actual cost of producing one additional item if the current production level is 20 items.
32. Find the actual cost of producing one additional item if the current production level is 30 items.

In Problems 33–37 let $C(x) = x^2 - 100x + 3{,}000$ be the total cost function for producing x items where $50 \le x \le 100$.

33. Derive the formula for the marginal cost of this function.
34. Find the marginal cost for a production level of 50 items.
35. Find the marginal cost for a production level of 60 items.
36. Find the actual cost of producing one additional item if the current production level is 50 items.
37. Find the actual cost of producing one additional item if the current production level is 60 items.

In Problems 38–40 suppose the number (in millions) of bacteria present in a culture at time t is given by the formula $N(t) = 2t^2 - 200t + 1{,}000$.

38. Derive the formula for the instantaneous rate of change of the number of bacteria with respect to time.
39. Find the instantaneous rate of change of the number of bacteria with respect to time at time $t = 3$.
40. Find the instantaneous rate of change of the number of bacteria with respect to time at the beginning of this experiment.
41. An object moving in a straight line travels d miles in t hours according to the formula $d(t) = \frac{1}{5}t^2 + 5t$. What is the object's velocity when $t = 3$?
42. Derive the formula for the instantaneous velocity in Problem 41.
43. An object moving in a straight line travels m miles in t minutes according to the formula $m(t) = t^2 + 10t$. What is the object's velocity when $t = 5$?
44. Derive the formula for the instantaneous velocity in Problem 43.

2.5
Differentiation Techniques, Part I

Differentiation is one of the most powerful and useful concepts in mathematics but it would indeed be cumbersome if we had to apply the definition of derivative as we did in the last section every time we wanted to use this concept. Luckily there are many shortcuts to the process, and these are presented as differentiation techniques in this and the following two sections. In stating and working with these derivative formulas it is helpful to use some alternative notations for derivative. We have already used y', f' and $f'(x)$. Some other notations include

$$\frac{dy}{dx} \qquad \frac{d}{dx}y \qquad \frac{d}{dx}f(x) \qquad D_x y \qquad D_x[f(x)]$$

These are usually pronounced "dee y, dee x," "derivative of y with respect to x," "derivative of f with respect to x," "derivative of y with respect to x," and "derivative of f of x with respect to x," respectively. We will use these notations interchangeably.

In this section we develop five very important derivative formulas, in the next section two more, and in Section 2.7 a derivative formula called the *chain rule*. We begin by finding a formula for the derivative of a power; that is, if $y = x^n$, then what is y'? In the last section (Example 1) we found that if $y = x^2$, then $f'(x) = 2x$. How about $y = x^3$?

EXAMPLE 1 If $f(x) = x^3$, find $f'(x)$.

Solution 1. $f(x) = x^3$
2. $f(x + h) = (x + h)^3 = x^3 + 3x^2h + 3xh^2 + h^3$

3. $f(x + h) - f(x) = x^3 + 3x^2h + 3xh^2 + h^3 - x^3$
$= 3x^2h + 3xh^2 + h^3$

4. $\dfrac{f(x + h) - f(x)}{h} = \dfrac{h(3x^2 + 3xh + h^2)}{h}$
$= 3x^2 + 3xh + h^2$

5. $\lim\limits_{h \to 0} \dfrac{f(x + h) - f(x)}{h} = \lim\limits_{h \to 0}(3x^2 + 3xh + h^2) = 3x^2$ ∎

If $y = x^4$, then, if we carry out the steps shown in Example 1, we will find $f'(x) = 4x^3$. The details are left as an exercise. Finally, look for a pattern:

$y = x^2 \quad \to \quad y' = 2x$
$y = x^3 \quad \to \quad y' = 3x^2$
$y = x^4 \quad \to \quad y' = 4x^3$
$y = x^5 \quad \to \quad ?$

Can you replace the question mark by the pattern you found?

$y = x^5 \quad \to \quad y' = 5x^4$
$y = x^6 \quad \to \quad y' = 6x^5$
$\vdots$
$y = x^n \quad \to \quad y' = nx^{n-1}$

We have simply demonstrated the plausibility of the following **power rule**. The general proof for any real number n requires algebra beyond the scope of this course, but we can look at several specific cases to get a feel for the proof. The proof for n a natural number is requested in Problem 69, Problem Set 2.5.

Power Rule If f is a differentiable function, and if $f(x) = x^n$, then
$$f'(x) = nx^{n-1}$$
for any real number n.

EXAMPLE 2 If $y = x$, then $n = 1$, and the power rule asserts that $y' = 1 \cdot x^{1-1} = x^0 = 1$. Verify this result by using the definition of a derivative.

Solution
1. $f(x) = x$
2. $f(x + h) = x + h$
3. $f(x + h) - f(x) = h$
4. $\dfrac{f(x + h) - f(x)}{h} = 1$
5. $\lim\limits_{h \to 0} \dfrac{f(x + h) - f(x)}{h} = \lim\limits_{h \to 0} 1 = 1$ ∎

EXAMPLE 3 Find derivatives of the following functions by using the power rule.

a. x^4 b. x^5 c. x^6 d. x^0 e. x^{-1}
f. x^{-2} g. $x^{1/3}$ h. $\sqrt{x}$ i. $x^{-2.5}$

Solution
a. If $y = x^4$, then $y' = 4x^3$.
b. If $y = x^5$, then $y' = 5x^4$.
c. If $y = x^6$, then $y' = 6x^5$.
d. If $y = x^0$, then $y' = 0x^{-1} = 0$.
e. If $y = x^{-1}$, then $y' = (-1)x^{-1-1} = -x^{-2}$ or $\frac{-1}{x^2}$.
f. If $y = x^{-2}$, then $y' = (-2)x^{-2-1} = -2x^{-3}$ or $\frac{-2}{x^3}$.
g. If $y = x^{1/3}$, then $y' = (\frac{1}{3})x^{1/3-1} = \frac{1}{3}x^{-2/3}$.
h. If $y = \sqrt{x}$, write (or think of it) as $y = x^{1/2}$; now apply the power rule: $y' = (\frac{1}{2})x^{1/2-1} = \frac{1}{2}x^{-1/2}$. Since the problem was given in radical notation, we state the answer in radical notation: $y' = \dfrac{1}{2\sqrt{x}}$.
i. If $y = x^{-2.5}$, then $y' = -2.5x^{-3.5}$. ∎

A constant rule can be easily derived using the derivative definition:

1. $f(x) = k$ is given.
2. $f(x + h) = k$
3. $f(x + h) - f(x) = 0$
4. $\dfrac{f(x + h) - f(x)}{h} = 0$
5. $\lim\limits_{h \to 0} \dfrac{f(x + h) - f(x)}{h} = \lim\limits_{h \to 0} \dfrac{k - k}{h} = \lim\limits_{h \to 0} 0 = 0$

Constant Rule

If $f(x) = k$ for some constant k, then
$$f'(x) = 0$$

EXAMPLE 4 Use the constant rule to find the derivatives of the given functions.

a. $y = 5$ b. $y = -18$ c. $y = \sqrt{3}$ d. $y = \pi$

Solution The derivative of any constant is 0; thus

a. $y' = 0$ b. $y' = 0$ c. $y' = 0$ d. $y' = 0$ ∎

The power rule and the constant rule for derivatives can be combined to prove the following result:

Constant Times a Function

If $y = kf(x)$ for a differentiable function f, then
$$y' = kf'(x)$$

2.5 DIFFERENTIATION TECHNIQUES, PART I

This derivative formula for a constant times a function is easy to derive using the definition of a derivative. (The procedure is left as an exercise). Example 5 illustrates how to use the formula.

EXAMPLE 5 Find the derivatives of the given functions.

a. $y = 3x^2$ b. $y = 6x^{-5}$ c. $y = -\frac{3}{5}x^{10}$ d. $y = -2\sqrt{x}$
e. $y = 5$

Solution
a. $y' = 3[(2)x^1]$
   $= 6x$
b. $y' = 6[(-5)x^{-6}]$
   $= -30x^{-6}$
c. $y' = -\frac{3}{5}[(10)x^9]$
   $= -6x^9$
d. $y' = -2[(\frac{1}{2})x^{1/2-1}]$
   $= -x^{-1/2}$

e. $y' = 0$ by the constant rule. However, note that we can write $y = 5 = 5x^0$ so $y' = 5 \cdot 0x^{-1} = 0$. This says that the power rule and the constant rule give the same results. ∎

The next two derivative formulas are for sums or differences of functions:

Sum Rule
Difference Rule

Suppose f and g are differentiable functions of x.

If $y = f + g$, then $y' = f' + g'$.
If $y = f - g$, then $y' = f' - g'$.

We assume that if f and g are differentiable at a point $x = a$, then the sum rule implies that the sum $f + g$ is also differentiable at $x = a$.

The following proof of the sum formula is optional, and you may skip over to Example 6 if you are not interested in deriving these formulas. We begin by writing y as $u(x)$ so that the sum formula can be written as $u(x) = f(x) + g(x)$. We want to show that $u'(x) = f'(x) + g'(x)$. We use the definition of derivative:

1. $u(x) = f(x) + g(x)$
2. $u(x + h) = f(x + h) + g(x + h)$
3. $u(x + h) - u(x) = [f(x + h) + g(x + h)] - [f(x) + g(x)]$
   $= [f(x + h) - f(x)] + [g(x + h) - g(x)]$
4. $\dfrac{u(x + h) - u(x)}{h} = \dfrac{[f(x + h) - f(x)] + [g(x + h) - g(x)]}{h}$
   $= \dfrac{f(x + h) - f(x)}{h} + \dfrac{g(x + h) - g(x)}{h}$
5. $\lim\limits_{h \to 0} \dfrac{u(x + h) - u(x)}{h} = \lim\limits_{h \to 0} \left[\dfrac{f(x + h) - f(x)}{h} + \dfrac{g(x + h) - g(x)}{h} \right]$
   $= \lim\limits_{h \to 0} \dfrac{f(x + h) - f(x)}{h} + \lim\limits_{h \to 0} \dfrac{g(x + h) - g(x)}{h}$

Thus $u'(x) = f'(x) + g'(x)$. The proof for the difference formula is identical to the one for the sum formula.

These last two derivative formulas, along with the ones already discussed, allow us to easily find the derivatives of polynomials, as shown in Example 6.

EXAMPLE 6 Find the derivatives of the given functions.

a. $y = 5x^2 - 3x + 15$ **b.** $y = 9x^3 - 4x^2 + 5x - 12$

Solution   **a.** $y' = 5(2)x - 3(1) + 0$ **b.** $y' = 9(3)x^2 - 4(2)x + 5(1) - 0$
$= 10x - 3$ $= 27x^2 - 8x + 5$ ∎

EXAMPLE 7 If $f(x) = 6x^8 - 5\sqrt{x} + \dfrac{4}{x}$, find $f'(x)$.

Solution Write (or think of) this as $f(x) = 6x^8 - 5x^{1/2} + 4x^{-1}$. Then

$$f'(x) = 6(8)x^7 - 5\left(\frac{1}{2}\right)x^{-1/2} + 4(-1)x^{-2}$$

$$= 48x^7 - \frac{5}{2\sqrt{x}} - \frac{4}{x^2}$$ ∎

EXAMPLE 8 If $f(x) = \dfrac{x^4 + 5x^3 + 1}{\sqrt{x}}$ find $f'(x)$.

Solution Write or think of this as

$$x^{-1/2}(x^4 + 5x^3 + 1) = x^{7/2} + 5x^{5/2} + x^{-1/2}$$

Thus, if $f(x) = x^{7/2} + 5x^{5/2} + x^{-1/2}$, then

$$f'(x) = \frac{7}{2}x^{5/2} + \frac{25}{2}x^{3/2} - \frac{1}{2}x^{-3/2}$$ ∎

EXAMPLE 9 Suppose the amount of calcium remaining in a person's bloodstream is given by the formula $A = t^{-3/2}$ where t is the number of days after the calcium was injected ($t \geq \frac{1}{2}$). How fast is the body removing calcium from the blood exactly 48 hours (or $t = 2$) after the calcium was injected?

Solution The rate of change (per day) of calcium in the blood is given by the derivative

$$A'(t) = \frac{dA}{dt} = -\frac{3}{2}t^{-5/2}$$

When $t = 2$, this rate is $-\frac{3}{2}(2)^{-5/2}$. This rate can be approximated on a calculator by pressing

| 2 | x^y | 2.5 | +/− | × | 1.5 | +/− | = | −0.2651650429

The amount of calcium in the blood is changing at the rate of about −.27 unit per day when $t = 2$. The negative sign tells us that the amount of calcium is decreasing.   ∎

Note in Example 9 that the variable is not x and that the symbol used is dA/dt. This symbol is read as "derivative of A with respect to t," or as "dee A, dee t." Even though the variables x and y are used most of the time, keep in mind that any variables may be substituted for x and y.

2.5 DIFFERENTIATION TECHNIQUES, PART I

The derivative rules of this section are summarized here for easy reference:

Differentiation Formulas

POWER RULE: If $y = x^n$, then $y' = nx^{(n-1)}$ for any real number n.
CONSTANT RULE: If $y = k$, then $y' = 0$.
CONSTANT TIMES A FUNCTION RULE: If $y = kf$, then $y' = kf'$.
SUM RULE: If $y = f + g$, then $y' = f' + g'$.
DIFFERENCE RULE: If $y = f - g$, then $y' = f' - g'$.

2.5
Problem Set

Find the derivatives of the functions in Problems 1–42 and simplify them algebraically.

1. $y = x^7$
2. $y = x^{16}$
3. $y = x^{12}$
4. $y = x^{-3}$
5. $y = x^{-5}$
6. $y = 25$
7. $y = -130$
8. $y = 5x^{-4}$
9. $y = -4x^{-8}$
10. $y = 5\sqrt{x}$
11. $y = -\frac{1}{2}\sqrt{x}$
12. $y = \frac{3}{4}\sqrt{x}$
13. $y = 5x^{-8}$
14. $y = -3x^{-2}$
15. $y - 12x^{5/4}$
16. $y = 5x^2 - 9$
17. $x = 3x^2 + x$
18. $y = 5x^3 - 9x^2$
19. $y = 2x^2 - 5x - 6$
20. $y = 5x^2 - 5x + 12$
21. $y = 5x^3 - 5x^2 + 4x - 5$
22. $y = 6x^3 - 25x^2 - 6x + 45$
23. $y = x^{-3} + x^2 + x^{-1}$
24. $y = x^{-5} - x^{-3} - x^{-1}$
25. $y = (x^4 + 2)^2$
26. $y = (x^5 - 1)^2$
27. $y = \frac{2}{x} + \frac{5}{x^2}$
28. $y = \frac{3}{x} - \frac{2}{x^3}$
29. $f(x) = -5x^7 + 2\sqrt{x} - 3x^{-1}$
30. $f(x) = 4x^{-3/4} - 5\sqrt{x} + 2x^{-3}$
31. $f(x) = \frac{x^3 + 4x^2 + 1}{x^2}$
32. $f(x) = \frac{3x^3 - 5x + 6}{2x^2}$
33. $f(x) = \frac{x^5 + 3x^3}{\sqrt{x}}$
34. $f(x) = \frac{2x^4 - 5x^2}{2\sqrt{x}}$
35. $\frac{d}{dx}\left(\sqrt[3]{x} + \frac{5}{8}\right)$
36. $\frac{d}{dy}(4y^{1/2} - 5)$
37. $\frac{d}{dD}(\pi D)$
38. $\frac{d}{dr}(\pi r^2)$
39. $\frac{dV}{dr}$ where $V = \frac{4}{3}\pi r^3$
40. $\frac{dV}{dR}$ where $V = \frac{10}{3}\pi R^2$
41. $\frac{dK}{dC}$ where $K = \frac{C^2}{4\pi}$
42. $\frac{dS}{dR}$ where $S = 4\pi R^2$

Find the slope of the tangent to the graph of f at the given point in Problems 43–48.

43. $f(x) = x^5$ at $(1, 1)$
44. $f(x) = x^5$ at $(2, 32)$
45. $f(x) = \frac{1}{\sqrt{x}}$ at $(1, 1)$
46. $f(x) = \frac{1}{\sqrt{x}}$ at $\left(4, \frac{1}{2}\right)$
47. $f(x) = 3x^2 - 4x + 10$ at $(0, 10)$
48. $f(x) = 3x^2 - 4x + 10$ at $(3, 25)$

Find the equation of the tangent line of f at the given point in Problems 49–54.

49. $f(x) = x^2 + 2x + 1$ at $(0, 1)$
50. $f(x) = x^2 + 2x + 1$ at $(-2, 1)$
51. $f(x) = 2 - \frac{2}{x^2}$ at $(1, 0)$
52. $f(x) = 2 - \frac{2}{x^2}$ at $(0, 2)$
53. $f(x) = \sqrt[3]{x}$ at $(1, 1)$
54. $f(x) = \sqrt[3]{x}$ at $(8, 2)$

APPLICATIONS

55. The cost of producing a certain type of boat is $C(x) = 20x^2 + 500x + 250{,}000$. What is the marginal cost?
56. The profit function for a certain item is $P(x) = 45x - 3x^3$. What is the marginal profit?
57. The relationship between the current and the time on an X-ray machine is given by the formula $m = 400/t$. Find dm/dt.
58. The relationship between the current and the distance on an X-ray machine is given by the formula $m = 256/s^2$. Find dm/ds.
59. The earnings (in thousands of dollars) of Amdex Corporation are a function of the number of years since its founding in 1986 according to the formula $A(t) = .05t^2 + 25t + 5$. At what rate are the earnings changing in 1992?

60. In a certain experiment, subjects are found to learn according to the model $N = 25\sqrt{x}$, where N is the number of tasks learned in x hours. How fast are the subjects learning at the end of the second hour?

61. Repeat Problem 60 except now determine how fast the subjects are learning at the end of the fifth hour.

62. In a wildlife reserve the population P of foxes depends on the population x of rabbits according to the formula $P = .0005x + .00001x^2$. What is the rate at which the population of foxes is changing when the number of rabbits is 100,000?

63. Repeat Problem 62 for a population of 1 million rabbits.

64. The number of gallons, g, of water pumped into a reservoir after t minutes is given by the formula $g(t) = 2t + \sqrt{t}$. At what rate is water flowing into the reservoir when $t = 10$?

65. Assume that the consumer price index (CPI) of an economy is described by the function

$$P(t) = -.05t^2 + 5t + 250$$

where t is the number of years after 1990 (i.e., $t = 0$ corresponds to 1990).
a. What was the average rate of change in the CPI from the years 1990 to 1994?
b. At what rate was the consumer price index changing in the year 1991?

66. Show that if $y = x^4$, then $y' = 4x^3$ by using the derivative definition.

67. Derive the formula for a constant times a function by using the derivative definition.

68. Prove the formula for the difference of functions by using the derivative definition.

69. Derive the power rule for a natural number n by using the binomial theorem.

2.6
Differentiation Techniques, Part II

There are two additional shortcut differentiation formulas that simplify work with calculus. These are the *product* and *quotient* formulas. Since the sum and difference formulas are easy to remember and are intuitively obvious, students often incorrectly try to generalize to product and quotient formulas. The derivative of a sum or difference is the sum or difference of the derivatives. But this is not true of the product and quotient formulas.

WARNING *The derivative of a product is not the product of the derivatives.*

The following simple example will illustrate that the derivative of a product is not the product of the derivatives. Let $f(x) = x^2$ and $g(x) = x^3$. Consider $y = f(x)g(x) = x^5$:

$$f'(x) = 2x \qquad g'(x) = 3x^2 \qquad y' = 5x^4$$

and

$$f'(x)g'(x) = 2x(3x^2) = 6x^3 \neq 5x^4$$

Note that $y' \neq f'g'$. The product formula tells us how to find the derivative of a product without the necessity of first multiplying them together.

Product Rule

If f and g are differentiable functions such that $y = f(x)g(x)$, then

$$y' = f(x)g'(x) + g(x)f'(x)$$

This formula can more easily be remembered in the following form: If $y = fg$, then

$$y' = fg' + gf' \qquad \text{First function times the derivative of the second plus second times the derivative of the first}$$

2.6 DIFFERENTIATION TECHNIQUES, PART II

This differentiation formula is proved in Example 4, but we will first work through a couple of examples to make sure you understand how it works.

EXAMPLE 1 If $f(x) = x^2$, $g(x) = x^3$, and $y = fg$, verify that $y' = fg' + gf' = 5x^4$.

Solution $f'(x) = 2x$ and $g'(x) = 3x^2$, so

$$fg' + f'g = (x^2)(3x^2) + (x^3)(2x)$$
$$= 3x^4 + 2x^4$$
$$= 5x^4$$

Thus $y' = fg' + f'g = 5x^4$. ∎

EXAMPLE 2 If $y = 2x^3(x^2 - 3x)$, find $\dfrac{dy}{dx}$ and simplify.

Solution You could multiply the factors together to obtain a polynomial, then use the results of the last section to find the derivative, or you can use the product rule. This example illustrates the product rule.

$$\dfrac{dy}{dx} = \dfrac{d}{dx}[\overbrace{2x^3}^{\text{first}}\ \overbrace{(x^2-3)}^{\text{second}}] = 2x^3 \overbrace{\dfrac{d}{dx}(x^2-3)}^{\text{first} \cdot \text{derivative of second}} + \overbrace{(x^2-3)\dfrac{d}{dx}(2x^3)}^{\text{second} \cdot \text{derivative of first}}$$

$$= 2x^3(2x) + (x^2-3)(6x^2)$$
$$= 4x^4 + 6x^4 - 18x^2$$
$$= 10x^4 - 18x^2$$
$$= 2x^2(5x^2 - 9)$$ ∎

EXAMPLE 3 If $y = (\sqrt{x}+1)(2\sqrt{x}-3)$, find dy/dx and simplify.

Solution $\dfrac{dy}{dx} = \dfrac{d}{dx}[\overbrace{(x^{1/2}+1)}^{\text{first}}\overbrace{(2x^{1/2}-3)}^{\text{second}}]$

$$= (x^{1/2}+1)\overbrace{\dfrac{d}{dx}(2x^{1/2}-3)}^{\text{derivative of second}} + (2x^{1/2}-3)\overbrace{\dfrac{d}{dx}(x^{1/2}+1)}^{\text{derivative of first}}$$

$$= (x^{1/2}+1)[2(\tfrac{1}{2})x^{-1/2}] + (2x^{1/2}-3)[\tfrac{1}{2}x^{-1/2}]$$
$$= (x^{1/2}+1)x^{-1/2} + (2x^{1/2}-3)[\tfrac{1}{2}x^{-1/2}]$$
$$= x^0 + x^{-1/2} + x^0 - \tfrac{3}{2}x^{-1/2}$$
$$= 2 - \tfrac{1}{2}x^{-1/2}$$
$$= 2 - \dfrac{1}{2\sqrt{x}} \cdot \dfrac{\sqrt{x}}{\sqrt{x}}$$
$$= 2 - \dfrac{\sqrt{x}}{2x}$$
$$= \dfrac{4x - \sqrt{x}}{2x}$$ ∎

WARNING Example 3 shows that the product rule is not always the most economical method. Notice that if you first multiply the factors to obtain

$$y = 2x - \sqrt{x} - 3$$

You can find the derivative by using the power rule:

$$y' = 2 - \frac{1}{2}x^{-1/2} = 2 - \frac{\sqrt{x}}{2x}$$

$$= \frac{4x - \sqrt{x}}{2x}$$

EXAMPLE 4 Prove the product rule. That is, if f and g are differentiable functions such that $y = f(x)g(x)$, then prove that

$$y' = f(x)g'(x) + g(x)f'(x)$$

Solution Use the definition of derivative:

$$y' = \lim_{h \to 0} \frac{f(x+h)g(x+h) - f(x)g(x)}{h}$$

Add and subtract $f(x)g(x+h)$ in the numerator:

$$y' = \lim_{h \to 0} \frac{f(x+h)g(x+h) - f(x)g(x) + f(x)g(x+h) - f(x)g(x+h)}{h}$$

$$= \lim_{h \to 0} \frac{[f(x+h)g(x+h) - f(x)g(x+h)] + [f(x)g(x+h) - f(x)g(x)]}{h}$$

$$= \lim_{h \to 0} \left[g(x+h) \cdot \frac{f(x+h) - f(x)}{h} + f(x) \cdot \frac{g(x+h) - g(x)}{h} \right]$$

$$= \lim_{h \to 0} \left[g(x+h) \cdot \frac{f(x+h) - f(x)}{h} \right] + \lim_{h \to 0} \left[f(x) \cdot \frac{g(x+h) - g(x)}{h} \right]$$

Since $g(x)$ is differentiable, it is continuous, and $\lim_{h \to 0} g(x+h) = g(x)$. Using this and the definition of derivative, we have

$$y' = g(x)f'(x) + f(x)g'(x)$$ ∎

WARNING The derivative formula for quotients also does *not* follow the pattern of simply taking the quotient of the derivatives. The quotient formula is stated in the following box, and you are asked to prove it in the problems. It is very easy to prove because you can write f/g as fg^{-1} and use the product rule together with the chain rule, which is discussed in the next section.

Quotient Rule

If f and g are differentiable functions such that $y = f(x)/g(x)$ then $y' = [g(x)f'(x) - f(x)g'(x)]/[g(x)]^2$. This formula can be more easily stated and remembered by using the following notation:

If $y = \dfrac{f}{g}$, then

$$y' = \frac{gf' - fg'}{g^2}$$

The *denominator* function times the derivative of the *numerator* minus the *numerator* times the derivative of the denominator all divided by the denominator squared
(Remember that *d* comes before *n* alphabetically.)
You can easily remember this by: "low dee high minus high dee low, over the square of the denominator must go."

EXAMPLE 5 If $y = \dfrac{2x^3}{3x+1}$, find $\dfrac{dy}{dx}$ and simplify.

Solution $y' = \dfrac{d}{dx}\left[\dfrac{2x^3}{3x+1}\right]$ ← top
← bottom

$$y' = \frac{\overbrace{(3x+1)}^{\text{bottom}}\overbrace{(2x^3)'}^{\text{derivative of top}} - \overbrace{(2x^3)}^{\text{top}}\overbrace{(3x+1)'}^{\text{derivative of bottom}}}{\underbrace{(3x+1)^2}_{\substack{\text{bottom}\\\text{squared}}}}$$

"Low dee high minus high dee low over the square of the denominator must go…"

$$= \frac{(3x+1)(6x^2) - (2x^3)(3)}{(3x+1)^2}$$

$$= \frac{18x^3 + 6x^2 - 6x^3}{(3x+1)^2}$$

$$= \frac{12x^3 + 6x^2}{(3x+1)^2}$$

Leave this in factored form—do *not* expand.

EXAMPLE 6 If $y = \dfrac{3x^2+2}{5x^2-1}$, find $\dfrac{dy}{dx}$ and simplify.

Solution $y' = \dfrac{(5x^2-1)(6x) - (3x^2+2)(10x)}{(5x^2-1)^2}$

$$= \frac{30x^3 - 6x - 30x^3 - 20x}{(5x^2-1)^2}$$

$$= \frac{-26x}{(5x^2-1)^2}$$

EXAMPLE 7   An apple factory can produce x thousand gallons of apple juice per week at an average weekly cost of

$$C(x) = \frac{x+15}{x}$$ per gallon on the domain $[1, 20]$.

Show that this average is always decreasing.

Solution   A function is decreasing when its derivative is negative. Thus we need to show the inequality $C'(x) < 0$ on $[1, 20]$. First find $C'(x)$.

Method I: Quotient rule

$$C'(x) = \frac{x(1) - (x+15)(1)}{x^2} = -\frac{15}{x^2}$$

Method II: Power rule. Write $C(x) = 1 + 15x^{-1}$. Then

$$C'(x) = -15x^{-2}$$

WARNING   Do not forget that quotients can often be written as products and that the problem may be easier to work as a product than as a quotient. However, in either case, you will obtain the same results if you do not make any mistakes. Since $C'(x) = -15x^{-2}$ we see that $C'(x) < 0$ for all nonzero x. Thus, the total cost is always decreasing on $[1, 20]$. ∎

2.6

Problem Set

Find the derivative for each of the functions in Problems 1–28 and simplify.

1. $f(x) = 5x^2(x^2 - 6)$
2. $f(x) = 3x^3(x^2 + 7)$
3. $f(x) = (x + 1)(x - 2)$
4. $f(x) = (2x - 5)(x + 7)$
5. $g(x) = (3x^2 + 5)(2x^2 - 5)$
6. $g(x) = (9x^2 - 8)(2x^2 - 5)$
7. $g(x) = 5x^4(2x^2 - 5x + 1)$
8. $g(x) = 9x^5(3x^2 - 2x + 15)$
9. $y = \dfrac{x}{x-3}$
10. $y = \dfrac{x}{3x-1}$
11. $y = \dfrac{x+5}{x-3}$
12. $y = \dfrac{3x+1}{15-x}$
13. $y = \dfrac{x^2+3}{x^2-5}$
14. $y = \dfrac{5-7x^3}{1+x^3}$
15. $y = \dfrac{x^2+x-6}{x^2-4}$
16. $y = \dfrac{2x^2-x}{x^2+6x}$
17. $f(x) = (2x^2 - 1)(x^3 + 2x^2 - 3)$
18. $f(x) = (3x^2 + 2)(x^3 - 5x^2 + 4)$
19. $f(x) = (4x^2 + x)(x^3 - 3x^2 + 13)$
20. $f(x) = (x^2 + 3x)(x^3 + 4x^2 + 25)$
21. $f(x) = (3x^3 - 2x + 5)(2x^4 + 5x - 9)$
22. $f(x) = (7x^4 - 8x^2 + 144)(5x^3 + 25)$
23. $f(x) = 5x^{2/3}(5x^{-1} + 3x)$
24. $f(x) = 3x^2(x^{-2} + x)$
25. $f(x) = 6\sqrt{x}(2x^2 - 5)$
26. $f(x) = 2(\sqrt{x} + 3x)(\sqrt{x} - x)$
27. $g(x) = \dfrac{2x}{5x^2 - 11x + 3}$
28. $g(x) = \dfrac{5x}{125 - 25x - 3x^2}$

Find the slope of the tangent to the graph of f at the value of x given in Problems 29–34.

29. $f(x) = \dfrac{5x^2 + 5x}{x - 5}$ at $x = 1$
30. $f(x) = \dfrac{x+2}{4x - 7x^2}$ at $x = -1$
31. $f(x) = x^{1/2}(x^2 + 3)$ at $x = 4$

32. $f(x) = x^{2/3}(x^2 - 5)$ at $x = 8$
33. $f(x) = \dfrac{x^2 - 5x + 1}{x^2 + 4x + 4}$ at $x = 0$
34. $f(x) = \dfrac{x^3 + 1}{x^3 - 1}$ at $x = 0$

Find an equation of the tangent line to the graph of f at the given value of x in Problems 35–40.

35. $f(x) = \dfrac{5x^2 + 5x}{x - 5}$ at $x = 0$
36. $f(x) = \dfrac{x + 2}{4x - 7x^2}$ at $x = 1$
37. $f(x) = x^{1/2}(x^2 + 3)$ at $x = 4$
38. $f(x) = x^{2/3}(x^2 - 5)$ at $x = 8$
39. $f(x) = \dfrac{x^2 - 5x + 1}{x^2 + 4x + 4}$ at $x = 1$
40. $f(x) = \dfrac{x^3 + 1}{x^3 - 1}$ at $x = 0$

APPLICATIONS

41. Classical economic theory predicts that in a free market the demand, D, for a commodity must decrease as the price, x, increases. Such a demand curve is drawn in Figure 2.15.

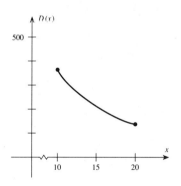

FIGURE 2.15
Demand curve
$$D(x) = \dfrac{100{,}000}{x^2 + 15x + 25}$$
on the domain $[10, 20]$

Find the rate of change of demand with respect to the price change for the demand curve shown in Figure 2.15.

42. If the demand curve is
$$D(x) = \dfrac{25{,}000}{x^2 + 3x + 20} \text{ on the domain } [10, 20]$$
find the rate of change of demand with respect to a price change.

43. Find the rate of change of demand for $x = 15$ using the demand curve in Figure 2.15, and interpret this answer in terms of the graph. Is the demand increasing or decreasing at that point?

44. Find the rate of change of demand for $x = 15$ using the demand curve in Problem 42 and interpret this answer. Is the demand increasing or decreasing at that point?

45. Dollars spent on advertising will influence the number of items purchased, up to some saturation point as shown by the graph in Figure 2.16.

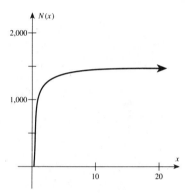

FIGURE 2.16
Number of items sold (in thousands) for dollars spent (in millions) according to the formula
$$N(x) = 1{,}500 - \dfrac{575}{x}$$

Find the rate of change of the number of items with respect to the number of dollars spent.

46. Find the rate of change of the number of items with respect to the number of dollars spent for the function
$$N(x) = 2{,}500 - \dfrac{400}{x}$$
where N is measured in thousands of items and x in thousands of dollars.

47. Find the rate of change of the number of items if $10,000,000 is spent for the function defined in Figure 2.16. Interpret your answer in terms of this graph. Is the number of items purchased increasing or decreasing at this point?

48. Find the rate of change of the number of items if $10,000 is spent for the function defined in Problem 46, and interpret this answer. Is the number of items purchased increasing or decreasing at this point?

49. Many drugs injected into the bloodstream have a dramatic effect over a very short period of time and then stabilize as shown in Figure 2.17.

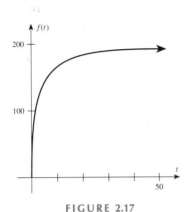

FIGURE 2.17
Amount of a drug (in number of milligrams) injected into the bloodstream after t minutes according to the formula
$$f(t) = \frac{t}{.01 + .005t}$$

Find the rate at which the amount of the drug is changing relative to the time t.

50. At what rate is the drug being absorbed into the system (in milligrams per minute) after 5 minutes for the drug shown in Figure 2.17?

51. Psychologists tell us that there is a certain interest value in learning a certain task, but after the task has been mastered, interest (and consequently, learning) decreases. Suppose an experiment is set up in which learning is tested by counting the number of items, N, learned by a subject in a given amount of time, t (in minutes), according to the formula given in Figure 2.18.

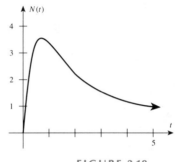

FIGURE 2.18
Learning curve
$$N(t) = \frac{10t}{2t^2 + 1}$$

What is the rate of change of number of items learned with respect to time, t?

52. Find the rate of learning for the learning curve shown in Figure 2.18 after 2 minutes. Interpret this answer in terms of the graph.

2.7
The Chain Rule

We now come to the last of the differentiation techniques we will consider, the **chain rule**. The chain rule, along with the other derivative theorems we have discussed, will enable us to efficiently find the derivative of almost every function we will consider. Remember, however, as you progress throughout this or other courses, that if you encounter a function that does not "fit" any of the differentiation formulas we have developed, you can resort to using the derivative definition.

Introduction to the Chain Rule

Suppose it is known that the carbon monoxide pollution in the air is .02 ppm (parts per million) for each person in a town whose population is growing at the rate of 1,000 people per year. To find the rate at which the level of pollution is increasing with respect to time, we form the product

(.02 ppm/person)(1,000 people/year) = 20 ppm/year

In this example, the level of pollution, L, is a function of the population, P, which is itself a function of time, t. Thus, L is a function of t, and

$$\begin{bmatrix} \text{RATE OF CHANGE OF } L \\ \text{WITH RESPECT TO } t \end{bmatrix} = \begin{bmatrix} \text{RATE OF CHANGE OF } L \\ \text{WITH RESPECT TO } P \end{bmatrix} \begin{bmatrix} \text{RATE OF CHANGE OF } P \\ \text{WITH RESPECT TO } t \end{bmatrix}$$

Expressing each of these rates in terms of an appropriate derivative, we obtain the following equation:

$$\frac{dL}{dt} = \frac{dL}{dP} \frac{dP}{dt}$$

These observations suggest the following differentiation rule.

Chain Rule

Suppose y is a differentiable function of u, and u, in turn, is a differentiable function of x. Then

$$\frac{dy}{dx} = \frac{dy}{du} \cdot \frac{du}{dx}$$

EXAMPLE 1 If $y = u^6$ and $u = 2x^3 - 5x$, find

a. $\dfrac{dy}{du}$ **b.** $\dfrac{du}{dx}$ **c.** $\dfrac{dy}{dx}$

Write your answer as a function of x.

Solution **a.** $\dfrac{dy}{du} = \dfrac{d}{du}(u^6) = 6u^5 = 6(2x^3 - 5x)^5$

b. $\dfrac{du}{dx} = \dfrac{d}{dx}(2x^3 - 5x) = 6x^2 - 5$

c. $\dfrac{dy}{dx} = \dfrac{dy}{du} \cdot \dfrac{du}{dx}$ This is the chain rule.
$= 6u^5(6x^2 - 5)$ Answers from parts **a** and **b**
$= 6(2x^3 - 5x)^5(6x^2 - 5)$ Write u in terms of x. ∎

EXAMPLE 2 If $y = 2u^5$ and $u = 2x^3 - 5x$, find $\dfrac{dy}{dx}$.

Solution When asking for the derivative of y with respect to x, it will be assumed that the answer should be given as a function of x, unless stated otherwise.

$\dfrac{dy}{dx} = \dfrac{dy}{du} \cdot \dfrac{du}{dx}$ This is the chain rule.
$= 10u^4(6x^2 - 5)$ Substitute $u = 2x^3 - 5x$ to write this as a
$= 10(2x^3 - 5x)^4(6x^2 - 5)$ function of x. ∎

More often than not, you will need to identify the function u, rather than being given the function u, as you were in Examples 1 and 2.

EXAMPLE 3 If $y = (4x + 1)^5$, find y'.

Solution Think: $u = 4x + 1$ so $y = \square^5$

$$y' = \frac{dy}{du} \cdot \frac{du}{dx} \qquad \text{Chain rule; remember } y' = \frac{dy}{dx}.$$

$$= 5(4x + 1)^4 \, (4) \qquad \text{Think: } y' = 5\,\square^4 \text{ times derivative of } \square.$$

$$= 20(4x + 1)^4$$

Because you will frequently use the chain rule, it is important that the process become natural and that you do not let the notation become too cumbersome. For this reason, the next example will show you in black what you would write down and in color what you should be thinking as you use the chain rule.

EXAMPLE 4 If $f(x) = (2x^2 + 3x - 5)^3$, find $f'(x)$.

Solution $f(x) = \overbrace{(2x^2 + 3x - 5)}^{u}{}^{\overbrace{3}^{n}}$ Think: $f(x) = \square^3$

$f'(x) = \overbrace{3}^{n} \overbrace{(2x^2 + 3x - 5)^2}^{u^{n-1}} \overbrace{(4x + 3)}^{u'} \qquad f'(x) = 3\,\square^2 \cdot \dfrac{d\,\square}{dx}$

$= 3(4x + 3)(2x^2 + 3x - 5)^2$

Differentiation of Composite Functions

The chain rule is actually a rule for differentiating composite functions. In particular, if $y = f(u)$ and $u = u(x)$, then y is the **composite function** $y = (f \circ u)(x) = f[u(x)]$ and the chain rule can be rewritten as follows.

Chain Rule for Composite Functions

> If u is differentiable at x and f is differentiable at $u(x)$, then the composite function $f \circ u$ is differentiable at x and
>
> $$(f \circ u)'(x) = f'(u)u'(x) \quad \text{or} \quad \frac{d}{dx}f[u(x)] = f'(u)\frac{du}{dx}$$

EXAMPLE 5 Differentiate $y = (3x^4 - 7x + 5)^3$.

Solution $y' = \boxed{3}(3x^4 - 7x + 5)^{\boxed{3-1}}(3x^4 - 7x + 5)'$ Think: $y = \square^3$ so $y' = 3\,\square^2 \cdot \dfrac{d\,\square}{dx}$

This step is usually done mentally. Note that we are thinking of $u(x) = 3x^4 - 7x + 5$. What you write down is:

$$y' = 3(3x^4 - 7x + 5)^2(12x^3 - 7)$$

You could, with a lot of work, have found the derivative in Example 5 without using the chain rule either by expanding the polynomial or by using the product rule. The answer would be the same, but would involve much more algebra. In order to compare these methods, however, consider the following problem with a simpler function.

EXAMPLE 6 Differentiate $y = (3x + 2)^2$.

a. By expansion
b. By the product rule
c. By the chain rule

Solution
a. $y = (3x + 2)^2 = 9x^2 + 12x + 4$
$y' = 18x + 12 = 6(3x + 2)$

b. $y = (3x + 2)^2 = (3x + 2)(3x + 2)$
$y' = (3x + 2)(3x + 2)' + (3x + 2)(3x + 2)'$
$= (3x + 2)(3) + (3x + 2)(3)$
$= 6(3x + 2)$

c. $y = (3x + 2)^2$
$y' = 2(3x + 2)(3) = 6(3x + 2)$ ∎

The chain rule allows us to find the derivative of functions that would otherwise be very difficult to handle. To efficiently use the chain rule, you must recognize when it is required and when it is not.

Chain rule not required	Chain rule required	Function of x	Function of u	Derivative
$y = 5x + 1$	$y = (5x + 1)^3$	$u = 5x + 1$	$y = u^3$	$y' = 3(5x + 1)^2(5)$
$y = 3x + 1$	$y = (3x + 1)^5$	$u = 3x + 1$	$y = u^5$	$y' = 5(3x + 1)^4(3)$
$y = 4x + 1$	$y = \sqrt{4x + 1}$	$u = 4x + 1$	$y = \sqrt{u}$	$y' = \frac{1}{2}(4x + 1)^{-1/2}(4)$
$y = 6x + 1$	$y = \dfrac{1}{6x + 1}$	$u = 6x + 1$	$y = \dfrac{1}{u}$	$y' = (-1)(6x + 1)^{-2}(6)$
$y = 2 - 7x$	$y = \sqrt[3]{(2 - 7x)^2}$	$u = 2 - 7x$	$y = \sqrt[3]{u^2}$	$y' = \dfrac{2}{3}(2 - 7x)^{-1/3}(-7)$
$y = \dfrac{3x + 1}{4x + 1}$	$y = \left(\dfrac{3x + 1}{4x + 1}\right)^2$	$u = \dfrac{3x + 1}{4x + 1}$	$y = u^2$	$y' = 2\left(\dfrac{3x + 1}{4x + 1}\right)\left[\dfrac{3x + 1}{4x + 1}\right]'$

$$= 2\left(\frac{3x + 1}{4x + 1}\right)\left[\frac{(4x + 1)(3) - (3x + 1)(4)}{(4x + 1)^2}\right]$$

$$= 2\left(\frac{3x + 1}{4x + 1}\right)\left[\frac{-1}{(4x + 1)^2}\right]$$

$$= \frac{-2(3x + 1)}{(4x + 1)^3}$$

Do you see the difference between the functions that do not require the chain rule and those that do? The chain rule is required when operating on a composite function.

The previous examples illustrate the most common type of composite function—namely, the one for which

$$y = [u(x)]^n$$

The rule for differentiating such functions is called the **generalized power rule**, and is a special case of the chain rule. In this course, this will be the most common application of the chain rule.

Generalized Power Rule

> Let u be some differentiable function of x and $y = [u(x)]^n$, where n is a rational number. Then
>
> $$y' = nu^{n-1}u' \qquad \text{or} \qquad \frac{dy}{dx} = n[u(x)]^{(n-1)}\frac{du}{dx}$$

EXAMPLE 7 Let $y = 15\sqrt{3x^2 + x}$. Find $\frac{dy}{dx}$ and simplify.

Solution
$$y' = 15\overbrace{(3x^2 + x)}^{u}{}^{\overbrace{1/2}^{n}}$$

$$= 15\overbrace{\left(\frac{1}{2}\right)}^{n}\overbrace{(3x^2 + x)^{-1/2}}^{u^{n-1}}\overbrace{(6x + 1)}^{u'}$$

$$= \frac{15(6x + 1)}{2\sqrt{3x^2 + x}}$$

You should notice that we simplified in Example 7. Sometimes it is as much of a problem to simplify as it is to use the chain rule and find the derivative. You might wish to review Appendix A.4, in particular simplifying by finding the common factor. The next two examples show that the chain rule can be used with the product and quotient rules as well as with the power rule.

EXAMPLE 8 Let $y = x^2(5x + 2)^3$. Find y' and simplify.

Solution This is a product, so begin with the product rule:

$$y' = x^2\overbrace{[(5x + 2)^3]'}^{\substack{\text{derivative} \\ \text{of second}}} + \overbrace{(5x + 2)^3}^{\substack{\text{second}}}\overbrace{[x^2]'}^{\substack{\text{derivative} \\ \text{of first}}}$$

Prime marks indicate that derivatives are yet to be found.

$$= x^2(3)(5x + 2)^2\underbrace{(5)}_{u'} + (5x + 2)^3(2x)$$

chain rule (generalized power rule)

$$= x(5x + 2)^2[x(3)(5) + 2(5x + 2)]$$ Remove common factors.

$$= x(5x + 2)^2(25x + 4)$$ Simplify expression in brackets.

2.7 THE CHAIN RULE

EXAMPLE 9 Let $y = \dfrac{(5x-3)^4}{-2x}$. Find $\dfrac{dy}{dx}$ and simplify.

Solution This is a quotient, so begin with the quotient rule:

$$\dfrac{dy}{dx} = \dfrac{-2x[(5x-3)^4]' - (5x-3)^4[-2x]'}{(-2x)^2} \quad \text{Prime marks indicate that derivatives are yet to be found.}$$

$$= \dfrac{-2x(4)(5x-3)^3(5) - (5x-3)^4(-2)}{(-2x)^2} \quad \text{Note the factor (5) from the generalized power rule.}$$

$$= \dfrac{-2(5x-3)^3[20x - (5x-3)]}{(-2x)^2} \quad \text{Common factor in numerator}$$

$$= \dfrac{-2(5x-3)^3(15x+3)}{4x^2} \quad \text{Simplify algebraically.}$$

$$= \dfrac{-(5x-3)^3 3(5x+1)}{2x^2}$$

$$= \dfrac{-3(5x-3)^3(5x+1)}{2x^2} \quad ■$$

EXAMPLE 10 Let $y = [(2x-1)(5x^3+2)]^9$. Find dy/dx.

Solution This is a power, so begin with the power rule:

$$y' = 9[(2x-1)(5x^3+2)]^8[(2x-1)(5x^3+2)]' \quad \text{Chain rule}$$
$$= 9[(2x-1)(5x^3+2)]^8[(2x-1)(15x^2) + (5x^3+2)(2)] \quad \text{Product rule}$$
$$= 9[(2x-1)(5x^3+2)]^8[30x^3 - 15x^2 + 10x^3 + 4] \quad \text{Simplify bracket.}$$
$$= 9[(2x-1)(5x^3+2)]^8[40x^3 - 15x^2 + 4]$$
$$= 9(40x^3 - 15x^2 + 4)[(2x-1)(5x^3+2)]^8 \quad ■$$

EXAMPLE 11 Find the standard form of the equation of the line tangent to the graph of f where $f(x) = \dfrac{24}{3x+2}$ at $x = 2$.

Solution We will use the point–slope form of the equation of a line—namely,

$$y - y_1 = m(x - x_1)$$

First, find (x_1, y_1), which is the point on the curve at $x = 2$:

$$f(2) = \dfrac{24}{3(2)+2} = 3; \quad \text{so the point is } (2, 3)$$

The slope of the tangent line at a point is the same as the value of the derivative at that point. Write $f(x) = 24(3x+2)^{-1}$.

$$f'(x) = -24(3x+2)^{-2}(3) \quad \text{Do not forget the factor 3 (because of the chain rule).}$$
$$= -72(3x+2)^{-2}$$

Thus,

$$f'(2) = -72[3(2) + 2]^{-2} = \frac{-72}{64} = \frac{-9}{8}$$

This value is the slope, m, in the equation:

$$y - 3 = \frac{-9}{8}(x - 2)$$
$$8y - 24 = -9x + 18$$
$$9x + 8y - 42 = 0$$

■

EXAMPLE 12 Find the point where the tangent line is horizontal for the function defined by $f(x) = \dfrac{x}{(3x + 2)^2}$.

Solution The tangent line will be horizontal when its slope is 0, and the slope of the tangent line is the same as the slope of the curve at a particular point.

$$f'(x) = \frac{(3x + 2)^2(1) - x(2)(3x + 2)(3)}{(3x + 2)^4}$$
$$= \frac{(3x + 2)[(3x + 2) - 6x]}{(3x + 2)^4}$$
$$= \frac{(3x + 2)(2 - 3x)}{(3x + 2)^4}$$
$$= \frac{2 - 3x}{(3x + 2)^3}$$

We need to find the value(s) of x for which

$$\frac{2 - 3x}{(3x + 2)^3} = 0$$
$$2 - 3x = 0 \qquad x \neq -\frac{2}{3}$$
$$x = \frac{2}{3}$$

We need to find the point on the curve for which $x = \frac{2}{3}$:

$$f\left(\frac{2}{3}\right) = \frac{\frac{2}{3}}{\left[3\left(\frac{2}{3}\right) + 2\right]^2}$$
$$= \frac{\frac{2}{3}}{16} = \frac{1}{24}$$

The point is $(\frac{2}{3}, \frac{1}{24})$.

■

2.7 Problem Set

Find the derivatives of the functions in Problems 1–30 and simplify.

1. $f(x) = (3x + 2)^3$
2. $f(x) = (6x + 5)^3$
3. $f(x) = (5x - 1)^4$
4. $f(x) = (4 - 2x)^5$
5. $y = (2x^2 + x)^3$
6. $y = (3x^2 - 2x)^3$
7. $y = (2x^2 - 3x + 2)^2$
8. $y = (3x^2 + 5x - 1)^2$
9. $g(x) = (x^3 + 5x)^4$
10. $g(x) = (2x^3 - 7x^2)^4$
11. $m(x) = (2x^2 - 5x)^{-2}$
12. $m(x) = (6x^2 - 3x)^{-3}$
13. $t(x) = (x^4 + 3x^3)^{-1}$
14. $t(x) = (5x^3 - 2x^2)^{-2}$
15. $y = 5(4x^3 + 3x^2)^3$
16. $y = 6(x^3 + 4x^2)^4$
17. $y = (x^2 - 3x)^{1/4}$
18. $y = (2x^2 + 5x)^{2/3}$
19. $y = \sqrt{x^2 + 16}$
20. $y = \sqrt{x^2 - 25}$
21. $y = 5\sqrt{x^3 + 8}$
22. $y = 9\sqrt{x^4 + 3x}$
23. $y = x\sqrt{3x + 1}$
24. $y = 2x\sqrt{5x - 3}$
25. $y = \sqrt[3]{2x + 5}$
26. $y = \sqrt[3]{7 - 5x}$
27. $y = \dfrac{1}{5x + 3}$
28. $y = \dfrac{1}{3x - 2}$
29. $y = \dfrac{1}{(x^2 + 3)^2}$
30. $y = \dfrac{1}{(x^3 - 2)^3}$

In Problems 31–36 find the standard form of the equation of the line tangent to the graph of f at the indicated value of x.

31. $f(x) = (3x^2 + 2)^2$ at $x = 1$
32. $f(x) = \dfrac{1}{2x - 3}$ at $x = 2$
33. $f(x) = \dfrac{8}{5 - 3x}$ at $x = -1$
34. $f(x) = x^2(2 - x)^3$ at $x = 1$
35. $f(x) = x\sqrt{x - 4}$ at $x = 5$
36. $f(x) = x\sqrt{x + 3}$ at $x = 1$

In Problems 37–42 find the value(s) of x for which the tangent line of the given function is horizontal.

37. $f(x) = x^2(x - 3)^2$
38. $f(x) = x^3(x - 2)^3$
39. $f(x) = \dfrac{x}{(2x + 3)^2}$
40. $f(x) = \dfrac{x - 1}{(x - 2)^2}$
41. $f(x) = \sqrt{x^2 - 5x + 20}$
42. $f(x) = \sqrt{x^2 + 2x + 3}$

In Problems 43–52
a. Find the "unsimplified" derivative.
b. Find the simplified version of the derivative.

43. $f(x) = (2x + 1)^2(3x + 2)^3$
44. $f(x) = (1 - 3x)^3(4 - x)^2$
45. $y = (5x + 1)^2(4x + 3)^{-1}$
46. $y = (9x + 2)^2(2x - 5)^{-1}$
47. $g(x) = \dfrac{(2x - 5)^2}{5x + 3}$
48. $g(x) = \dfrac{(7x + 11)^3}{x^2 + 3}$
49. $t(x) = \dfrac{(x + 5)^4}{(2x - 5)^2}$
50. $t(x) = \dfrac{(x^2 - 3x + 1)^2}{(5x + 2)^2}$
51. $f(x) = \dfrac{3x}{\sqrt{x^2 + 1}}$
52. $f(x) = \dfrac{-2x}{\sqrt{4 - x^2}}$

APPLICATIONS

53. A colony of bacteria that has been sprayed by a bacterial agent in a controlled situation can be measured in terms of N, the number of viable bacteria remaining after t minutes, according to the formula
$$N(t) = 10^9(50 - 2t)^3$$
What is the rate of decrease at the end of 10 minutes?

54. A psychologist finds that after t minutes ($0 \leq t \leq 10$) a person is able to score
$$s(t) = \dfrac{100t^2}{(2t + 10)^2}$$
on a performance test. Is the person's score increasing or decreasing at time $t = 10$ minutes?

55. If $10,000 is invested at an annual rate r compounded annually, the amount in the account at the end of 10 years is
$$A = 10,000(1 + r)^{10}$$
Find the rate of change of A with respect to r (i.e., find $\dfrac{dA}{dr}$).

56. If the interest in Problem 55 is compounded quarterly (instead of annually), the formula is
$$A = 10,000\left(1 + \dfrac{r}{4}\right)^{40}$$
Find $\dfrac{dA}{dr}$.

57. If you are to receive $10,000 in 10 years, the present value of this amount, if it is invested at an annual rate r, is given by the formula
$$P = \dfrac{10,000}{(1 + r)^{10}}$$
Find the rate of change of P with respect to r (i.e., find $\dfrac{dP}{dr}$).

58. If the amount in Problem 57 is compounded monthly instead of annually, the formula is

$$P = \frac{10{,}000}{(1 + \frac{r}{12})^{120}}$$

Find $\dfrac{dP}{dr}$.

59. The registrar of an eastern agricultural college estimates that the total student enrollment t years from now is given by the formula

$$N = 8{,}000 - \frac{4{,}000}{\sqrt{1 + .2t}} \quad (t \geq 0)$$

a. What is the enrollment now?
b. What is the enrollment in 10 years?
c. What is the rate of change in enrollment with respect to time?
d. How fast is the enrollment changing today?
e. How fast is the enrollment changing in 10 years?

60. The psychologist L. L. Thurstone studied learning by asking subjects to memorize a list of n words. One such formula he used to predict learning time, T, was

$$T = 4n\sqrt{n - 3}$$

a. Find the rate of change in time with respect to the length of the list n.
b. Find $\dfrac{dT}{dn}$ for $n = 20$ and interpret your result.

61. Prove the quotient rule.

*2.8
Review

The material of this chapter is reviewed in the following list of objectives. After each objective there are some practice questions. For a sample test select the first question of each set and check your answers. The second question for each objective has no answer given. If you are having trouble with a particular type of problem, look back at the indicated section in the text. When you are finished reviewing these objectives, a sample examination is given at the end of this section.

[2.1]
Objective 2.1: Given a function defined by a graph, find a limit as $x \to c$.

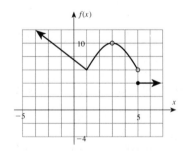

1. $\lim\limits_{x \to 3} f(x)$ **2.** $\lim\limits_{x \to 1} f(x)$
3. $\lim\limits_{x \to 5} f(x)$ **4.** $\lim\limits_{x \to \infty} f(x)$

Objective 2.2: Find limits by using a calculator and filling in values on a table.

5. $\lim\limits_{x \to 3} \dfrac{5x + 1}{3 - x}$

6. $\lim\limits_{x \to 0} \dfrac{5x + 1}{3 - x}$

7. $\lim\limits_{|x| \to \infty} \dfrac{5x + 1}{3 - x}$

8. $\lim\limits_{|x| \to \infty} \dfrac{145 - 2{,}000x + 15x^3}{2x^2 - 3x^3}$

Objective 2.3: Evaluate limits.

9. $\lim\limits_{x \to 1} \dfrac{x^2 + 5x - 6}{x - 1}$

10. $\lim\limits_{x \to 1} \dfrac{x^2 + 7x + 6}{x - 1}$

11. $\lim\limits_{x \to -1} \dfrac{x^2 + 7x + 6}{x - 1}$

12. $\lim\limits_{x \to \infty} \dfrac{3x^2 - 5x + 10}{8x^2 + 2x - 5}$

* Optional section.

[2.2]
Objective 2.4: *From a graph, find all suspicious points and tell which of them are points of discontinuity.*

13.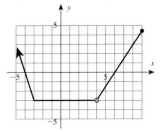

14.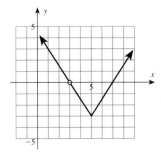

15.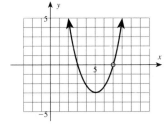

16.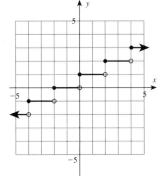

Objective 2.5: *Decide whether a given applied situation describes a continuous or a discontinuous function.*

17. The number of bacteria in a culture as a function of time
18. The distance that a skydiver falls as a function of the time since leaving the aircraft
19. The odometer reading on an automobile as a function of the distance traveled
20. Distance a skydiver has fallen as shown on frames of a motion picture film

Objective 2.6: *Given a function defined over a certain interval, determine whether the function is continuous at all points in the interval. Give the points of discontinuity.*

21. $f(x) = \dfrac{x^2 - 15x + 56}{x - 8}, \quad -5 \leq x \leq 5$

22. $f(x) = \dfrac{x^2 - 15x + 56}{x - 8}, \quad 0 \leq x \leq 10$

23. $f(x) = \begin{cases} \dfrac{x^2 - 15x + 56}{x - 8} & 0 \leq x \leq 10, x \neq 8 \\ 4 & x = 8 \end{cases}$

24. $f(x) = \begin{cases} \dfrac{x^2 - 15x + 56}{x - 8} & 0 \leq x \leq 10, x \neq 8 \\ 1 & x = 8 \end{cases}$

[2.3]
Objective 2.7: *Find the average rate of change for a given function on some interval.*

25. $y = 3x^2 + 4x$ for $x = 1$ to $x = 3$
26. $y = 3 + 2x - x^2$ for $x = -1$ to $x = 1$
27. $y = \sqrt{3x}$ for $x = 0$ to $x = 6$
28. $y = \dfrac{1}{x} - 5$ for $x = 1$ to $x = 4$

Objective 2.8: *Find the instantaneous rate of change for a given function at some point.*

29. $y = 3x^2 + 4x$ at $x = 1$
30. $y = 3 + 2x - x^2$ at $x = -1$
31. $y = \sqrt{3x}$ at $x = 3$
32. $y = \dfrac{1}{x} - 5$ at $x = 1$

[2.4]
Objective 2.9: *Know the definition of derivative and also use the definition to find a derivative.*

33. In your own words, state the definition of derivative.
34. $y = 3 - 8x^2$
35. $y = \dfrac{1}{x - 5}$
36. $y = \sqrt{x^2 + 1}$

[2.5]
Objective 2.10: *Use the power rule, constant rule, constant times a function rule, sum rule, and difference rule to find the derivative of a given function.*

37. $y = x^{14}$
38. $y = x^{-8}$
39. $y = x^{-7/9}$
40. $y = 2x$
41. $y = 150$
42. $y = -\dfrac{23}{25}$
43. $y = 2x^3 - 5x^2 + 12$
44. $f(x) = 45 - 13x^2 - 5x^3$

[2.6]
Objective 2.11: *Use the product and quotient rules to find derivatives.*

45. $y = (1 - 3x)(2 + 9x)$
46. $y = 5x^3(x^2 - 3x + 9)$
47. $y = \sqrt{x}(x - 1)^{-1}$
48. $y = 5x\sqrt{9 - x}$
49. $y = \dfrac{1}{(3x^2 + 1)}$
50. $y = \dfrac{x + 10}{x - 5}$
51. $y = \dfrac{\sqrt{x}}{5x - 3}$
52. $y = \dfrac{x}{\sqrt{x + 3}}$

[2.7]
Objective 2.12: *Find the derivative of a function using the generalized power rule.*

53. $f(x) = (5x + 9)^4$
54. $f(x) = (4 - 3x)^8$
55. $y = (5x^2 - 3x)^{1/5}$
56. $y = \dfrac{(3x + 7)^5}{x^2 - 5}$

SAMPLE TEST
The following sample test (45 minutes) is intended to review the main ideas of this chapter.

Find the limits in Problems 1–4.

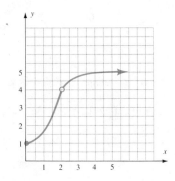

1. $\lim\limits_{x \to 2} f(x)$
2. $\lim\limits_{x \to 2} \dfrac{2 - 5x + 2x^2}{x - 2}$
3. $\lim\limits_{x \to 2} \dfrac{6x + 1}{3 - x}$
4. $\lim\limits_{|x| \to \infty} \dfrac{5x^2 - 3x + 1}{8x^2 - 5x + 4}$
5. Find the point(s) of discontinuity, if any.

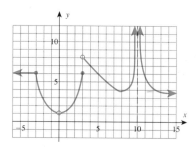

6. Find the point(s) of discontinuity, if any.
$$f(x) = \begin{cases} \dfrac{2x^2 - 7x - 15}{x - 5} & \text{if } -10 \leq x \leq 10,\ x \neq 5 \\ 13 & \text{if } x = 5 \end{cases}$$

7. Find the average rate of change for the function $y = 5 + x - 2x^2$ for $x = 1$ to $x = 3$.
8. Find the instantaneous rate of change for the function in Problem 7 at $x = 1$.
9. State the definition of derivative.
10. Use the definition of derivative to find the derivative of $y = 1/x$.

Find dy/dx for the functions in Problems 11–18.

11. $y = \dfrac{29}{125}$
12. $y = 3x^2 - 5x + 3$
13. $y = 5x^2 - 2x^{-1}$
14. $y = 2(x - 5)(3x + 2)$
15. $y = \dfrac{15}{4x - 5}$
16. $y = (3x^2 - 5x)^{1/4}$

17. $y = \dfrac{(x-5)^3}{1-2x}$

18. $y = \dfrac{5x^3}{\sqrt{x-5}}$

19. Suppose that the profit, P, for a manufacturer is a function of the number of units produced and behaves according to the model

$$P = \dfrac{1{,}000 - x^2}{100 - x}$$

If the current level of production is 20 units, what is the per unit change in profit if production is increased from 20 to 25 units?

20. Find the marginal profit for the function given in Problem 19.

3
Additional Derivative Topics

CHAPTER OVERVIEW
This chapter enhances and amplifies the idea of the derivative. The discussion should strengthen your understanding of the concept of the derivative by emphasizing that it is a rate of change.

PREVIEW
First, we are introduced to finding a derivative without actually solving for y; this is called *implicit differentiation*. Then meaning is given to dy and dx in the concept called a *differential*. Differentials, derivatives, and implicit differentiation are used in building some business models. The chapter closes by using derivatives to relate rates.

PERSPECTIVE
The concept of a derivative is such an important idea in mathematics that we need time to develop an appreciation of some of the many different places it can be used. This chapter forms a bridge between the definition and manipulation involved in finding a derivative (developed in the last chapter) and the applications of derivatives (introduced in the next chapter).

3.1 Implicit Differentiation
3.2 Differentials
3.3 Business Models Using Differentiation
3.4 Related Rates
3.5 Chapter 3 Review
    Chapter Objectives
    Sample Test

MODELING APPLICATION 3

Publishing: An Economic Model

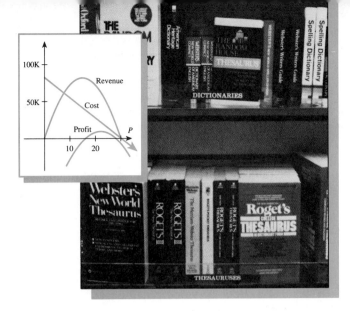

Karlin Press sells its *World Dictionary* at a list price of $20 and presently sells 5,000 copies per year. Suppose you are being considered for an editorial position and are asked to do an analysis of the price and sales of this book in order to assess your competency for the position.

After you have finished this chapter, write a paper based on this modeling application. This Modeling Application is continued on page 147.

APPLICATIONS

Management (*Business, Economics, Finance, and Investments*)
Rate of change of price with respect to number of items (3.1, Problems 38–39)
Relationship between sales and advertising cost (3.2, Problem 31)
Average cost (3.2, Problem 32)
Changes in revenue and profit (3.2, Problems 37–38)
Demand equation (3.3, Problem 22)
Revenue, cost, or profit functions (3.3, Problems 23–24, 30, 37–40; 3.5, Problem 25; 3.5, Test Problem 10)
Marginal revenue, cost, or profit (3.3, Problems 25–29, 31, 33–35; 3.5, Problems 25–27)
Average marginal revenue, cost, or profit (3.3, Problems 16–18, 32, 36)
Profit and loss (3.3, Problem 37)
Price elasticity of demand (3.3, Problems 41–44)
Rate of change of profit (3.4, Problems 10–11)
Positioning of a robot arm in an assembly-line conveyor belt (3.4, Problem 13)

Management (*continued*)
Batching process (3.4, Problems 14–15)
Rate of change of wholesale price of apples (3.4, Problem 25)
Maximize monthly revenue from an apartment complex (3.5, Test Problem 9)

Life Sciences (*Biology, Ecology, Health, and Medicine*)
Concentration of alcohol in the bloodstream (3.2, Problem 33)
Average adult pulse rate as a function of a person's height (3.2, Problem 34)
Area of a circular oil slick (3.2, Problem 35)
Area of a dilated pupil (3.2, Problem 36)
Blood velocity (Poiseuille's Law) (3.4, Problems 16–19)
Oil spill (3.4, Problem 22)
The rate of rabies as a function of the skunk population (3.4, Problem 26)
Treatment of a stomach disorder (3.4, Problems 27–28)
The level of carbon monoxide in a city (3.4, Problem 30)

Social Sciences (*Demography, Political science, Population, Psychology, Society, and Sociology*)
Effect of advertising on voting (3.2, Problem 39)
Effect on learning Spanish vocabulary by changing study time (3.2, Problem 40)
Urban sprawl (3.4, Problem 20)

General Interest
Seismological application with concentric circles (3.1, Problem 40)
Distance of one car from another (3.4, Problem 12)
Ripples in a pond (3.4, Problems 23–24)
Height of a weather balloon (3.4, Problem 29)

Modeling Application—
Publishing: An Economic Model

3.1
Implicit Differentiation

Sometimes we are given a function or an equation in terms of two or more variables, say, x and y. In order to use the derivative formulas developed in Chapter 2 it is necessary to **explicitly** solve for y. For example, consider the equation

$$3x^2 - 2x + 5y - 10 = 0$$

Find the slope of this curve at the point $(4, -6)$. To do this, we need to find the derivative. We solve for y:

$$5y = -3x^2 + 2x + 10$$

$$y = -\frac{3}{5}x^2 + \frac{2}{5}x + 2$$

Therefore

$$\frac{dy}{dx} = -\frac{3}{5}(2)x + \frac{2}{5} = -\frac{6}{5}x + \frac{2}{5}$$

The value at the point $(4, -6)$ is found by substitution of the x value ($x = 4$):

$$y' = -\frac{6}{5}(4) + \frac{2}{5} = -\frac{22}{5}$$

Another way to find the derivative is to use the chain rule to find it **implicitly**—that is, without first solving for y. For this example,

$$3x^2 - 2x + 5y - 10 = 0$$

Think of y as a function of x so that

$$D_x x = 1 \quad \text{and} \quad D_x y = \frac{dy}{dx}$$

Now differentiate both sides (with respect to x) to obtain

$$D_x(3x^2 - 2x + 5y - 10) = D_x(0)$$

$$\underline{D_x(3x^2)} + \underline{D_x(-2x)} + \underline{D_x(5y)} + \underline{D_x(-10)} = \underline{D_x(0)}$$

$$6x \quad + \quad (-2) \quad + \quad 5\frac{dy}{dx} \quad + \quad 0 \quad = \quad 0$$

Chain rule since y
is a function of x

$$6x - 2 + 5\frac{dy}{dx} = 0$$

Solve for $\dfrac{dy}{dx}$:

$$5\frac{dy}{dx} = -6x + 2$$

$$\frac{dy}{dx} = -\frac{6}{5}x + \frac{2}{5}$$

We can now find the slope at $(4, -6)$ as before. You might ask, why would we want to find the derivative implicitly? Why not always solve for y and then find the derivative? Sometimes it is inconvenient, or even impossible to solve for y, but nevertheless we can find the derivative. The idea is to take the derivative of both sides of the equation, thus treating y as a function of x and using the chain rule. Remember, the derivative of x with respect to x is 1, but the derivative of y with respect to x is $\dfrac{dy}{dx}$.

EXAMPLE 1 Find $\dfrac{dy}{dx}$ for $x^2 + y^2 = 25$ both implicitly and explicitly and find the equation of the tangent line at $(3, 4)$.

Solution First, implicitly (without solving for y):

$D_x(x^2 + y^2) = D_x(25)$ Take the derivative of both sides with respect to x.

$D_x(x^2) + D_x(y^2) = D_x(25)$ Derivative of a sum

$2x + 2y\dfrac{dy}{dx} = 0$ Derivative of x^2 is $2x$.

Derivative of y^2 is $2y\dfrac{dy}{dx}$ by the chain rule (since we are assuming that y is some, yet unfound, function of x).

$2y\dfrac{dy}{dx} = -2x$ Subtract $2x$ from both sides.

$\dfrac{dy}{dx} = -\dfrac{x}{y}$ Divide both sides by $2y$, $y \neq 0$.

Next, find the derivative explicitly (that is, solve for y first):

$y^2 = 25 - x^2$ Subtract x^2 from both sides.

$y = \sqrt{25 - x^2}$ Square root property; select positive value for y since y is positive for the given point, $(3, 4)$.

$\dfrac{dy}{dx} = \dfrac{1}{2}(25 - x^2)^{-1/2}(-2x)$ Do not forget the chain rule when finding the derivative.

$= -\dfrac{x}{\sqrt{25 - x^2}}$ Notice this is the same as the answer from the first part since $y = \sqrt{25 - x^2}$.

Finally, the equation of the tangent line at $(3, 4)$ is found by finding the slope at $(3, 4)$ using the derivative. The graph is shown in the margin.

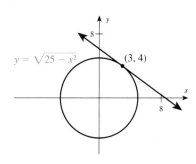

$\dfrac{dy}{dx} = -\dfrac{x}{y} = -\dfrac{3}{4}$ Evaluate at $(3, 4)$.

Use the point–slope form of the equation of a line, namely

$y - y_1 = m(x - x_1)$

or

$$y - 4 = -\frac{3}{4}(x - 3)$$

In standard form this is

$$3x + 4y - 25 = 0$$ ∎

Notice from Example 1 that *y is not necessarily a function of x*. When working with implicit differentiation it is often the case for a given value of x that there may be more than one value for y, which violates the definition of a function. We get around this difficulty by saying that if a segment (part) of a graph can be represented by a differentiable function in some vicinity of a point (x, y), then dy/dx will have meaning in that vicinity.*

We now summarize the procedure for implicit differentiation.

Implicit Differentiation

Given an equation involving x and y, where y is assumed to be a differentiable function of x *in some vicinity* of a point (x, y), we can find dy/dx as follows:

1. Take the derivative of both sides with respect to x.
2. Solve for dy/dx:
   a. Collect all terms involving dy/dx on the left side and all terms not involving dy/dx on the right side.
   b. Factor dy/dx from the left side.
   c. Solve for dy/dx by dividing both sides of the equation by the left-hand factor that does not contain dy/dx.

EXAMPLE 2 Find $\dfrac{dy}{dx}$ where $x^5 - y^5 = 211$.

Solution

$D_x(x^5 - y^5) = D_x(211)$ 1. Take the derivative of both sides with respect to x.

$5x^4 - 5y^4 \dfrac{dy}{dx} = 0$ Do not forget chain rule for $D_x(y^5)$.

$-5y^4 \dfrac{dy}{dx} = -5x^4$ 2. Solve for $\frac{dy}{dx}$.

$\dfrac{dy}{dx} = \dfrac{-5x^4}{-5y^4}$

$= \dfrac{x^4}{y^4}$ ∎

Note that $\dfrac{dy}{dx}$ is stated in terms of both x and y, so if you were asked for the slope of the curve at some point you would need to know both components of the point,

---

* This is, of course, not a definition. The idea of a *vicinity*, or as it is sometimes called, a *neighborhood*, of a point requires a rather precise definition, but the intuitive idea should be clear enough.

not just the x value as when we found $\dfrac{dy}{dx}$ explicitly. In Example 2 the slope of $x^5 - y^5 = 211$ at $(3, 2)$ is
$$\frac{dy}{dx} = \frac{x^4}{y^4} = \frac{81}{16}$$

WARNING Remember that all the usual derivative procedures apply when doing implicit differentiation, especially when using the product rule. Consider Example 3 carefully.

EXAMPLE 3 If $x^2 y^3 = 1$, find $\dfrac{dy}{dx}$.

Solution Use implicit differentiation; do not forget to use the product rule:
$$D_x(\underbrace{x^2 y^3}_{\text{Product rule}}) = D_x(1)$$

$$\overbrace{x^2 D_x(y^3) + y^3 D_x(x^2)}^{} = 0$$

$$x^2 \left(3y^2 \frac{dy}{dx}\right) + y^3(2x) = 0$$

$$3x^2 y^2 \frac{dy}{dx} = -2xy^3$$

$$\frac{dy}{dx} = -\frac{2xy^3}{3x^2 y^2} = -\frac{2y}{3x}$$

EXAMPLE 4 Find the line tangent to $x^2 + xy + y^2 - 7 = 0$ at $(1, 2)$.

Solution First find the slope of the curve at $(1, 2)$, which means find the derivative at that point. Carry out implicit differentiation:
$$2x + x\frac{dy}{dx} + y + 2y\frac{dy}{dx} + 0 = 0$$

$$x\frac{dy}{dx} + 2y\frac{dy}{dx} = -2x - y$$

$$\frac{dy}{dx}(x + 2y) = -2x - y$$

$$\frac{dy}{dx} = \frac{-2x - y}{x + 2y}$$

At $(1, 2)$:
$$\frac{dy}{dx} = \frac{-2(1) - 2}{1 + 2(2)} = \frac{-4}{5}$$

Finally, the equation of the tangent line is found by using the point–slope form:
$$y - y_1 = m(x - x_1)$$
$$y - 2 = \frac{-4}{5}(x - 1)$$
$$5y - 10 = -4x + 4$$
$$4x + 5y - 14 = 0$$

3.1
Problem Set

Suppose that x and y are related by the equations in Problems 1–10. Find dy/dx using both implicit and explicit differentiation.

1. $y + 5x^2 + 12 = 0$
2. $3x^3 - y + 15 = 0$
3. $xy = 5$
4. $5xy = x^2$
5. $x^2 - 3xy = 50$
6. $10x = xy$
7. $x^2 + y^2 = 4, y > 0$
8. $x^3 - y^3 = 9$
9. $2x^4 - 5y^3 = 7$
10. $x^5 y^3 = -1$

Suppose that x and y are related by the equations in Problems 11–20 and use implicit differentiation to find dy/dx.

11. $x^2 - xy + y^2 = 1$
12. $4x^2 - 3xy + 9y^2 = 4$
13. $(2x + 1)^2 + (3y - 5)^2 = 41$
14. $(5x - 2)^2 - (2y + 1)^2 = 30$
15. $\dfrac{x^2 - y^2}{2x + y^2} = 10$
16. $\dfrac{3x^2 + y^2}{x + y} = 100$
17. $3x^2 y^3 - 3xy^2 + 5xy = 2$
18. $5x^5 y^2 - 5xy^2 - 8x = 100$
19. $x^3 + 2x^2 + xy - 4 = 0$
20. $x^3 - 3x^2 + 2xy + 3 = 0$

Find the standard form of the equation of the line tangent to the curves in Problems 21–31 at the indicated point.

21. $x^2 + y^2 - 4 = 0$ at $(0, 2)$
22. $x^3 - y^3 - 9 = 0$ at $(2, -1)$
23. $2x^4 - 5y^3 - 7 = 0$ at $(-1, -1)$
24. $(2x + 1)^2 + (3y - 5)^2 = 41$ at $(2, 3)$
25. $x^2 + xy + y^2 - 7 = 0$ at $(1, -3)$
26. $x^2 + xy + y^2 - 12 = 0$ at $(2, -4)$
27. $(x - 2)^2 + (y - 1)^2 = 9$ at $x = 2$
28. $(x - 3)^2 + (y - 1)^2 = 16$ at $x = 3$
29. $x^2 - 2xy + y^2 = 0$ at $x = -1$
30. $\dfrac{(x-1)^2}{9} + \dfrac{(y+1)^2}{16} = 1$ at $x = 1$
31. $\dfrac{(x+1)^2}{25} + \dfrac{(y-1)^2}{4} = 1$ at $x = -1$

APPLICATIONS
Many graphs are famous enough to have been named for a variety of different applications. Problems 32–37 are examples of such curves. Find the slope of the tangent line at the indicated point.

32. Circle: $x^2 + y^2 = 5x + 4y$

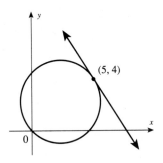

33. Semicubical parabola: $y^2 = 4x^3$

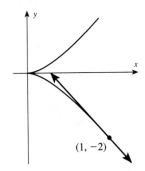

34. Bifolium: $(x^2 + y^2)^2 = 4x^2 y$

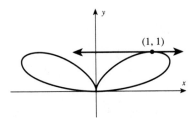

35. Folium of Descartes: $2(x^3 + y^3) = 9xy$

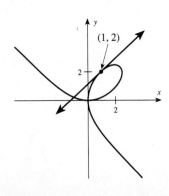

36. Two-leaved rose: $8(x^2 + y^2)^2 = 100(x^2 - y^2)$

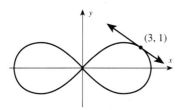

37. Lemniscate of Bernoulli: $12(x^2 + y^2)^2 = 625xy$
(Give slope to nearest tenth.)

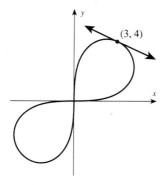

38. Find the rate of change of the price, p, in dollars, with respect to the number of items, x, if

$$x = \sqrt{5{,}000 - p^2}$$

39. Find the rate of change of the price, p, in dollars, with respect to the number of items, x, if

$$x = p^2 - 5p + 500$$

40. Seismologists sometimes use equations of circles around seismological stations. Show that the circles with the equations

$$x^2 + y^2 - 12x - 6y + 25 = 0 \quad \text{and}$$
$$x^2 + y^2 + 2x + y - 10 = 0$$

have tangent lines with the same slope at $x = 2$.

3.2
Differentials

Suppose we consider dx and dy from the derivative dy/dx as two separate quantities. In Chapter 1 we used the symbols Δx and Δy to mean the change in x and the change in y, as follows. Let (x_1, y_1) and (x_2, y_2) be any two points on some curve: then

$$\Delta x = x_2 - x_1 \quad \text{and} \quad \Delta y = y_2 - y_1$$

We *define* dx, called the **differential of x**, to be an independent variable equal to the change in x. That is, we define dx to be Δx. Then, if f is differentiable at x, we *define* dy, called the **differential of y**, according to the formula

$$dy = f'(x)\, dx$$

Thus, if $dx \neq 0$, then

$$\frac{dy}{dx} = f'(x)$$

There is a very clear geometrical representation of what we have done here, shown in Figure 3.1 on page 126. Note that $dx = \Delta x$ (the change in x), $\Delta y =$ the change in y that occurs for a change of Δx, and $dy =$ the rise of the tangent line relative to $\Delta x = dx$.

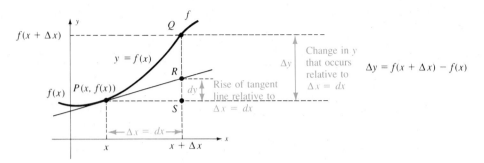

FIGURE 3.1
Geometrical definition of *dx* and *dy*

EXAMPLE 1   Use Figure 3.1 to describe Δx, Δy, dx, and dy.

Solution   Since $\Delta x = dx$, we see that these are the length of the segment PS; these quantities represent the change in x. The number Δy represents the change in y, which is the length of the segment QS. Finally, the distance dy is defined to be the rise of the tangent line at the point P, so dy is the length of the segment RS. ∎

EXAMPLE 2   Find Δy and dy for $f(x) = 6x - 2x^2$ when $x = 2$ and $\Delta x = .1$.

Solution   For this example, $y = f(x)$, so

$$x_1 = 2 \quad \text{and} \quad x_2 = x_1 + \Delta x = 2 + .1 = 2.1$$
$$y_1 = f(x_1) = f(2) = 6(2) - 2(2)^2 = 4$$
$$y_2 = f(x_2) = f(2.1) = 6(2.1) - 2(2.1)^2 = 3.78$$

Now

$$dy = f'(x)\,dx$$
$$= (6 - 4x)(.1) \quad \text{Since } f'(x) = 6 - 4x \text{ and } \Delta x = .1$$

When $x = 2$,

$$dy = [6 - 4(2)](.1) = -.2$$

To find Δy we know that

$$\Delta x = x_2 - x_1 \quad \text{and} \quad x_2 = x_1 + \Delta x$$

so that

$$\Delta y = f(x_2) - f(x_1)$$
$$= f(x_1 + \Delta x) - f(x_1) \quad \text{Substitute.}$$
$$= f(2 + .1) - f(2)$$
$$= f(2.1) - f(2)$$
$$= 3.78 - 4$$
$$= -.22$$

∎

Approximations Using Differentials*

Note in Example 2 that the numerical values for Δy and dy are almost the same. Since it is generally easier to calculate dy than Δy, it is often convenient to approximate a change in y by using dy. Example 3 illustrates this idea.

EXAMPLE 3 Suppose the demand for a certain product is a function of its price, x, and can be specified according to the formula

$$p(x) = -3x^3 - 2x^2 + 1{,}000$$

How would the demand change as the price changes from \$2 to \$2.10?

Solution The question asks for Δy when $x = 2$ and $\Delta x = .10$. Since

$$\Delta y = p(x + \Delta x) - p(x)$$

you want to find

$$\begin{aligned}
\Delta y &= p(2.10) - p(2) \\
&= [-3(2.10)^3 - 2(2.10)^2 + 1{,}000] - [-3(2)^3 - 2(2)^2 + 1{,}000] \\
&= 963.397 - 968 \quad \text{By calculator} \\
&= -4.603
\end{aligned}$$

This means that you would expect the demand to go down about 5 units. Instead of doing this calculation, however, you might approximate Δy by using dy:

$$\begin{aligned}
dy &= p'(x)\,dx \\
&= (-9x^2 - 4x)\,dx
\end{aligned}$$

Now substitute the values $x = 2$ and $\Delta x = .10$:

$$\begin{aligned}
dy &= [-9(2)^2 - 4(2)](.10) \\
&= -4.4
\end{aligned}$$

You would estimate that the demand would decrease by four units. ∎

In Example 3 the error introduced by using dy instead of Δy is one unit, but there is a lot less "work" involved. Closer examination shows that the error is only about .203. Example 4 shows how we can estimate the amount of error introduced in an approximation.

---

* Many books introduce approximations using differentials with problems such as approximating $\sqrt{26}$. To do this, let $f(x) = \sqrt{x}$ and let $x = 25$, $\Delta x = 1$ so

$$\begin{aligned}
f(x + \Delta x) &= f(x) + \Delta y & f(26) &= f(25 + 1) = f(25) + \Delta y \\
&\approx f(x) + dy & &\approx f(25) + dy \\
&= f(x) + f'(x)\,dx & &= f(25) + f'(25)\,dx
\end{aligned}$$

Since the functions $f(25)$ and $f'(25)$ are easy to evaluate, and since $dx = \Delta x = 1$, we see that differentials can be used to approximate $\sqrt{26}$. However, with the widespread use of calculators this does not seem like a good motivational problem for using differentials, so we will concentrate on other important applications of the differential.

EXAMPLE 4   What is the error introduced by using dy instead of Δy in Example 3 for an increase in price from 2 to $2 + h$ dollars?

Solution   For an increase from 2 to $2 + h$,

$$\Delta y = p(x + \Delta x) - p(x)$$
$$= [-3(2 + h)^3 - 2(2 + h)^2 + 1{,}000] - [-3(2)^3 - 2(2)^2 + 1{,}000]$$

This is a formidable calculation (the details are left for you). After simplification we find that

$$\Delta y = -44h - 20h^2 - 3h^3$$

However, if we approximate this by dy, we see that

$$dy = p'(x)\,dx$$
$$= (-9x^2 - 4x)\,dx$$

At $x = 2$ and $dx = h$,

$$dy = [-9(2)^2 - 4(2)]h = -44h$$

By comparing these two calculations we see that the error introduced by using dy rather than Δy is

$$-20h^2 - 3h^3$$

If h is small (such as .1), we see that the error will be small ($-.203$ in this case). ■

These examples lead us to the following summary:

Differential Approximation

If $f'(x)$ exists, then, for small Δx,

$$\Delta y \approx dy$$

and

$$f(x + \Delta x) = f(x) + \Delta y$$
$$\approx f(x) + dy$$
$$= f(x) + f'(x)\,dx$$

EXAMPLE 5   Colortex manufactures pigments for tinting various paints. Its profit function is given by

$$P(x) = 10x - \frac{x^2}{1{,}000} - 3{,}000$$

where x is the number of tubes of pigment produced. What is the expected change in profit if production is changed from 2,000 to 2,010 tubes of pigment?

Solution   The change in profit is ΔP and the change in production is Δx. The change in production, $\Delta x = 10$, is given, and since it is relatively small we will approximate

ΔP by dP:

$$dP = P'(x)\,dx$$
$$= \left(10 - \frac{2x}{1{,}000}\right)dx$$
$$= (10 - .002x)\,dx$$

Since $\Delta x = dx = 10$ at a production level of 2,000, we have

$$dP = [10 - .002(2{,}000)](10) = 60$$

This means that we would expect the profit to increase by \$60 if production were increased by 10 tubes of pigment. ∎

Differential Formulas

Each of the derivative rules derived in the last chapter has a corresponding differential form. For example, the generalized power rule states that

$$\text{If} \quad y = u^n \quad \text{then} \quad \frac{dy}{dx} = nu^{n-1} \cdot \frac{du}{dx}$$

By multiplying both sides by dx we obtain a differential form of this same derivative formula:

$$\text{If} \quad y = u^n \quad \text{then} \quad dy = nu^{n-1} \cdot du$$

All of the derivative rules previously discussed are now summarized in differential form:

Differential Rules

If u and v are differentiable functions and c is a constant, then

CONSTANT RULE:	$dc = 0$
POWER RULE:	$du^n = nu^{n-1} \cdot du$
SUM RULE:	$d(u + v) = du + dv$
DIFFERENCE RULE:	$d(u - v) = du - dv$
PRODUCT RULE:	$d(uv) = u\,dv + v\,du$
QUOTIENT RULE:	$d\left(\dfrac{u}{v}\right) = \dfrac{v\,du - u\,dv}{v^2}$

EXAMPLE 6 **a.** If $y = \sqrt{x}$, then $\dfrac{dy}{dx} = \tfrac{1}{2}x^{-1/2}$, so, in differential form,

$$dy = \frac{1}{2\sqrt{x}}\,dx$$

b. If $u = x^2 + 1$, then $\dfrac{du}{dx} = 2x$, so, in differential form,

$$du = 2x\,dx$$

∎

3.2
Problem Set

Find dy for the functions in Problems 1–20.
1. $y = 5x^3$
2. $y = 35x^4$
3. $y = 10x^{-1}$
4. $y = 100x^{-2}$
5. $y = 100x^3 - 50x + 10$
6. $y = 500x^5 - 40x^2 + 5,000$
7. $y = 3\sqrt{x-1}$
8. $y = 5\sqrt{x^2 + 4}$
9. $y = 5x$
10. $y = 6x$
11. $y = (5x - 3)^2(x^2 - 3)$
12. $y = (2x^2 + 1)(3x - 1)^3$
13. $y = \dfrac{3x + 1}{x - 2}$
14. $y = \dfrac{5x + 2}{4x + 3}$
15. $y = x^2\left(1 - \dfrac{1}{x} + \dfrac{3}{x^2}\right)$
16. $y = x^3\left(1 - \dfrac{5}{x} + \dfrac{10}{x^2}\right)$
17. $y = \left(5 - \dfrac{1}{x^2}\right)\left(1 + \dfrac{1}{x}\right)$
18. $y = \left(6 - \dfrac{10}{x^3}\right)\left(5x - \dfrac{1}{x}\right)$
19. $y = \dfrac{x^2 - 5x + 1}{x^2 + 3x - 2}$
20. $y = \dfrac{2x^2 + 3x - 5}{3x^2 - 2x + 1}$

Evaluate dy and Δy for the values in Problems 21–24.
21. $y = f(x) = x^2 - 2x + 5$, $x = 10$, $\Delta x = .1$
22. $y = f(x) = 2x^2 - 3x$, $x = 5$, $\Delta x = .2$
23. $y = f(x) = \sqrt{5x}$, $x = 5$, $\Delta x = .15$
24. $y = f(x) = \sqrt{3x - 2}$, $x = 100$, $\Delta x = 2$

Estimate Δy by using dy in Problems 25–30.
25. $y = \dfrac{2x - 3}{5x + 2}$, $x = 100$, $\Delta x = 3$
26. $y = \dfrac{x^2 + 1}{x - 3}$, $x = 30$, $\Delta x = 1$
27. $y = 20\left(3 - \dfrac{1}{x^2}\right)$, $x = 10$, $\Delta x = .02$
28. $y = \dfrac{1 + 3x}{\sqrt{x}}$, $x = 20$, $\Delta x = .1$
29. $y = \dfrac{450(x + 300)}{\sqrt{x - 50}}$, $x = 1,000$, $\Delta x = 10$
30. $y = \dfrac{1,000(20 - 30x)}{\sqrt{x^2 + 20}}$, $x = 50$, $\Delta x = 5$

Use differential approximations in Problems 31–40.

APPLICATIONS
31. Suppose that sales, S, can be expressed as a function of advertising cost (c, in thousands of dollars) according to the formula
$$S(c) = 500c - c^2$$
Estimate the increase in sales that will result by increasing the advertising budget from \$100,000 to \$110,000.

32. The average cost (in dollars) to manufacture x items is
$$A(x) = .05x^3 + .1x^2 + .5x + 10$$
Approximate the change in the average cost as x changes from 10 to 11.

33. The concentration of alcohol in the bloodstream x hours after drinking 1 ounce of alcohol is approximately
$$A(x) = \dfrac{3x}{100 + x^2}$$
Estimate the change in concentration as x changes from 3 to 3.5.

34. The average adult pulse rate, b, in beats per minute can be expressed as a function of the person's height, h (in inches), according to the formula
$$b = \dfrac{600}{\sqrt{h}}$$
Approximate the change in pulse rate for a change in height from 72 to 73.5 inches.

35. Find the approximate increase in the area of a circular oil slick as its radius increases from 2 to 2.1 miles.

36. The pupil of a patient's eye is nearly circular and will dilate when the patient is given a certain drug. Estimate the increase in the area of a patient's pupil if the radius increases from 4 to 4.1 millimeters.

37. A company manufactures and sells x items per day. If the cost and revenue equations (in dollars) are
$$C(x) = 400 + 30x \quad \text{and} \quad R(x) = 60x$$
find the approximate changes in revenue and profit if production is increased from 100 to 110 items.

38. A company produces and sells x items per day (in hundreds). The cost and revenue equations (in thousands of dollars) are given on the domain [0, 35]:
$$C(x) = 3x^2 - 40x + 200 \quad \text{and} \quad R(x) = 500x - \dfrac{x^2}{10}$$
Find the approximate changes in revenue and profit if production is increased from 10 to 11 items.

39. The number of people, N, expected to vote in the next election is a function of number of hours, x, of television

advertising according to the formula

$$N(x) = 25,000 + \frac{x^2}{10} - \frac{x^3}{50,000}$$

Find the approximate change in N as x changes from 1,000 to 1,100 hours.

40. A student learns y Spanish vocabulary words in x hours according to the formula

$$y = 50\sqrt{x}$$

What is the approximate increase in the number of words learned as x changes from 3 to 3.5 hours?

3.3
Business Models Using Differentiation

Marginal Analysis

The application of derivatives, that is, rates of change, to cost, revenue, and profit is an important component in the decision-making process for business executives and managers. The word *marginal* is used in business and economics to mean rate of change. Some of these ideas have appeared in examples in previous parts of this book, but for completeness and easy reference they are summarized below:

Marginal Analysis

Suppose that x is the number of units sold in some time interval at a price of p dollars and that

$$C(x) = \text{TOTAL COST} \quad \text{and} \quad R(x) = \text{TOTAL REVENUE}$$
$$= (\text{price per item})(\text{number sold})$$
$$= px$$

then

$$P(x) = R(x) - C(x) = \text{TOTAL PROFIT}$$
$$p(x) = \text{PRICE (or DEMAND) function}$$

Note that capital P and lowercase p denote different functions.

$$C'(x) = \text{MARGINAL COST}$$
$$R'(x) = \text{MARGINAL REVENUE}$$
$$P'(x) = \text{MARGINAL PROFIT}$$

Marginal can be related to the definition of *derivative* as follows:

Marginal cost, also called the *marginal propensity*, is the rate of change in cost per unit change in production at a given output level. This derivative approximates the *extra cost* of producing one additional unit.

Marginal revenue is the rate of change in revenue per unit change in production at a given output level. This derivative approximates the *extra* revenue for selling one additional unit.

Marginal profit is the rate of change of profit per unit change in production at a given output level. This derivative approximates the *extra profit* for selling one additional unit.

In practice, the smallest level of change in production is one unit. This was denoted in the last section by $\Delta x = dx = 1$. The actual changes in cost, revenue, and profit are denoted by ΔC, and ΔR, and ΔP, respectively, and are approximated by dC, dR, and dP. Recall from the last section that

$$dC = C'(x)\,dx$$
$$dR = R'(x)\,dx$$
$$dP = P'(x)\,dx$$

EXAMPLE 1 Suppose it is known that the total cost (in dollars) for a product is given by the equation

$$C(x) = 18{,}500 + 8.45x$$

where x is the number of items sold. Find the marginal cost.

Solution The marginal cost is $C'(x)$:

$$C'(x) = 8.45$$

This means that it costs an additional \$8.45 to produce one more item at all production levels. ∎

The marginal cost is not always a constant. If

$$C(x) = 500 + .045x^2$$

then $C'(x) = .09x$, which is a function of x. This means that the marginal cost depends on the production level of x units. For example, if present production is 1,000 items and if we want to know the cost of producing one additional item, then $x = 1{,}000$, $\Delta x = 1$, and

$$\begin{aligned} dC &= C'(x)\,dx \\ &= .09(1{,}000)(1) \qquad \text{Since } C'(x) = .09x \\ &= 90 \end{aligned}$$

This means that to increase production by one unit would lead to an approximate increase in total cost of \$90.

EXAMPLE 2 Suppose that the market demand for the product in Example 1 is linear and that it has been found that at a selling price of \$15 a company will sell 7,500 items per month, but at a \$25 selling price, sales would drop to 7,000 items per month. Graph the demand function. Also, find the marginal profit.

Solution The first step is to decide on the variables. Apparently, the variables are the price, p, and the number of items per month, x. In mathematics it is customary to put the independent variable (which in this problem is p) on the horizontal axis, and the dependent variable (in this problem it is x) on the vertical axis. This is exactly the opposite of the way that economists do it. In 1890, a book by Alfred Marshall, *Principles of Economics*, the classic that is one of the foundation stones of modern price theory, made the break with mathematics and put the dependent variable, x, on the horizontal axis.* Also note that an economist would have called

---

* It is unfortunate that mathematicians and economists disagree about how to draw the graphs shown in Figure 3.2. Many mathematics professors will want to put p on the horizontal axis, but the simple fact is that Marshall's scheme is now used by everybody, although mathematicians must wonder about the odd ways of economists.

3.3 BUSINESS MODELS USING DIFFERENTIATION

the number of items n instead of x, but by calling the number of items x instead of n, we are set to draw these graphs the same way that they are drawn in economics and at the same time make them more agreeable to mathematicians by using x for the horizontal axis.

DEMAND: It is linear and passes through the points (7500, 15) and (7000, 25). Use the slope–intercept form: $y - y_1 = m(x - x_1)$, or in terms of the variables of this problem:

$$p - p_1 = m(x - x_1) \quad \text{for ordered pairs } (x, p)$$

First, find m: $\quad m = \dfrac{y_2 - y_1}{x_2 - x_1} \quad$ Slope formula

$\qquad\qquad\quad = \dfrac{p_2 - p_1}{x_2 - x_1} \quad$ Slope formula for the variables in this problem

$\qquad\qquad\quad = \dfrac{25 - 15}{7{,}000 - 7{,}500} = \dfrac{10}{-500} = -.02$

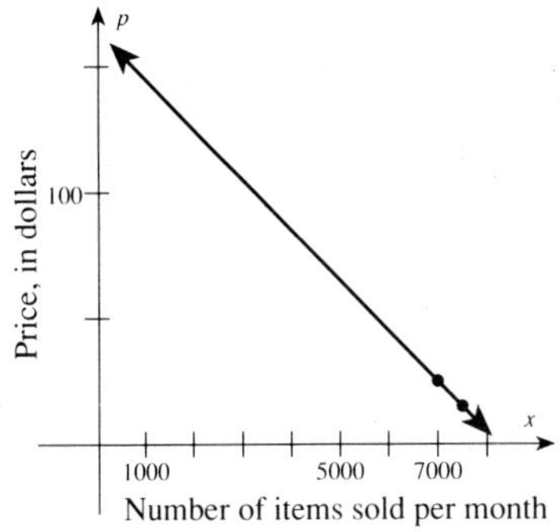

Thus, the equation is

$$p - 15 = -.02(x - 7{,}500) \quad \text{Substitute (7500, 15) for } (x_1, p_1).$$
$$p - 15 = -.02x + 150$$
$$p = -.02x + 165$$

We can draw the demand function as shown in Figure 3.2.

$$\text{TOTAL REVENUE} = xp = x(-.02x + 165) = -.02x^2 + 165x$$
$$\text{TOTAL COST} = 18{,}500 + 8.45x \quad \text{From Example 1}$$

so

$$\text{TOTAL PROFIT} = \text{TOTAL REVENUE} - \text{TOTAL COST}$$
$$P(x) = (-.02x^2 + 165x) - (18{,}500 + 8.45x)$$
$$= -.02x^2 + 156.55x - 18{,}500$$

The marginal profit is

$$P'(x) = -.04x + 156.55$$

The marginal profit is the rate of change of profit with respect to changes in the number of items. But what does this really mean, and how could it be used? Consider Table 3.1 (page 134) which gives some values for prices from $100 to $80.

Notice that as price decreases, the production increases. This makes sense because the higher the price, the less the demand. This production number is determined from the price according to the formula $x = -50p + 8{,}250$.* The profit is calculated from the profit function $P(x) = -.02x^2 + 156.55x - 18{,}500$. Finally take a close look at the profit and marginal profit columns. The maximum profit occurs when the product is priced at $87. Also notice that as long as the marginal profit was positive the profit was increasing, and when the marginal profit became

FIGURE 3.2

---

* Solve $p = -.02x + 165$ for x.

TABLE 3.1
Comparison of Prices, Production Levels, Profit, and Marginal Profit

p Price	x Production	P(x) Profit	P'(x) Marginal profit
100	3,250	$279,037.50	26.55
99	3,300	$280,315.00	24.55
98	3,350	$281,492.50	22.55
97	3,400	$282,570.00	20.55
96	3,450	$283,547.50	18.55
95	3,500	$284,425.00	16.55
94	3,550	$285,202.50	14.55
93	3,600	$285,880.00	12.55
92	3,650	$286,457.50	10.55
91	3,700	$286,935.00	8.55
90	3,750	$287,312.50	6.55
89	3,800	$287,590.00	4.55
88	3,850	$287,767.50	2.55
87	3,900	$287,845.00	.55
86	3,950	$287,822.50	−1.45
85	4,000	$287,700.00	−3.45
84	4,050	$287,477.50	−5.45
83	4,100	$287,155.00	−7.45
82	4,150	$286,732.50	−9.45
81	4,200	$286,210.00	−11.45
80	4,250	$285,587.50	−13.45

negative, the profit started to decline. It looks like the "turning point" is when the marginal profit is 0. We will show in the next chapter that the maximum value for profit occurs when the marginal profit is zero. ∎

EXAMPLE 3 Find the break-even points for the model described in Examples 1–2. Draw graphs for the revenue and the cost. Label the portions of the graph showing both profit and loss, and interpret Table 3.1 in terms of the graph.

Solution The break-even point (from Chapter 1) is when the revenue and cost are equal. Thus

$$-.02x^2 + 165x = 18{,}500 + 8.45x$$
$$.02x^2 - 156.55x + 18{,}500 = 0$$

$$x = \frac{156.55 \pm \sqrt{(-156.55)^2 - 4(.02)(18{,}500)}}{2(.02)}$$

$$\approx \frac{156.55 \pm 151.75}{.04}$$

$$\approx 120 \text{ and } 7{,}707$$

Substitute $x = 120$ into either the revenue or cost equation:

$$18{,}500 + 8.45(120) = 19{,}514; \text{ the point is } (120, 19514)$$
$$18{,}500 + 8.45(7{,}707) = 83{,}628; \text{ the point is } (7707, 83628)$$

The graphs are shown in Figure 3.3.

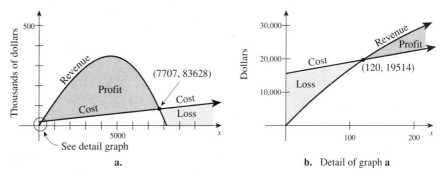

FIGURE 3.3
Cost, revenue, and profit functions

Notice from Figure 3.3 that the profit is positive for $120 < x < 7{,}707$. The maximum profit (from Table 3.1) occurs when $p = 87$ and the production is 3,900. This seems reasonable when we look at Figure 3.3a since the largest profit is the part with the biggest bulge. If fewer than 120 or more than 7,707 items are produced, then the profit is negative; this is called a *loss*.

Average Marginal Analysis

Sometimes marginal analysis is carried out relative to the average cost, average revenue, or average profit. *Average* here means *cost per unit, revenue per unit*, or *profit per unit*. The average of a number x is denoted by $\bar{x}$.

Average Marginal Analysis

If x is the number of units of a product produced in some time interval, then

$$\bar{C}(x) = \frac{C(x)}{x} = \text{cost per unit} = \text{AVERAGE TOTAL COST}$$

$\bar{C}'(x) = $ MARGINAL AVERAGE COST

$$\bar{R}(x) = \frac{R(x)}{x} = \text{revenue per unit} = \text{AVERAGE TOTAL REVENUE}$$

$\bar{R}'(x) = $ MARGINAL AVERAGE REVENUE

$$\bar{P}(x) = \frac{P(x)}{x} = \text{profit per unit} = \text{AVERAGE TOTAL PROFIT}$$

$\bar{P}'(x) = $ MARGINAL AVERAGE PROFIT

EXAMPLE 4 Find the marginal average cost, marginal average revenue, and marginal average profit for the functions

$$C(x) = 18{,}500 + 8.45x$$
$$R(x) = -.02x^2 + 165x$$
$$P(x) = -.02x^2 + 173.45x - 18{,}500$$

Solution $\bar{C}(x) = \dfrac{C(x)}{x} = \dfrac{18{,}500 + 8.45x}{x} = 18{,}500x^{-1} + 8.45$ This is the average cost or cost per unit.

Marginal average cost: $\bar{C}'(x) = -18{,}500x^{-2}$

$\bar{R}(x) = \dfrac{R(x)}{x} = \dfrac{-.02x^2 + 165x}{x} = -.02x + 165$ This is the average revenue, or revenue per unit.

Marginal average revenue: $\bar{R}'(x) = -.02$

$\bar{P}(x) = \dfrac{P(x)}{x} = \dfrac{-.02x^2 + 173.45x - 18{,}500}{x} = -.02x + 173.45 - 18{,}500x^{-1}$ This is the average profit, or profit per unit.

Marginal average profit: $\bar{P}'(x) = -.02 + 18{,}500\, x^{-2}$

Note that $\bar{P}'(x) = \bar{R}'(x) - \bar{C}'(x)$. ∎

3.3 Problem Set

The cost to produce x items is

$$C(x) = 200 + 6x - x^2 + x^3$$

Find the marginal cost for the given value of x in Problems 1–4.
1. $x = 1$
2. $x = 0$
3. $x = 4$
4. $x = 2$

The revenue function for a product is

$$R(x) = 9x - .001x^2$$

Find the marginal revenue for the given value of x in Problems 5–8.
5. $x = 0$
6. $x = 10$
7. $x = 50$
8. $x = 100$

The profit function for a product is

$$P(x) = x^3 - 8x^2 + 2x + 50$$

Find the marginal profit for the given value of x in Problems 9–12.
9. $x = 2$
10. $x = 1$
11. $x = 4$
12. $x = 6$

Find the marginal average cost for the functions given in Problems 13–15.
13. $C(x) = 200 + 6x - x^2 + x^3$
14. $C(x) = .25x^2 + 45x + 225$
15. $C(x) = 5{,}000 + .4x^2$

Find the marginal average revenue for the functions given in Problems 16–18.
16. $R(x) = 9x - .001x^2$
17. $R(x) = 50x - .5x^2$
18. $R(x) = 10x - .01x^2$

Find the marginal average profit for the functions given in Problems 19–21.
19. $P(x) = x^3 - 8x^2 + 2x + 50$
20. $P(x) = 5x - .05x^2$
21. $P(x) = x^3 - 50x^2 + 5x + 200$

APPLICATIONS

In Problems 22–27 suppose that the sales of a company are presently 10,000 items per year and that at a list price of $15 the company will sell 15,000 items but at a list price of $25 sales will drop to 5,000 items. Use (x, p) where x is the number of items and p is the price.

22. What is the demand equation (assuming it is linear)?
23. What is the revenue function?
24. What is the cost function in terms of price if

$$C(x) = 15{,}000 + 8.5x$$

where x is the total number of items sold?

25. Find and interpret the marginal revenue function.
26. Find and interpret the marginal cost function.
27. Find and interpret the marginal profit function.

Suppose the demand equation for a product is

$$p = 200 - .04x$$

Use (x, p) where x is the total number of items produced and p is the price. Also suppose the cost equation is

$$C(x) = 50{,}000 + 50x$$

Use this information in Problems 28–37.

28. What is the marginal cost in terms of x?
29. What is the marginal average cost in terms of x?
30. What is the revenue equation in terms of x?
31. What is the marginal revenue in terms of x?
32. What is the marginal average revenue in terms of x?
33. Find $R'(1{,}000)$ and $R'(2{,}500)$ and interpret.
34. What is the marginal profit in terms of x?
35. Find $P'(1{,}000)$ and $P'(2{,}500)$ and interpret.
36. What is the marginal average profit in terms of x?
37. Graph the cost and revenue functions on the same coordinate system and show the regions of profit and loss.
38. The total cost to produce x items is

    $$C(x) = 1{,}000 + 5x - x^2 + x^3$$

    Find the marginal cost.
39. The total revenue for x items is

    $$R(x) = 5x - .001x^2$$

    Find the marginal revenue.

40. The profit in dollars from the sale of x items is

    $$P(x) = x^3 - 3x^2 + 5x + 10$$

    Find the marginal profit.

The laws of supply and demand can be used to predict the direction of changes in price and quantity in response to various shifts in supply and demand. However, it is often not enough to know whether quantity rises or falls in response to a change in price; it is also important to know by how much. The relative responsiveness of consumers to a change in the price of an item is called the **price elasticity of demand**. If $p(x)$ is a differentiable demand function, then the price elasticity of demand is denoted by the Greek letter η (eta) and is defined by the following equation:

$$\eta = \frac{\text{Percentage change in quantity demanded}}{\text{Rate of change in price}}$$

$$= \frac{\dfrac{p}{x}}{\dfrac{dp}{dx}}$$

For a given price, if $|\eta| < 1$, the demand is **inelastic**, and if $|\eta| > 1$, the demand is **elastic**. Determine whether the demand functions in Problems 41–44 are elastic, inelastic, or neither at the indicated x-value.

41. $p(x) = 500 - 4x$; where $x = 50$
42. $p(x) = 10 - .005x$; where $x = 5{,}000$
43. $p(x) = 100 - .5x^2$; where $x = 20$
44. $p(x) = 500(x + 1)^{-1}$; where $x = 24$

*3.4
Related Rates

The concept of derivative has been defined and interpreted. We now use the idea that the derivative represents a rate of change of one variable with respect to another variable. That is, we focus on the fact that

$$\frac{dy}{dx} = \text{the rate of change of } y \text{ with respect to } x$$

$$\frac{dP}{dx} = \text{the rate of change of } P \text{ with respect to } x$$

$$\frac{dz}{dt} = \text{the rate of change of } z \text{ with respect to } t$$

and so on.

* Optional section.

If a formula is given as an equation, we know that we often solve that equation for an unknown value. However, sometimes the information we have at hand is the *rate* at which the variables are changing. We now explore how to use a given formula to find unknown quantities when we *know* the rates of change. For example, if the profit P is given by the equation

$$P = 75x - \frac{x^3}{50{,}000} - 100{,}000$$

for sales of a total of x items, and if you know the rate at which the profit is changing *with respect to time*, you can transform this profit formula into one involving rates by using implicit differentiation and taking the derivative of both sides with respect to time:

$$\frac{dP}{dt} = \frac{d}{dt}\left(75x - \frac{x^3}{50{,}000} - 100{,}000\right)$$

$$= 75\frac{dx}{dt} - \frac{3}{50{,}000}x^2\frac{dx}{dt}$$

Notice that this formula now relates

$$\frac{dP}{dt} = \text{the rate of change of } P \text{ with respect to time } t$$

$$\frac{dx}{dt} = \text{the rate of change of } x \text{ with respect to time } t$$

Since the formula relates rates, the problem is called a **related rate** problem. When working a related rate problem you must distinguish between the general situation and the specific situation. The **general situation** comprises properties that are true at *every* instant of time, while the **specific situation** refers to those properties that are true only at the *particular* instant of time that the problem investigates. When working related rate problems you should first formulate a model that incorporates all the facts that are true at every instant of time (the general situation). Then substitute into the equation from the general situation the facts that represent the particular instant under investigation (the specific situation). Finally, solve the equation from the specific situation for the desired unknown. This process is illustrated by the following examples.

EXAMPLE 1   The total profit, P, is related to the number of items produced, x, by the formula

$$P = 75x - \frac{x^3}{50{,}000} - 100{,}000$$

Suppose that production is increasing by 100 units per week. How fast is the profit increasing when production is at 1,000 items?

Solution   *The general situation:*   The formula for the general situation is given. We take the derivative of both sides of the given equation with respect to t in order to transform the equation into one on related rates:

$$\frac{dP}{dt} = 75\frac{dx}{dt} - \frac{3}{50,000}x^2\frac{dx}{dt}$$

This equation is true for every instant of time in this problem.

The specific situation: At the instant under consideration in this problem (namely, when production is at 1,000 items), we are given that production is increasing at 100 units per week. That is, we are given $dx/dt = 100$. Thus, when production is 1,000 items, we have

$$\frac{dP}{dt} = 75(100) - \frac{3}{50,000}(1,000^2)(100)$$
$$= 1,500$$

This means that the profit is increasing at \$1,500 per week when production is 1,000 items and when production is increasing at 100 items per week. ∎

EXAMPLE 2 Two conveyor belts are moving away from an assembly point at right angles to one another, as shown in Figure 3.4. If two items simultaneously leave the assembly table, how fast is the distance between them changing after 2 minutes?

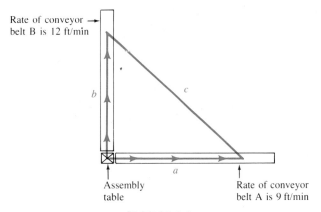

FIGURE 3.4
Assembly point configuration

Solution *The general situation:* Since a, b, and c in Figure 3.4 represent the sides of a right triangle, they are related by the formula

$$c^2 = a^2 + b^2$$

Convert this formula into one involving rates by taking the derivative of both sides with respect to time t:

$$2c\frac{dc}{dt} = 2a\frac{da}{dt} + 2b\frac{db}{dt}$$

The specific situation: At the instant under consideration (namely, at $t = 2$ minutes), you know that $da/dt = 9$ and $db/dt = 12$. Find dc/dt when $t = 2$. When $t = 2$, $a = 18$, $b = 24$, and

$$c^2 = 18^2 + 24^2$$
$$c = 30$$

Substitute these values into the formula relating the rates:

$$2(30)\frac{dc}{dt} = 2(18)(9) + 2(24)(12)$$

$$\frac{dc}{dt} = 15$$

The items are moving apart at 15 feet per minute. ∎

EXAMPLE 3 A cylindrical vat with dimensions given in Figure 3.5 is being filled at the rate of 1,000 cubic feet per minute. How fast is the surface rising when the depth is 12 feet?

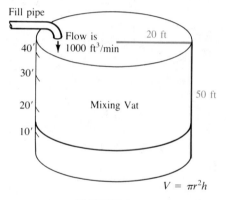

FIGURE 3.5
Mixing vat in an industrial process

Solution *The general situation:* The necessary relationship from Figure 3.5 requires a geometric formula from your previous courses. This is the formula for the volume of a cylinder: $V = \pi r^2 h$. In this problem, $r = 20$ feet (constant), so that the formula we will use is $V = 400\pi h$. Convert this to a formula involving rates:

$$\frac{dV}{dt} = \frac{d}{dt}(400\pi h)$$

$$= 400\pi \frac{dh}{dt}$$

The specific situation: The instant under investigation is when the depth is 12 feet. At this instant, $dV/dt = 1,000$ and you want to find dh/dt. Substitute these values into the formula from the general situation:

$$1{,}000 = 400\pi \frac{dh}{dt}$$

$$\frac{dh}{dt} = \frac{1{,}000}{400\pi} \approx .796$$

That is, the height is rising at .796 feet per minute (this is about $9\frac{1}{2}$ inches per minute). ∎

The procedure for solving related rate problems is summarized in the box.

Procedure for Solving Related Rate Problems

THE GENERAL SITUATION

1. Draw a figure if appropriate. Use letters to describe the data in the problem. Since rates involve variables and not constants, be careful not to label a quantity with a number unless it *never* changes in the problem.
2. Find a formula relating the variables. (The appropriate formula for most of the problems in this section is given.)
3. Differentiate the equation: usually implicitly and usually with respect to time.

THE SPECIFIC SITUATION

4. List the known quantities; list as unknown the quantity you want to find. If there are other variables or quantities in the problem use the given formula to eliminate those "extra" variables. Substitute in all the values in the formula. The only remaining variable should be the unknown. Solve for the unknown.

EXAMPLE 4 Illustrate steps 3 and 4 of the above procedure to find and interpret dy/dt where $x^3 + 5y^2 = 84$ and $dx/dt = 10$ where $x = 4$. Assume that both x and y are positive.

Solution *The general situation (step 3):* You are asked to find the rate at which y is changing with respect to time at *that particular instant* when $x = 4$ if you know that x is changing at a rate of 10 units per unit of time. This means that $dx/dt = 10$ is part of the general situation (it does not change throughout the problem). On the other hand, $x = 4$ and dy/dt are part of the specific situation (this is the instant with which we are concerned). Differentiate the formula implicitly:

$$3x^2 \frac{dx}{dt} + 10y \frac{dy}{dt} = 0$$

Known: $dx/dt = 10$ To find: dy/dt

The specific situation (step 4): The formula still involves both an x and a y, but we are interested in finding dy/dt at the instant when $x = 4$ (x is not a constant in this problem), so we use the *given* formula to find y *at that instant*:

$$x^3 + 5y^2 = 84$$
$$4^3 + 5y^2 = 84$$
$$5y^2 = 20$$
$$y^2 = 4$$
$$y = 2, -2 \quad \text{Reject } y = -2 \text{ since } y \text{ must be positive.}$$

Now substitute all values into the formula; the only unknown value should be dy/dt:

$$3x^2 \frac{dx}{dt} + 10y \frac{dy}{dt} = 0$$

$$3(4)^2(10) + 10(2) \frac{dy}{dt} = 0$$

$$\frac{dy}{dt} = -24$$

The result says that y is decreasing at the rate of 24 units per unit of time. ∎

EXAMPLE 5 The blood in a blood vessel flows faster toward the center of the blood vessel and slower toward the outside of the blood vessel. This flow is described by a formula called *Poiseuille's law*:

$$V = \frac{p}{4Lv}(R^2 - r^2)$$

where V is the velocity of the blood, R is the radius of the blood vessel, r is the distance of the blood from the center of the blood vessel, and p, L, and v are constants related to the blood pressure, length of the blood vessel, and the viscosity of the blood vessel. If a person goes from a warm house into a cold winter night the person's blood vessels will contract at a rate of

$$\frac{dR}{dt} = -.0025 \text{ mm/min}$$

at a place where $R = .01$. Also assume that $r = .005$, $p = 100$, $L = 1$ mm, and $v = .05$ are constants. Find the rate of change of velocity with respect to time at the location where $R = .01$.

Solution *The general situation:* The necessary relationship is given with Poiseuille's law, which is (with the necessary constants)

$$V = \frac{100}{4(1)(.05)}(R^2 - .005^2)$$

Convert this to a formula involving rates:

$$\frac{dV}{dt} = 500\left(2R \cdot \frac{dR}{dt}\right) = 1{,}000R \frac{dR}{dt}$$

The specific situation: The instant under investigation is when $R = .01$ and $\frac{dR}{dt} = -.0025$ is

$$\frac{dV}{dt} = 1{,}000(.01)(-.0025) = -.025$$

The velocity is decreasing at the rate of .025 mm/min. ■

3.4

Problem Set

In Problems 1–9 find the indicated rate, given the other information. Assume that both x and y are positive.

1. Find dy/dt where $x^2 + y^2 = 25$ and $dx/dt = 4$ when $x = 3$.
2. Find dx/dt where $x^2 + y^2 = 25$ and $dy/dt = 2$ when $x = 4$.
3. Find dy/dt where $5x^2 - y = 100$ and $dx/dt = 10$ when $x = 10$.
4. Find dx/dt where $4x^2 - y = 100$ and $dy/dt = -6$ when $x = 1$.
5. Find dx/dt where $y = 2\sqrt{x - 9}$ and $dy/dt = 5$ when $x = 9$.
6. Find dy/dt where $y = 5\sqrt{x + 9}$ and $dx/dt = 2$ when $x = 7$.
7. Find dy/dt where $xy = 10$ and $dx/dt = -2$ when $x = 5$.
8. Find dy/dt where $5xy^2 = 10$ and $dx/dt = -2$ when $x = 1$.
9. Find dx/dt where $x^2 + xy - y^2 = 20$ and $dy/dt = 5$ when $x = 4$ and $y > 0$.

APPLICATIONS

10. The profit, P, in dollars, is related to the number of items produced, x, by the formula

$$P = 125x - \frac{x^2}{200} - 500 \quad \text{where } 0 \le x \le 1{,}000$$

Suppose that production is increasing at five units per week. How fast is the profit changing when production is at 200 items per week?

11. Suppose that production in Problem 10 is decreasing by one unit per week. How fast is the profit changing when production is at 200 items?

12. Two cars start driving from the same point. One goes north at a rate of 30 mph and the other west at 40 mph. How fast is the distance between them changing after 5 hours?

13. A product is moving along a conveyor belt at the rate of 3 feet per second. A robot arm is suspended 8 feet above the belt. At what rate is the distance between the product and the robot arm changing when the product is 9 feet from the base of the arm?

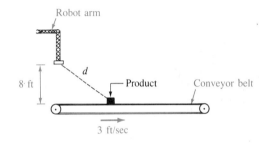

14. How fast is the surface of the vat described in Example 3 rising when the depth is 20 feet?

15. How fast is the surface of the vat described in Example 3 rising if it is being filled at a rate of 500 cubic feet per minute and the other details remain the same?

16. Find the rate of change of blood velocity with respect to time in Example 5 at the instant where $R = .02$.

17. Find the rate of change of blood velocity with respect to time in Example 5 if

$$\frac{dR}{dt} = -.0015 \text{ mm/min}$$

and all the other details of the example remain the same.

18. A person with heart problems needs to take a nitroglycerin tablet which dilates the blood vessels. Suppose the constants are the same as in Example 5 and the nitroglycerin dilates the blood vessel at the rate of

$$\frac{dR}{dt} = .003 \text{ mm/min}$$

What is the rate of change of blood velocity at the instant where $R = .01$?

19. A person with heart problems needs to take a nitroglycerin tablet which dilates the blood vessels. Suppose the constants are the same as in Example 5 and the nitroglycerin

dilates the blood vessel at the rate of

$$\frac{dR}{dt} = .003 \text{ mm/min}$$

What is the rate of change of blood velocity at the instant where $R = .02$?

20. Urban sprawl for a certain city is increasing in a circular manner in such a way that the radius r is increasing at the rate of 3 miles per year. At the moment where $r = 25$ miles, how fast is the area increasing?

21. A supplier services a circular area in such a way that its radius r is increasing at the rate of 1.5 miles per year. How fast is the area increasing at the instant when $r = 4$ miles?

22. Assume that oil is spilled from a ruptured tanker and forms a circular oil slick whose radius R is increasing at a constant rate of one-half a foot per minute ($dR/dt = .5$). How fast is the area of the spill increasing when the radius of the spill is 100 feet?

23. A pebble is thrown into a still pond and the result is a circular ripple. If the radius of this circle is expanding at 1 foot per second, how fast is the area changing when the radius is 10 feet?

24. How fast is the circumference of the circular ripple in Problem 23 changing?

25. The wholesale price of apples is d dollars per ton, and the daily supply x is related to the price by the formula

$$d = \frac{-5x}{x + 1{,}000} + 70$$

Suppose that there are 2,000 tons available today and that the supply is decreasing at 200 tons per day. At what rate is the price changing?

26. The number of cases of rabies, r, in an area is related to the number of skunks, s, according to the formula

$$r = .0005s^{2/3}$$

A program to eliminate the skunk population is instituted and it is estimated that 200 skunks per day are destroyed and that the present skunk population is 5,000. At what rate is the number of cases of rabies changing?

27. A certain medical procedure requires that a spherical balloon be inserted into the stomach and then inflated. If the radius of the balloon is increasing at the rate of .3 centimeter per minute, how fast is the volume changing when the radius is 4 centimeters? (Use $V = \frac{4}{3}\pi r^3$.)

28. How fast is the surface area of the sphere in Problem 27 increasing? (Use $S = 4\pi r^2$.)

29. A weather balloon is rising vertically at the rate of 10 feet per second. An observer is standing on the ground 300 feet from the point where the balloon was released. At what rate is the distance between the observer and the balloon changing when the balloon is 400 feet high?

30. The level of carbon monoxide (in parts per million, ppm) in a city can be predicted by considering it as a function of the number of registered automobiles in that city according to the formula

$$P = 1 + .2x + .001x^2$$

If the number of automobiles is increasing at 4,000 per year, how is the level of carbon monoxide changing when the city has exactly 25,000 cars?

*3.5
Review

The material of this chapter is reviewed in the following list of objectives. After each objective there are some practice questions. For a sample test select the first question of each set and check your answers. The second question for each objective has no answer given. If you are having trouble with a particular type of problem, look back at the indicated section in the text. When you are finished reviewing these objectives, a sample examination is given at the end of this section.

[3.1]
Objective 3.1: *Find the derivative of y with respect to x implicitly.*

1. $x^5 + 2y^2 + y + 10 = 0$
2. $3x^8y^4 = 1$
3. $(x - 3)^2 - (y + 1)^2 = 1$
4. $x^3 - 3y^2 + 40 = 0$

Objective 3.2: *Find the slope of a tangent line at a given point.*

5. A semicubical parabola
   $y^2 = 8x^3$ at (2, 8)

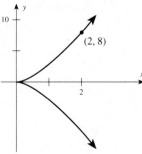

* Optional section.

6. A semicubical parabola

$$y = x^{2/3} \text{ at } (8, 4)$$

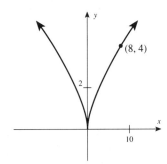

7. $y = 5x^2(6x - 5)^3$ at $x = 1$
8. $x^3 - 3y^2 + 40 = 0$ at $(2, 4)$

Objective 3.3: *Find the standard form equation of a line tangent to a given curve at a given point.*

9. A semicubical parabola

$$y^2 = 8x^3 \text{ at } (2, 8)$$

10. A semicubical parabola

$$y = x^{2/3} \text{ at } (8, 4)$$

11. $y = 5x^2(6x - 5)^3$ at $x = 1$
12. $x^3 - 3y^2 + 40 = 0$ at $(2, 4)$

[3.2]
Objective 3.4: *Find dy for the given functions.*

13. $y = 6\sqrt{3x^2 + 5}$
14. $y = \dfrac{2x + 1}{x - 3}$
15. $y = \dfrac{2x^2 + x - 1}{3x^2 + x - 4}$
16. $y = \left(1 - \dfrac{1}{x^2}\right)\left(2 - \dfrac{1}{x}\right)$

Objective 3.5: *Find dy and Δy for the indicated values.*

17. $y = f(x) = 5x^2$, $x = 10$, $\Delta x = .1$
18. $y = f(x) = 2x^2 + 5x - 3$, $x = 5$, $\Delta x = .1$
19. $y = f(x) = \sqrt{6x}$, $x = 6$, $\Delta x = .2$
20. $y = f(x) = \dfrac{x + 1}{x - 5}$, $x = 1$, $\Delta x = .01$

Objective 3.6: *Estimate Δy by using dy.*

21. $y = 5\left(10 - \dfrac{1}{x^3}\right)$, $x = 10$, $\Delta x = .03$
22. $y = \dfrac{1 - 2x}{\sqrt{x}}$, $x = 50$, $\Delta x = .1$
23. $y = \dfrac{x^2 - 1}{x + 2}$, $x = 20$, $\Delta x = 1$

24. $y = \dfrac{50(10 - 8x)}{\sqrt{x^2 + 10}}$, $x = 40$, $\Delta x = 2$

[3.3]
Objective 3.7: *Solve applied problems based on the preceding objectives. For specific examples of the types of applications look at the list of applications in this chapter on page 119.*

25. **a.** *Profit function.* Suppose that the profit, P, for a manufacturer is a function of the number of units produced and behaves according to the model

$$P(x) = \dfrac{200 - x^3}{10 - x^2}$$

If the current level of production is 10 units, what is the per unit increase in profit if production is increased from 10 to 15 units?

b. *Marginal profit.* Find the marginal profit for the function given in part **a**.

26. *Cost function.* Suppose that the cost, C, in dollars for producing x items is given by the formula

$$C(x) = 20x^3(5x - 100)^2$$

Find the average rate of change of cost as x increases from 50 to 75 units.

27. *Marginal cost.* Find the marginal cost for the function given in Problem 26.

28. *Rate of change.* An object moving in a straight line travels d centimeters in t minutes according to the formula $d(t) = .005t^3 + 20t$. What is the object's velocity?

[3.4]
Objective 3.8: *Find the indicated rate, given the other information. Assume that both x and y are positive.*

29. Find $\dfrac{dy}{dt}$ where $4x^2 + 9y^2 = 36$ and $\dfrac{dx}{dt} = 3$ when $x = 2$ and $y \geq 0$.

30. Find $\dfrac{dx}{dt}$ where $y = 4x^2$ and $\dfrac{dy}{dt} = -5$ when $x = 2$.

31. Find $\dfrac{dy}{dt}$ where $x^2 + 2xy + y^2 = 49$ and $\dfrac{dx}{dt} = 5$ when $x = 4$.

32. Find $\dfrac{dx}{dt}$ where $xy = 4$ and $\dfrac{dy}{dt} = -4$ when $x = 1$.

SAMPLE TEST

The following sample test (45 minutes) is intended to review the main ideas of this chapter.

Find dy/dx in Problems 1–3 by using implicit differentiation.
1. $y - 10x^2 + 6x - 15 = 0$
2. $6xy = 20x$

3. $x^4 + 5xy - 3x^2y + 9xy^2 - 155 = 0$
4. Find the standard form of the equation of the line tangent to the circle $(x - 2)^2 + y^2 = 25$ at the point $(5, 4)$.

Find the indicated rate, given the information in Problems 5–8.

5. Find dy/dt where $9x^2 + 16y^2 = 145$ and $dx/dt = 4$ when $x = 3$.
6. Find dy/dt where $xy = 9$ and $dx/dt = -8$ when $x = 1$.
7. Find dx/dt where $y = 25x^2$ and $dy/dt = -4$ when $x = 3$.
8. Find dx/dt where $x^2 + 2xy + y^2 = 64$ and $dy/dt = 3$ when $x = 4$.
9. A property management company manages 100 apartments renting for $500 with all the apartments rented. For each $50 per month increase in rent there will be 2 vacancies with no possibility of filling them. If x represents the number of $50 price increases, find the marginal revenue.
10. Suppose that for a company manufacturing lawn chairs, the cost and revenue functions are

$$C = 15{,}000 + 45x$$

$$R = 100x - \frac{x^2}{2{,}000}$$

where the production output is x chairs per week. If production is increasing at a rate of 50 chairs per week when production output is 2,000 chairs per week, find the rate of increase (decrease) in **a.** cost **b.** revenue **c.** profit.

MODELING APPLICATION 3

Publishing: An Economic Model

Karlin Press sells its *World Dictionary* at a list price of $20 and presently sells 5,000 copies per year. Suppose you are being considered for an editorial position and are asked to do an analysis of the price and sales of this book in order to assess your competency for the position. The annual costs associated with this book are summarized in the following table. The dictionary presently has a net cost of 80% of the list (selling) price.

Costs Associated with Publishing
World Dictionary

Cost	Amount, $
Advertising	1,750.00
Author's royalty	0.00
Binding	1.85 per book
Composition	3,600.00
Computer services	750.00
Investment return	3,000.00
Operating overhead	9,000.00*
Printing	4.90 (per book)
Set-up charges	400.00
Storage	.50 (per book)
Taxes	1.20 (per book)

*Salaries and offices are prorated for this title.

Analyze the price and sales of the dictionary. Your analysis should reach a conclusion about pricing the book for maximum revenue and/or maximum profit. Some consideration should also be given to market demand (which is assumed to be linear). Suppose you do some additional market research to find that at a list price of $15, the company will sell 7,500 copies, but at $25 sales would drop to 2,500. You might also want to give some thought to inventory and reprint schedule. To this end you find that the annual inventory (for maximum profit) should be 6,885, but 7,000 copies are allowed because they need some for office and sample copies. For general guidelines about writing this essay, see the commentary accompanying Modeling Application 1 on page 51.

4
Applications and Differentiation

CHAPTER OVERVIEW
The calculus of Chapters 2 and 3 is put to work in this chapter by discussing some important applications of the derivative. The applications demonstrate some of the power and versatility of differential calculus.

PREVIEW
Graphing, aided by the idea of a derivative, leads to curve sketching and the determination of relative maximums and minimums. We then turn to optimization (finding the absolute maximum or minimum), an extremely important concept in the business world.

PERSPECTIVE
We continue to explore the usefulness of calculus in a variety of different real-world settings. The only reason for introducing the derivative in this book is to apply it as a real-world model and to use it to derive additional mathematics to use in real-world models.

4.1 First Derivatives and Graphs
4.2 Second Derivatives and Graphs
4.3 Curve Sketching—Relative Maximums and Minimums
4.4 Absolute Maximum and Minimum
4.5 Chapter 4 Review
   Chapter Objectives
   Sample Test

MODELING
APPLICATION 4

Health Care Pricing

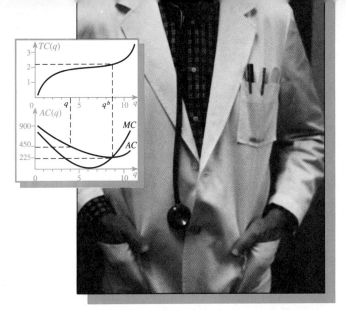

"Use of surgery and hospital costs vary greatly by region ... many doctors are baffled by this finding. And some fear that politicians and reimbursement officials will seize on data that nobody understands to rationalize budget cuts for Medicare and other health-insurance programs."—*The Wall Street Journal*, March 5, 1986, page 33

The figure in the margin shows a total cost function for New York hospitals, based on an article, "Financial Management and DRGS," by Steven Ullmann, *Health Services Research*, 1984. Set up another coordinate axis, and carefully construct the graphs of the average cost function and of the marginal cost function. As an example, show how to find the average cost and marginal cost for 175,000 patient-days per year.*

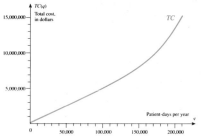

Write a paper that develops a mathematical model associated with the pricing of health care. For general guidelines about writing this essay, see the commentary for Modeling Application 1 on page 51.

*This modeling application is from Yves Nievergelt, "Graphic Differentiation Clarifies Health Care Pricing," *UMAP Journal*, Volume 9, No. 1. Copyright 1988 by COMAP, Inc.

APPLICATIONS

Management (*Business, Economics, Finance, and Investments*)
Sales as a function of advertising
  (4.1, Problem 43)
Point of diminishing returns
  (4.2, Problems 41–48)
Determining when sales are increasing
  (4.2, Problems 49–50)
Marginal revenue
  (4.2, Problem 51; 4.3, Problem 36; 4.5, Problem 29)
Cost–benefit model
  (4.3, Problem 35; 4.5, Problem 30)
Average cost (4.3, Problems 37–38)
Maximum profit
  (4.4, Problems 21–27, 34–37)
Minimal average cost
  (4.4, Problems 28, 32–33)

Management (*continued*)
Maximize volume of a box (4.4, Problem 31)
Minimize cost (4.4, Problem 39)
Worker efficiency (4.4, Problem 40)
Largest area in an enclosure
  (4.4, Problems 41–42)
Tour pricing for group fares
  (4.4, Problems 43–44)
Property management (4.5, Problem 31)

Life Sciences (*Biology, Ecology, Health, and Medicine*)
Concentration of a drug in the bloodstream
  (4.3, Problem 39)
Fish swimming to spawn (4.3, Problem 40)
Air pollution (4.4, Problems 29, 38)
Maximize yield per acre (4.4, Problem 45)

Social Sciences (*Demography, Political science, Population, Psychology, Society, and Sociology*)
Learning curve (4.1, Problem 44)
Timing of a political campaign
  (4.1, Problem 45)
Voting patterns (4.4, Problem 30)

General Interest
Maximum volume of a box (4.4, Problem 31)
Largest area in an enclosure
  (4.4, Problems 41–42)
Minimize cardboard for a poster
  (4.5, Problem 32)

Modeling Application—
Health Care Pricing

4.1
First Derivatives and Graphs

We continue our study of the derivatives and their applications in this chapter. Many of the phenomena we will study can be better understood by looking at a graph of the relationship between two variables, so our first application is to apply the idea of the derivative to the graph of a function.

Increasing and Decreasing Functions

If the graph of a function rises from left to right, the function is said to be *increasing*, and if it drops, it is said to be *decreasing*, as is shown in Figure

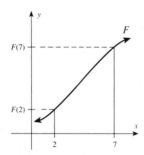
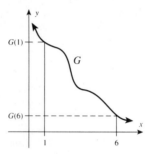
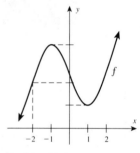

Function F is an increasing function on $(2, 7)$; $F'(x) > 0$ for all x in $(2, 7)$.

Function G is a decreasing function on $(1, 6)$; $G'(x) < 0$ for all x in $(1, 6)$.

Function f is increasing on $(-2, -1)$, decreasing on $(-1, 1)$, and increasing on $(1, \infty)$.

FIGURE 4.1

Notice that functions are classified as increasing on an open interval (a, b) if for every x_1 and x_2 on the interval such that $x_2 > x_1$, then $f(x_2) > f(x_1)$. If a function is increasing, then we say its graph is rising. Furthermore, if $f'(x) > 0$ for some x, then we say that f is increasing at the point x. Similarly, we say that f is decreasing on some interval I if for every x_1 and x_2 on the interval such that $x_2 > x_1$, then $f(x_2) < f(x_1)$, and is decreasing at a point x if $f'(x) < 0$. If a function is decreasing, then we say its graph is falling. These features of functions are generalized graphically in Table 4.1.

Increasing and Decreasing Functions

If $f'(x) > 0$ for all x on an interval I, then f is **increasing** over I.
If $f'(x) < 0$ for all x on an interval I, then f is **decreasing** over I.

Graphing Parabolas Using Calculus

We will now apply these ideas to parabolas whose equations have the form $y = ax^2 + bx + c$. Section 1.4 promised that this type of parabola would be discussed

TABLE 4.1
Increasing and Decreasing Functions on the Interval (a, b)

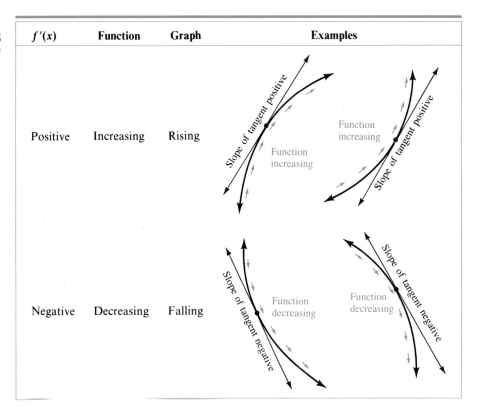

$f'(x)$	Function	Graph	Examples
Positive	Increasing	Rising	
Negative	Decreasing	Falling	

after we learned some calculus techniques. Now recall that parabolas of this form are functions and open upward if $a > 0$ and downward if $a < 0$. Furthermore the vertex of the parabola gives the extreme values for the function—it is a high point if $a < 0$ and a low point if $a > 0$. The slope of the parabola at the vertex must be zero. These possibilities are summarized in Figure 4.2.

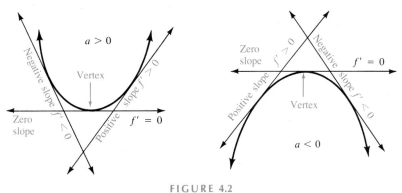

FIGURE 4.2
Graphs of parabolas showing slope

EXAMPLE 1 Graph $y = 3x^2 + 12x + 9$ and indicate the intervals for which the function is increasing and for which it is decreasing. Also compare the methods of finding the vertex by completing the square (as we did in Chapter 1) with that of using calculus (using the derivative).

Solution The function is increasing or decreasing depending on where the derivative is positive and where it is negative. If $f(x) = y = 3x^2 + 12x + 9$, then

$$f'(x) = 6x + 12$$
$$= 6(x + 2)$$

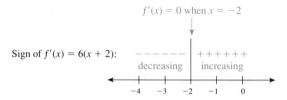

If a parabola is decreasing on $(-\infty, -2)$ and increasing on $(-2, \infty)$ with a horizontal tangent line at $x = -2$, it must open upward. Now we will compare the methods for finding the vertex:

Complete the square (precalculus)	*Use calculus*
$y = 3x^2 + 12x + 9$	$y = 3x^2 + 12x + 9$
$y - 9 = 3(x^2 + 4x)$	$y' = 6x + 12$
$y - 9 + 12 = 3(x^2 + 4x + 2^2)$	$y' = 0$ if $x = -2$
$y + 3 = 3(x + 2)^2$	If $x = -2$,
Vertex is $(-2, -3)$	then $y = 3(-2)2 + 12(-2) + 9 = -3$
	Vertex is $(-2, -3)$

The graph is shown in the margin.

EXAMPLE 2 Graph $y = 10x - x^2$.

Solution First, find the vertex: $y' = 10 - 2x$
Then, find the horizontal tangent: If $y' = 0$, then $10 - 2x = 0$
$$10 = 2x$$
$$5 = x$$
If $x = 5$, then
$$y = 10(5) - (5)^2 = 25$$
The vertex is the point $(5, 25)$.
The parabola opens downward ($a = -1$ is less than zero), has the y-intercept at $(0, 0)$, and is symmetric about the vertical line through the vertex. The graph is shown in the margin.

Maximums and Minimums

Since we will be graphing functions other than parabolas, some new terminology will be helpful. If f is a continuous function on an interval $[a, b]$, then there will be

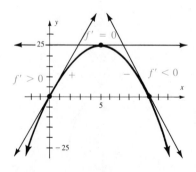

Increasing on $(-\infty, 5)$
Decreasing on $(5, \infty)$

some point c_5 in that interval so that

$$f(c_5) \geq f(x) \text{ for all } x \text{ on } [a, b]$$

The value $f(c_5)$ is called an **absolute maximum** on $[a, b]$, and we say that the function f has an absolute maximum at $x = c_5$. Similarly, there is another point a on $[a, b]$, not necessarily different from c_5, such that

$$f(a) \leq f(x) \text{ for all } x \text{ on } [a, b]$$

The value $f(a)$ is called an **absolute minimum** on $[a, b]$, and we say that the function f has an absolute minimum at $x = a$.

In addition to these absolute extremes, there may also be points that are higher or lower than the surrounding points. The values of the function at these points are called *local* (or *relative*) *maximums and minimums*. Several of these points are shown in Figure 4.3 and are defined in the following box.

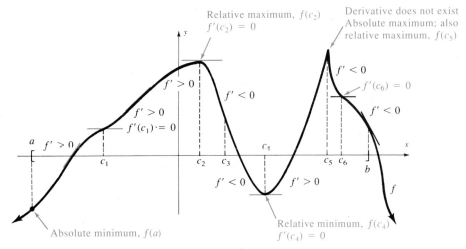

FIGURE 4.3
Absolute and Relative Extremes

Relative Maximum — A function f has a **local** (or **relative**) **maximum** at $x = c$ if there exists some interval (a, b) around c so that $f(x) < f(c)$ for all x (except c) in the interval (a, b). The value M for which $f(c) = M$ is the **relative maximum**.

Relative Minimum — A function f has a **local** (or **relative**) **minimum** at $x = c$ if there exists some interval (a, b) around c so that $f(x) > f(c)$ for all x (except c) in the interval (a, b). The value m for which $f(c) = m$ is the **relative minimum**.

We focus on absolute maximums and minimums in Section 4.3. In this section we use calculus to locate the relative extremes. Notice in Figure 4.3 that each and every relative maximum and relative minimum occurs at values of x for which $f'(x) = 0$ (horizontal tangent lines) or else where $f'(x)$ does not exist (places with sharp points on the graph). In Appendix A the term **critical value** was used when solving quadratic inequalities to mean a value that caused a factor or an expression to be zero. In the context of derivatives, a critical value is not only a value that causes a factor of the derivative to be zero (to have a horizontal tangent line), but also a value for which the derivative does not exist.

Critical Value

> A **critical value** for a function is an interior point c of its domain at which the function has a horizontal tangent, or at which the derivative does not exist. That is, c is a critical value if
>
> $f'(c) = 0$ or $f'(c)$ does not exist.

Of what significance are the critical values? Suppose f is a continuous function defined over an interval $[a, b]$, with critical values $c_1, c_2, \ldots, c_n$. Then everything that is interesting about the curve f will occur either at the endpoints a, b, or at one of these critical values. Everywhere else, the curve is either increasing or decreasing. We call this the **function behavior principle**. Be careful about the way you read this principle. It does *not* say that if c is a critical value, then $f(c)$ is a relative maximum or a relative minimum. It does say that $f(c)$ is a candidate to be a relative maximum or a relative minimum. We must, therefore, develop a strategy that will lead us to identify any relative maximums and minimums from the list of candidates (i.e., critical values). We use two derivative tests for this purpose. We first discuss a derivative test that works in all cases. In the next section, we discuss an easier, second-derivative test, which will work most of the time. For those cases in which this easier second-derivative test does not work, we will need to revert back to the first-derivative test discussed in this section.

First-Derivative Test

The first test, appropriately called the **first-derivative test**, makes use of the fact that the derivative gives the slope of a tangent line of a curve at a particular point. Suppose that c is a critical value. Let c^- be a point to the left of c (but to the right of any other critical value) and let c^+ be a point to the right of c (but to the left of any other critical value). There are then four possibilities as shown in Figure 4.4.

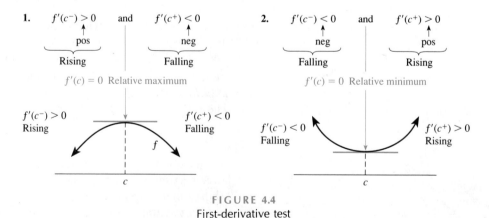

FIGURE 4.4
First-derivative test

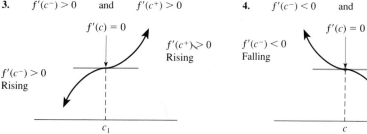

3. $f'(c^-) > 0$ and $f'(c^+) > 0$

4. $f'(c^-) < 0$ and $f'(c^+) < 0$

Not a relative maximum or minimum

Not a relative maximum or minimum

FIGURE 4.4 (continued)
First-derivative test

First-Derivative Test

1. Find the critical values:
   a. Find all values c such that $f'(c) = 0$.
   b. Find all values c for which the derivative does not exist but $f(c)$ is defined.
2. a. $f(c)$ is a **relative maximum** if

   $$f'(c^-) > 0 \quad \text{(rising)} \quad \text{and} \quad f'(c^+) < 0 \quad \text{(falling)}$$

   b. $f(c)$ is a **relative minimum** if

   $$f'(c^-) < 0 \quad \text{(falling)} \quad \text{and} \quad f'(c^+) > 0 \quad \text{(rising)}$$

EXAMPLE 3 Find the critical values for $f(x) = 5x^3 + 4x^2 - 12x - 25$.

Solution Critical values are values in the domain for which $f'(x) = 0$ or $f'(x)$ is not defined.

$$f'(x) = 15x^2 + 8x - 12$$

We see that $f'(x)$ is defined for all x, so we look for x values for which

$$f'(x) = 0$$
$$15x^2 + 8x - 12 = 0$$
$$(3x - 2)(5x + 6) = 0$$

The critical values are $x = \frac{2}{3}$ and $x = -\frac{6}{5}$. ∎

If we want to graph the curve whose equation is given in Example 3 we can use the first-derivative test. We want to know where $f'(x)$ is positive and where it is negative.

Look at the factored form of $f'(x)$:

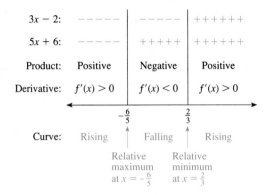

By plotting some points you can graph this curve, as shown in Figure 4.5.

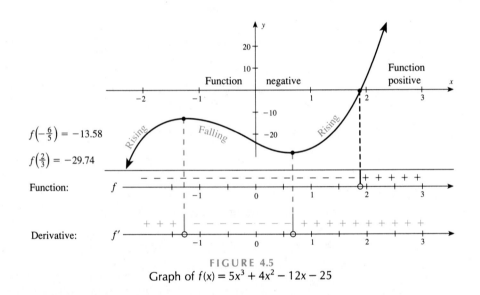

FIGURE 4.5
Graph of $f(x) = 5x^3 + 4x^2 - 12x - 25$

$f(x)$ values found by calculator

x	$f(x)$	$f'(x)$	
$-\frac{6}{5}$	-13.58	0	Relative maximum
$\frac{2}{3}$	-29.74	0	Relative minimum
2	7	>0	Rising ⎫ Same for
-2	-25	>0	Rising ⎬ each interval
0	-25	<0	Falling ⎭ by function behavior principle

EXAMPLE 4 Find the critical values for $f(x) = |x+1|$.

Solution If $x > -1$:
$$f(x) = x + 1 \quad \text{and} \quad f'(x) = 1$$
If $x < -1$:
$$f(x) = -x - 1 \quad \text{and} \quad f'(x) = -1$$

At $x = -1$,
$$\lim_{h \to -1^+} \frac{|-1+h+1| - |-1+1|}{h} = \lim_{h \to -1^+} 1 = 1$$
$$\lim_{h \to -1^-} \frac{|-1+h+1| - |-1+1|}{h} = \lim_{h \to -1^-} -1 = -1$$

Thus the derivative does not exist at $x = -1$ and $x = -1$ is in the domain, so $x = -1$ is a critical value.

Apply the first-derivative test. Let
$$f'(c^-) = -1 \quad \text{since} \quad c^- < -1 \quad \text{(falling)}$$
$$f'(c^+) = 1 \quad \text{since} \quad c^+ > -1 \quad \text{(rising)}$$

There is a relative minimum at $f(-1) = 0$. The graph is shown in Figure 4.6. ■

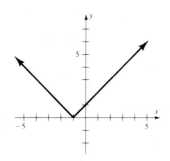

FIGURE 4.6
Graph of $f(x) = |x+1|$

4.1
Problem Set

Graph the curves in Problems 1–16 by using the derivative to find the vertex.

1. $f(x) = 8x^2$
2. $f(x) = -12x^2$
3. $f(x) = -20x^2$
4. $2x^2 + 5y = 0$
5. $5y + 15x^2 = 0$
6. $5x^2 + 4y = 20$
7. $f(x) = 5x^2 - 20x + 2$
8. $f(x) = 9 + 24x - 12x^2$
9. $x^2 + 4y - 3x + 1 = 0$
10. $x^2 - 4x + 10y + 13 = 0$
11. $9x^2 + 6x + 18y - 23 = 0$
12. $9x^2 + 6y + 18x - 23 = 0$
13. $y = 5x^2 - 3x + 1$
14. $y = 2x^2 + 5x - 8$
15. $y = 7x^2 + 2x - 3$
16. $y = 2x^2 - 5x + 3$

Identify the intervals for which the functions in Problems 17–22 are increasing, decreasing, or constant. Also identify the places where there is a horizontal tangent.

17.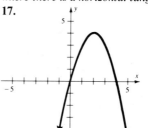

18.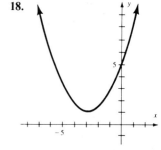

158 CHAPTER FOUR APPLICATIONS AND DIFFERENTIATION

19.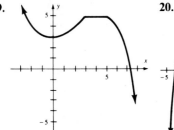

20.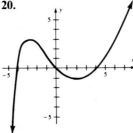

21.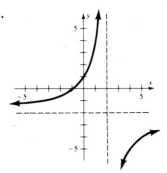

22.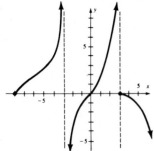

23. The graph of a function f is shown. Draw the tangent lines and measure the slope of f at the points labeled A, B, C, D, and E. Then set up another coordinate system and plot the corresponding points to draw the graph of f'.

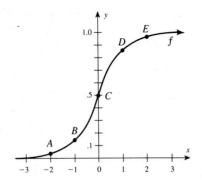

24. The graph of a function f is shown. Draw the tangent lines and measure the slope of f at the points labeled A, B, C, D, and E. Then set up another coordinate system and plot the corresponding points to draw the graph of f'.

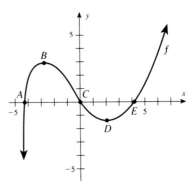

25. The graph of the derivative g' is shown, but it does not show the function g. Sketch the general shape of a function g that may have as its derivative the given graph g'.

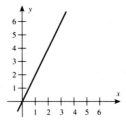

26. The graph of the derivative g' is shown, but it does not show the function g. Sketch the general shape of a function g that may have as its derivative the given graph g'.

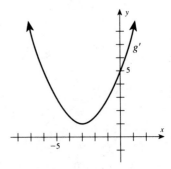

For the functions in Problems 27–42, **(a)** find the critical values and **(b)** determine the intervals for which the functions are increasing and decreasing.

27. $f(x) = 5 + 10x - x^2$
28. $f(x) = 10 + 6x - x^2$
29. $f(x) = x^2 - 12x + 6$
30. $f(x) = x^2 + 5x + 6$

31. $g(x) = 2x^3 - 3x^2 - 36x + 4$
32. $g(x) = x^3 + 3x^2 - 24x - 4$
33. $y = x^3 + 5x^2 + 8x - 5$
34. $y = \frac{8}{3}x^3 - 5x^2 - 3x + 10$
35. $y = 3x^4 + 2x^3 - 9x^2 + 12$
36. $y = 3x^4 + 8x^3 - 18x^2 + 5$
37. $f(x) = 3x^5 - 25x^3 - 540x + 90$
38. $f(x) = 3x^5 - 85x^3 + 240x - 200$
39. $y = \dfrac{x+1}{x-2}$
40. $y = \dfrac{x^2}{x-1}$
41. $f(x) = 1 + \dfrac{1}{x} + \dfrac{1}{x^2}$
42. $f(x) = x - 4\sqrt{x}$

APPLICATIONS

43. If a company has sales of
$$S(x) = 10{,}000 + 5{,}000x - 25x^2 - x^3$$
where x is the amount spent on advertising in thousands of dollars, when are the sales increasing?

44. The time t, in minutes, that it takes to learn a list of x items is given by the formula
$$t(x) = 5x\sqrt{x - 10} \qquad \text{for } x \geq 10$$
For what values of x is t increasing?

45. Suppose you are the campaign manager for a candidate, and also suppose that a function showing the percentage of people who are aware of your candidate is
$$f(x) = \dfrac{10x}{x^2 + 50} + .1$$
where x is the number of months after you begin campaigning. If the election is held on November 2, when should you begin campaigning if you want the maximum number of people to be aware of your candidate at election time?

4.2
Second Derivatives and Graphs

Higher-Order Derivatives

We have used the derivative to find out where f is increasing and where it is decreasing. Since the derivative of f is also a function we can repeat the process and use the derivative of f' to determine where f' is increasing and where it is decreasing. The derivative of a derivative is called its **second derivative** and is denoted by f''. Taking still another derivative gives a function f''' called the **third derivative**. Successive derivatives are denoted by

$$f', f'', f''', f^{(4)}, f^{(5)}, \ldots, f^{(n)}$$

Note the use of parentheses for fourth derivatives and higher. This is to avoid confusion with powers f^4, f^5, and so on.

The graph of a function and its successive derivatives is shown in Figure 4.7.

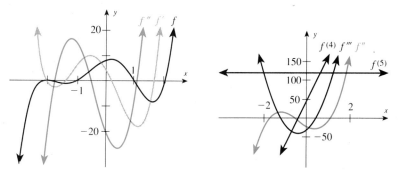

FIGURE 4.7
Comparison of graphs of a function and its derivatives

COMPUTER COMMENT

Use the *Function plotter* on the software accompanying this book to draw the graph of a function. You can also graph relations using the *Implicit relation plotter*.

EXAMPLE 1 Find all successive derivatives of $f(x) = x^5 + 2x^4 - 5x^3 - 10x^2 + 4x + 8$.

Solution
$f'(x) = 5x^4 + 8x^3 - 15x^2 - 20x + 4$
$f''(x) = 20x^3 + 24x^2 - 30x - 20$
$f'''(x) = 60x^2 + 48x - 30$
$f^{(4)}(x) = 120x + 48$
$f^{(5)}(x) = 120$
$f^{(6)}(x) = f^{(7)}(x) = \cdots = 0$ ∎

Another commonly used notation for higher derivatives uses the dy/dx notation as follows:

$$\frac{dy}{dx}, \frac{d^2y}{dx^2}, \frac{d^3y}{dx^3}, \frac{d^4y}{dx^4}, \ldots, \frac{d^ny}{dx^n}$$

EXAMPLE 2 If $y = 8x^{1/2}$, find the first three derivatives.

Solution
$$\frac{dy}{dx} = 8\left(\frac{1}{2}\right)x^{1/2-1} = 4x^{-1/2}$$

$$\frac{d^2y}{dx^2} = \frac{d}{dx}\left[\frac{dy}{dx}\right] = \frac{d}{dx}[4x^{-1/2}] = 4\left(-\frac{1}{2}\right)x^{-1/2-1} = -2x^{-3/2}$$

$$\frac{d^3y}{dx^3} = \frac{d}{dx}\left[\frac{d^2y}{dx}\right] = \frac{d}{dx}[-2x^{-3/2}] = -2\left(-\frac{3}{2}\right)x^{-3/2-1} = 3x^{-5/2}$$ ∎

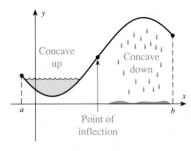

FIGURE 4.8
Concavity and inflection points

Concavity

Now we can use the second derivative to develop a second-derivative test to inform us about the shape of a graph. If a curve lies above its tangent line at each point on an interval (a, b), it is **concave upward** over (a, b); if a curve lies below its tangent line at each point on the interval (a, b), it is **concave downward**. A point on a graph that separates a concave downward portion of a curve from a concave upward portion is called an **inflection point**. Inflection points can occur where $f''(x) = 0$ or where $f''(x)$ is undefined. Figure 4.8 illustrates these ideas.

We can now use the second derivative to obtain additional information about a graph. We consider two possibilities: $f''(x) > 0$ and $f''(x) < 0$.

Case I: If $f''(x) > 0$ on (a, b), then f' is increasing on (a, b).
What are the possibilities for f when f' is increasing?

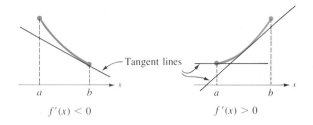
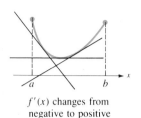

f' is increasing and is negative; the curve is falling
f' is increasing and is positive; the curve is rising
f' is increasing and changes from negative to positive

Curves having the shape illustrated by $f''(x) > 0$ are concave upward.

Case II: If $f''(x) < 0$ on (a, b), then f' is decreasing on (a, b).
What are the possibilities for f when f' is decreasing?

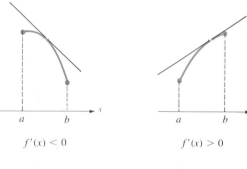
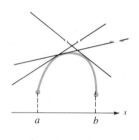

f' is decreasing and negative; the curve is falling
f' is decreasing and positive; the curve is rising
f' is decreasing and changes from positive to negative

Curves having the shape illustrated by $f''(x) < 0$ are concave downward.

EXAMPLE 3 Given: $4x^3 - 5x^2 - 8x - 2y + 20 = 0$

a. Find the critical values.
b. Find the intervals for which the function is increasing or decreasing.
c. Find the interval for which the function is concave upward or concave downward.

Solution a. You could first solve for y and then find the derivative, but instead you find y' implicitly:

$$12x^2 - 10x - 8 - 2y' = 0$$
$$2y' = 12x^2 - 10x - 8$$
$$y' = 6x^2 - 5x - 4$$

Set $y' = 0$ and solve for the critical values.

$$6x^2 - 5x - 4 = 0$$
$$(3x - 4)(2x + 1) = 0$$

The critical values are $x = \frac{4}{3}$ and $x = -\frac{1}{2}$.

b. The intervals on which the function is increasing or decreasing can be found by finding where the derivative is positive or negative. The procedure is identical to that used when solving inequalities (see Appendix A.6).

$$6x^2 + 5x - 4 = (3x - 4)(2x + 1)$$

Plot the critical values and solve on a number line.

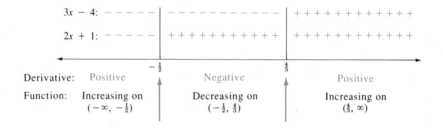

c. For the concavity, look at the second derivative:

$$y'' = 12x - 5$$

Set $y'' = 0$ and solve.

$$12x - 5 = 0$$
$$x = \frac{5}{12}$$

The first derivative is not zero at $x = \frac{5}{12}$, so

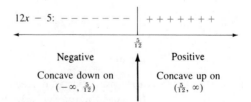

4.2 SECOND DERIVATIVES AND GRAPHS   163

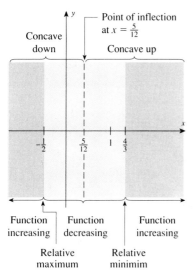

In preparation for graphing such curves we summarize all of this information on a coordinate grid as follows:

Color screen shows portions where the function is decreasing.
Gray screen shows portions where the function is increasing.
These screens are separated by points on the graph that are relative maximums or relative minimums.

Concavity is also marked, and places where concavity changes are marked by a point of inflection.

COMPUTER COMMENT

All of the work of this example can be summarized using the *Function plotter*.

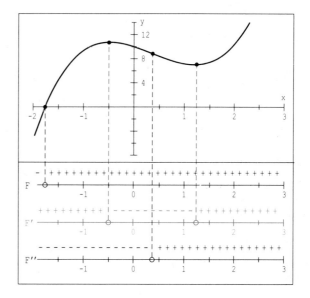

Notice that F shows where the function is positive and where it is negative. F′ shows where the function is increasing and where it is decreasing. F″ shows where the function is concave down and where it is concave up.
Where the curve crosses the axis the value of F is zero;
where the curve reaches a relative maximum or minimum, the value of F′ is zero;
where the curve changes concavity, the value of F″ is zero.

EXAMPLE 4 Identify the intervals for which the function is increasing, the intervals for which it is decreasing, where it is concave upward, where it is concave downward, as well as places where there is a horizontal tangent line. Identify the points that give relative maximum and relative minimum values for the function.

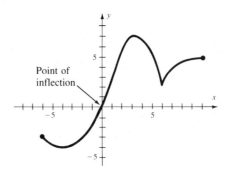

Solution

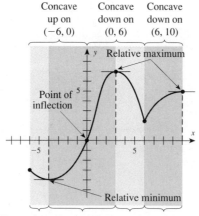

The horizontal tangents are at $x = -4$, $x = 3$, and $x = 10$.

(There is no horizontal tangent at $x = 6$, since there is not even a derivative defined at $x = 6$.)

Second-Derivative Test

For most of your work you will use what is called the **second-derivative test**. Figure 4.9 illustrates the two possibilities that give a relative maximum or a relative minimum.

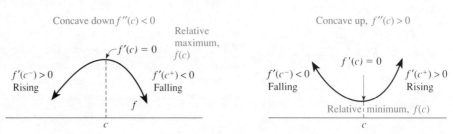

FIGURE 4.9

If a curve has a horizontal tangent line at $x = c$ and is concave upward ($f'' > 0$) in an interval containing the point of tangency, then $f(c)$ is a relative minimum. On the other hand, if the curve is concave downward ($f'' < 0$), then $f(c)$ is a relative maximum. This observation leads us to the second-derivative test:

Second-Derivative Test

> If f is a continuous function on an interval (a, b),
>
> 1. Find the critical values of f. Suppose that c is a critical value of f.
> 2. **a.** If $f''(c) > 0$, then $f(c)$ is a relative minimum.
>   *Note:* Greater than zero, concave *up*; thus a relative minimum.
>   **b.** If $f''(c) < 0$, then $f(c)$ is a relative maximum.
>   *Note:* Less than zero, concave *down*; thus a relative maximum.
>   **c.** If $f''(c) = 0$, then the second-derivative test fails and gives no information, so the first-derivative test must be used.

EXAMPLE 5 Graph the parabola defined by $f(x) = x^2 - 8x + 7$ by using the second-derivative test.

Solution
$f(x) = x^2 - 8x + 7$
$f'(x) = 2x - 8$; critical value $x = 4$
$f''(x) = 2$; thus, there is a relative minimum at $x = 4$ since $f''(4) = 2 > 0$. This minimum value is

$$f(4) = 16 - 8(4) + 7 = -9$$

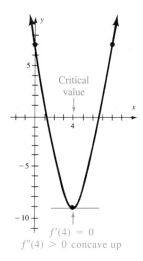

$f'(4) = 0$
$f''(4) > 0$ concave up

EXAMPLE 6 Graph $f(x) = 3x^4 + 8x^3 - 6x^2 - 24x + 6$.

Solution You need to keep track of the function for points on the curve, the derivative for slope and critical values, and the second derivative for concavity and points of inflection. We will keep track of this with the following table.

Function	First derivative	Second derivative
$f(x) = 3x^4 + 8x^3 - 6x^2 - 24x + 6$	$f'(x) = 12x^3 + 24x^2 - 12x - 24$	$f''(x) = 36x^2 + 48x - 12$
	Critical values: This is a polynomial, so the derivative is defined everywhere. $12x^3 + 24x^2 - 12x - 24 = 0$ $x^3 + 2x^2 - x - 2 = 0$ $x^2(x + 2) - (x + 2) = 0$ $(x^2 - 1)(x + 2) = 0$ $(x - 1)(x + 1)(x + 2) = 0$ $x = 1, -1, -2$	
Evaluate function to plot points: $f(1) = 3(1)^4 + 8(1)^3 - 6(1)^2 - 24(1) + 6$ $= -13$ Point $(1, -13)$ $f(-1) = 3 - 8 - 6 + 24 + 6$ $= 19$ Point $(-1, 19)$ $f(-2) = 14$ Point $(-2, 14)$	Check $x = 1$ Check $x = -1$ Check $x = -2$	**Second-derivative test:** $f''(1) = 36(1)^2 + 48(1) - 12$ $= 72 > 0$ Relative minimum $f(1) = 13$ at $x = 1$ $f''(-1) = 36 - 48 - 12$ $= -24 < 0$ Relative maximum $f(-1) = 19$ at $x = -1$ $f''(-2) > 0$ Relative minimum $f(-2) = 14$ at $x = -2$
		Check concavity: $36x^2 + 48x - 12 = 0$ $3x^2 + 4x - 1 = 0$ $x = \dfrac{-2 \pm \sqrt{7}}{3}$ $\approx .22, -1.55$
$f(.22) \approx .52$	$f'(.22) \approx -25$	Check $x = \dfrac{-2 + \sqrt{7}}{3}$ $\approx .22$
Point $(.22, .52)$	decreasing	Point of inflection at $x \approx .22$
$f(-1.55) \approx 16.31$	$f'(-1.55) \approx 7.6$	Check $x = \dfrac{-2 - \sqrt{7}}{3}$ ≈ -1.55
Point $(-1.55, 16.31)$	increasing	Point of inflection at $x \approx -1.55$

If there were any other relative maximums or minimums, they would have shown up in the list of critical values (which they did not), so it is easy to complete the graph. The graph is shown in Figure 4.10.

FIGURE 4.10
Graph of $f(x) = 3x^4 + 8x^3 - 6x^2 - 24x + 6$

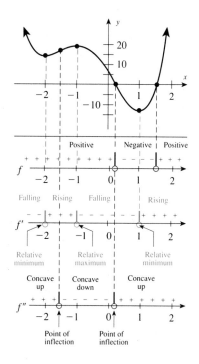

EXAMPLE 7 Graph $f(x) = x^4$.

Solution

Function	First derivative	Second derivative
$f(x) = x^4$	$f'(x) = 4x^3$	$f''(x) = 12x^2$
	Critical values: $4x^3 = 0$ $x = 0$	
$f(0) = 0$ Point $(0, 0)$	Check $x = 0$	**Second-derivative test:** $f''(x) = 0$ Test fails.
$f(-1) = 1$ Point $(-1, 1)$ $f(1) = 1$ Point $(1, 1)$ Point $(0, 0)$ Plot some points: $f(2) = 16$ $f(-2) = 16$	**First-derivative test:** Choose $c^- = -1$ $f'(-1) < 0$ decreasing Choose $c^+ = 1$ $f'(1) > 0$ increasing relative minimum at $x = 0$	

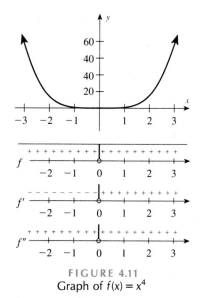

FIGURE 4.11
Graph of $f(x) = x^4$

4.2

Problem Set

Find the indicated derivatives for the functions in Problems 1–12.

1. $f(x) = 2x^5 - 3x^4 + x^3 - 5x^2 + 19x - 120$; $f'''(x)$
2. $f(x) = 25 + x - 3x^2 + 4x^3 - x^4 + 3x^5$; $f^{(5)}(x)$
3. $y = \sqrt{5x}$; $\dfrac{d^4y}{dx^4}$
4. $y = 18\sqrt{3x}$; $\dfrac{d^3y}{dx^3}$
5. $g(x) = 4x^3 - 2x^{-1}$; $g^{(4)}(x)$
6. $g(x) = 5x^{-2} - 3x$; $g'''(x)$
7. $y = 3x^{5/3}$; $\dfrac{d^3y}{dx^3}$
8. $y = -4x^{-3/2}$; $\dfrac{d^4y}{dx^4}$
9. $y = \dfrac{3}{x-1}$; $\dfrac{d^3y}{dx^3}$
10. $y = \dfrac{-5}{2x+1}$; $\dfrac{d^3y}{dx^3}$
11. $y = \dfrac{x^2-1}{x+4}$; $\dfrac{d^2y}{dx^2}$
12. $y = \dfrac{3x^2+5}{x-9}$; $\dfrac{d^2y}{dx^2}$

Find all relative maximums and minimums in Problems 13–30 by using either the first- or the second-derivative test as appropriate. You do not need to graph these functions.

13. $f(x) = (2x+1)^4$
14. $f(x) = (3x-5)^4$
15. $f(x) = \dfrac{x^2+9}{x}$
16. $f(x) = \dfrac{x^2+1}{x}$
17. $f(x) = \sqrt{x^2+1}$
18. $f(x) = \sqrt{x^2+4}$
19. $f(x) = 8x^2 - x^4$
20. $f(x) = 2x^2 - x^4$
21. $f(x) = x^3 + 5x^2 - 8x + 10$
22. $f(x) = 6x^3 - 21x^2 - 36x + 15$
23. $y = 4x^3 - 27x^2 - 30x - 6$
24. $y = 2x^3 + 7x^2 - 40x + 5$
25. $g(x) = x + \dfrac{2}{x}$
26. $g(x) = x + \dfrac{4}{x}$
27. $g(x) = \dfrac{x^2}{x-1}$
28. $g(x) = \dfrac{x^2}{x-9}$
29. $f(x) = (x+1)^{2/3}$
30. $f(x) = (x-3)^{2/3}$

For the functions in Problems 31–40, **(a)** find the critical values; **(b)** determine the intervals for which the functions are increasing and for which they are decreasing; and **(c)** determine the intervals for which the functions are concave up or concave down.

31. $f(x) = 3 + 12x - x^2$
32. $f(x) = 8 + 6x - x^2$
33. $g(x) = x^3 - 7x^2 - 5x + 8$
34. $g(x) = x^3 - 5x^2 - 8x + 10$
35. $y = x^3 + 11x^2 - 45x + 125$
36. $y = x^3 - 2x^2 - 15x - 75$
37. $12x^3 - 5x^2 - 4x - 2y + 14 = 0$
38. $2x^3 - 12x^2 - 30x - 6y + 5 = 0$
39. $f(x) = (2-x)^3 - 8$
40. $f(x) = (1+x)^3 - 1$

APPLICATIONS

If $S(x)$ is the number of units of a product sold after spending x dollars on advertising, then the **point of diminishing returns** is that point of inflection for S at which the rate of change of sales changes from positive to negative.

41. A company estimates that it will sell $S(x)$ units of a product after spending x thousands of dollars on advertising, according to the formula

$$S(x) = 4{,}000 - x^3 + 45x^2 + 60x$$

for $10 \leq x \leq 35$. What is the point of diminishing returns?

42. A company estimates that it will sell $S(x)$ units of a product after spending x thousands of dollars on advertising, according to the formula

$$S(x) = 20{,}000 - 4x^3 + 180x^2 - 2{,}400x$$

for $10 \leq x \leq 30$. What is the point of diminishing returns?

43. For the application in Problem 41, when is the rate of change of sales per unit change in advertising increasing?
44. For the application in Problem 42, when is the rate of change of sales per unit change in advertising increasing?
45. What are the sales corresponding to the maximum rate of change for the sales in Problem 41?
46. What are the sales corresponding to the maximum rate of change for the sales in Problem 42?
47. Graph both S and S' for Problem 41.
48. Graph both S and S' for Problem 42.
49. The marketing research department found that the demand for a product can be approximated by

$$p = 1{,}156 - \dfrac{1}{12}x^2 \quad \text{for } 0 \leq x \leq 100$$

Find the relative maximums and minimums for the revenue function.

50. Over which intervals is the graph of the revenue function in Problem 49 concave upward? Concave downward?

51. A manufacturer has determined that the demand of selling items is given by the formula

$$p(x) = 5 - \left(\frac{x}{100}\right)^2$$

a. Find an expression for the revenue.
b. Find the marginal revenue.
c. Is the marginal revenue increasing or decreasing when $x = 100$ items?

4.3
Curve Sketching—Relative Maximums and Minimums

One of the most important tools in mathematics is the ability to quickly sketch a wide variety of functions. In this section we combine the graphing techniques of calculus with those we used in Chapter 1.*

Table 4.2 presents procedures for graphing functions. It is, of course, not necessary to use all the procedures to graph every function. And, if any of the procedures are too difficult, it is possible to omit that particular procedure. The more you know about a curve, the fewer points you need to plot.

TABLE 4.2 Graphing Strategy

Step	Procedure
Simplify	First, simplify, if possible, the function you wish to graph. That is, combine similar terms, reduce fractions, and simplify radical expressions.
Second-derivative test	Use the second-derivative test to find the relative maximum or minimum values: 1. Find the critical values, c. 2. $f''(c) > 0$, relative minimum $f(c)$ at $x = c$; concave up $\quad f''(c) < 0$, relative maximum $f(c)$ at $x = c$; concave down $\quad f''(c) = 0$, test fails
First-derivative test	Use the first-derivative test if the second-derivative test fails: 1. Relative minimum $f(c)$ if $f'(c^-) < 0$ and $f'(c^+) > 0$ 2. Relative maximum $f(c)$ if $f'(c^-) > 0$ and $f'(c^+) < 0$ The curve is increasing if $f'(x) > 0$ The curve is decreasing if $f'(x) < 0$
Asymptotes	If $f(x) = P(x)/D(x)$, where $P(x)$ and $D(x)$ are polynomial functions with no common factors, then: 1. *Vertical asymptote:* $x = r$ if $D(r) = 0$ 2. *Horizontal asymptote:* If $$\lim_{x \to \infty} f(x) = L$$ then $y = L$ is a horizontal asymptote.
Intercepts	1. *x-intercept:* Set $y = 0$ and solve for x. 2. *y-intercept:* Set $x = 0$ and solve for y.
Plot points	Plot any additional points necessary to draw the graph.

---

* If you omitted Chapter 1, it is now a good idea to look at Section 1.5.

EXAMPLE 1 Graph $f(x) = \dfrac{x^2}{x-2}$.

Solution

Function	First derivative	Second derivative
$f(x) = \dfrac{x^2}{x-2}$	$f'(x) = \dfrac{(x-2)(2x) - x^2(1)}{(x-2)^2}$ $= \dfrac{2x^2 - 4x - x^2}{(x-2)^2}$ $= \dfrac{x^2 - 4x}{(x-2)^2} = \dfrac{x(x-4)}{(x-2)^2}$	$f''(x) = \dfrac{(x-2)^2(2x-4) - x(x-4)(2)(x-2)}{(x-2)^4}$ $= \dfrac{8}{(x-2)^3}$ There were several simplification steps left out; can you fill in the details?
	Critical values: $\dfrac{x(x-4)}{(x-2)^2} = 0$ $x(x-4) = 0$ $x = 0, 4$ Critical values	
$f(0) = \dfrac{0^2}{0-2} = 0$ Point $(0, 0)$ $f(4) = \dfrac{4^2}{4-2} = 8$ Point $(4, 8)$	Check $x = 0$ Check $x = 4$	**Second-derivative test** $f''(0) = \dfrac{8}{(0-2)^3} = -1 < 0$ Relative maximum at $x = 0$ $f''(4) = \dfrac{8}{(4-2)^3} = 1 > 0$ Relative minimum at $x = 4$
Intercepts: $x = 0$ found above $y = 0$: $0 = \dfrac{x^2}{x-2}$ $0 = 0$ $(0, 0)$ is only intercept.	**Increasing and Decreasing** x: $--$ $+++$ $+++$ $+++$ $x-4$: $--$ $---$ $---$ $+++$ $(x-2)^2$: $++$ $+++$ $+++$ $+++$ $\longleftarrow\!\!\longrightarrow$ 0 2 4 Derivative: Pos. Neg. Neg. Pos. Function: Rising Falling Falling Rising	Curve rising Curve falling Curve falling Curve rising Relative minimum at $x = 4$ Relative maximum at $x = 0$ $x = 2$

Since this is a rational function, we check asymptotes. There is a vertical asymptote $x = 2$. Check for a horizontal asymptote:

$\lim\limits_{|x| \to \infty} \dfrac{x^2}{x - 2}$ does not exist; no horizontal asymptotes.

$\left[\text{Note: Check both } \lim\limits_{x \to \infty} \dfrac{x^2}{x - 2} \text{ and } \lim\limits_{x \to -\infty} \dfrac{x^2}{x - 2}. \right]$

Draw the asymptote, and plot another value, say $x = 8$, to draw the graph shown in Figure 4.12.

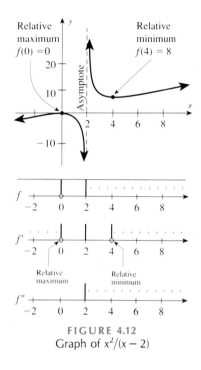

FIGURE 4.12
Graph of $x^2/(x - 2)$

EXAMPLE 2 Graph $f(x) = x + \dfrac{9}{x}$.

Solution The domain excludes $x = 0$.

Function	First derivative	Second derivative
$f(x) = x + 9x^{-1}$	$f'(x) = 1 - 9x^{-2}$	$f''(x) = 18x^{-3}$
	Critical values: $1 - 9x^{-2} = 0$ $x^2 = 9$ $x = \pm 3$ Also, derivative does not exist at $x = 0$, but this value is not in the domain for f.	

(continued)

Function	First derivative	Second derivative
$f(3) = 3 + 9(3)^{-1}$ $= 6$ Point $(3, 6)$ $f(-3) = -3 + 9(-3)^{-1}$ $= -6$ Point $(-3, -6)$	Check $x = 3$ Check $x = -3$	**Second-derivative test:** $f''(3) = 18(3)^{-3} = \frac{2}{3} > 0$ Relative minimum at $x = 3$ $f''(3) = 18(-3)^{-3} = -\frac{2}{3} < 0$ Relative maximum at $x = -3$
Intercepts: $x = 0$: not in the domain; no y-intercept $y = 0$: $0 = x + 9x^{-1}$ $x^2 + 9 = 0$ No solution; no x-intercept **Asymptotes:** Horizontal: $x = 0$ Vertical: $\lim\limits_{|x| \to \infty} f(x)$ does not exist. No vertical asymptotes	**Increasing and decreasing:** Note: $f'(x) = 1 - 9x^{-2}$ $= \dfrac{x^2 - 9}{x^2}$ $= \dfrac{(x-3)(x+3)}{x^2}$ Factors: $x - 3$: $--\ \|\ --\ \|\ --\ \|\ +++$ $x + 3$: $--\ \|\ ++\ \|\ ++\ \|\ +++$ x^2: $\ \ ++\ \|\ ++\ \|\ ++\ \|\ +++$ $\qquad\quad\ -3\quad\ \ 0\quad\ \ 3$ Derivative: Pos. Neg. Neg. Pos. Function: Rising Falling Falling Rising f'': Neg. Neg. Pos. Pos. Concavity: Down Down Up Up	**Concavity:** $f''(x) = \dfrac{18}{x^3}$ **Points of inflection:** $f''(x) \neq 0$; none

One additional observation that might help with this problem has to do with the limit we considered when looking for a horizontal asymptote:

$$\lim_{|x| \to \infty} x + \frac{9}{x}$$

Although it is true that this limit does not exist, you might also notice that, for large values of $|x|$, the term $9/x$ is negligible, so that the values of f are almost identical to the values on the line $y = x$. This line is called a **slant asymptote**. If you draw the line $y = x$ you can use this line to help you sketch f as shown in Figure 4.13.

4.3 CURVE SKETCHING—RELATIVE MAXIMUMS AND MINIMUMS 173

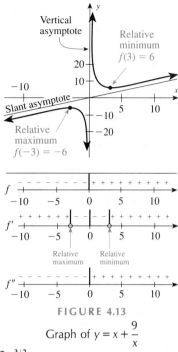

FIGURE 4.13

Graph of $y = x + \dfrac{9}{x}$

EXAMPLE 3 Graph $f(x) = 12x^{1/2} - 2x^{3/2}$.

Solution The domain is all nonnegative real numbers.

Function	First derivative	Second derivative
$f(x) = 12x^{1/2} - 2x^{3/2}$	$f'(x) = 6x^{-1/2} - 3x^{1/2}$ $= 3x^{-1/2}(2 - x)$	$f''(x) = -3x^{-3/2} - \tfrac{3}{2}x^{-1/2}$ $= -\tfrac{3}{2}x^{-3/2}(2 + x)$
	Critical values: First, $f'(x) = 0$ if $x = 2$ Not defined if $x = 0$	
$f(2) = 12(2)^{1/2} - 2(2)^{3/2}$ $= 12\sqrt{2} - 4\sqrt{2}$ $= 8\sqrt{2}$ Point $(2, 8\sqrt{2})$	Check $x = 2$	**Second-derivative test:** $f''(2) = -\tfrac{3}{2}(2)^{-3/2}(2 + 2)$ < 0 Relative maximum at $x = 2$
$f(0) = 12(0)^{1/2} - 2(0)^{3/2}$ $= 0$ Point $(0, 0)$	Check $x = 0$	$f''(x) = -3(0)^{-3/2} - \tfrac{3}{2}(0)^{-1/2}$ $= 0$ Test fails
	First-derivative test: Values to the left of 0 not defined	

(continued)

Function	First derivative	Second derivative
There are no asymptotes.	$f(c) > 0$ to the right of critical value $x = 0$; this is a relative minimum	
Intercepts: y-intercept ($x = 0$): $12x^{1/2} - 2x^{3/2}$ $= 12(0)^{1/2} - 2(0)^{3/2}$ $= 0$ Point $(0, 0)$ x-intercept ($y = 0$): $12x^{1/2} - 2x^{3/2} = 0$ $2x^{1/2}(6 - x) = 0$ $x = 0, 6$ These are the points $(6, 0)$ and $(0, 0)$	**Increasing and Decreasing:** $2 - x$: Not in domain $\|$ + + $\|$ - - - x: + + $\|$ + + + $\quad\quad\quad 0 \quad\quad 2$ Derivative: Positive Negative Function: Rising Falling f'': Negative Negative Concavity: Down Down $f'(6) = 3(6)^{-1/2}(2 - 6)$ < 0 Falling at $x = 6$	FIGURE 4.14 Graph of $12x^{1/2} - 2x^{3/2}$

The graph is shown in Figure 4.14.

4.3
Problem Set

In Problems 1–6, draw a curve on the interval (a, b) satisfying the stated conditions.

1. $f'(x) > 0$ and $f''(x) > 0$
2. $f'(x) > 0$ and $f''(x) < 0$
3. $f'(x) < 0$ and $f''(x) > 0$
4. $f'(x) < 0$ and $f''(x) < 0$
5. Passes through $(-2, 3)$, $(0, 5)$, and $(2, 7)$; $f'(-2) = 0$ and $f'(2) = 0$; $f''(x) > 0$ if $x < 0$, $f''(x) < 0$ if $x > 0$, and $f''(0) = 0$
6. Passes through $(2, 8)$, $(5, 6)$, and $(8, 4)$; $f'(2) = 0$ and $f'(8) = 0$; $f''(x) < 0$ if $x < 5$, $f''(x) > 0$ if $x > 5$, and $f''(5) = 0$

Graph the functions in Problems 7–34.

7. $y = 6x - x^2$
8. $y = 20x - 5x^2$
9. $y = 2x^2 - 3x + 5$
10. $3x^2 + 4x - 2y + 8 = 0$
11. $y = x^3 + 3x^2 - 9x + 5$
12. $y = x^3 - 3x^2 - 24x + 10$
13. $y = x^3 - 48x + 50$
14. $y = x^3 - 75x + 1$
15. $f(x) = x^3 + 1$
16. $f(x) = x^5$
17. $g(x) = x^4 - 1$
18. $g(x) = (x - 1)^4$
19. $f(x) = x^4 - 2x^2$
20. $f(x) = x^4 - 8x^2$

21. $g(x) = x + \dfrac{4}{x}$ 22. $g(x) = x + \dfrac{1}{x}$

23. $f(x) = 3x^4 + 8x^3 - 6x^2 - 24x - 5$

24. $f(x) = 3x^4 + 20x^3 - 24x^2 - 240x + 20$

25. $y = \dfrac{x^2}{x-1}$ 26. $y = \dfrac{x^2}{x+1}$

27. $y = \dfrac{x+1}{x-1}$ 28. $y = \dfrac{2x+1}{x-1}$

29. $y = \dfrac{x-5}{x+5}$ 30. $y = \dfrac{x-2}{x+2}$

31. $y = 3x^{2/3}$ 32. $y = -3x^{1/3}$

33. $f(x) = 6x^{1/2} - 4x^{3/2}$

34. $f(x) = 6x^{1/2} - 12x^{3/2} + 5$

APPLICATIONS

35. The cost of removing $p\%$ of the pollutants from the atmosphere is given by the formula

$$C(p) = \dfrac{20{,}000p}{100 - p}$$

Graph $C(p)$ on $[0, 100]$.

36. A manufacturer has determined that the demand of selling x items is given by the formula

$$p(x) = 5 - \left(\dfrac{x}{100}\right)^2$$

Sketch the marginal revenue function.

37. The cost of manufacturing a certain item is

$$C(x) = 5{,}000 + \dfrac{1}{2}x^2$$

where x is the number of units produced. Graph the average cost function and the marginal cost function on the same coordinate axes.

38. The cost of manufacturing a product is

$$C(x) = 10{,}000 + .1x^2$$

where x is the number of units produced. Graph the average cost function and the marginal cost function on the same coordinate axes.

39. The concentration C of a drug in the bloodstream t minutes after injection is given by the formula

$$C(t) = \dfrac{10t}{8 + t^3}$$

Graph $C(t)$.

40. The number of fish swimming upstream to spawn is approximated by the formula

$$S(x) = 13{,}200x - 2x^3 - 45x^2 - 10{,}000$$

where x is the temperature of the water in degrees Fahrenheit ($30 \leq x \leq 70$). Graph $S(x)$.

4.4

Absolute Maximum and Minimum

We can now use the derivative in one of its most powerful applications for management, life sciences, and social sciences—that of finding the absolute maximum and absolute minimum of a given function. All of the necessary techniques have now been developed, so the examples in this section illustrate the wide variety of ways these ideas can be applied. Absolute maximum and minimum were defined in the last section, and we only need now to be aware that the **absolute maximum** of a continuous function will occur at the critical value giving the largest value of the function, or else it will occur at an endpoint of a closed interval. On the other hand, the **absolute minimum** will occur at the critical value giving the smallest value of the function or at an endpoint. These situations were illustrated in Figure 4.3, which is repeated at the top of page 176 for easy reference.

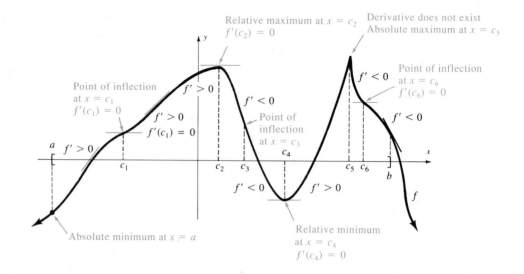

CALCULATOR COMMENT

Graphing calculators and the [trace] key allow you to look at a function on some interval and then to approximate the maximum and minimum values to any reasonable desired degree of accuracy.

Procedure for Finding the Absolute Maximum and Absolute Minimum

Given a continuous function f over an interval $[a, b]$, there exists an absolute maximum and minimum for f on the interval. To find these absolute extremes, carry out the following steps:

1. Find the endpoints and the critical values: $a, c_1, c_2, \ldots, c_n, b$. Remember, the critical values are values that cause the derivative to be either zero or undefined.
2. Evaluate $f(a), f(c_1), f(c_2), \ldots, f(b)$.
3. The **absolute maximum** of f is the largest of the values found in step 2.
4. The **absolute minimum** of f is the smallest of the values found in step 2.

EXAMPLE 1 Find the absolute maximum and minimum for

$$f(x) = 3x^5 - 50x^3 + 135x + 20$$

on $[-2, 4]$.

Solution   **1.** $f'(x) = 15x^4 - 150x^2 + 135$
2. Critical values:

$$15x^4 - 150x^2 + 135 = 0$$
$$x^4 - 10x^2 + 9 = 0$$
$$(x^2 - 9)(x^2 - 1) = 0$$
$$(x - 3)(x + 3)(x - 1)(x + 1) = 0$$
$$x = 3, -3, 1, -1$$

3. List: $-2, \underbrace{-1, 1, 3,}_{} 4$ ← Endpoints

Critical values
$x = -3$ is not in
the interval $[-2, 4]$,
so do not test $x = -3$.

4. Test each of these values:

$$f(-2) = 3(-2)^5 - 50(-2)^3 + 135(-2) + 20 = 54$$
$$f(-1) = 3(-1)^5 - 50(-1)^3 + 135(-1) + 20 = -68$$
$$f(1) = 3(1)^5 - 50(1)^3 + 135(1) + 20 = 108$$
$$f(3) = 3(3)^5 - 50(3)^3 + 135(3) + 20 = -196$$
$$f(4) = 3(4)^5 - 50(4)^3 + 135(4) + 20 = 432$$

CALCULATOR COMMENT

When graphing a function on a calculator, the most difficult part is determining the correct domain and range. For example, on a TI81 if you input the function in Example 1 using the $\boxed{Y=}$ key and then graph using the standard domain and range you obtain:

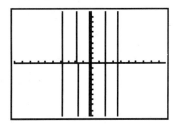

Now, use the $\boxed{\text{RANGE}}$ key to set Xmin $= -2$; Xmax $= 10$; then use the $\boxed{\text{TRACE}}$ key to find X $= 4.0631579$ Y $= 536.84$ and X $= 2.963158$ Y $= -194.1321$ so that you use the $\boxed{\text{RANGE}}$ key to set Ymin $= -200$;

(continued)

Ymax = 550. The GRAPH now shows:

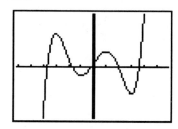

Now the TRACE gives an approximation for the extreme values: Min: X = 2.9894737, Y = −195.9604 and the Max: X = 4 Y = 432

As noted, the real value of finding absolute maximum and minimum values is in the applications to which these ideas can be applied. The remainder of this section concerns such applications.

EXAMPLE 2 Westel Corporation manufactures telephones and has developed a new cellular phone. Production analysis shows that its price must not be less than $50; if x units are sold, then the demand is given by the formula

$$p(x) = 150 - x$$

The total cost of producing x units is given by the formula

$$C(x) = 2{,}500 + 30x$$

Find the maximum profit, and determine the price that should be charged to maximize the profit.

Solution The total profit, $P(x)$, is found by

$P(x) =$ TOTAL REVENUE − TOTAL COST

$= $ (number of items)(price per item) − cost

$= x \cdot p(x) - C(x)$

$= x(150 - x) - (2{,}500 + 30x)$

$= 150x - x^2 - 2{,}500 - 30x$

$= -x^2 + 120x - 2{,}500$

Find the absolute maximum value of P:

1. Critical points: $P'(x) = -2x + 120$ and $-2x + 120 = 0$ when $x = 60$. There are no values for which the derivative is not defined, so there is one critical value: $x = 60$.
2. Endpoints: Given $p(x) \geq 50$, we see that

$$150 - x \geq 50$$
$$100 \geq x$$
$$x \leq 100$$

so one endpoint is 100. The other endpoint is merely implied, namely, 0 phones, so that the domain is [0, 100].
3. Find the absolute maximum of P by checking $x = 0, 60$, and 100:

$$P(0) = -2{,}500$$
$$P(60) = -(60)^2 + 120(60) - 2{,}500 = 1{,}100$$
$$P(100) = -(100)^2 + 120(100) - 2{,}500 = -500$$

4. The absolute maximum of P is 1,100 and occurs when $x = 60$.

The answer to the question asked might not be the absolute maximum or minimum value, and you must be careful to answer the question asked. For example, this problem asks for the price. The price needs to be found for $x = 60$. Use the demand function:

$$p(60) = 150 - 60 = 90$$

The phones should be priced at $90 per unit. ∎

The analysis in Example 2 uses the techniques of this chapter, but we were solving this type of problem in Chapter 1. Two additional techniques will simplify the work shown in Example 2. The first relates second-degree problems to the material of this chapter. It is called the **second-derivative test for absolute maximum and minimum**:

Second-Derivative Test for Absolute Maximum and Minimum

If f is a continuous function with only *one* critical value c and if $f''(c)$ exists and is not zero [that is, if $f''(c) = 0$, then this test fails], and if

$f''(c) > 0$ then there is an *absolute minimum* at $x = c$
$f''(c) < 0$ then there is an *absolute maximum* at $x = c$

This test tells us in Example 2 that since there was only one critical value at $x = 60$ and since $P''(x) = -2$, $x = 60$ *must* give the absolute maximum value of P, so we did not need to go through the evaluation of P at 0, 60, and 100.

The second technique involves assuming that the business in question has known cost and revenue functions $C(x)$ and $R(x)$, as shown in Figure 4.15 on page 180. Profit is positive when $R(x)$ exceeds $C(x)$. The points a and b are the break-even points.

The maximum profit occurs at a critical point c_1 of $P(x)$. Assume that $P'(x)$ exists for all x in some interval, usually $[0, \infty)$ and that the critical point c_1 occurs at some x in this interval, so that

$$P'(x) = 0$$

Then

$$P(x) = R(x) - C(x)$$
$$P'(x) = R'(x) - C'(x)$$
$$0 = R'(x) - C'(x)$$
$$R'(x) = C'(x)$$

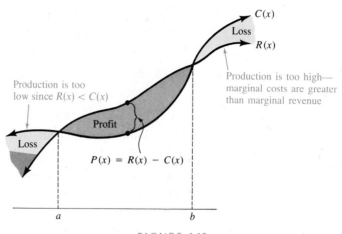

FIGURE 4.15
Maximum value of a profit function

Maximum Profit

> The maximum profit is achieved when the marginal revenue and marginal cost are equal.

Turning, one final time, to Example 2, we see that

$$R(x) = x(150 - x) = 150x - x^2 \quad \text{and} \quad C(x) = 2{,}500 + 30x$$

so

$$R'(x) = 150 - 2x \qquad C'(x) = 30$$

Thus, the maximum profit is found when

$$150 - 2x = 30$$
$$120 = 2x$$
$$60 = x$$

It is well known in economics that the minimal average cost of a product occurs when the average cost is equal to the marginal cost. Although this principle will not be proved here, it is illustrated in Example 3.

EXAMPLE 3 A product has a total cost function given by

$$C(x) = .125x^2 + 20{,}000$$

where x is the number of units produced. Show that the minimal average cost occurs when the average cost is equal to the marginal cost.

Solution First, find the average cost: $\bar{C}(x) = \dfrac{C(x)}{x} = .125x + 20{,}000x^{-1} \qquad x > 0$

Next, find the minimal average cost: $\bar{C}'(x) = .125 - 20{,}000x^{-2}$
Critical values:

$$\bar{C}'(x) = 0$$
$$.125 - 20{,}000x^{-2} = 0$$
$$.125 = 20{,}000x^{-2}$$
$$.125x^2 = 20{,}000$$
$$x^2 = 160{,}000$$
$$x = 400,\ -400$$
$$\uparrow$$
Reject since $x > 0$.

There is only one critical value, $c = 400$ (see Figure 4.16). Use the second-derivative test.

$$\bar{C}''(x) = 40{,}000x^{-3}$$

and $\bar{C}''(400) > 0$, so by the second-derivative test for absolute maximum and minimum, the absolute minimum occurs at $x = 400$: $\bar{C}(400) = .125x + 20{,}000(400)^{-1} = 100$.

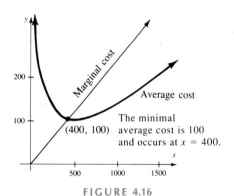

FIGURE 4.16
The minimal average cost of a product occurs when the average cost is equal to the marginal cost.

Now, find the marginal cost: $C'(x) = .25x$. Find the value for which the average cost is equal to the marginal cost:

$$\dfrac{C(x)}{x} = C'(x)$$

Then

$$125x + 20{,}000x^{-1} = .25x$$
$$20{,}000x^{-1} - .125x = 0$$
$$20{,}000 - .125x^2 = 0 \quad \text{Multiply both sides by } x \ (x > 0).$$
$$160{,}000 - x^2 = 0 \quad \text{Divide both sides by .125.}$$
$$(400 - x)(400 + x) = 0$$
$$x = 400, \ -400$$
$$\uparrow$$
$$\text{Reject since } x > 0$$

EXAMPLE 4 As more and more industrial areas are constructed, there is a growing need for standards ensuring the control of the pollutants released into the air. Suppose that the air pollution at a particular location is based on the distance from the source of the pollution according to the following principle:

For distances greater than or equal to 1 mile, the concentration of particulate matter (in parts per million, ppm) decreases as the reciprocal of the distance from the source.

This means that if you live 3 miles from a plant emitting 60 ppm, the pollution at your home is $\frac{60}{3} = 20$ ppm. On the other hand, if you live 10 miles from the plant, the pollution at your home is $\frac{60}{10} = 6$ ppm.

Suppose that two plants 10 miles apart are releasing 60 and 240 ppm, respectively, and you want to know the location between them at which the pollution is a minimum.

Solution

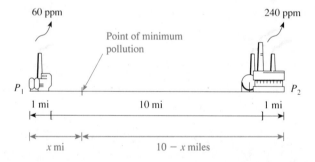

Let x be the distance from plant P_1. The domain is $[1, 9]$ since the given formula implies that you cannot be closer than 1 mile to either plant.

$$\text{Pollution from } P_1 = \frac{60}{x}$$

$$\text{Pollution from } P_2 = \frac{240}{10 - x}$$

The total pollution at any point is the sum of the pollution from the plants:

$$P(x) = \frac{60}{x} + \frac{240}{10 - x}$$

4.4 ABSOLUTE MAXIMUM AND MINIMUM

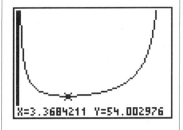

CALCULATOR COMMENT

Graph $\boxed{Y=}$
$60/X + 240/(10-X)$
$\boxed{\text{RANGE}}$
Xmin = 0
Xmax = 10
Ymin = 0
Ymax = 300

X = 3.3684211 Y = 54.002976

$\boxed{\text{TRACE}}$ shows:
X = 3.3684211 Y = 54.002976

Find the absolute minimum value of P on $[1, 9]$:

1. Critical points:
   a. Values for which $P'(x) = 0$:

   $$P(x) = 60x^{-1} + 240(10-x)^{-1}$$
   $$P'(x) = -60x^{-2} - 240(10-x)^{-2}(-1)$$
   $$ = -60x^{-2} + 240(10-x)^{-2}$$

   Set $P'(x) = 0$ and solve:

   $$\frac{-60}{x^2} + \frac{240}{(10-x)^2} = 0 \quad \text{Multiply by } x^2(10-x)^2.$$
   $$-60(10-x)^2 + 240x^2 = 0 \quad \text{Divide by 60.}$$
   $$-(100 - 20x + x^2) + 4x^2 = 0$$
   $$3x^2 + 20x - 100 = 0$$
   $$(3x - 10)(x + 10) = 0$$
   $$x = \frac{10}{3}, -10$$

   Reject this value since it is not in the domain.

   b. Values for which $P'(x)$ do not exist: $x = 0$ and $x = 10$.
      Reject these since they are not in the domain $[1, 9]$.
2. Endpoints: $x = 1$, $x = 9$
3. Find the absolute minimum of P by checking $x = 1$, $\frac{10}{3}$, and 9:

   $$P(1) = \frac{60}{1} + \frac{240}{9} \approx 87 \text{ ppm}$$
   $$P\left(\frac{10}{3}\right) = \frac{60}{\frac{10}{3}} + \frac{240}{\frac{20}{3}} = 54 \text{ ppm}$$
   $$P(9) = \frac{60}{9} + \frac{240}{1} \approx 247 \text{ ppm}$$

The minimum pollution is found at $3\frac{1}{3}$ miles from plant P_1. ∎

EXAMPLE 5 The voting patterns of a geographical region show that the percent, P, of voters in a national election varies according to age, x, and fits the model predicted by the formula.

$$P(x) = .002x^3 - .195x^2 + 6x \qquad 18 \leq x \leq 50$$

Graph P and discuss its possible meaning, and then find the absolute maximum and minimum percentage of the population voting in a national election.

Solution $P'(x) = .006x^2 - .390x + 6$

Set $P'(x) = 0$ and solve to find the critical values:

$$.006x^2 - .390x + 6 = 0$$
$$x^2 - 65x + 1{,}000 = 0 \quad \text{Divide by .006.}$$
$$(x - 25)(x - 40) = 0$$

$P''(x) = .012x - .390$, so

$P''(25) = .012(25) - .390 < 0$ Relative maximum at $x = 25$
$P''(40) = .012(40) - .390 > 0$ Relative minimum at $x = 40$

Find some values of P (which are needed both for the graph and for the absolute maximum and absolute minimum):

$P(25) = 59.375$
$P(40) = 56$
$P(18) = 56.484$
$P(50) = 62.5$

From ages 18 to 25 the percentage of the population voting is increasing (derivative positive), and from 25 to 40 it is decreasing, at which time it begins to increase again. From the graph in Figure 4.17 and the calculations we see that the absolute maximum percentage of the population voting is at age 50, the endpoint of the interval [18, 50]. ∎

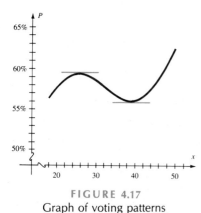

FIGURE 4.17
Graph of voting patterns

EXAMPLE 6 A shipping box is to be constructed from a piece of corrugated cardboard 17 inches long and 11 inches wide. A square is to be cut out of each corner so that the sides can be folded up to form the box. What is the maximum volume (to the nearest in.³) for the finished box?

Solution Let x be the size of the cutout square. The volume, V, can be calculated as a function of x:

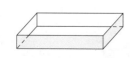

$$V(x) = lwh = (17 - 2x)(11 - 2x)x$$
$$= (187 - 56x + 4x^2)x$$
$$= 4x^3 - 56x^2 + 187x$$

The domain is [0, 5.5] since x cannot be negative, nor can $11 - 2x$. Take the derivative and find the critical values:

$$V'(x) = 12x^2 - 112x + 187$$

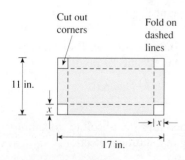

$$V'(x) = 0 \quad \text{when} \quad x = \frac{112 \pm \sqrt{(-112)^2 - 4(12)(187)}}{2(12)}$$
$$\approx 7.1555308, \; 2.1778026$$

This first value is not in the domain, so $x \approx 2.18$ inches. The length of the box is $17 - 2x \approx 12.644395$ and the width is $11 - 2x \approx 6.6443948$. The volume is

$$V \approx 2.1778026 \times 12.644395 \times 6.6443948 \approx 182.96668$$

In practical measurements, you should cut $2\frac{1}{8}$ in. from each corner to form a box with dimensions

$$12\frac{3}{4} \text{ in. by } 6\frac{3}{4} \text{ in. by } 2\frac{1}{8} \text{ in.}$$

with a volume of 183 in.^{3}. ∎

4.4

Problem Set

Find the absolute maximum and minimum for the functions in Problems 1–20 and also give the x-values for those extrema.

1. $f(x) = x^3 - 3x^2$ on $[-1, 4]$
2. $f(x) = x^3 - 12x$ on $[0, 4]$
3. $f(x) = 3x^5 - 50x^3 + 135x + 750$ on $[-5, 0]$
4. $f(x) = 3x^5 - 50x^3 + 135x + 750$ on $[-3, 3]$
5. $g(x) = x^3 - 2x^2 - 15x + 42$ on $[0, 5]$
6. $g(x) = x^3 + 5x^2 + 3x + 20$ on $[-5, 0]$
7. $h(x) = 3x^5 - 25x^3 + 60x - 200$ on $[0, 3]$
8. $h(x) = 3x^5 - 25x^3 + 60x - 200$ on $[-3, 3]$
9. $f(x) = 1 + x^{1/3}$ on $[-1, 8]$
10. $g(x) = 1 - x^{2/3}$ on $[-1, 8]$
11. $f(x) = \sqrt{x} + x$ on $[0, 4]$
12. $g(x) = x - \sqrt{x}$ on $[0, 9]$
13. $f(x) = (x + 1)^2(x - 3)$ on $[0, 5]$
14. $g(x) = (x - 1)(x + 4)^3$ on $[-5, 1]$
15. $f(x) = 2(x - 3)^{-1}$ on $[0, 5]$
16. $f(x) = 2x(x - 1)^{-1}$ on $[0, 3]$
17. $f(x) = |x|$ on $[-3, 3]$
18. $f(x) = |x + 2|$ on $[-3, 3]$
19. $f(x) = \dfrac{x - 1}{(x + 1)^2}$ on $[0, 3]$
20. $f(x) = \dfrac{x}{(x - 1)^2}$ on $[-2, 0]$

APPLICATIONS

In Problems 21–26 revenue and cost functions are given. Find the number of items, x, which produces a maximum profit. Also find the maximum profit.

21. $R(x) = 35x - .03x^2$; $C(x) = 6{,}000 + 5x$
22. $R(x) = 2{,}000x - .05x^2$; $C(x) = 120{,}000 + 150x$
23. $R(x) = 5{,}000x - .1x^2$; $C(x) = 80{,}000 + 200x$
24. $R(x) = 2x + \dfrac{x}{x - 16}$; $C(x) = \dfrac{2}{3} + x$ on $[0, 15]$
25. $R(x) = 5x + \dfrac{x}{x - 64}$; $C(x) = 10 + x$ on $[0, 62]$
26. $R(x) = 6x + \dfrac{x}{x - 245}$; $C(x) = 50 + x$ on $[0, 240]$
27. Rework Example 2 except that the price must not be less than $100.
28. Rework Example 3 for $C(x) = .25x^2 + 12{,}100$.
29. Rework Example 4, except assume that the plants are 20 miles apart.
30. Rework Example 5, except assume that the voting formula is

    $$P(x) = .002x^3 - .207x^2 + .48x$$

31. Rework Example 6 for a piece of cardboard 10 inches long and 5 inches wide.
32. Show the result of Problem 28 graphically.
33. A manufacturer has the following costs in producing x items ($0 \le x \le 200$): unit cost, \$25; fixed costs, \$500; repairs, $x^2/20$ dollars.
    a. What is the average cost $\bar{C}(x)$ per item if x items are produced?
    b. Find the critical values for $\bar{C}(x)$. Where is this average increasing and where is it decreasing?
    c. What is the minimal average cost?
34. A consulting firm determines that the demand (price equation) for a certain product is $p = 1{,}000 - 10x$ dollars, where x is the number of items produced. The cost function is $C(x) = 5{,}000 + 500x$. Find the maximum profit and determine the price that should be charged to make the maximum profit. [*Note:* Price should be nonnegative.]

35. The price equation for a new product is $50 - x$ dollars, where x is the number of items produced. The cost function is $C(x) = 100 + 20x$. Find the maximum profit and determine the price that should be charged to make the maximum profit. [*Note:* Price must be nonnegative.]

36. Suppose a tax of $2 per item is imposed on the product described in Problem 35. Find the maximum profit and determine the price that should be charged to make the maximum profit.

37. Suppose a tax of $4 per item is imposed on the product described in Problem 35. Find the maximum profit and determine the price that should be charged to make the maximum profit.

38. Suppose that the air pollution at a particular location is based on the distance from the source of the pollution according to the principle that for distances greater than or equal to one mile, the concentration of particulate matter (in parts per million, ppm) decreases as the reciprocal of the distance from the source. Suppose that two plants 50 miles apart are releasing 180 ppm and 300 ppm, respectively. What is the location between the plants where pollution is a minimum?

39. A retailer has determined that the cost C for ordering and storing x units of a product is

$$C(x) = 5x + \frac{50{,}000}{x} \text{ on } [1, 200]$$

Find the order size that will minimize the cost if the delivery truck can bring a maximum of 125 per order.

40. An assembly line worker can memorize $p\%$ of a given list of tasks in x continuous hours according to the formula

$$p(x) = 95x - 25x^2 \text{ on } [0, 3]$$

How long should it take a worker to memorize the maximum percentage? What is the maximum percentage?

41. A farmer has 1,000 feet of fence and wishes to enclose a rectangular plot of land. The land borders a river and no fence is required on that side. (Refer to the figure at the top of the next column.) What should the length of the side parallel to the river be in order to include the largest possible area?

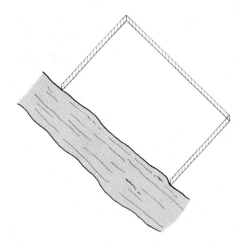

42. A fence must be built around a rectangular area of 1,600 ft². The fence along three sides is to be made of a material that costs $12 per foot. The material for the fourth side costs $4 per foot. Find the dimensions (rounded to the nearest foot) of the rectangle that would be the least expensive to build, provided that no side is less than 10 ft long.

43. A tour agency is booking a tour and has 100 people signed up. The price of a ticket is $2,000 per person. The agency has booked a plane seating 150 people at a cost of $125,000. Additional costs to the agency are incidental fees of $500 per person. For each $5 that the price is lowered, a new person will sign up. How much should the price be lowered to maximize the profit for the tour agency?

44. Suppose that the tour agency described in Problem 43 is able to obtain a booking for a plane that will seat 225 people at the same price for the charter as the first airline. Answer the question under these conditions.

45. A viticulturist estimates that if 50 grapevines are planted per acre, each grapevine will produce 150 pounds of grapes. For each additional grapevine planted per acre (up to 20), the average yield per vine drops by 2 pounds. How many grapevines should be planted to maximize the yield per acre?

*4.5

Review

The material of this chapter is reviewed in the following list of objectives. After each objective there are some practice questions. For a sample test select the first question of each set and check your answers. The second question for each objective has no answer given. If you are having trouble with a particular type of problem, look back at the indicated section in the text. When you are finished reviewing these objectives, a sample examination is given at the end of this section.

* Optional section.

[4.1]
Objective 4.1: *Graph a parabola by using calculus to find its vertex and how it opens.*
1. $y = 2x^2 - 8x + 5$
2. $x^2 - 6x + 3y - 4 = 0$
3. $x^2 + 4x + 2y + 3 = 0$
4. $2x^2 - 16x - 3y + 23 = 0$

Objective 4.2: *Graph a curve by finding the critical values, finding when it is increasing or decreasing, and by deciding upon the concavity.*
5. $f(x) = x^3 - 27x$
6. $f(x) = 2x^3 - 3x^2 - 36x$
7. $g(x) = x^3 + 3x^2 - 9x + 5$
8. $g(x) = 4x^3 + 5x^2 + 2x + 3$

[4.2]
Objective 4.3: *Find successive derivatives of a given function.*
9. Find all derivatives of $y = 3x^4 - x^3 + 5x^2 + 79$.
10. Find all derivatives of $y = 1 - 3x^3 + x^5$.
11. Find the first four derivatives of $f(x) = \dfrac{1}{\sqrt{x}}$.
12. Find the first four derivatives of $g(x) = 3x^5 - 3x^{-1}$.

[4.3]
Objective 4.4: *Find all relative maximums and minimums for a given function.*
13. $f(x) = x^3 - x^2 - 5x$
14. $g(x) = x - \dfrac{2}{x}$
15. $t(x) = 4x^2 - x^4$
16. $s(x) = x^3 - 2x^2 - 4x - 8$

Objective 4.5: *Graph a given function.*
17. $f(x) = 4x^3 - 30x^2 + 48x$
18. $g(x) = 3x^5 - 50x^3 + 135x + 12$
19. $y = x^3 - 1$
20. $y = \dfrac{x-2}{x+1}$

[4.4]
Objective 4.6: *Find the absolute maximum and minimum for a function defined on a closed interval.*
21. $y = 3x^5 - 85x^3 + 240x$ on $[-2, 5]$
22. $y = 2x + \dfrac{32}{x}$ on $[1, 9]$
23. $y = (x-1)(x+3)^3$ on $[-5, 5]$
24. $y = 1 - x^{3/2}$ on $[0, 4]$

Objective 4.7: *If you are given revenue and cost functions, find the number of items to produce a maximum profit.*
25. $R(x) = 50x - .05x^2$; $C(x) = 5{,}000 + 2x$
26. $R(x) = 10x - 3x^2 + .01x^3$; $C(x) = 15{,}000 + x$; $x \geq 100$
27. $R(x) = 4x + .345x^2 - .005x^3$; $C(x) = 4{,}000 + x$
28. $R(x) = 10x + \dfrac{x}{(x-49)}$; $C(x) = 3{,}000 + x$ on $[0, 48]$

Objective 4.8: *Solve applied problems based on the preceding objectives.*
29. *Marginal revenue.* A manufacturer has determined that the demand is given by the formula
$$p(x) = 25 - \left(\dfrac{x}{500}\right)^2$$
Is the marginal revenue increasing or decreasing when $x = 100$ items?

30. *Cost–benefit model.* The cost–benefit model relating the cost, C, of removing $p\%$ of the pollutants from the atmosphere is
$$C(p) = \dfrac{50{,}000p}{110 - p}$$
where the domain of p is $[0, 100]$. Find the minimum cost if the EPA requires 85% of the pollutants to be removed.

31. *Property management.* A property management company manages 100 apartments renting for $500 with all the apartments rented. For each $50 per month increase in rent there will be two vacancies with no possibility of filling them. What rent per apartment will maximize the monthly revenue?

32. A rectangular cardboard poster is to have 208 sq in. for printed matter. It is to have a 3-inch margin at the top and a 2-inch margin at the sides and bottom.

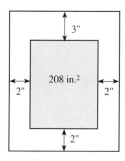

Find the length and width (to the nearest inch) of the poster so that the amount of cardboard used is minimized.

SAMPLE TEST

The following sample test (45 minutes) is intended to review the main ideas of this chapter.

1. **a.** Find all nonzero derivatives of $y = 1 - 3x^3 + x^5$.
   b. Find the first four derivatives of $g(x) = 3x^5 - 3x^{-1}$.

Graph the functions in Problems 2–5.

2. $x^2 - 6x + 3y - 4 = 0$
3. $g(x) = x^3 + 3x^2 - 9x + 5$
4. $y = x + \dfrac{4}{x}$
5. $y = \dfrac{x-2}{x+1}$

Find all the relative maximums and minimums for the functions in Problems 6 and 7.

6. $f(x) = 8x^2 - 2x^4$
7. $g(x) = x + 8/x$
8. Find the absolute maximum and minimum for
$$y = (x-1)(x+3)^3 \text{ on } [-5, 5]$$

9. A manufacturer has determined that the price of selling x items is given by the formula
$$p(x) = 50 - (x/1{,}000)^2$$
Is the marginal revenue increasing or decreasing when $x = .100$ items?

10. A manufacturer needs to package a product in a closed rectangular box. If the sides of the box cost $4.00 per square foot and the base and top (which are squares) cost $8.00 per square foot, what are the dimensions of the box with greatest volume that can be constructed for $96?

CUMULATIVE REVIEW
for chapters 2–4

Evaluate the limits in Problems 1–5.

1. Find $\lim\limits_{x \to 3} f(x)$ of the accompanying graph.

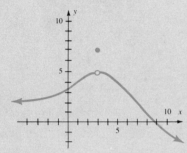

2. Find $\lim\limits_{x \to -4} \dfrac{x+8}{x-4}$.

3. Find $\lim\limits_{x \to 5} \dfrac{2x^2 - 7x - 15}{x - 5}$.

4. Find $\lim\limits_{x \to 2} \dfrac{3x^2 - 5x - 2}{x - 2}$.

5. Find $\lim\limits_{x \to \infty} \dfrac{(3x+1)(5x-2)}{x^2}$.

Find the points of discontinuity over the given domain for the functions in Problems 6–8.

6. $f(x) = \dfrac{x^2 - 15x + 56}{x - 8}$

7. $f(x) = \begin{cases} \dfrac{x^2 - 15x + 56}{x - 8} & \text{if } 0 \le x \le 10,\ x \ne 8 \\ 7 & \text{if } x = 8 \end{cases}$

8. $f(x) = \begin{cases} \dfrac{x^2 - 15x + 56}{x - 8} & \text{if } 0 \le x \le 10,\ x \ne 8 \\ 1 & \text{if } x = 8 \end{cases}$

9. Find the instantaneous rate of change for $y = 4 - \dfrac{1}{x}$ at $x = 1$.

10. State the definition of derivative.

Find the derivative of the functions in Problems 11–16.

11. $y = 25 - 12x^2 - 5x^3$
12. $y = 5x\sqrt{9 - x}$
13. $y = (2 - 5x)^5$
14. $y = \dfrac{(2x + 5)^8}{x^2 - 5}$
15. $3x^5 y^4 = 10$
16. $2x^2 + xy + 3y^2 = 100$

17. Find the equation of the line tangent to $f(x) = 2x - 6x^3$ at the point $(1, -4)$.

18. Graph $2x^2 + 24x - 3y - 3 = 0$ using calculus.

19. Find all derivatives of $y = 2x^4 - x^3 + 3x^2 + 25$.

20. Find all relative maximums and minimums of $f(x) = 4x^2 - x^4$.

21. Find the absolute maximum and minimum of $y = \sqrt{x} - x$ on $[0, 9]$.

22. Find the maximum profit when $R(x) = 20x + .11x^2 - .01x^3$ and $C(x) = 10{,}000 + x$.

APPLICATIONS

23. A manufacturer has determined that the price of an item is determined by the number of items sold according to the following price formula for x items sold:

    $$p(x) = 25 - \dfrac{x}{500}$$

    Is the marginal revenue increasing or decreasing when $x = 100$ items?

24. Find the marginal cost for the function $C(x) = 20x^3(5x - 100)^2$.

25. An artist's print is 20 in.² and is to be framed so that it has 2 in. of matting on each side, and 4 in. on the top and bottom. What are the dimensions of the frame enclosing the smallest area possible? (That is, the frame should enclose the least area; around your answer to the nearest inch.)

5
Exponential and Logarithmic Functions

5.1 Exponential Functions
5.2 Logarithmic Functions
5.3 Logarithmic and Exponential Equations
5.4 Derivatives of Logarithmic and Exponential Functions
5.5 Chapter 5 Review
Chapter Objectives
Sample Test

CHAPTER OVERVIEW
There are many applications involving growth or decay that cannot be modeled with only the algebraic functions considered thus far. Two additional functions—exponential and logarithmic—are now added to our repertoire. Each is defined and graphed, and we learn how to find their derivatives.

PREVIEW
We are first introduced to exponential functions and then to logarithmic functions. Next, we solve equations involving these functions, and, finally, we learn how to differentiate them.

PERSPECTIVE
Persons needing calculus for management (business, economics, finance, or investments), life sciences (biology, ecology, health, or medicine), or social sciences (demography, political science, population, psychology, society, or sociology) need to understand how things grow (as in money or populations), and how things decay (as in forgetting or inflation). In order to understand these concepts, two of the most useful functions are introduced in this chapter: exponential and logarithmic functions. The material of this chapter will be particularly important in Section 9.2 when solving differential equations.

MODELING
APPLICATION

World Running Records

World record running rates for mile run (meters/minute)

World records for footraces at all distances have improved consistently ever since records have been kept. For example, if we consider the mile run, the magic 4-minute mile was broken in 1954 by Roger Bannister of the United Kingdom. Since then, the record has decreased steadily to the present time when the world record is under 3 minutes 47 seconds!

After you have finished this chapter, write a paper that develops a mathematical model to answer the question, Will a 3-minute mile ever be run? Is there an "ultimate" time for a mile run, and, if so, what should we expect that time to be? For general guidelines about writing this essay, see the commentary for Modeling Application 1 on page 51.

*This modeling application is from Joseph Brown, "Predicting Future Improvements in Footracing," MATYC Journal, Fall 1980, pp. 173–179.

APPLICATIONS

Management (*Business, Economics, Finance, and Investments*)
Finding the future value in an interest-bearing account
 (5.1, Problems 40–45)
Number of units sold as a function of advertising expenditures
 (5.2, Problems 40, 41; 5.5, Test Problem 18)
Determining the time for an interest-bearing account
 (5.3, Problems 47–52; 5.5, Problem 46)
Monthly payment on an installment loan
 (5.3, Problems 55–56)
Maximum value of a demand equation
 (5.4, Problem 41)
Marginal cost (5.4, Problem 42)
Rate at which a money supply is increasing
 (5.4, Problem 45)
Sales growth pattern (5.4, Problem 48)
Price–demand equation
 (5.4, Problem 49; 5.5, Test Problem 19)

Management (*continued*)
Price–supply equation (5.4, Problem 50)
Amount spent on advertising
 (5.5, Problem 48)
Equilibrium for supply/demand
 (5.5, Problem 51)

Life Sciences (*Biology, Ecology, Health, and Medicine*)
Determining the pH of a substance
 (5.2, Problems 46, 48)
Petroleum reserves and time until they are exhausted (5.3, Problems 59–60)
Spread of a disease in a town
 (5.4, Problems 43–44)
Blood pressure of the aorta artery
 (5.4, Problem 46)
Amount of drug present in the body after being administered
 (5.4, Problem 47)
Half-life (5.5, Problem 47)

Social Sciences (*Demography, Political science, Population, Psychology, Society, and Sociology*)
Learning curve for a typing test
 (5.2, Problems 42–45)
World growth rate (5.3, Problems 44–46)
Forgetting curve in psychology
 (5.3, Problems 53–54)
Dating an artifact in archaeology
 (5.3, Problems 57–58; 5.5, Problem 47)
Percentage who will hear a presidential announcement
 (5.5, Problem 52; Test Problem 20)

General Interest
Insurance policy (5.1, Problem 50)
Magnitude of an earthquake on the Richter scale (5.2, Problems 47, 49)
Inflation (5.5, Problem 45)

Modeling Application—
World Running Records

5.1
Exponential Functions

Linear, quadratic, polynomial, and rational functions are examples of what are all called **algebraic functions**. An algebraic function is a function that can be expressed in terms of algebraic operations alone. If a function is not algebraic, it is called a **transcendental function**. In this chapter two examples of transcendental functions— *exponential* and *logarithmic* functions—are discussed.

Definition of Exponential Function

Exponential Function

The function f is an **exponential function** if

$$f(x) = b^x$$

where b is a positive constant other than 1 and x is any real number. The number x is called the **exponent** and b is called the **base**.

Laws of Exponents

Recall that if n is a natural number, then

$$b^n = \underbrace{b \cdot b \cdot b \cdot \cdots \cdot b}_{n \text{ factors}}$$

Furthermore, if $b \neq 0$, then $b^0 = 1$, $b^{-n} = 1/b^n$, and $b^{1/n} = \sqrt[n]{b}$ (for $b \geq 0$ when n is even). Also, $b^{m/n} = (b^{1/n})^m$. These definitions are used in conjunction with five **laws of exponents**:

Laws of Exponents

For a and b positive real numbers and rational numbers p and q, there are five rules that govern the use of exponents. The form 0^0 and division by zero, as well as even roots of negative numbers, are excluded whenever they occur.

First law: $b^p \cdot b^q = b^{p+q}$

Second law: $\dfrac{b^p}{b^q} = b^{p-q}$

Third law: $(b^p)^q = b^{pq}$

Fourth law: $(ab)^p = a^p b^p$

Fifth law: $\left(\dfrac{a}{b}\right)^p = \dfrac{a^p}{b^p}$

EXAMPLE 1 Simplify the following expressions.

a. $16^{1/2} = (4^2)^{1/2}$
$= 4^1$
$= 4$

b. $-16^{1/2} = -(4^2)^{1/2}$
$= -4$

c. $(-16)^{1/2}$ is not defined since b must be greater than or equal to zero for square roots ($x = \frac{1}{2}$ is a square root)

d. $343^{2/3} = (7^3)^{2/3}$
$= 7^2$
$= 49$

e. $25^{-3/2} = (5^2)^{-3/2}$
$= 5^{-3}$
$= 1/5^3$
$= 1/125$ ∎

EXAMPLE 2 Use the ordinary rules of algebra to simplify the following expressions.

a. $x(x^{2/3} + x^{1/2}) = x^1 x^{2/3} + x^1 x^{1/2}$
$= x^{1+2/3} + x^{1+1/2}$
$= x^{5/3} + x^{3/2}$

b. $(x^{1/2} + y^{1/2})(x^{1/2} - y^{1/2}) = x^{1/2}x^{1/2} - x^{1/2}y^{1/2} + x^{1/2}y^{1/2} - y^{1/2}y^{1/2}$
$= x - y$ ∎

Graphing Exponential Functions

The next step is to enlarge the domain of x to include all real numbers. This is done with the help of the following property:

Squeeze Theorem for Exponents

> Suppose b is a real number greater than 1. Then for any real number x there is a unique real number b^x. Moreover, if p and q are any two positive, rational numbers such that $p < x < q$, then
>
> $b^p < b^x < b^q$

The squeeze theorem gives meaning to expressions such as $2^{\sqrt{3}}$. Since

$$1.732 < \sqrt{3} < 1.733$$

the squeeze theorem says that

$$2^{1.732} < 2^{\sqrt{3}} < 2^{1.733}$$

This means that even though only rational exponents were previously defined, we can extend the definition to any real number exponent by using the squeeze theorem to give us any desired degree of accuracy. The case where $0 < b < 1$ is considered in Problem Set 5.1.

We can now consider an exponential function since this extended definition allows exponents to be any real number.

EXAMPLE 3 Graph the function $f(x) = 2^x$.

Solution Begin by plotting points, as shown in the table and Figure 5.1.

x	$y = f(x) = 2^x$
-3	$2^{-3} = \frac{1}{8}$
-2	$2^{-2} = \frac{1}{4}$
-1	$2^{-1} = \frac{1}{2}$
0	$2^0 = 1$
1	$2^1 = 2$
2	$2^2 = 4$
3	$2^3 = 8$

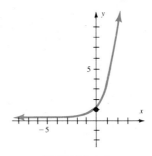

FIGURE 5.1
Graph of $y = 2^x$

Connect these points with a smooth curve to obtain the graph shown in Figure 5.1. Note that the squeeze theorem allows us to draw this as a smooth curve.

EXAMPLE 4 Sketch $f(x) = (\frac{1}{2})^x$.

Solution Notice that $y = (\frac{1}{2})^x = (2^{-1})^x$ or 2^{-x}.

x	$y = f(x) = (\frac{1}{2})^x$
-3	$(2^{-1})^{-3} = 8$
-2	$(2^{-1})^{-2} = 4$
-1	$(2^{-1})^{-1} = 2$
0	$(2^{-1})^0 = 1$
1	$(\frac{1}{2})^1 = \frac{1}{2}$
2	$(\frac{1}{2})^2 = \frac{1}{4}$
3	$(\frac{1}{2})^3 = \frac{1}{8}$

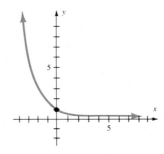

FIGURE 5.2
Graph of $f(x) = (\frac{1}{2})^x$

Connect the points shown in the table with a smooth graph, as shown in Figure 5.2.

The exponential function can be used as a model for certain types of growth or decay. If $b > 1$ (as in Example 3), then the exponential function is called a **growth function**, and if $0 < b < 1$ (as in Example 4), then it is called a **decay function**.

Interest

One of the most common examples of growth is the way money grows in a bank account. This is a very important concept not only in business but in personal money management. If a sum of money, called the **principal** or **present value**, is denoted by P and invested at an annual interest rate of r for t years, then the *future amount* of money present is denoted by A and is found by

$$A = P + I$$

where I denotes the amount of interest. **Interest** is an amount of money paid for the use of another's money. **Simple interest** is found by multiplication:

$$I = Prt$$

For example, $1,000 invested for 3 years at 15% simple interest generates interest of

$$I = \$1,000(.15)(3) = \$450$$

so the future amount in 3 years is

$$A = \$1,000 + \$450 = \$1,450$$

Most businesses, however, pay interest on the interest as well as on the principal, and when this is done, it is called **compound interest**. For example, $1,000 invested at 15% annual interest compounded annually for 3 years can be found as follows:

First year: $\quad A = P + I$
$\quad\quad\quad\quad\quad = P \cdot 1 + Pr \quad\quad$ $I = Prt$ and $t = 1$
$\quad\quad\quad\quad\quad = P(1 + r) \quad\quad\quad$ For this example, $A = \$1,000(1 + .15)$
$\quad\quad\quad\quad\quad\quad\quad\quad\quad\quad\quad\quad\quad\quad\quad\quad\quad = \$1,150$

Second year: $\quad A = P(1 + r) + I \quad\quad$ The total amount from the first year becomes the principal for the second year.
$\quad\quad\quad\quad\quad\quad = P(1 + r) \cdot 1 + P(1 + r) \cdot r$
$\quad\quad\quad\quad\quad\quad = P(1 + r)(1 + r)$
$\quad\quad\quad\quad\quad\quad = \mathbf{P(1 + r)^2} \quad\quad$ For this example, $A = \$1,000(1 + .15)^2$
$\quad\quad\quad\quad\quad\quad\quad\quad\quad\quad\quad\quad\quad\quad\quad = \$1,322.50$

Third year: $\quad A = P(1 + r)^2 \cdot 1 + P(1 + r)^2 \cdot r$
$\quad\quad\quad\quad\quad = P(1 + r)^2(1 + r)$
$\quad\quad\quad\quad\quad = P(1 + r)^3 \quad\quad$ For this example, $A = \$1,000(1 + .15)^3$
$\quad\quad\quad\quad\quad\quad\quad\quad\quad\quad\quad\quad\quad = \$1,520.88$

The pattern illustrated in Example 3 can be generalized:

Compound Interest or Future Value Formula

> If a principal of P dollars is invested at an interest rate of i per period for a total of N periods, then the future amount A is given by the formula
>
> $$A = P(1 + i)^N$$

Compound interest is usually stated in terms of an annual interest rate r and a given number of years t. The frequency of compounding (that is, the number of compoundings per year) is denoted by n where

$\quad\quad\quad n = 1 \quad\quad$ if compounded annually
$\quad\quad\quad n = 2 \quad\quad$ if compounded semiannually
$\quad\quad\quad n = 4 \quad\quad$ if compounded quarterly

$n = 12$ if compounded monthly
$n = 360$ if compounded daily (ordinary interest)
$n = 365$ if compounded daily (exact interest)

Therefore, $i = r/n$ and $N = nt$, as illustrated in Examples 5 and 6.

EXAMPLE 5   If $12,000 is invested for 5 years at 18% compounded semiannually, what is the amount present at the end of 5 years?

Solution $P = \$12{,}000;\ r = .18;\ i = \frac{.18}{2} = .09;\ t = 5;\ N = 2(5) = 10$; thus

$$A = P(1 + i)^N$$
$$= \$12{,}000(1 + .09)^{10}$$
$$\approx \$28{,}408.36$$

Use a calculator and round answers involving money to the nearest cent; however, do not round until you are ready to state your *final* answer.   ∎

EXAMPLE 6   Rework Example 5 except now the interest is compounded monthly.

Solution $P = \$12{,}000;\ i = \frac{.18}{12} = .015;\ N = 12(5) = 60$. Thus

$$A = \$12{,}000(1 + .015)^{60}$$
$$\approx \$29{,}318.64$$ ∎

Examples 5 and 6 show how the amount increases as the number of times it is compounded increases. A reasonable extension is to ask what happens if the interest is compounded even more frequently than monthly. Can we compound daily, hourly, every minute, or every split second? The answer is yes; in fact, money can be compounded **continuously**, which means that at every instant the newly accumulated interest is used as part of the principal for the next instant. In order to understand these concepts consider the following contrived example. Suppose $1 is invested at 100% interest for 1 year compounded at different intervals. The compound interest formula for this example is

$$A = \left(1 + \frac{1}{n}\right)^n$$

The calculations of this formula for different values of n are shown in Table 5.1.

TABLE 5.1
Effect of Compound Interest on a $1 Investment

Number of periods	Formula	Amount
Annually, $n = 1$	$(1 + \frac{1}{1})^1$	$2.00
Semiannually, $n = 2$	$(1 + \frac{1}{2})^2$	$2.25
Quarterly, $n = 4$	$(1 + \frac{1}{4})^4$	$2.44
Monthly, $n = 12$	$(1 + \frac{1}{12})^{12}$	$2.61
Daily, $n = 360$	$(1 + \frac{1}{360})^{360}$	$2.715
Hourly, $n = 8{,}640$	$(1 + \frac{1}{8{,}640})^{8{,}640}$	$2.7181

If these calculations are continued for even larger n, you will obtain the following:

$n = 10{,}000$	then, the formula yields	2.718145926
$n = 100{,}000$		2.718268237
$n = 1{,}000{,}000$		2.718280469
$n = 10{,}000{,}000$		2.718281828
$n = 100{,}000{,}000$		2.718281828

The calculator cannot distinguish values of $(1 + \frac{1}{n})^n$ for larger n. These values are approaching a particular number. This number, it turns out, is an irrational number, so it does not have a convenient decimal representation. (That is, its decimal representation does not terminate and does not repeat.) Mathematicians have agreed to denote this number by the symbol e, which is defined as a limit:

The Number e

$$\lim_{n \to \infty} \left(1 + \frac{1}{n}\right)^n = e$$

EXAMPLE 7 Find e, e^2, and e^{-3}.

Solution On a calculator, locate a key labeled e^x.

$$e \approx 2.718281828$$
$$e^2 \approx 7.389056099$$
$$e^{-3} \approx .0497870684$$

For interest *compounded continuously*, the following formula is used:

$$A = Pe^{rt}$$

You can find e, as well as powers of e, by using a calculator.

EXAMPLE 8 Find the future value of the principal in Example 5 if the interest is compounded continuously.

Solution Since $P = \$12{,}000$, $r = .18$, and $t = 5$,

$$A = \$12{,}000 e^{(.18)(5)}$$
$$\approx \$29{,}515.24$$

You should memorize at least the first six digits of e:

$$e \approx 2.71828$$

5.1

Problem Set

Simplify the expressions in Problems 1–18. Eliminate negative exponents from your answers.

1. $25^{1/2}$
2. $-25^{1/2}$
3. $(-25)^{1/2}$
4. $(-27)^{1/3}$
5. $-27^{1/3}$
6. $27^{1/3}$
7. $7^{1/3} \cdot 7^{2/3}$
8. $8^{4/3} \cdot 8^{-1/3}$
9. $1{,}000^{-1/3}$
10. $.001^{-2/3}$
11. $100^{-3/2}$
12. $.01^{-3/2}$
13. $\dfrac{2^3 \cdot 2^{-4}}{2^5 \cdot 2^{-2}}$
14. $\dfrac{3^{-2} \cdot 3^3}{3^5 \cdot 3^{-4}}$
15. $\dfrac{9^{1/2} \cdot 27^{2/3}}{3^2 \cdot 81}$
16. $\dfrac{4^{1/2} \cdot 8^{2/3}}{16 \cdot 4^3}$
17. $2^{-1} + 3^{-1}$
18. $\dfrac{4^{-1} + 9^{-1}}{36^{-1}}$

Evaluate the expressions in Problems 19–24.
19. e^3
20. e^{-2}
21. $e^{.05}$
22. $e^{-.2}$
23. $e^{.045}$
24. $e^{-4.85}$

Simplify the expressions in Problems 25–30.
25. $x^{1/2}(x^{1/2} + x^{1/2})$
26. $x(x^{1/2} + x^{-1/2})$
27. $x^{2/3}(x^{-2/3} + x^{1/3})$
28. $x^{1/4}(x^{3/4} + x^{-1/4})$
29. $(x^{1/2} + y^{1/2})^2$
30. $(x^{1/2} - y^{1/2})^2$

Sketch the graph of each function in Problems 31–39.
31. $y = 3^x$
32. $y = 4^x$
33. $y = (\tfrac{1}{3})^x$
34. $y = e^x$
35. $y = e^{-x}$
36. $y = -e^{-x}$
37. $y = e^{x-2}$
38. $y = 10^{x-1}$
39. $y - 2 = 2^{x+1}$

APPLICATIONS

40. If $1,000 is invested at 12% compounded semiannually, how much money will there be in 10 years?

41. If $1,000 is invested at 16% compounded continuously, how much money will there be in 25 years?

42. If $1,000 is invested at 14% compounded continuously, how much money will there be in 10 years?

43. If $8,500 is invested at 18% compounded monthly, how much money will there be in 4 years?

44. If $3,600 is invested at 15% compounded daily, how much money will there be in 7 years? (Use a 365-day year; this is *exact interest*.)

45. If $9,400 is invested at 14% compounded daily, how much money will there be in 6 months? (Use a 360-day year; this is *ordinary interest*.)

46. Find formulas for present value P for interest compounded n times per year and for interest compounded continuously.

47. If you want to have $10,000 in the bank five years from now and make a deposit compounded semiannually at 11.5%, how much do you need to deposit?

48. Repeat Problem 47 except deposit in a bank paying interest daily. (Assume a 365-day year.)

49. Repeat Problem 47 except deposit in a bank paying continuous interest.

50. Suppose an insurance agent offers you an insurance policy that will pay your beneficiary $25,000 if you die in the next 25 years, *and* will pay you $25,000 in 25 years if you live. What is the present value of the $25,000 you will receive if you live? Assume that the annual inflation rate will be 12% compounded annually for the next 25 years.

51. The hypothesis for the squeeze theorem for exponents requires $b > 1$. What can you say about the case when $0 < b < 1$?

5.2

Logarithmic Functions

Definition of Logarithmic Functions

Suppose the exponent for an exponential function is the unknown. For example, the compound interest formula

$$A = P(1 + i)^N$$

was used in the last section to find A. Now suppose that A, P, and i are known, but the length of time it will take for P to grow to A is not known. If the exponent in an equation is the unknown value, the equation is called an **exponential equation**. Consider

$$A = b^x$$

where $b > 1$. How can we solve this exponential equation for x? Notice that

x **is the exponent of base b that yields the value A**

This can be rewritten as

$x = $ **exponent of base b to get A**

It appears that the equation is now solved for x, but we have simply changed the notation. The expression "exponent of b to get A" is called, for historical reasons, "the log of A to the base b." That is,

$x = \log A$ **to base** b

This phrase is shortened to the notation

$$x = \log_b A$$

The term *log* is an abbreviation for **logarithm**, which means **exponent**.

Logarithm

The logarithm

$$x = \log_b A$$

is the **exponent** on a base b that yields the value A. In symbols

$$x = \log_b A \iff b^x = A$$

Logarithmic Function

The function f defined for $x > 0$ by

$$f(x) = \log_b x$$

where $b > 0$, $b \neq 1$, is called the **logarithmic function with base b**.

Remember the definition of logarithm is nothing more than a notational change.

$M = b^N$ is equivalent to $N = \log_b M$ or $Q = \log_b S$ is equivalent to $S = b^Q$

(Exponent / Base labels)

WARNING Remember: *logarithm means exponent.*

EXAMPLE 1 Change the following expressions from exponential form to logarithmic form.

a. $5^2 = 25$ The base is 5 and the exponent is 2: $\log_5 25 = 2$.
b. $3^2 = 9$ This is the same as $\log_3 9 = 2$.
c. $\frac{1}{8} = 2^{-3}$ In logarithmic form: $\log_2 \frac{1}{8} = -3$.
d. $\sqrt{16} = 4$ Logarithmic form: $\log_{16} 4 = \frac{1}{2}$. (Remember $\sqrt{16} = 16^{1/2}$.) ∎

EXAMPLE 2 Change the following expressions from logarithmic form to exponential form.

a. $\log_{10} 100 = 2$ The base is 10 and the exponent is 2: $10^2 = 100$.
b. $\log_{10} 1/1{,}000 = -3$ This is the same as $10^{-3} = 1/1{,}000$.
c. $\log_3 1 = 0$ Exponential form: $3^0 = 1$. ■

Evaluate Logarithms

To **evaluate a logarithm** means to find a numerical value for the given logarithm. The first ones you are asked to evaluate use a property of exponents that says if two bases on two logarithms are the same and the numbers are equal, then the exponents must be equal.

Exponential Property of Equality

For positive real b ($b \neq 1$)

$b^x = b^y$ if and only if $x = y$

EXAMPLE 3 Evaluate the given logarithms.

a. $\log_2 64$ **b.** $\log_3 \frac{1}{9}$ **c.** $\log_9 27$ **d.** $\log_{10} 1$ **e.** $\log_{10} 100$ **f.** $\log_{10} 10$

Solution **a.** Since it is usually necessary to supply a variable to convert to exponential form, we use N in these examples. That is,

$$N = \log_2 64$$

We write this in exponential form: $2^N = 64$
$$2^N = 2^6$$
$$N = 6 \quad \text{Use the exponential property of equality.}$$

Thus $\log_2 64 = 6$.

b. Let $N = \log_3 \frac{1}{9}$, so $3^N = \frac{1}{9}$
$$3^N = 3^{-2}$$
$$N = -2$$

Thus $\log_3 \frac{1}{9} = -2$.

c. $N = \log_9 27$, so $9^N = 27$
$$3^{2N} = 3^3$$
$$2N = 3$$
$$N = \frac{3}{2}$$

Thus $\log_9 27 = \frac{3}{2}$.

d. $\log_{10} 1 = 0$. Can you do this mentally?
e. $\log_{10} 100 = 2$.
f. $\log_{10} 10 = 1$. ■

Suppose you cannot use the exponential property of equality. For example, suppose you want to find $\log_{10} 5.03$. Since 5.03 is between 1 and 10 and

$$10^0 = 1$$
$$10^x = 5.03 \quad \text{You want to find this } x.$$
$$10^1 = 10$$

The number x should be between 0 and 1, by the squeeze theorem for exponents. There are tables that show approximations for these exponents. However, calculators have, to a large extent, eliminated the need for extensive log tables. Remember, however, that calculator answers are approximate. The key for $\log_{10}$ is simply labeled *log*. Base 10 is fairly common, and if the logarithm is to the base 10 it is called a **common logarithm** and is written without the subscript 10.

EXAMPLE 4 Evaluate the following logarithms correct to four decimal places.

a. log 7.68 **b.** log 852 **c.** log .00728

Solution **a.** DISPLAY: 0.88536122

To four decimal places, $x = .8854$

b. DISPLAY: 2.930439595

To four decimal places, 2.9304.

c. DISPLAY: -2.137868621

To four decimal places, log .00728 = −2.1379.

Keep in mind that a logarithm is an exponent. That is, for Example 4a, the answer .8854 is the *exponent* on the base of 10 that gives the given number, 7.68. That is

$$10^{.8854} \approx 7.68$$

In addition to common logarithms, another logarithm is frequently encountered. This is a logarithm to the base e called a **natural logarithm**. Natural logarithms are denoted by

$$\log_e x = \ln x$$

and this is sometimes pronounced as "lawn x."

EXAMPLE 5 Find ln 3.49 to two decimal places.

Solution BY CALCULATOR: If your calculator has a $\boxed{\log}$ key, chances are it also has a $\boxed{\ln}$ key:

DISPLAY: 1.249901736

EXAMPLE 6 Find ln .403.

Solution DISPLAY: -0.908818717

Thus ln .403 ≈ −.909. Always keep in mind that a logarithm is nothing more than an exponent, so this calculator display means that

$$e^{-.909} \approx .403$$

Graphing Logarithmic Functions

To graph a logarithmic function you use the definition of logarithm and then construct a table of values.

EXAMPLE 7 Graph $y = \log_2 x$.

Solution From the definition of logarithm, $2^y = x$. Use this equation to construct a table of values:

y	$x = 2^y$
-3	$2^{-3} = \frac{1}{8}$
-2	$2^{-2} = \frac{1}{4}$
-1	$2^{-1} = \frac{1}{2}$
0	$2^0 = 1$
1	$2^1 = 2$
2	$2^2 = 4$
3	$2^3 = 8$

The graph is shown in Figure 5.3.

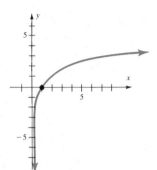

FIGURE 5.3
Graph of $y = \log_2 x$

Exponential and Logarithmic Functions Are Inverses

Compare your answer in Example 7 with the graph of $y = 2^x$ shown in Figure 5.4.

Functions whose graphs are symmetric with respect to the line $y = x$ as shown in Figure 5.4 are said to be **inverse functions**. This relationship is needed in order to find e on several brands of calculators. If a calculator has

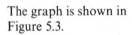

keys, but no e^x key, you can still use the calculator to find e^x. Since

$$y = \ln x \quad \text{and} \quad y = e^x$$

are inverse functions, to find e (or e^1) press

| 1 | INV | ln x | DISPLAY: 2.718281828

These two keys give the inverse of the ln x function; that is, they give e^x when the x value is input just prior to pressing these keys.

If, for example, you need $e^{5.2}$ on such a calculator, press

| 5.2 | INV | ln x | DISPLAY: 181.2722419

FIGURE 5.4
Graphs of $y = 2^x$, $y = \log_2 x$, and $y = x$

EXAMPLE 8 What is the sequence of keys to press to evaluate $\ln e$?*

Solution | 1 | INV | ln x | ln x | The display is, of course, 1.

This gives e.   This gives ln e.

*Answers may vary depending on brand and model.

5.2

Problem Set

Write the equations in Problems 1–6 in logarithmic form.
1. $64 = 2^6$
2. $125 = 5^3$
3. $9 = (\frac{1}{3})^{-2}$
4. $\frac{1}{2} = 4^{-1/2}$
5. $a = b^c$
6. $m = n^p$

Write the equations in Problems 7–12 in exponential form.
7. $\log_4 2 = \frac{1}{2}$
8. $\log_2 \frac{1}{8} = -3$
9. $\log .01 = -2$
10. $\log 10{,}000 = 4$
11. $\ln e^2 = 2$
12. $\ln x = .03$

Use the definition of logarithm or a calculator to evaluate the expressions in Problems 13–30.
13. $\log_b b^2$
14. $\log_t t^3$
15. $\log_\pi \sqrt{\pi}$
16. $\ln e^4$
17. $\log 1{,}000$
18. $\log_2 8$
19. $\log_{16} 1$
20. $\log 4.27$
21. $\log 1.08$
22. $\log 8.43$
23. $\log 9{,}760$
24. $\log .042$
25. $\log .321$
26. $\log .0532$
27. $\ln 2.27$
28. $\ln 16.77$
29. $\ln 2$
30. $\ln 13$

Graph the functions in Problems 31–39.
31. $y = \log_3 x$
32. $y = \log_{1/2} x$
33. $y = \ln x$
34. $y = \log x$
35. $y = \ln \sqrt{x}$
36. $y = \log_\pi x$
37. $y = \ln(x + 2)$
38. $y + 2 = \ln x$
39. $y - 1 = \log(x - 2)$

APPLICATIONS

40. An advertising agency conducts a survey and finds that the number of units sold, N, is related to the amount a spent on advertising (in dollars) according to the following formula:
$$N = 1{,}500 + 300 \ln a \quad a \geq 1$$
   a. How many units are sold after spending $1,000?
   b. How many units are sold after spending $50,000?

41. Graph the sales curve (see Problem 40)
$$N = 1{,}500 + 300 \ln a$$

42. Psychologists have found that in many learning situations a person's rate of learning is rapid at first and then slows down. For the learning curve for learning the touch system of typing, where t is measured in months and P is the number of words typed per minute,
$$P(t) = 80(1 - e^{-.2t})$$
   a. What is the expected number of words typed per minute after one month?
   b. What is the expected number of words typed per minute after six months?

43. A learning curve describes the rate at which a person learns certain tasks. If a person sets a goal of typing N words per minute (wpm), the length of time, t (in days), to achieve this goal is given by
$$t = -62.5 \ln(1 - N/80)$$
   a. How long would it take to learn to type 30 wpm?
   b. If we accept this formula, is it possible to learn to type 80 wpm?

44. Graph the learning curve (see Problem 42)
$$P(t) = 80(1 - e^{-.2t})$$

45. Graph the learning curve (see Problem 43)
$$t = -62.5 \ln(1 - N/80)$$

46. The pH of a substance measures its acidity or alkalinity. It is found by the formula
$$pH = -\log[H^+]$$
where $[H^+]$ is the concentration of hydrogen ions in an aqueous solution given in moles per liter.
   a. What is the pH (to the nearest tenth) of a lemon for which $[H^+] = 2.86 \times 10^{-4}$?
   b. What is the pH (to the nearest tenth) of rain water for which $[H^+] = 6.31 \times 10^{-7}$?

47. The Richter scale for measuring earthquakes, developed by Gutenberg and Richter, relates the energy E (in ergs) to the magnitude of the earthquake, M, by the formula
$$M = \frac{\log E - 11.8}{1.5}$$
   a. A small earthquake is one that releases 15^{15} ergs of energy. What is the magnitude of such an earthquake on the Richter scale?
   b. A large earthquake is one that releases 10^{25} ergs of energy. What is the magnitude of such an earthquake on the Richter scale?

48. Graph the pH formula in Problem 46
$$pH = -\log[H^+]$$
for $1 \leq [H^+] \leq 10^{14}$.
(*Hint:* Let the scale on the x-axis be powers of 10.)

49. Graph the magnitude of an earthquake

$$M = \frac{\log E - 11.8}{1.5}$$

for $10^{12} \leq E \leq 10^{17}$.
(*Hint:* Let the scale on the *x*-axis be powers of 10.)

50. Prove that $\log_b 1 = 0$ for all $b > 0$, $b \neq 1$.
51. Prove that $\log_b b^x = x$ for all x, where $b > 0$, $b \neq 1$.
52. Prove that $\ln e^x = x$ for all x.
53. Prove that $b^{\log_b x} = x$ for all $x > 0$ where $b > 0$, $b \neq 1$.

5.3
Logarithmic and Exponential Equations

Logarithmic Equations

We now discuss solving two types of equations involving transcendental functions. The first type is called a **logarithmic equation**. All logarithmic equations fall into one of four categories, as illustrated by Examples 1–4.

EXAMPLE 1 Solve $\log_2 \sqrt{2} = x$ for x.

Solution *The exponent is the unknown*; apply the definition of logarithm:

$$2^x = \sqrt{2}$$
$$2^x = 2^{1/2}$$
$$x = \frac{1}{2} \quad \text{Exponential property of equality}$$

EXAMPLE 2 Solving $\log_x 25 = 2$.

Solution *The base is the unknown*; apply the definition of logarithm:

$$x^2 = 25$$
$$x = \pm 5$$

Be sure the values you obtain are permissible values for the definition of a logarithm. In this case, $x = -5$ is not a permissible value since a logarithm with a negative base is not defined. Therefore the solution is $x = 5$.

EXAMPLE 3 Solve $\ln x = 5$.

Solution *The power itself is the unknown*; apply the definition of logarithm:

$$e^5 = x$$
$$x \approx 148.41 \quad \text{DISPLAY:} \quad 148.4131591$$

The first three categories of logarithmic equation problems all use the definition of logarithm. The last category requires a property that follows from the exponential property of equality.

Log of Both Sides Theorem

If A, B, and b are positive real numbers with $b \neq 1$, then

$$\log_b A = \log_b B \quad \text{is equivalent to} \quad A = B$$

EXAMPLE 4 Solve $\log_5 x = \log_5 72$.

Solution Use the log of both sides theorem: $x = 72$. ∎

When solving a logarithmic equation, the goal is to write the logarithmic equation with a *single* logarithmic function on either one or both sides of the equation. If the logarithmic equation has the form

$$\log_b A = \log_b B$$

you can use the log of both sides theorem to find the unknown, as Example 4 shows. On the other hand, if the equation has the form

$$\log_b A = N$$

(a log on one side only), then you apply the definition of logarithm to solve the equation, as Examples 1–3 show. In either case, you must first algebraically simplify in order to put the equations into one of these forms. To do this you need some additional theorems about logarithms.

Since logarithms are exponents, we can rewrite the laws of exponents in logarithmic form. For example,

$$b^M \cdot b^N = b^{M+N}$$

This means that if $b^M = A$ and $b^N = B$, then in logarithmic form

$$M = \log_b A \quad \text{and} \quad N = \log_b B$$

But if we apply the definition of logarithm to $b^M \cdot b^N = b^{M+N}$, we obtain

$$M + N = \log_b b^M b^N$$

or, by substitution,

$$\log_b A + \log_b B = \log_b AB$$

This is the logarithmic form for the first law of exponents. The second and third laws of exponents can also be written in logarithmic form, as summarized here. You are asked to derive the second and third laws of logarithms in the problem set.

First Law of Logarithms	$\log_b AB = \log_b A + \log_b B$	The log of the product of two numbers is the sum of the logs of those numbers.
Second Law of Logarithms	$\log_b \dfrac{A}{B} = \log_b A - \log_b B$	The log of the quotient of two numbers is the log of the numerator minus the log of the denominator.
Third Law of Logarithms	$\log_b A^p = p \log_b A$	The log of the pth power of a number is p times the log of that number.

Examples 5 and 6 use these laws of logarithms to solve a logarithmic equation that reduces to the type with a logarithm on both sides of the equation. This means that the last step will be to use the log of both sides theorem.

EXAMPLE 5 Solve $\log_8 3 + \frac{1}{2}\log_8 25 = \log_8 x$ for x.

Solution
$$\log_8 3 + \frac{1}{2}\log_8 25 = \log_8 x$$

$\log_8 3 + \log_8 25^{1/2} = \log_8 x$	Third law of logarithms
$\log_8 3 + \log_8 5 = \log_8 x$	$25^{1/2} = \sqrt{25} = 5$
$\log_8(3 \cdot 5) = \log_8 x$	First law of logarithms
$15 = x$	Log of both sides theorem

The solution is 15. ∎

EXAMPLE 6 Solve $\ln x - \frac{1}{2}\ln 2 = \frac{1}{2}\ln(x + 4)$ for x.

Solution
$$\ln x - \frac{1}{2}\ln 2 = \frac{1}{2}\ln(x + 4)$$
$$2\ln x - \ln 2 = \ln(x + 4)$$
$$2\ln x = \ln 2 + \ln(x + 4)$$
$$\ln x^2 = \ln 2(x + 4)$$
$$x^2 = 2x + 8$$
$$x^2 - 2x - 8 = 0$$
$$(x - 4)(x + 2) = 0$$
$$x = 4, -2$$

Since $\ln(-2)$ is not defined, $x = -2$ is an extraneous root. Therefore, the solution is 4. ∎

Exponential Equations

The second type of equations are **exponential equations** and they fall into one of three categories.

	Common log	Natural log	Arbitrary base
Base:	10	e	b
Example:	$10^x = 5$	$e^x = 3.456$	$7^x = 3$

Examples 7–9 illustrate each of these types.

EXAMPLE 7 Solve $10^{5x+3} = 195$ for x.

Solution This is a *common log* problem. Use the definition of logarithm to write it in logarithmic form: $\log 195 = 5x + 3$.
Now solve this equation for x: $\log 195 - 3 = 5x$
$$\frac{\log 195 - 3}{5} = x$$

For an approximate answer use a calculator.   DISPLAY: -0.1419930777

∎

EXAMPLE 8 Solve $e^{-.000425t} = \frac{1}{2}$ for t.

Solution This is a natural logarithm problem. Use the definition to write

$$\ln .5 = -.000425t$$

$$t = \frac{\ln .5}{-.000425}$$

Use a calculator to find $t \approx 1{,}630.934543$; this is about $1{,}600$. ■

Since you do not have tables or calculator keys for bases other than 10 or e, you must proceed differently for an *arbitrary base b*. You should use the log of both sides theorem and the procedure for solving equations. Generally you can choose either base 10 or base e for solving these equations.

EXAMPLE 9 Solve $7^x = 3$ for x.

Solution We will work this two ways: base 10 and base e so it is clear that both produce exactly the same answer.

Base e	Base 10	
$\ln 7^x = \ln 3$	$\log 7^x = \log 3$	Log of both sides
$x \ln 7 = \ln 3$	$x \log 7 = \log 3$	Third law of exponents
$x = \dfrac{\ln 3}{\ln 7}$	$x = \dfrac{\log 3}{\log 7}$	Divide both sides by the coefficient of x.
$\approx \dfrac{1.099}{1.946}$	$\approx \dfrac{.4771}{.8451}$	Evaluate by calculator.
$\approx .5646$	$\approx .5646$	Divide. ■

WARNING $\dfrac{\log 3}{\log 7} \neq \log \dfrac{3}{7}$

There is another method for solving the exponential equation of Example 9. This one directly applies the definition of logarithm:

$$7^x = 3 \quad \text{is the same as} \quad \log_7 3 = x$$

If there were a logarithm base 7 key on a calculator or a log base 7 table, you would have the answer. Since there are not, you need one final logarithm theorem that changes logarithms from one base to another.

Change of Base Theorem

$$\log_a x = \frac{\log_b x}{\log_b a} \quad \text{In particular,} \quad \log_a x = \frac{\log x}{\log a} = \frac{\ln x}{\ln a}.$$

Notice that to change from base a to another (possibly more familiar) base b, you simply change the base on the given logarithm from a to b and then divide by the logarithm to the base b of the old base a. The proof of this theorem is identical to the steps outlined in Example 9. That is, if $a^N = x$, then $N = \log_a x$ and

$$\log_b a^N = \log_b x \qquad \text{Take the } \log_b \text{ of both sides.}$$
$$N \log_b a = \log_b x \qquad \text{Third law of exponents}$$
$$N = \frac{\log_b x}{\log_b a} \qquad \text{Divide both sides by } \log_b a.$$
$$\text{Thus, } \log_a x = \frac{\log_b x}{\log_b a} \qquad \text{Substitute } N = \log_a x.$$

EXAMPLE 10 Change $\log_7 3$ to logarithms with base 10 and evaluate.

Solution
$$\log_7 3 = \frac{\log 3}{\log 7}$$
$$\approx \frac{.4771}{.8451} \approx .5646$$

EXAMPLE 11 Solve $6^{3x+2} = 200$ for x.

Solution
$$\log_6 200 = 3x + 2 \qquad \text{Use the definition of logarithm.}$$
$$\log_6 200 - 2 = 3x \qquad \text{Solve for } x.$$
$$x = \frac{\log_6 200 - 2}{3}$$
$$= \frac{\frac{\log 200}{\log 6} - 2}{3}$$

By calculator, $x \approx .3190157417 \approx .32$.

Applications

An important application of exponential equations involves the calculation of human population growth. Human populations grow according to the exponential equation

$$P = P_0 e^{rt}$$

where P_0 is the size of the initial population, r is the growth rate, t is the length of time, and P is the size of the population after time t.

EXAMPLE 12 On April 3, 1987, newspaper headlines proclaimed that the world population reached 5 billion. If the annual growth rate is 2%, when will the world population reach 6 billion?

Solution The initial population P_0 is 5 (billion), P is 6 (billion), $r = .02$, and t is the unknown. Substitute the known values into the population growth formula:

$$6 = 5e^{.02t}$$

$$\frac{6}{5} = e^{.02t}$$

$.02t = \ln 1.2$ Use the definition of logarithm; note that $\frac{6}{5} = 1.2$.*

$$t = \frac{\ln 1.2}{.02} \approx 9.11607784$$

We would expect the world's population to reach 6 billion about 9 years after it reached 5 billion; this would be in 1996. ∎

EXAMPLE 13 An almanac lists the 1970 population of San Antonio, Texas, as 654,153, and the 1980 population as 783,296. What was the growth rate of San Antonio for this period?

Solution   Since $P_0 = 654{,}153$, $P = 783{,}296$, and $t = 10$, we have

$$P = P_0 e^{rt}$$

$$\frac{P}{P_0} = e^{rt}$$

$$rt = \ln\left(\frac{P}{P_0}\right) \quad \text{Use the definition of logarithm.*}$$

$$r = \frac{1}{t} \ln \frac{P}{P_0}$$

For this example,

$$r = \frac{1}{10} \ln \frac{783{,}296}{654{,}153} \approx .0180169389$$

The growth rate is about 1.8%. ∎

EXAMPLE 14 How long (to the nearest month) will it take for your money to double if you deposit it at a credit union paying 12.5% interest compounded quarterly?

Solution   The compound interest formula is

$$A = P(1 + i)^N$$

where $i = \frac{r}{n} = \frac{.125}{4} = .03125$, $A = 2P$, and the unknown is $N = nt = 4t$, where t is the time (in years) to be found to the nearest month.

$$2P = P(1.03125)^{4t}$$

$2 = 1.03125^{4t}$ Divide both sides by P ($P \neq 0$).

$4t = \log_{1.03125} 2$ Use the definition of logarithm.*

$t = \dfrac{1}{4} \log_{1.03125} 2$ Divide both sides by 4.

≈ 5.6313765 By calculator.

---

* Many people like to solve equations like these by "taking the log of both sides." This is an acceptable way of proceeding, but unnecessary. It dates back to the time when calculators were not available. The method we use here assumes that you have a calculator, and reinforces the definition of a logarithm as an exponent.

This is 5 years and some months. To find the number of months, subtract 5 and multiply by 12 (to convert .6313765 of a year to months). The answer (to the nearest month) is 5 years 8 months. ∎

5.3

Problem Set

Solve Problems 1–43.

1. $\log_5 25 = x$
2. $\log_2 128 = x$
3. $\log(\frac{1}{10}) = x$
4. $\log_x 84 = 2$
5. $\log_x 28 = 2$
6. $\log x = 2$
7. $\ln x = 3$
8. $\ln x = \ln 14$
9. $\ln 9.3 = \ln x$
10. $\log_3 x^2 = \log_3 125$
11. $\ln x^2 = \ln 12$
12. $\log_2 8\sqrt{2} = x$
13. $\log_3 27\sqrt{3} = x$
14. $\log_x 1 = 0$
15. $\log_x 10 = 0$
16. $\log x = 5$
17. $2^x = 128$
18. $8^x = 32$
19. $125^x = 25$
20. $(\frac{2}{3})^x = \frac{9}{4}$
21. $3^{4x-3} = \frac{1}{9}$
22. $27^{2x+1} = 3$
23. $(\frac{3}{2})^x = 10$
24. $1 = 2\log x - \log 1{,}000$
25. $\ln x - \ln e^3 = 1$
26. $\log_a 102 + \log_a 4 - \log_a 3.1 = 1$
27. $\log_b 6 - \log_b 2.8 + \log_b 3.9 = 1$
28. $\log_8 5 + \frac{1}{2}\log_8 9 = \log_8 x$
29. $\log_7 x - \frac{1}{2}\log_7 4 = \frac{1}{2}\log_7(2x - 3)$
30. $\ln x - \frac{1}{2}\ln 3 = \frac{1}{2}\ln(x + 6)$
31. $\frac{1}{2}\ln x = 3\ln 5 - \ln x$
32. $\log_x(5x - 4) = 2$
33. $\log_x(x + 6) = 2$
34. $\frac{1}{2}\log_2 x = 3\log_2 3 - \log_2 x$
35. $\ln 10 - \frac{1}{2}\ln 25 = \ln x$
36. $10^{5-3x} = .041$
37. $10^{2x-1} = 515$
38. $5^{-x} = 8$
39. $4^x = .82$
40. $e^{2x} = 10$
41. $e^{5x} = \frac{1}{4}$
42. $e^{1-2x} = 3$
43. $e^{1-5x} = 15$

APPLICATIONS

44. The world population reached 4 billion on March 18, 1976 and 5 billion on April 3, 1987. What was the growth rate for this period of time (to the nearest hundredth of a percent)?

45. Use the data given in Example 12 to estimate the year the world population will reach 7 billion.

46. Use the data given in Example 12 to estimate the year the world population will reach 8 billion.

47. If $1,000 is invested at 12% compounded semiannually, how long will it take (to the nearest half-year) for the money to double?

48. Repeat Problem 47, except compound daily and give your answer to the nearest day (365-day year).

49. Repeat Problem 47, except compound continuously and give your answer to the nearest day (365-day year).

50. How long will it take for $1,000 to triple if it is invested at 12% compounded quarterly? (Give your answer to the nearest quarter.)

51. Repeat Problem 50, except compound continuously and give your answer to the nearest day.

52. If $1,000 is invested at 12% interest compounded quarterly, how long will it take (to the nearest quarter) for the money to reach $2,500?

53. Psychologists are concerned with forgetting. In an experiment, students were asked to remember a set of nonsense syllables, such as "nem." They then had to recall the syllables after t seconds. The model used to describe forgetting in this experiment is

$$R = 80 - 27\ln t \qquad t > 1$$

where R is the percentage of students who remember the syllables after t seconds.
   a. What percentage of the students remembered the syllables after 3 seconds?
   b. In how many seconds would only 10% of the students remember the syllables?

54. Solve the forgetting curve formula in Problem 53 for t.

55. If P dollars are borrowed for n months at a monthly interest rate of i, then the monthly payment m is found by the formula

$$m = \frac{Pi}{1 - (1 + i)^{-n}}$$

Use this formula to find the monthly car payment after a down payment of $2,487 on a new car costing $12,487. The car is financed for 4 years at 12%. (*Hint:* $P = $10,000 and $i = .01$.)

56. A home loan is made for $110,000 at 12% interest for 30 years. What is the monthly payment and what is the

total amount of interest paid over the life of the loan? (*Hint:* Use the formula given in Problem 55.)

57. A formula used for carbon-14 dating in archaeology is

$$A = A_0 \left(\frac{1}{2}\right)^{t/5,700} \quad \text{or} \quad P = \left(\frac{1}{2}\right)^{t/5,700}$$

where P is the percentage of carbon-14 present after t years. Solve for t in terms of P.

58. Some bone artifacts found at the Lindemeir site in northeastern Colorado were tested for their carbon-14 content. If 25% of the original carbon-14 was still present, what is the probable age of the artifacts? Use the formula in Problem 57.

59. In 1975 ($t = 0$), the world use of petroleum, P_0, was 19,473 million barrels of oil. If the world reserves at that time were estimated to be 584,600 million barrels and the growth rate for the use of oil is k, then the total amount A used during a time interval $t > 0$ is given by

$$A = \frac{P_0}{k}(e^{kt} - 1)$$

How long will it be before the world reserves are depleted if $k = 8\%$? What does your answer mean?

60. Solve the formula in Problem 59 for t.

61. Prove that $\log_b \frac{A}{B} = \log_b A - \log_b B$.

62. Prove that $\log_b A^p = p \log_b A$.

5.4

Derivatives of Logarithmic and Exponential Functions

Derivative of Logarithmic Function

Rates of change and graphs of logarithmic and exponential functions are fairly common applications in management and the life and social sciences. In Chapters 2 and 3 the notion of a derivative as a limit was defined; that is, if $y = f(x)$, then

$$y' = \lim_{h \to 0} \frac{f(x+h) - f(x)}{h}$$

provided this limit exists. We found a variety of derivatives by using this formula and derived some derivative formulas that made the process very efficient, so that in some problems we no longer had to resort to the definition of derivative. Now we will find the derivatives of the logarithmic and exponential functions. Our first attempt might be to try to apply some of our derivative formulas, but since the logarithmic and exponential functions are transcendental and not algebraic, none of the derivative formulas applies. Whenever we need to find the derivative of a new class of functions we must go back to the definition of derivative.

We begin by finding the derivative of the natural logarithm; that is, we let $y = \ln x$. We need to find

$$\lim_{h \to 0} \frac{\ln(x+h) - \ln x}{h} = \lim_{h \to 0} \frac{\ln\left(\frac{x+h}{x}\right)}{h} \quad \text{This is a property of logarithms; do you see which one?}$$

$$= \lim_{h \to 0} \frac{1}{h} \ln\left(\frac{x+h}{x}\right)$$

$$= \lim_{h \to 0} \ln\left(\frac{x+h}{x}\right)^{1/h} \quad \text{Another property of logarithms}$$

$$= \lim_{h \to 0} \ln\left(1 + \frac{h}{x}\right)^{1/h}$$

Let $m = \dfrac{x}{h}$ so that $\dfrac{h}{x} = \dfrac{1}{m}$ and $\dfrac{1}{h} = \dfrac{m}{x}$. Also, as $h \to 0$, $m \to \infty$. Now substitute these changes into the derivative formula:

$$\lim_{h \to 0} \ln\left(1 + \dfrac{h}{x}\right)^{1/h} = \lim_{m \to \infty} \ln\left(1 + \dfrac{1}{m}\right)^{m/x}$$

$$= \lim_{m \to \infty} \ln\left[\left(1 + \dfrac{1}{m}\right)^{m}\right]^{1/x}$$

$$= \lim_{m \to \infty} \dfrac{1}{x} \ln\left(1 + \dfrac{1}{m}\right)^{m}$$

$$= \dfrac{1}{x} \lim_{m \to \infty} \ln\left(1 + \dfrac{1}{m}\right)^{m}$$

$$= \dfrac{1}{x} \ln e \qquad \text{Remember the definition of } e.$$

$$= \dfrac{1}{x} \qquad \ln e = 1$$

This turns out to be a very pleasing result because of its simplicity! We can also obtain a more general result by applying the chain rule:

Derivative of Logarithm

If $y = \ln|x|$, then $y' = \dfrac{1}{x}$.

If $y = \ln|u|$, where u is a function of x, then $y' = \dfrac{u'}{u}$.

The absolute value bars have been added because the domain of $\ln x$ includes only positive values of x. We are not differentiating $|x|$ here but are simply restricting the domain of x.

EXAMPLE 1 If $y = \ln|3x|$, find y'.

Solution $u(x) = 3x$, so $u'(x) = 3$. Thus

$$y' = \dfrac{3}{3x} = \dfrac{1}{x}$$

∎

EXAMPLE 2 If $y = \ln|5x^3 + 3x^2 - 4|$, find y'.

Solution $u(x) = 5x^3 + 3x^2 - 4$, so $u'(x) = 15x^2 + 6x$.   (This step is usually done mentally.)

$$y' = \dfrac{15x^2 + 6x}{5x^3 + 3x^2 - 4}$$

∎

5.4 DERIVATIVES OF LOGARITHMIC AND EXPONENTIAL FUNCTIONS

EXAMPLE 3 If $y = (3x^2 + 2x)\ln|5x - 7|$, find y'.

Solution Use the product rule:

$$y' = (3x^2 + 2x)\left(\frac{5}{5x-7}\right) + (6x+2)\ln|5x-7|$$

$$= \frac{5x(3x+2)}{5x-7} + 2(3x+1)\ln|5x-7|$$

Properties of logarithms can sometimes simplify the process of finding derivatives. This is especially true when quotients are involved, as shown in Example 4.

EXAMPLE 4 If $y = \ln\left(\frac{x}{x+1}\right)$ find y' as follows:

a. By taking the derivative of the logarithm
b. By using properties of logarithms before taking the derivative

Solution

a. $\dfrac{d}{dx}\left[\ln\left(\dfrac{x}{x+1}\right)\right] = \dfrac{1}{\frac{x}{x+1}}\left[\dfrac{x}{x+1}\right]' = \dfrac{x+1}{x}\left[\dfrac{(x+1)(1) - x(1)}{(x+1)^2}\right] = \dfrac{1}{x(x+1)}$

b. $\dfrac{d}{dx}\left[\ln\left(\dfrac{x}{x+1}\right)\right] = \dfrac{d}{dx}[\ln x - \ln(x+1)] = \dfrac{1}{x} - \dfrac{1}{x+1}$

$= \dfrac{x+1-x}{x(x+1)} = \dfrac{1}{x(x+1)}$

WARNING *Take note of the proper notation for logarithms.* The notation of logarithms can be confusing. Remember, $\ln x$ means $\log_e x$, $\log x$ means $\log_{10} x$. Also, be sure you distinguish between $\ln x^2$, which means $\ln(xx)$, and $(\ln x)^2$, which means $(\ln x)(\ln x)$.

EXAMPLE 5 If $y = \log x$ find $\dfrac{dy}{dx}$.

Solution The derivative formula we derived applies only to natural logarithms. In order to find the derivative for a common logarithm, we use the change of base rule:

$y = \log x$

$= \dfrac{\ln x}{\ln 10}$

$= \dfrac{1}{\ln 10}\ln x$ *Note:* $\dfrac{1}{\ln 10}$ is a constant.

$\dfrac{dy}{dx} = \dfrac{d}{dx}\left(\dfrac{1}{\ln 10}\ln x\right)$ Take the derivative of both sides.

$= \dfrac{1}{\ln 10}\dfrac{1}{x}$ $\dfrac{d}{dy}\ln x = \dfrac{1}{x}$

EXAMPLE 6

The demand equation of a commodity is

$$p = \frac{200 \ln(x+1)}{x}$$

and the cost of producing x units is $C(x) = 2x$. Show that the profit function has a relative maximum and not a minimum point.

Solution The profit function $P(x)$ is found by $P(x) = R(x) - C(x)$. We are given the cost function, but we need to find $R(x)$. The revenue function is found by multiplying the number of units times the demand, so $R(x) = xp = 200 \ln(x+1)$.

$$P(x) = 200 \ln(x+1) - 2x$$

$$P'(x) = 200 \frac{1}{x+1} - 2$$

$$= 200(x+1)^{-1} - 2$$

Set $P'(x) = 0$ and solve for x:

$$200(x+1)^{-1} - 2 = 0$$

$$200 = 2(x+1)$$

$$99 = x$$

To determine if this is a relative maximum or minimum, check the second derivative:

$$P''(x) = -200(x+1)^{-2}$$

This function is negative for all values of x, so the profit function has a relative maximum at $x = 99$ and no relative minimum. ∎

CALCULATOR COMMENT

When graphing P in Example 6 using a function using a standard domain and range, nothing shows on the screen. Using the TRACE, it appears that the largest value for x is
X=98.947368 Y=723.03401
This leads us to input the following values in the
RANGE:
Xmin=0
Xmax=200
Ymin=0
Ymax=1000·
Graph shows:

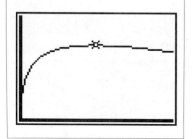

Derivative of Exponential Function

Now we turn our attention to finding the derivative of the exponential function $y = e^x$. To do this, we use the definition of logarithm to write $x = \ln y$. We can now use the derivative rule for $\ln y$ (found above) by using implicit differentiation on the equation $\ln y = x$:

$$\frac{1}{y} \cdot \frac{dy}{dx} = 1$$

We solve for dy/dx:

$$\frac{dy}{dx} = y$$

Since $y = e^x$, we see that the derivative of e^x is e^x; what could be easier than this! This result shows one of the reasons why base e rather than base 10 is used almost exclusively in more advanced work. Again, this result is generalized using the chain rule:

WARNING

$$\frac{d}{dx}(e^x) \neq xe^{x-1}$$

This is a common mistake.

Derivative of Exponential

If	$y = e^x$,	then $y' = e^x$.
If	$y = e^u$,	where u is a function of x, then $y' = u' \cdot e^u$.

EXAMPLE 7 If $y = e^{10x}$, then $y' = 10e^{10x}$.

EXAMPLE 8 If $y = e^{4x^2+5}$, then $y' = 8xe^{4x^2+5}$.

EXAMPLE 9 If $y = x^3 e^{4x}$, then $y' = x^3(4e^{4x}) + 3x^2 e^{4x}$ Use the product rule first.
$$= 4x^3 e^{4x} + 3x^2 e^{4x} \quad \text{or} \quad x^2 e^{4x}(4x + 3)$$

EXAMPLE 10 If $y = \dfrac{e^x}{\ln|x|}$, then $y' = \dfrac{\ln|x| \cdot e^x - e^x \cdot (1/x)}{\ln^2|x|}$ Use the quotient rule first. Note that $[\ln(x)]^2$ is written more simply as $\ln^2 x$.

$$= \dfrac{\dfrac{e^x}{x}(x\ln|x| - 1)}{\ln^2|x|}$$

$$= \dfrac{e^x(x\ln|x| - 1)}{x\ln^2|x|}$$

EXAMPLE 11 Suppose you study the spread of a disease introduced into a small town of 2,000 persons. Assume that everyone in the town has an equal chance of contracting the disease. A model for predicting the number of people, N, contracting the disease is given by

$$N(t) = \dfrac{2,000}{1 + 1,999e^{-.5t}}$$

where t is the number of days since the disease was introduced into the community. What is the rate at which members of the community are contracting the disease?

Solution   The requested function is dN/dt. Write $N(t) = 2,000(1 + 1,999e^{-.5t})^{-1}$. Then,

$$\dfrac{dN}{dt} = 2,000(-1)(1 + 1,999e^{-.5t})^{-2}(-.5)1,999e^{-.5t}$$

$$= \dfrac{2,000(-1)(-.5)1,999e^{-.5t}}{(1 + 1,999e^{-.5t})^2}$$

$$= \dfrac{1,999,000e^{-.5t}}{(1 + 1,999e^{-.5t})^2}$$

This means, for example, that on day 0, $N(0) = 2,000/(1 + 1,999) = 1$ person has the disease, but dN/dt at $t = 0$ is

$$\dfrac{1,999,000}{2,000^2} = .49975$$

so the rate means that nearly one person is contracting the disease every 2 days. On the other hand, after 10 days, $N(10) \approx 138$, so about 138 people have the disease and the rate at which new people are now contracting the disease is dN/dt at $t = 10$:

$$\dfrac{1,999,000e^{-5}}{(1 + 1,999e^{-5})^2} \approx 64.33598929$$

which means that at the time $t = 10$, new people are contracting the disease at the rate of about 64 people per day.

Graphing Techniques for Logarithmic and Exponential Functions

EXAMPLE 12 Graph $f(x) = 10x - 5x \ln x$.

Solution Domain is $x > 0$.

Function	First derivative	Second derivative
$f(x) = 10x - 5x \ln x$	$f'(x) = 5 - 5 \ln x$ Detail of work: $f'(x) = 10 - (5x \cdot \frac{1}{x} + 5 \ln x)$ $= 10 - 5 - 5 \ln x$ $= 5 - 5 \ln x$ Critical values: $5 - 5 \ln x = 0$ $\ln x = 1$ $x = e$	$f''(x) = -5x^{-1}$
$f(e) = 10e - 5e \ln e$ $= 10e - 5e$ $= 5e$ ≈ 13.59	Check $x = e$ If $x < e$, $f'(x) > 0$ increasing If $x > e$, $f'(x) < 0$ decreasing	$f''(e) = -5e^{-1} < 0$ Relative maximum at $x = e$ No inflection point Concave downward for $x > 0$

The graph is shown in Figure 5.5.

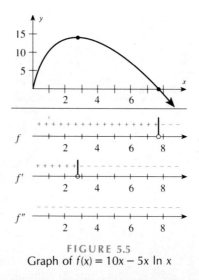

FIGURE 5.5
Graph of $f(x) = 10x - 5x \ln x$

EXAMPLE 13 Graph $f(x) = xe^x$.

Solution Domain is $(-\infty, \infty)$.

Function	First derivative	Second derivative
$f(x) = xe^x$	$f'(x) = e^x(x + 1)$ Detail of work: $f'(x) = x\dfrac{d}{dx}e^x + e^x\dfrac{d}{dx}(x)$ $ = xe^x + e^x$ $ = e^x(x + 1)$ Critical values: $e^x(x + 1) = 0$ Since $e^x \neq 0$, the only critical value is $x = -1$.	$f''(x) = e^x(x + 2)$
$f(-1) = -e^{-1}$	If $x < -1$, $f'(x) < 0$ decreasing If $x > -1$, $f'(x) > 0$ increasing	$f''(-1) = e^{-1} > 0$ Relative minimum at $x = -1$
$f(0) = 0$		Points of inflection: $f''(x) = 0$ if $x = -2$ If $x < -2$ Concave downward If $x > -2$ Concave upward

The graph is shown in Figure 5.6.

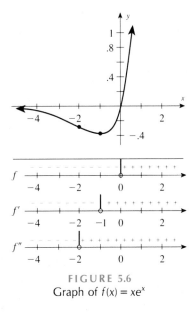

FIGURE 5.6
Graph of $f(x) = xe^x$

5.4

Problem Set

Find the derivatives of the functions in Problems 1–34.

1. $y = e^{2x}$
2. $y = e^{8x}$
3. $y = e^{-x}$
4. $y = e^{-3x}$
5. $f(x) = xe^x$
6. $f(x) = x^2 e^{3x}$
7. $f(x) = (2x^3 + 1)e^{5x^2}$
8. $f(x) = (5x^2 - x)e^{-3x^2}$
9. $y = -2e^{5x}(3x^2 + 5)$
10. $y = (3x + 2e^{5x})^4$
11. $y = \ln|5 - x|$
12. $y = \ln|3 - x^2|$
13. $y = (2x^3 + e^{5x})^4$
14. $y = (e^{6x} - 5x^4)^3$
15. $y = \ln x^4$
16. $y = (\ln x)^4$
17. $y = \ln \sqrt{x}$
18. $y = \sqrt{\ln|x|}$
19. $y = \dfrac{\ln x^3}{e^x}$
20. $y = \dfrac{e^{3x}}{\ln x^2}$
21. $f(x) = \ln\left|\dfrac{x}{x+2}\right|$
22. $f(x) = \ln\left|\dfrac{x-3}{x^2}\right|$
23. $f(t) = e^{t^2}(1 - e^{-t})$
24. $f(t) = e^{(t+1)/t}$
25. $y = (e^x - e^{-x})^2$
26. $y = (e^{-x} + e^x)^3$
27. $y = \dfrac{\ln|x|}{5x + 2}$
28. $y = \dfrac{3x^2 - 7}{\ln|x|}$
29. $f(x) = \dfrac{e^x}{\ln|3x|}$
30. $f(x) = \dfrac{e^{5x}}{\ln|5x|}$
31. $y = \dfrac{2{,}000}{1 + 6e^{.3x}}$
32. $y = \dfrac{500}{1 - 30e^{.2x}}$
33. $y = \dfrac{100 \ln|x|}{1 + 40e^{-.2x}}$
34. $y = \dfrac{900 \ln|x|}{1 - 70e^{-.3x}}$

Graph the functions in Problems 35–40.

35. $f(x) = 2x - 5x \ln x$
36. $f(x) = x - \ln x$
37. $f(x) = \dfrac{\ln x}{x}$, $x \geq 1$
38. $f(x) = \dfrac{\ln x^2}{x^2}$, $x \geq 1$
39. $y = x^2 e^{3x}$
40. $y = e^x - e^{-x}$

APPLICATIONS

41. The demand equation of a commodity is

$$d = \dfrac{500 \ln(x + 10)}{x^2}$$

where x is the number of units produced ($1 \leq x \leq 20$). The cost of producing x units is $C(x) = 2x^2$. Find the marginal revenue.

42. Find the marginal cost for the information in Problem 41.

43. The spread of a disease in a town of 5,000 people is described by the model

$$N(t) = \dfrac{5{,}000}{1 + 4{,}999e^{-.1t}}$$

where N is the number of people contracting the disease after t days.
   a. How many people have the disease on day 0?
   b. What is the rate at which people are getting the disease on day 0?

44. a. How many people in Problem 43 have the disease on the tenth day?
   b. What is the rate at which people are getting the disease on the tenth day?

45. The amount of money, P, invested at 12% interest and compounded continuously for t years, has a future value A according to the formula $A = Pe^{.12t}$. At what rate is the amount of money A increasing after 1 year? After 5 years?

46. The blood pressure in the aorta changes between beats according to the formula

$$P(t) = e^{-kt}$$

for an appropriate constant k, where t is the time in milliseconds since the last beat. What is the rate at which the pressure is changing with respect to time after 3 milliseconds if $k = .025$?

47. The amount of a drug present in the body is a function of the amount administered and the length of time t (in hours) since it was administered. For 5 milliliters of a certain drug, the amount of drug present behaves according to the formula

$$A(t) = 5e^{-.03t}$$

What is the rate at which the amount of drug present is changing expressed as a function of time?

48. Sales often follow a growth pattern of rapid initial growth with some leveling off after a period of time. This pattern is described by the formula

$$S(t) = 25{,}000 - 10{,}000e^{-.2t}$$

   a. How many items will be sold initially?
   b. What is the rate of change of sales initially?
   c. What is the rate of change of sales after t years?

49. The price–demand equation for x units of a commodity is given by

$$d(x) = 500e^{-.1x}$$

Find the marginal revenue.

50. The price–supply equation for x units of the commodity described in Problem 49 is given by

$$s(x) = 50e^{.05x}$$

Find the marginal supply.

51. If $y = \log x$, show that

$$y' = \frac{\log e}{x}$$

52. If $y = \log_b x$, show that

$$y' = \frac{\log_b e}{x}$$

53. If $y = \log_b u$, where u is a function of x, show that

$$\frac{dy}{dx} = \frac{\log_b e}{u} \cdot \frac{du}{dx}$$

*5.5
Review

The material of this chapter is reviewed in the following list of objectives. After each objective there are some practice questions. For a sample test select the first question of each set and check your answers. The second question for each objective has no answer given. If you are having trouble with a particular type of problem, look back at the indicated section in the text. When you are finished reviewing these objectives, a sample examination is given at the end of this section.

[5.1]
Objective 5.1: Simplify expressions with positive, negative, and fractional exponents.
1. $125^{2/3}$
2. $2^{1/2} \cdot 3^{1/3}$
3. $\dfrac{27^{2/3}}{27^{1/2}}$
4. $(x^{1/2} - y^{1/2})(x^{1/2} + y^{1/2})$

Objective 5.2: Sketch the graph of exponential functions.
5. $y = (\tfrac{1}{2})^x$
6. $y = -2^x$
7. $y = 2^{-x}$
8. $y = e^{-x/2}$

Objective 5.3: Evaluate expressions with the natural base.
9. e^1
10. e^4
11. $e^{1.05}$
12. $e^{-.005}$

[5.2]
Objective 5.4: Write an exponential equation in logarithmic form.
13. $10^{.5} = \sqrt{10}$
14. $e^0 = 1$
15. $9^3 = 729$
16. $(\sqrt{2})^3 = 2\sqrt{2}$

Objective 5.5: Write a logarithmic equation in exponential form.
17. $\log 1 = 0$
18. $\ln \tfrac{1}{e} = -1$
19. $\log_2 64 = 6$
20. $\log_\pi \pi = 1$

Objective 5.6: Evaluate common and natural logarithms.
21. $\ln 3$
22. $\log 3$
23. $\log .0021$
24. $\ln .013$

Objective 5.7: Graph logarithmic functions.
25. $y = \log x$
26. $y = \ln x$
27. $y = \log_5 x$
28. $y = \log_{1/5} x$

[5.3]
Objective 5.8: Solve logarithmic equations.
29. $\log_5 25 = x$
30. $\log_x(x + 6) = 2$
31. $3 \log 3 - \tfrac{1}{2} \log 3 = \log \sqrt{x}$
32. $2 \ln \tfrac{e}{\sqrt{7}} = 2 - \ln x$

Objective 5.9: Solve exponential equations.
33. $10^{x+2} = 125$
34. $5^{2x-3} = .5$
35. $e^{4-3x} = 15$
36. $10^{-x^2} = .45$

[5.4]
Objective 5.10: Find the derivatives of exponential functions.
37. $y = 4.9e^{-.05x^2}$
38. $y = 3{,}500x - 500e^{3x}$
39. $y = x^3 e^{x^4}$
40. $y = \dfrac{1{,}000}{x - 999e^{-.2x}}$

Objective 5.11: Find the derivatives of logarithmic functions.
41. $y = \ln x^2$
42. $y = \ln|x^4(x^2 - 5)|$
43. $y = \dfrac{\ln x^2}{x + 3}$
44. $y = 250 \ln|3x^2 - 5|$

* Optional section.

Objective 5.12: *Solve applied problems based on the preceding objectives.*

45. If a person's present salary is $30,000 per year, use the formula $A = Pe^{rt}$ to determine the salary necessary to equal this salary in 15 years if you assume the 1991 annual inflation rate of 8.2% compounded continuously.

46. If $5,500 is invested at 13.5% compounded daily, how long will it take for this to grow to $10,000? (Use a 365-day year.)

47. *Archaeology.* The half-life formula for carbon-14 dating is
$$A = 10\left(\frac{1}{2}\right)^{t/5{,}700}$$
where A is the amount present (in milligrams) after t years. Solve this equation for t.

48. *Advertising.* An advertising agency conducted a survey and found that the number of units sold, N, is related to the amount spent on advertising (in dollars) by the formula
$$A = 1{,}500 + 300 \ln a \qquad (a \geq 1)$$
What is the rate at which N is changing at the instant that $a = \$10{,}000$?

49. The atmospheric pressure P in pounds per square inch (psi) is given by
$$P = 14.7e^{-.21a}$$
where a is the altitude above sea level (in miles). As a hot-air balloon is rising the pressure is constantly changing. How fast is the pressure changing when the balloon is one mile above sea level?

50. When a satellite has an initial radioisotope power supply of 50 watts, its power output in watts is given by the equation
$$P = 50e^{-t/250}$$
where t is the time in days. Solve for t.

51. *Equilibrium point.* Suppose the price–demand and price–supply equations for x thousands of units of a commodity are given by
DEMAND: $p(x) = 300e^{-.3x}$
SUPPLY: $s(x) = 30e^{.1x}$
Find the value of x for the equilibrium point.

52. Suppose the president makes a major policy announcement. A model to predict the percentage, N, of the population that will have heard the announcement is a function of the time, t (in days), after the announcement. Suppose a suitable model is given by the formula
$$N = 1 - e^{-2.5t}$$
How long will it take for 90% of the population ($N = .9$) to hear of the announcement?

SAMPLE TEST
The following sample test (45 minutes) is intended to review the main ideas of this chapter.

Solve Problems 1–8.
1. $\log_6 36 = x$
2. $\log_x(2x + 15) = 2$
3. $2\log 2 - \frac{1}{2}\log 2 = \log\sqrt{x}$
4. $2\ln\frac{e}{\sqrt{5}} = 1 - \ln x$
5. $10^{x-1} = 250$
6. $3^{1-2x} = .5$
7. $e^{2x+3} = 10$
8. $10^{-x^2} = .75$

Find the derivatives in Problems 9–16.
9. $y = 5.5e^{-.5x^2}$
10. $y = (x^2 - 1)e^{3x}$
11. $y = \ln x^5$
12. $y = (\ln x)^5$
13. $y = \ln|x^2(4 - x^3)|$
14. $y = e^x \ln x$
15. $y = \dfrac{\ln x^2}{4 - x}$
16. $y = \dfrac{1{,}000 \ln|x^3|}{2e^x}$

17. The inflation formula (continuous)
$$A = Pe^{rt}$$
gives the future value after t years of an inflation rate of r. Solve this equation for r.

18. An advertising agency conducts a survey that finds that the number of units sold, N, is related to the amount spent on advertising (in dollars) by the formula
$$N = 2{,}500 + 200 \ln a \qquad (a \geq 1)$$
What is the rate at which N is changing at the instant that $a = \$20{,}000$?

19. The price–demand and price–supply equations for x thousands of units of a commodity are given by

Demand: $p(x) = 200e^{-.2x}$

Supply: $s(x) = 20e^{.1x}$

Find the number of units that should be manufactured in order to achieve equilibrium.

20. The president makes a major policy announcement. The percentage of the population that will have heard about the announcement is a function of the time t (in days) after the announcement according to the formula

$$N = 1 - e^{-1.5t}$$

where N is the percentage written as a decimal. How long will it take for 90% of the population to hear about the announcement?

6 The Integral

CHAPTER OVERVIEW
The last main idea of calculus is introduced in this chapter—the idea of an integral and a procedure called integration.

PREVIEW
The integral is first introduced as an antiderivative, and then an indefinite integral is defined. The idea of a definite integral is applied to finding areas, and finally we give a formal definition of the definite integral and state the Fundamental Theorem of Calculus.

PERSPECTIVE
The skills you learn in this chapter will be put to use in the next two chapters when applications of integration are considered. As you will see in this chapter, the ability to "reverse" the process of differentiation is important since you quite often know the derivative but need to know the original function. The indefinite integral is used for that purpose.

6.1 The Antiderivative
6.2 Integration by Substitution
6.3 The Definite Integral
6.4 Area Between Curves
6.5 The Fundamental Theorem of Calculus
6.6 Numerical Integration
6.7 Chapter 6 Review
Chapter Objectives
Sample Test

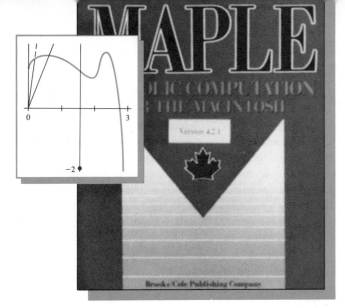

MODELING
APPLICATION 6

Computers in Mathematics

In the 1980s, mathematics instruction was forever altered with the advent of powerful, inexpensive calculators and computer programs that would do symbolic manipulation and function graphing. This new software allows both instructors and students to focus on concepts and processes rather than on tedious and complicated calculations involved in realistic problems.

DERIVE, the successor to MUMATH, is a powerful computer algebra system for PC-compatible computers. It is available from The Soft Warehouse, 3615 Harding Avenue, Suite 505, Honolulu, HI 96816.

MATHEMATICA is advertised as more than a program—it is a *system* for doing mathematics. It is available from Wolfram Research, Inc., P.O. Box 6059, Champaign, IL 61821. It runs on both Macintosh and IBM-compatible computers.

MAPLE also does algebraic manipulation in an easy-to-use format for Macintosh computers. It is available from Brooks/Cole Publishing, 555 Forest Lodge Road, Pacific Grove, CA 93950.

After you have finished this chapter, write a paper on a computer program that does symbolic manipulation. Your paper should describe the program, as well as evaluate it.

APPLICATIONS

Management (*Business, Economics, Finance, and Investments*)
Finding the cost function given the marginal cost (6.1, Problems 37–40)
Finding the profit function for Turbo Plus, Inc. (6.1, Problem 43)
Finding the cost of producing 100,000 processors (6.1, Problem 44)
Determining the output after t hours of work (6.2, Problem 42)
Finding the profit when the marginal profit is known (6.2, Problems 43–44)
Finding the sales (6.2, Problem 45)
Determining the accumulated sales (6.3, Problems 45–46)
Finding the Gini index to measure money flow (6.3, Problems 47–48)
Accumulated cost and revenue (6.3, Problem 49; 6.4, Problems 42–43)
Total cost of production (6.4, Problem 36)
Automobile sales from a graph in the *Wall Street Journal* (6.5, Problems 37–38)
Total paper and paperboard production in the U.S. (6.5, Problem 39)

Management (*continued*)
Light output vs. time for a flashbulb (6.6, Problem 38)
Treasury bond rate (6.7, Problem 60)
Determining if the eucalyptus tree grows fast enough to make commercial growing a profitable enterprise (6.7, Test Problem 17)

Life Sciences (*Biology, Ecology, Health, and Medicine*)
Infection by a new strain of influenza (6.1, Problems 45–46)
Flu epidemic (6.2, Problem 41)
Predicting oil consumption (petroleum) given present production and rate of consumption (6.2, Problems 48–49; 6.3, Problems 35–39; 6.7, Problem 19)
Toxic dumping (6.3, Problems 40–44; 6.7, Problem 59)
Rate of healing (6.3, Problem 50)
Amount of pollutants dumped into the Russian River (6.4, Problem 37)
Chemical pollution (6.4, Problem 39)

Life Sciences (*continued*)
Growing rate of a culture (6.4, Problem 40)
Rabbit population (6.5, Problem 41)

Social Sciences (*Demography, Political science, Population, Psychology, Society, and Sociology*)
Predicting population given growth rate (6.1, Problems 41–42; 6.2, Problem 47)
Predicting the world population (6.2, Problem 46; 6.5, Problem 42)
Growth rate of Chicago, Illinois (6.7, Test Problem 18)

General Interest
Velocity of an object thrown off Hoover Dam (6.1, Problems 47–48)
Temperature variations (6.5, Problems 43–44)
Estimating areas of swimming pools and parcels (6.5, Problems 45–46; 6.6, Problem 37)

Modeling Application—
Computers in Mathematics

6.1 The Antiderivative

Calculus is divided into two broad categories. The first, **differential calculus**, involves the definition of derivative and related applications of instantaneous rates of change, marginal costs, profits, and curve sketching, as well as finding maximums and minimums. In this chapter, we begin our study of the second part of calculus, **integral calculus**, which deals with the definition of another limit—the **definite integral**. Integral calculus is used to find areas, volumes, and functions when we have information about the function's rate of change. For example, if we know the present world population and the rate at which it is growing, we can produce a formula (subject to certain assumptions) that predicts the population size at any future time or estimates its size at some time in the past.

EXAMPLE 1 Multiplex Corporation knows that the marginal cost (in thousands of dollars) to produce x items (in thousands) is given by the formula

$$f(x) = 2x + 4$$

Find the cost of producing 50,000 items ($x = 50$) if the fixed costs are $25,000.

Solution In Section 3.3 marginal cost was defined to be the rate of change in cost per unit change in production at an output level of x units. This means that if F is the cost function, then the marginal cost, f, is a function that can be found by taking the derivative of F at x:

$$F'(x) = f(x)$$

Find a function F so that $F'(x) = f(x)$:

$$F'(x) = 2x + 4$$

By trial and error (we will develop better methods very shortly), we see that

$$F(x) = x^2 + 4x$$

is such a function, but this function is not unique:

If $F_1(x) = x^2 + 4x + 6$, then $F'_1(x) = 2x + 4$;
if $F_2(x) = x^2 + 4x - 100$, then $F'_2(x) = 2x + 4$;
if $F_3(x) = x^2 + 4x + 14.8$, then $F'_3(x) = 2x + 4$;...

There are many functions that have $2x + 4$ as a derivative, but they all differ by a constant since the derivative of a constant is 0. Thus

$$F(x) = x^2 + 4x + C$$

for any constant C. The cost of producing 50,000 items is given by $F(50)$, but first we need to find C, the **constant of integration**. The fixed costs are the costs that are present even if $x = 0$. This means that $F(0) = 25$, so

$$F(0) = 0^2 + 4(0) + C = 25$$
$$C = 25$$

Thus

$$F(x) = x^2 + 4x + 25$$

is the cost equation for our original problem, so
$$F(50) = 50^2 + 4(50) + 25 = 2{,}725$$
The cost of producing 50,000 items is $2,725,000. ∎

In Example 1 we went through a (trial-and-error) process that might be called the process of **antidifferentiation**. A more common name for the process of antidifferentiation is **integration**. If $F(x)$ is an antiderivative of $f(x)$, then $F(x) + C$ is called the **indefinite integral** of $f(x)$. The adjective *indefinite* is used because the constant C is arbitrary or indefinite.

Indefinite Integral

If F is an antiderivative of f, then we write
$$\int f(x)\, dx = F(x) + C$$
and say, "The indefinite integral of $f(x)$ with respect to x is $F(x) + C$." The symbol $\int$ is called the **integral symbol**, $f(x)$ is called the **integrand**, and C is called the **constant of integration**.

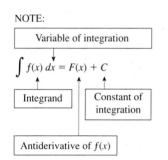

NOTE:

We will say more about the symbol dx later, but for now remember that the dx in the indefinite integral identifies the independent variable of integration, as illustrated by Example 2.

EXAMPLE 2 $\int 3x\, dx$ is read "the indefinite integral of $3x$ with respect to x."

$\int 9y^2\, dy$ is read "the indefinite integral of $9y^2$ with respect to y."

$\int (5z^2 + 3z + 2)\, dz$ is read "the indefinite integral of $5z^2 + 3z + 2$ with respect to z."

$\int x^2 y^3\, dx$ is read "the indefinite integral of $x^2 y^3$ with respect to x." ∎

EXAMPLE 3   Show that $F(x) = x^6$ is an antiderivative of $f(x) = 6x^5$.

Solution   Use the power rule for derivatives:
$$F(x) = x^6$$
$$F'(x) = 6x^5$$ ∎

EXAMPLE 4   Find an antiderivative of $f(x) = x^5$.

Solution   By trial and error (the power rule for derivatives in reverse),
$$F(x) = \frac{x^6}{6} + C$$
Check: $F'(x) = \dfrac{6x^5}{6} + 0 = x^5$ ∎

Integral notation makes it easy to state some integration formulas. For example, the power rule for derivatives says if $f(x) = x^n$, then $f'(x) = nx^{n-1}$, so if this process is reversed we find a corresponding integration formula to be

$$\text{If} \quad f(x) = x^n, \quad \text{then} \quad F(x) = \int x^n \, dx = \frac{x^{n+1}}{n+1} + C \quad n \neq -1$$

($n \neq -1$ keeps the denominator from being 0).

To prove this formula, find the derivative of

$$F(x) = \frac{x^{n+1}}{n+1} + C \quad n \neq -1$$

$$F'(x) = \frac{(n+1)x^{n+1-1}}{n+1} + 0$$

$$= x^n$$

EXAMPLE 5

a. $\displaystyle\int x^8 \, dx = \frac{x^9}{9} + C$

b. $\displaystyle\int x^3 \, dx = \frac{x^4}{4} + C$

c. $\displaystyle\int \sqrt{x} \, dx = \int x^{1/2} \, dx$

$$= \frac{x^{3/2}}{\frac{3}{2}} + C = \frac{2}{3} x^{3/2} + C$$

d. $\displaystyle\int dx = \int x^0 \, dx = \frac{x^1}{1} + C = x + C$

e. $\displaystyle\int \frac{dx}{x^3} = \int x^{-3} \, dx$

$$= \frac{x^{-2}}{-2} + C = -\frac{1}{2} x^{-2} + C \quad \blacksquare$$

As you see, if the derivatives of two functions are equal, then the functions differ by at most a constant. This fact is summarized by adding C in our answers for Example 5. This result is summarized in the following box.

Functions with Equal Derivatives

If F and G are differentiable functions on an interval (a, b) such that

$$F'(x) = G'(x)$$

then

$$F(x) = G(x) + C$$

for some constant C.

If we quickly review the derivative formulas from Chapter 2, we should be able to restate each as an integration formula.

Review of Derivative Formulas

POWER RULE:	If $y = x^n$, then $y' = nx^{n-1}$.
CONSTANT RULE:	If $y = k$, then $y' = 0$.
CONSTANT TIMES A FUNCTION RULE:	If $y = kf$, then $y' = kf'$.
SUM RULE:	If $y = f + g$, then $y' = f' + g'$.
DIFFERENCE RULE:	If $y = f - g$, then $y' = f' - g'$.
LOGARITHMIC RULE:	If $y = \ln\|x\|$, then $y' = \dfrac{1}{x}$.
EXPONENTIAL RULE:	If $y = e^x$, then $y' = e^x$.

(We will consider the product, quotient, and chain rules later.)

These derivative formulas can be divided into two types: two *direct integration forms* and three *procedural types*. For now, all our examples can be reduced to one of the direct integration types by first using the procedural types.

Integration Formulas

DIRECT INTEGRATION FORMULAS

$$\int x^n \, dx = \begin{cases} \dfrac{x^{n+1}}{n+1} + C & n \neq -1 \\ \ln|x| + C & n = -1 \end{cases}$$

Power rule (This includes the constant rule as the case where $n = 0$.)

Logarithmic rule

$$\int e^x \, dx = e^x + C$$

Exponential rule

PROCEDURAL INTEGRATION FORMULAS

$$\int kf(x) \, dx = k \int f(x) \, dx$$

The **integral of a constant times a function** is that constant times the integral.

$$\int [f(x) + g(x)] \, dx = \int f(x) \, dx + \int g(x) \, dx$$

The **integral of a sum** is the sum of the integrals.

$$\int [f(x) - g(x)] \, dx = \int f(x) \, dx - \int g(x) \, dx$$

The **integral of a difference** is the difference of the integrals.

EXAMPLE 6
$$\int (3x^2 - 4x + 5)\,dx = \int 3x^2\,dx - \int 4x\,dx + \int 5\,dx \qquad \text{Sum and difference formulas}$$
$$= 3\int x^2\,dx - 4\int x\,dx + 5\int dx \qquad \text{Constant formula}$$
$$= 3\left(\frac{x^3}{3} + C_1\right) - 4\left(\frac{x^2}{2} + C_2\right) + 5(x + C_3) \qquad \text{Power formula}$$
$$= x^3 - 2x^2 + 5x + (3C_1 - 4C_2 + 5C_3)$$
$$= x^3 - 2x^2 + 5x + C$$

In Example 6 since $3C_1 - 4C_2 + C_3$ represents a sum of arbitrary constants, they can be combined into a single constant, C. Also, the steps in Example 6 are shown in great detail so you can relate them to the proper formulas, but in practice your work would look like this:

$$\int (3x^2 - 4x + 5)\,dx = \frac{3x^3}{3} - \frac{4x^2}{2} + 5x + C$$
$$= x^3 - 2x^2 + 5x + C$$

When we carry out the process of writing an integral in functional notation (or as a numerical value), we say that we are **evaluating** the integral.

EXAMPLE 7 Evaluate the given integrals.

a. $\int (5x^4 + 2x^3 + \sqrt[3]{x})\,dx = \frac{5x^5}{5} + \frac{2x^4}{4} + \frac{x^{4/3}}{\frac{4}{3}} + C \qquad \text{Remember, } \sqrt[3]{x} = x^{1/3}$

$$= x^5 + \frac{1}{2}x^4 + \frac{3}{4}x^{4/3} + C$$

b. $\int \frac{1}{x}\,dx = \int x^{-1}\,dx = \ln|x| + C$

c. $\int (x - e^x)\,dx = \frac{x^2}{2} - e^x + C$

d. $\int \frac{x^3 + x + 1}{x}\,dx = \int \left(\frac{x^3}{x} + \frac{x}{x} + \frac{1}{x}\right)dx \qquad \text{Simplify first, then integrate.}$

$$= \int (x^2 + 1 + x^{-1})\,dx$$
$$= \frac{x^3}{3} + x + \ln|x| + C$$

As shown in Example 1, it is possible to find C if some value of the original function is known. The most commonly known value is the one when the variable is 0. That is, if

$$F'(x) = f(x)$$

and $f(0)$ is known, then this known value is called the **initial value**.

EXAMPLE 8 The population of Santa Rosa, California, is growing at the rate $3{,}000 + 500t^{1/4}$, where t is measured in years. The present population is 92,500. Predict the population in 5 years.

Solution The population function for time t is found by knowing
$$P'(t) = 3{,}000 + 500t^{1/4}$$
Thus, by integrating with respect to t,
$$P(t) = \int (3{,}000 + 500t^{1/4})\,dt = 3{,}000t + \frac{500t^{1/4+1}}{\frac{5}{4}} + C$$
$$= 3{,}000t + 400t^{5/4} + C$$
The initial condition ($t = 0$) is 92,500, so
$$P(0) = 3{,}000(0) + 400(0)^{5/4} + C = 92{,}500$$
$$C = 92{,}500$$
In 5 years,
$$P(5) = 3{,}000(5) + 400(5)^{5/4} + 92{,}500$$
$$\approx 110{,}490.6976$$
The population in 5 years will be about 110,000. ∎

6.1 Problem Set

Evaluate the indefinite integrals in Problems 1–32. You can check each answer by differentiating your answers.

1. $\int x^7\,dx$
2. $\int x^{10}\,dx$
3. $\int 4x^3\,dx$
4. $\int 12x^5\,dx$
5. $\int 3\,dx$
6. $\int 8\,dx$
7. $\int (5x + 7)\,dx$
8. $\int (5 - 2x)\,dx$
9. $\int dx$
10. $\int (9x^2 - 4x + 3)\,dx$
11. $\int (18x^2 - 6x + 5)\,dx$
12. $\int \dfrac{dx}{5}$
13. $\int \dfrac{dx}{9}$
14. $\int \dfrac{3\,dx}{x}$
15. $\int \dfrac{5\,dx}{x^2}$
16. $\int -3e^x\,dx$
17. $\int (1 - e^x)\,dx$
18. $\int (x - \sqrt{x})\,dx$
19. $\int (\sqrt[3]{x} + \sqrt{2})\,dx$
20. $\int (\sqrt[3]{5} + \sqrt{2})\,dx$
21. $\int \dfrac{3x - 1}{\sqrt{x}}\,dx$
22. $\int \dfrac{5x^2 + 2x + 3}{\sqrt{x}}\,dx$
23. $\int \dfrac{x}{\sqrt[3]{x}}\,dx$
24. $\int \dfrac{x^2}{\sqrt[3]{x^2}}\,dx$
25. $\int \dfrac{x^2 + x + 1}{x}\,dx$
26. $\int \dfrac{3x^2 + 2x + 1}{x}\,dx$
27. $\int y^3 \sqrt{y}\,dy$
28. $\int (3y^{-2/3} - 4y^{1/2})\,dy$
29. $\int (1 + z^2)^2\,dz$
30. $\int (3z + 1)^2\,dz$
31. $\int (3u^2 - u^{-1} + e^u)\,du$
32. $\int (5u^4 - 3u^{-1} + 4e^u)\,du$

In Problems 33–36 find the antiderivative $F(x)$ for the given function that satisfies the initial conditions.

33. $f(x) = 3x^2 + 4x + 1$; $F(0) = 10$
34. $f(x) = 3x^2 + 4x + 1$; $F(1) = -8$
35. $f(x) = \dfrac{5x + 1}{x}$; $F(1) = 5{,}000$
36. $f(x) = \dfrac{3x^2 + 2x + 1}{x}$; $F(1) = -50$

APPLICATIONS

In Problems 37–40 find the cost function for the given marginal cost functions.

37. $C'(x) = 4x + 8$; fixed cost = $5,000
38. $C'(x) = 128x + 300$; fixed cost = $32,000
39. $C'(x) = .009x^2$; fixed cost = $18,500
40. $C'(x) = x + 100e^x$; fixed cost = $1,900
41. The population of New Haven, Connecticut, is growing at the rate of $450 + 600\sqrt{t}$, where t is measured in years, and the present population is 420,000. Predict the population in 5 years.
42. The population of Charlotte, North Carolina, is growing at the rate of $1{,}000 + 500t^{3/4}$, where t is measured in years, and the present population is 330,000. Predict the population in 10 years.
43. The marginal profit of Turbo Plus is

$$P'(x) = 200x - 10{,}000$$

where x is the sales in thousands of items. Find the profit function if there is a $50,000 loss (−$50,000 profit) when no items are produced.

44. The marginal cost for producing x new 32-bit microprocessors is

$$C'(x) = 100x^{-2/3} + .00001\,x$$

If the fixed costs are $120,000, what is the cost of producing 100,000 processors?

45. A new strain of influenza is known to be spreading at the rate of $240t - 3t^2$ cases per day, where t is the number of days measured from the first recorded outbreak. How many people will be affected on the tenth day, if there were 50 cases on the first day?
46. Repeat Problem 45 for the fifth day, if there were five cases on the first day.
47. It is known that the acceleration of a freely falling body is a constant, 32 feet per second per second. Use the formula

$$a(t) = -32$$

to derive a formula for the velocity (at time t) for an object thrown from the top of Hoover Dam (726 feet) with an initial velocity of -72 feet per second. [*Hint:* If v is the velocity function, then $v'(t) = a(t)$; it is negative because the object was thrown downward.]

48. Derive a formula for the height of the object thrown from Hoover Dam in Problem 47. [*Hint:* If h is the height function, then $h'(t) = v(t)$.]
49. The slope of the tangent line to a curve is given by

$$f'(x) = 3x^2 + 5$$

If the point $(1, 2)$ is on the curve, find the equation of the curve.

50. Find the equation of a curve whose slope is $\sqrt{x}$ if the point $(9, 19)$ is on the curve.

6.2

Integration by Substitution

One of the most important integration techniques comes from the *generalized power rule*. For example, consider

$$\int x(x^2 + 1)^2\, dx$$

We could proceed by writing

$$\int x(x^4 + 2x^2 + 1)\, dx = \int (x^5 + 2x^3 + x)\, dx = \int \frac{x^6}{6} + \frac{2x^4}{4} + \frac{x^2}{2} + C$$

$$= \frac{1}{6}x^6 + \frac{1}{2}x^4 + \frac{1}{2}x^2 + C$$

However, it is often not practical (or possible) to multiply [as with $x(x^2 + 1)^{12}$, for example]. We will proceed, instead, by *substitution*. The complicating ingredient is the extra factor as a result of the chain rule (Section 2.7). In order to handle this factor, we need to use differentials (discussed in Section 3.2).

Let us take another look at the preceding example. This time, instead of multiplying, we will make a substitution. Look at the expression raised to a power, namely, $x^2 + 1$. Let

$$u = x^2 + 1$$

then

$$\frac{du}{dx} = 2x$$

Now, solve for dx:

$$du = 2x\,dx$$

$$dx = \frac{du}{2x} \quad (x \neq 0)$$

If substitution eliminates the variable x (leaving only the variable u), the integrand may be in a simpler, more easily integrable, form.

$$\int x(x^2 + 1)^2\,dx = \int xu^2 \frac{du}{2x}$$

$$= \frac{1}{2}\int u^2\,du \qquad \text{Note: The } x\text{'s cancel and } \tfrac{1}{2} \text{ is a constant.}$$

$$= \frac{1}{2}\frac{u^3}{3} + C \qquad \text{Use the power rule on } u$$

$$= \frac{(x^2 + 1)^3}{6} + C \qquad \text{Write } u \text{ in terms of } x \text{ for the final answer.}$$

By using substitution we are able to integrate this function without a lot of multiplication. This has tremendous advantages, as shown in Example 1.*

EXAMPLE 1 Evaluate $\int x(x^2 + 1)^{12}\,dx$.

Solution Let $u = x^2 + 1$; then $\dfrac{du}{dx} = 2x$, so $dx = \dfrac{du}{2x}$ $(x \neq 0)$. Substitute to obtain

$$\int x\underbrace{(x^2 + 1)^{12}}_{u}\,\underbrace{dx}_{\frac{du}{2x}} = \int xu^{12}\frac{du}{2x} = \frac{1}{2}\int u^{12}\,du$$

Bring constant out in front of integration.

$$= \frac{1}{2}\frac{u^{13}}{13} + C = \frac{1}{26}\underbrace{(x^2 + 1)^{13}}_{} + C$$

Replace u with original variable. ∎

---

* If you did expand, you would see that the result agrees with that shown here.

$$\frac{1}{6}(x^6 + 3x^4 + 3x^2 + 1) + C = \frac{1}{6}x^6 + \frac{1}{2}x^4 + \frac{1}{2}x^2 + \frac{1}{6} + C$$

EXAMPLE 2 Evaluate $\int x^2 \sqrt{x^3 - 2} \, dx$.

Solution Let $u = x^3 - 2$; $\dfrac{du}{dx} = 3x^2$, so $dx = \dfrac{du}{3x^2}$. Substitute

$$\int x^2 \sqrt{x^3 - 2} \, dx = \int x^2 \sqrt{u} \, \frac{du}{3x^2}$$

$$= \frac{1}{3} \int u^{1/2} \, du$$

$$= \frac{1}{3} \frac{u^{3/2}}{\frac{3}{2}} + C$$

$$= \frac{2}{9} u^{3/2} + C$$

$$= \frac{2}{9} (x^3 - 2)^{3/2} + C \qquad \blacksquare$$

The procedure is generalized below:

Integration by Substitution

1. Make a choice for u, say, $u = f(x)$.
   (u is usually inside parentheses, or an exponent, or under a radical.)
2. Find $\dfrac{du}{dx} = f'(x)$.
3. Make the substitution*
   $$u = f(x) \quad \text{and} \quad du = f'(x) \, dx$$

WARNING Note ⟶ Step 4.

4. *Everything* must be in terms of u with no remaining x's. If this is not possible, try another substitution.
5. Evaluate the integral.
6. Replace u by $f(x)$ so the final answer is in terms of x.

When using the substitution method, make sure that every x is eliminated. If this cannot be done, try a different choice, but remember that not every function *can* be integrated by this technique. This means that *after* substitution and simplification, you *must* have one of the following forms:

$$\int u^n \, du = \begin{cases} \dfrac{u^{n+1}}{n+1} + C & n \neq -1 \\ \ln|u| + C & n = -1 \end{cases}$$

or

$$\int e^u \, du = e^u + C$$

* In making the substitution $du = f'(x) \, dx$ it is often easiest to assume that values that cause division by 0 are excluded from the domain, and then solve $du = f'(x) \, dx$ for dx before making the substitution. If you do this, pay particular attention to the WARNING on step 4.

6.2 INTEGRATION BY SUBSTITUTION

EXAMPLE 3 Evaluate $\int \dfrac{x^2\, dx}{(x^3 - 2)^5}$.

Solution Let u be the value in parentheses; that is, let $u = x^3 - 2$. Then

$$du = 3x^2\, dx \quad \text{so} \quad dx = \dfrac{du}{3x^2}$$

$$\int \dfrac{x^2\, dx}{(x^3 - 2)^5} = \int \dfrac{x^2}{u^5} \cdot \dfrac{du}{3x^2} = \int \dfrac{du}{3u^5}$$

> All x's must be eliminated:
> $(x^3 - 2)^5 = u^5$ and $dx = \dfrac{du}{3x^2}$

$$= \dfrac{1}{3} \int u^{-5}\, du = \dfrac{1}{3} \dfrac{u^{-4}}{-4} + C$$

$$= \dfrac{-(x^3 - 2)^{-4}}{12} + C$$

EXAMPLE 4 Evaluate $\int e^{3x+5}\, dx$.

Solution Let u be the exponent; that is, $u = 3x + 5$, so

$$du = 3\, dx \quad \text{and} \quad dx = \dfrac{du}{3}$$

Substitute:

$$\int e^{3x+5}\, dx = \int e^u \dfrac{du}{3}$$

$$= \dfrac{1}{3} \int e^u\, du$$

$$= \dfrac{1}{3} e^u + C$$

$$= \dfrac{1}{3} e^{3x+5} + C$$

EXAMPLE 5 Evaluate $\int x^3 \sqrt{3x^2 - 1}\, dx$.

Solution Let u be the expression under the radical; that is, $u = 3x^2 - 1$, so

$$du = 6x\, dx \quad \text{and} \quad dx = \dfrac{du}{6x}$$

Substitute:

$$\int x^3 \sqrt{u}\, \frac{du}{6x} = \frac{1}{6} \int x^2 \sqrt{u}\, du$$

You can handle this.

WARNING
x^2 is a variable, so do not try to treat it like a constant. That is, do not bring out the x^2 in front of the integral sign.

Only constants may be moved across an integral.

All x's must be eliminated; note:
$$u = 3x^2 - 1$$
$$u + 1 = 3x^2$$
$$\frac{u+1}{3} = x^2$$

$$= \frac{1}{6} \int \frac{u+1}{3} \sqrt{u}\, du$$

$$= \frac{1}{18} \int (u^{3/2} + u^{1/2})\, du$$

$$= \frac{1}{18} \left(\frac{u^{5/2}}{\frac{5}{2}} + \frac{u^{3/2}}{\frac{3}{2}} \right) + C$$

$$= \frac{1}{45} u^{5/2} + \frac{1}{27} u^{3/2} + C$$

$$= \frac{1}{45} (3x^2 - 1)^{5/2} + \frac{1}{27} (3x^2 - 1)^{3/2} + C \quad \blacksquare$$

Different parts of a problem may require different substitutions.

EXAMPLE 6 Evaluate $\int \left(\frac{x^2}{1-x^3} + xe^{x^2} \right) dx$.

Solution

$$\int \left(\frac{x^2}{1-x^3} + xe^{x^2} \right) dx = \int \frac{x^2}{1-x^3}\, dx + \int xe^{x^2}\, dx$$

$u = 1 - x^3$
$du = -3x^2\, dx$

$v = x^2$
$dv = 2x\, dx$

$$= \int \frac{x^2}{u} \frac{du}{-3x^2} + \int xe^v \frac{dv}{2x}$$

$$= -\frac{1}{3} \int u^{-1}\, du + \frac{1}{2} \int e^v\, dv$$

$$= -\frac{1}{3} \ln|u| + \frac{1}{2} e^v + C$$

$$= -\frac{1}{3} \ln|1 - x^3| + \frac{1}{2} e^{x^2} + C \quad \blacksquare$$

EXAMPLE 7 The TexRite Company has found that the marginal profit for a product is

$$P'(x) = \frac{1{,}600x}{\sqrt[3]{(8x^2 - 33{,}344)^2}}$$

where x is the number of units sold on the domain $[75, 5000]$. If the break-even point is 100 units [that is, $P(100) = 0$], what is the approximate total profit for 1,000 items?

Solution

$$P(x) = \int \frac{1{,}600x\,dx}{\sqrt[3]{(8x^2 - 33{,}344)^2}} = \int \frac{1{,}600x}{\sqrt[3]{u^2}} \frac{du}{16x}$$

Let $u = 8x^2 - 33{,}344$
$du = 16x\,dx$

$$= \int 100 u^{-2/3}\,du$$

$$= 100 \frac{u^{1/3}}{\frac{1}{3}} + C$$

$$= 300(8x^2 - 33{,}344)^{1/3} + C$$

Now, $P(100) = 0$, so

$$300(8 \cdot 100^2 - 33{,}344)^{1/3} + C = 0$$
$$C = -10{,}800$$
$$P(x) = 300(8x^2 - 33{,}344)^{1/3} - 10{,}800$$
$$P(1{,}000) \approx 49{,}116.52389$$

The profit is about $49,000.

6.2
Problem Set

Evaluate the indefinite integrals in Problems 1–32.

1. $\int (5x + 3)^3\,dx$
2. $\int (1 - 5x)^4\,dx$
3. $\int \frac{5}{(5 - x)^4}\,dx$
4. $\int \frac{6}{(x + 8)^2}\,dx$
5. $\int \sqrt{3x + 5}\,dx$
6. $\int \sqrt[5]{9 - 4x}\,dx$
7. $\int 6x(3x^2 + 1)\,dx$
8. $\int 2x(4 - 5x^2)\,dx$
9. $\int \frac{2x + 5}{\sqrt{x^2 + 5x}}\,dx$
10. $\int \frac{2x + 3}{\sqrt{x^2 + 3x}}\,dx$
11. $\int \frac{dx}{6 + 5x}$
12. $\int \frac{dx}{1 - 4x}$
13. $\int \frac{x\,dx}{1 - 3x^2}$
14. $\int \frac{2x\,dx}{x^2 + 5}$
15. $\int e^{5x}\,dx$
16. $\int e^{1-3x}\,dx$
17. $\int 5x^2 e^{4x^3}\,dx$
18. $\int 3xe^{x^2+5}\,dx$
19. $\int \frac{2x - 1}{(4x^2 - 4x)^2}\,dx$
20. $\int \frac{2x - 1}{(x - x^2)^3}\,dx$
21. $\int \frac{4x^3 - 4x}{x^4 - 2x^2 + 3}\,dx$
22. $\int \frac{x^3 - x}{(x^4 - 2x^2 + 3)^2}\,dx$
23. $\int \frac{\ln|x|}{x}\,dx$
   Hint: Let $u = \ln|x|$.
24. $\int \frac{\ln|x + 1|}{x + 1}\,dx$
   Hint: Let $u = \ln|x + 1|$.
25. $\int \ln e^x\,dx$
26. $\int \ln e^{x^2}\,dx$

27. $\int \left(\dfrac{3x}{4x^2+1} + x^2 e^{x^3}\right) dx$

28. $\int \left(\dfrac{4x+6}{\sqrt{x^2+3x}} + xe^{3x^2}\right) dx$

29. $\int x\sqrt[3]{(x^2+1)^2}\, dx$

30. $\int (x^2 - 4x + 4)^{2/5}\, dx$

31. $\int \dfrac{x\, dx}{\sqrt{x+1}}$

32. $\int x^2 \sqrt{5-x}\, dx$

In Problems 33–36 consider the following integrals.

a. $\int (x^2+1)^n\, dx$
b. $\int (x^2+1)^n x\, dx$
c. $\int (x^2+1)^n x^2\, dx$
d. $\int (x^2+1)^n x^3\, dx$

33. Integrate each part for $n = 1$.
34. Integrate each part for $n = 2$.
35. Integrate each part for $n = 3$.
36. Which parts can you easily integrate for an arbitrary n? What causes these parts to be comparatively easy to integrate and why?

In Problems 37–40 consider the following integrals.

a. $\int (x^3+1)^n\, dx$
b. $\int (x^3+1)^n x\, dx$
c. $\int (x^3+1)^n x^2\, dx$
d. $\int (x^3+1)^n x^3\, dx$

37. Integrate each part for $n = 1$.
38. Integrate each part for $n = 2$.
39. Integrate each part for $n = 3$.
40. Which parts can you easily integrate for an arbitrary n? What causes these parts to be comparatively easy to integrate and why?

APPLICATIONS

41. A flu epidemic is spreading at the rate of

$$P'(t) = 40t(5 - t^2)^2$$

people per day, where t is the number of days since the first outbreak. If P is a function representing the number of sick people, and if $P(0) = 10$, find P.

42. Let $P(t)$ be the total output after t hours of work. Find P if the rate of production at time t ($1 < t < 40$) is

$$P'(t) = \dfrac{50}{50-t}$$

and production (to the nearest unit) is 71 when $t = 1$.

43. If the marginal profit (in dollars) for a product is

$$P'(x) = \dfrac{100x}{\sqrt[3]{(x^2-36)^2}}$$

where x is the number of units sold ($x \geq 7$), find the profit for 100 items if the break-even point is 10 units.

44. The marginal profit for a product is

$$P'(x) = \dfrac{20x}{\sqrt{x^2-16}}$$

where x is the number of units sold ($x \geq 5$). If the break-even point is 5 units, what is the profit for 25 items?

45. Sales are changing at a rate given by the formula

$$S'(t) = 2{,}000 e^{-.2t}$$

where t is the time in years and $S(t)$ is the sales. What are the sales in 3 years if initial sales are 15,000 units?

46. In 1976 the world population reached 4 billion and was growing at a rate approximated by the formula

$$P'(t) = .072 e^{.018t}$$

where t is measured in years since 1976 and $P(t)$ gives the population. What is the predicted population in the year 2000?

47. In 1985 the growth rate of San Antonio, Texas, was given by the formula

$$P'(t) = 14{,}000 e^{.0175t}$$

where t is measured in years since 1980 and $P(t)$ gives the population. If the population in 1984 was 842,779, predict the population in the year 2000.

48. Between 1980 and 1985 worldwide oil consumption dropped from 2,400 million barrels to 1,950 million barrels. This rate of consumption is given by the formula

$$R(t) = 78 e^{-.04t}$$

where t is measured in years since 1985. The total consumption since 1985 is given by $T(t)$, where $T'(t) = R(t)$. Find the consumption from 1985 to the year 2000.

49. If we consider worldwide oil consumption from 1925 to 1985, the rate of consumption is given by the formula

$$R(t) = 32.5 e^{.048t}$$

where t is measured in years since 1985. If the worldwide consumption from 1925 to 1985 was 1,950 million barrels, predict the consumption from 1985 to the year 2000 if the total consumption is given by $T(t)$, where $T'(t) = R(t)$.

6.3 The Definite Integral

We will introduce the definite integral with an example, and our approach will be informal and intuitive. These ideas will be formalized in Section 6.5 when we introduce a notion called Riemann sums and the Fundamental Theorem of Calculus.

The Definite Integral

In 1986 world oil prices plummeted. The price drop was related to overproduction and declining consumption. Suppose that between 1980 and 1985 the rate of consumption (in billions of barrels) is given by the formula

$$R(t) = 78e^{-.04t}$$

where t is measured in years after 1985.* If we use this consumption rate, we can predict the amount of oil that will be consumed from 1987 to 1995. The total consumption since 1985 is given by $T(t)$, where $T'(t) = R(t)$. Then

$$T(t) = \int T'(t)\,dt = \int R(t)\,dt$$
$$= \int (78e^{-.04t})\,dt$$
$$= \frac{78e^{-.04t}}{-.04} + C$$
$$= -1{,}950 e^{-.04t} + C$$

We can find C because if $t = 0$, then oil consumption is also zero. Thus

$$T(0) = -1{,}950 e^{-.04t} + C = 0$$
$$-1{,}950 + C = 0$$
$$C = 1{,}950$$

Thus

$$T(2) = -1{,}950 e^{-.04(2)} + 1{,}950 \approx 150 \qquad \text{Total consumption from 1985 to 1987}$$

$$T(10) = -1{,}950 e^{-.04(10)} + 1{,}950 \approx 643 \qquad \text{Total consumption from 1985 to 1995}$$

We can find the total consumption from 1987 to 1995 by subtraction:

$$T(10) - T(2) = 643 - 150 = 493$$

We would predict approximately 493 billion barrels to be consumed, assuming the rate of consumption does not change. However, if prices continue to drop, we would expect the rate of consumption to again turn upward (see Problems 35–39 in Problem Set 6.3).

* Equations like this are derived in Chapter 9.

Let us take a closer look at what we have done. Since $T(t)$ is an antiderivative of $R(t)$ over the interval from 1987 to 1995, we see that $T(10) - T(2)$ is the *net change* of the function T over this interval. In general, if $F(x)$ is any function and a and b are real numbers with $a < b$, then the *net change of $F(x)$ over the interval $[a,b]$* is the number

$$F(b) - F(a)$$

The quantity $F(b) - F(a)$ is often abbreviated by the symbol

$$F(x)\Big|_a^b$$

This discussion suggests the following formula:

Definite Integral

Let f be a function defined over the interval $[a,b]$. Then the **definite integral** of f over this interval is denoted by

$$\int_a^b f(x)\,dx$$

and is the net change of an antiderivative of f over that interval. Thus, if $F(x)$ is an antiderivative of $f(x)$, then

$$\int_a^b f(x)\,dx = F(x)\Big|_a^b = F(b) - F(a)$$

WARNING The definite integral $\int_a^b f(x)\,dx$ is a real number. The indefinite integral $\int f(x)\,dx$ is a set of functions—namely, all of those functions whose derivative is f(x).

The function f is called the **integrand** and the constants a and b are called the **limits of integration**. The number x is called a **dummy variable** because the definite integral is a fixed number and not a function of x.

Evaluating Definite Integrals

Both the integral sign and the symbol dx are dropped when the antiderivative is found. Also, when using the definite integral it is not necessary to take into account a constant of integration. [Remember, in the introductory example, the constant of integration (1,950) had no bearing on the final answer since it was first added and then subtracted.] The process of finding the value of a definite integral is called **evaluating the integral**.

EXAMPLE 1 Evaluate the given integrals.

a. $\int_2^3 6x^2\,dx$ **b.** $\int_{-2}^2 x^3\,dx$ **c.** $\int_1^{10} \frac{dx}{x}$

Solution **a.** $\int_2^3 6x^2\,dx = \frac{6x^3}{3}\Big|_2^3 = \frac{6(3)^3}{3} - \frac{6(2)^3}{3} = 2(27) - 2(8) = 38$

— Evaluate at 3 first.
Antiderivative of $6x^2$
Subtract the value of the antiderivative at 2.

b. $\int_{-2}^{2} x^3 \, dx = \frac{x^4}{4}\bigg|_{-2}^{2} = \frac{2^4}{4} - \frac{(-2)^4}{4} = 0$

c. $\int_{1}^{10} \frac{dx}{x} = \ln|x|\bigg|_{1}^{10} = \ln|10| - \ln|1| = \ln 10 - 0 \approx 2.3$

Remember, $\ln 1 = 0$.

Properties of the Definite Integral

There are many properties of the definite integral, several of which are analogous to those we stated for the indefinite integral:

Properties of the Definite Integral

1. $\int_{a}^{a} f(x) \, dx = 0$ where a is in the domain of f

2. $\int_{a}^{b} dx = b - a$

3. $\int_{a}^{b} f(x) \, dx = -\int_{b}^{a} f(x) \, dx$

4. $\int_{a}^{b} f(x) \, dx = \int_{a}^{c} f(x) \, dx + \int_{c}^{b} f(x) \, dx$ for any points a, b, and c in the closed interval $[a, b]$

5. $\int_{a}^{b} kf(x) \, dx = k\int_{a}^{b} f(x) \, dx$

6. $\int_{a}^{b} [f(x) \pm g(x)] \, dx = \int_{a}^{b} f(x) \, dx \pm \int_{a}^{b} g(x) \, dx$

EXAMPLE 2 Evaluate the given integrals.

a. $\int_{10}^{10} (3x^2 - \sqrt{x}) \, dx$ **b.** $\int_{4}^{3} (3x^2 - \sqrt{x}) \, dx$

c. $\int_{-2}^{\pi/\sqrt{2}} e^t \, dt + \int_{\pi/\sqrt{2}}^{2} e^t \, dt$ (correct to the nearest tenth)

Solution **a.** $\int_{10}^{10} (3x^2 - \sqrt{x}) \, dx = 0$

b. $\int_{4}^{3} (3x^2 - \sqrt{x}) \, dx = -\int_{3}^{4} (3x^2 - \sqrt{x}) \, dx = -\left[\frac{3x^3}{3} - \frac{x^{3/2}}{\frac{3}{2}}\right]\bigg|_{3}^{4}$

$= -(4^3 - \frac{2}{3} \cdot 4^{3/2}) + (3^3 - \frac{2}{3} \cdot 3^{3/2})$

$= -(64 - \frac{16}{3}) + (27 - 2\sqrt{3})$

$= \frac{16}{3} - 37 - 2\sqrt{3}$

$= \frac{-95 - 6\sqrt{3}}{3}$

c. $\int_{-2}^{\pi/\sqrt{2}} e^t \, dt + \int_{\pi/\sqrt{2}}^{2} e^t \, dt = \int_{-2}^{2} e^t \, dt = e^t \bigg|_{-2}^{2} = e^2 - e^{-2} \approx 7.2$

Substitution with the Definite Integral

Always be careful about the evaluation of the variable when you use substitution in order to find the antiderivative.

EXAMPLE 3 Evaluate $\int_{1}^{\sqrt{5}} \frac{6x\,dx}{\sqrt{1+3x^2}}$.

Solution

$$\int_{1}^{\sqrt{5}} \frac{6x\,dx}{\sqrt{1+3x^2}} = \int_{x=1}^{x=\sqrt{5}} \frac{du}{\sqrt{u}}$$

Note that we need to say $x = \sqrt{5}$ and $x = 1$ are the limits of integration so that they are not confused with the variable u.

$$= \int_{x=1}^{x=\sqrt{5}} u^{-1/2}\,du$$

$$= 2u^{1/2} \Big|_{x=1}^{x=\sqrt{5}}$$

Also notice: $u = 1 + 3x^2$
$du = 6x\,dx$

$$= 2\sqrt{1+3x^2} \Big|_{1}^{\sqrt{5}}$$

$$= 2(4) - 2(2)$$

$$= 4$$

Instead of substituting back to the original variable, we can change the limits of integration. For Example 3, since $u = 1 + 3x^2$, we can substitute $x = 1$ (lower limit):

$$u = 1 + 3(1)^2 = 4$$

and $x = \sqrt{5}$ (upper limit):

$$u = 1 + 3(\sqrt{5})^2 = 16$$

Then we can write

— If $x = \sqrt{5}$, then $u = 16$.

— Note the change in the limits of integration.

$$\int_{1}^{\sqrt{5}} \frac{6x\,dx}{\sqrt{1+3x^2}} = \int_{4}^{16} u^{-1/2}\,du$$

If $x = 1$, then $u = 4$.

$$= 2u^{1/2} \Big|_{4}^{16}$$

$$= 2(\sqrt{16} - \sqrt{4})$$

$$= 4$$

Substitution with the Definite Integral

If f is continuous on the set of values taken on by g, and g' is continuous on $[a, b]$, and if f has an antiderivative on that interval, then

$$\int_{a}^{b} f[g(x)]g'(x)\,dx = \int_{g(a)}^{g(b)} f(u)\,du \qquad u = g(x), \quad du = g'(x)\,dx$$

provided these integrals exist.

EXAMPLE 4   Evaluate $\int_1^2 e^{-3t}\,dt$ correct to three decimal places.

Solution $\int_1^2 e^{-3t}\,dt = -\dfrac{1}{3}\int_{-3}^{-6} e^u\,du = -\dfrac{1}{3}e^u\Big|_{-3}^{-6} = -\dfrac{1}{3}(e^{-6} - e^{-3}) \approx .016$

> Let $u = -3t$; if $t = 2$, then $u = -6$
> $du = -3\,dt$; if $t = 1$, then $u = -3$

Applications

If you are given some function and want to find the rate at which the function is changing at some point, then we know that the derivative will give us this rate. On the other hand, it is not uncommon to know a function which *itself* represents the rate of change. For example, you might know the divorce rate, or rate of flow of blood, or velocity (which is the rate of change of distance with respect to time), or rate of discharge of pollutants into a water supply, or marginal cost (rate of change of cost with respect to number of items). We could cite many other examples in which you might know the rate and wish to find the original function. It should be clear that to find the original function, whose rate of change is known, requires an integral, but it might not be clear what this integral represents. Some examples are given in Table 6.1.

TABLE 6.1
Examples of Given Rates and Their Associated Integrals

Given rate	Associated definite integral
Divorce rate over some interval of time	Total accumulated number of divorces for the given time interval
Rate of flow of blood over some time interval	Total accumulated flow of blood for the given time interval
Velocity	Distance (or position) function
Rate of discharge over some time interval	Total amount of discharge over the given time interval
Marginal cost	Total cost function
Marginal revenue	Total revenue function
Marginal profit	Total profit function
Rate of depreciation with respect to time, t	The value of the item after t years
Rate of consumption after some time, t	The total consumption after t years

EXAMPLE 5   A factory has been dumping pollutants into a lake since 1985. The rate of change in the concentration of the pollutant (in parts per million) at time t is given by the formula

$$P(t) = 150t^{3/2}$$

where t is the number of years since 1985. Ecologists have established that when the total pollution level reaches 24,000 ppm (parts per million), all bass in the lake will be killed. How much longer could the factory operate before all bass in the lake are killed?

Solution We know the rate and wish to find the total accumulation. This is a process that leads to a definite integral. The total accumulation of pollution dumped into the lake in x years since 1985 is given by

$$\int_0^x 150 t^{3/2}\, dt = \frac{150 t^{5/2}}{\frac{5}{2}}\bigg|_0^x = 60 t^{5/2}\bigg|_0^x = 60 x^{5/2}$$

Now, we need to solve the equation $60 x^{5/2} = 24{,}000$:

$$60 x^{5/2} = 24{,}000$$
$$x^{5/2} = 400$$
$$(x^{5/2})^{2/5} = (400)^{2/5}$$
$$x \approx 10.985605$$

The bass in the lake will all be dead approximately 11 years after 1985—that is, in 1996. ∎

6.3

Problem Set

Evaluate the definite integrals in Problems 1–30.

1. $\int_2^3 x^2\, dx$
2. $\int_{-2}^3 x^3\, dx$
3. $\int_1^4 \sqrt{x}\, dx$
4. $\int_1^{100} dx$
5. $\int_{-4}^{-1} \sqrt{1 - 2x}\, dx$
6. $\int_{-3}^1 \sqrt{1 - x}\, dx$
7. $\int_0^2 e^{2x}\, dx$
8. $\int_0^3 e^{-x}\, dx$
9. $\int_1^{10} x^{-1}\, dx$
10. $\int_1^{e^3} \frac{dx}{x}$
11. $\int_1^{32} x^{-2/5}\, dx$
12. $\int_4^9 x^{3/2}\, dx$
13. $\int_3^8 (x - 2)^{-1}\, dx$
14. $\int_5^{10} (x - 3)^{-1}\, dx$
15. $\int_{-1}^2 x(1 + x^4)\, dx$
16. $\int_{-1}^1 x^3(1 + x^4)\, dx$
17. $\int_2^2 x(1 + x)^4\, dx$
18. $\int_{-1}^1 x^2(1 + x)^4\, dx$
19. $\int_0^1 \frac{dx}{(2x + 1)^2}$
20. $\int_{-1}^0 \frac{dx}{(1 - 3x)^3}$
21. $\int_1^2 \left(x^{-2} - \frac{2}{x} + x^3\right) dx$
22. $\int_1^2 \left(x^2 + \frac{5}{x} - \sqrt{x}\right) dx$
23. $\int_6^6 \left(x^2 + \frac{\sqrt{5x}}{3} - 5\right) dx$
24. $\int_0^2 \left(x^2 + \frac{\sqrt{5x}}{3} - 5\right) dx$
25. $\int_{-1}^1 \frac{x\, dx}{\sqrt{1 + 3x^2}}$
26. $\int_{-1}^1 x\sqrt{1 + 3x^2}\, dx$
27. $\int_{-1}^0 5x^2(x^3 + 1)^{10}\, dx$
28. $\int_{-1}^0 3x^4(1 - x^5)^3\, dx$
29. $\int_1^e 3x^{-1} \ln^3 x\, dx$
30. $\int_1^e x^{-1} \ln^2 x\, dx$

31. If f and g are continuous on $[-3, 2]$ and
$$\int_{-3}^{0} f(x)\,dx = 4, \quad \int_{0}^{2} f(x)\,dx = 10,$$
and $\int_{-3}^{2} g(x) = -5,$ find

a. $\int_{-3}^{2} f(x)\,dx$
b. $\int_{2}^{-3} g(x)\,dx$
c. $\int_{-3}^{2} [2f(x) - 3g(x)]\,dx$

32. If f and g are continuous on $[5, 10]$ and
$$\int_{5}^{8} f(x)\,dx = -3, \quad \int_{8}^{10} f(x)\,dx = 4,$$
and $\int_{5}^{10} g(x)\,dx = 2,$ find

a. $\int_{5}^{10} f(x)\,dx$
b. $\int_{7}^{7} [f(x) + g(x)]\,dx$
c. $\int_{10}^{5} g(x)\,dx$

33. If f is continuous on $[1, 5]$ and
$$\int_{1}^{3} f(x)\,dx = 9 \quad \text{and} \quad \int_{1}^{5} f(t)\,dt = 15, \quad \text{find} \int_{3}^{5} f(y)\,dy.$$

34. If g is continuous on $[-3, 2]$ and
$$\int_{-3}^{2} g(x)\,dx = 7 \quad \text{and} \quad \int_{0}^{2} g(t)\,dt = 4, \quad \text{find} \int_{-3}^{0} g(y)\,dy.$$

APPLICATIONS

35. Suppose that the rate of oil consumption (in billions of barrels) since 1985 is given by the formula
$$R(t) = 78e^{-.04t}$$
where t is measured in years since 1985. Predict the total number of barrels of oil consumed between 1987 and 2000.

36. Suppose that because of the 1986 crash in oil prices the demand for oil (in billions of barrels) changed so that it conforms to the formula
$$R(t) = 32.5e^{.048t}$$
where t is measured in years since 1986. Predict the total number of barrels consumed between 1988 and 1996 using 1986 as $t = 0$.

37. Repeat Problem 36 for the years 1987 to 2000.

38. If the known oil reserves in 1985 were 670.3 billion barrels, use the information in Problem 35 to estimate the length of time before all known reserves are depleted.

39. If the known oil reserves in 1985 were 670.3 billion barrels, use the information in Problem 36 to estimate the length of time before all known reserves are depleted.

40. What is the total amount of pollutants dumped into the lake in Example 5 between the years 1985 and 1990? Recall that the formula for the concentration of the pollutants (in ppm) at time t is given by the formula
$$P(t) = 150t^{3/2}$$

41. A factory has been dumping pollutants into a river since 1990. The rate of concentration of the pollutants (in ppm) at time t is given by the formula
$$P(t) = 350t^{5/2}$$
where t is the number of years since 1990. What is the total amount of pollutants dumped into the river between 1990 and 1995?

42. What is the total amount of pollutants dumped into the lake in Example 5 in 1992? Recall that the formula for the concentration of the pollutants (in ppm) at time t is given by the formula
$$P(t) = 150t^{3/2}$$

43. What is the total amount of pollutants dumped into the river in Problem 41 in 1992? Recall that the formula for the concentration of the pollutants (in ppm) at time t is given by the formula
$$P(t) = 350t^{5/2}$$

44. It is known that all life within 20 miles of a dumping site will die if the total concentration of the pollutants reaches 28,000 ppm. If dumping started in 1990, when will the pollution level be high enough to kill all life within the 20-mile radius of the dumping site? Assume that the formula for the concentration of pollutants (in ppm) at time t is given by $P(t) = 350t^{5/2}$.

45. The sales of Thornton Publishing are continuously growing according to the function
$$S(t) = 50e^{t}$$
where $S(t)$ is the rate of sales in dollars in the tth year. What are the accumulated sales after 10 years?

46. When will the accumulated sales in Problem 45 pass $1 million?

47. In economics, the Gini index, G, can be used to measure how money is distributed among the population according to the formula
$$G = 1 - \int_{0}^{1} 2f(x)\,dx$$

The closer G is to zero, the more the money is spread evenly throughout the population. Find G for the function $f(x) = x^3$.

48. Find the Gini index, G, for the function $f(x) = x^{3/2}$. The Gini index is a measure for the amount of money spread throughout the population (see Problem 47) and is given by

$$G = 1 - \int_0^1 2f(x)\,dx$$

The closer this number is to zero, the more the money is spread evenly throughout the population.

49. The total accumulated cost and revenue for a piece of business machinery is

$$C'(t) = \frac{t}{20} \quad \text{and} \quad R'(t) = 10te^{-t^2}$$

where t is the time in years. Find the cost and revenue functions if the initial cost is $25,000 and the initial revenue is $0.

50. The rate of healing for a wound is given by the formula

$$A'(t) = -.85e^{-.1t}$$

where A is the number of square centimeters of unhealed skin after t days. How much will the area change in the first week if the initial wound has an area of 10 cm²?

6.4
Area Between Curves

Area Function

One of the most common applications of definite integrals involves the area under a curve.

Let f be a continuous nonnegative function on $[a, b]$. Let $A(x)$ be the **area function** that denotes the area of the region of $[a, b]$ bounded by f on top and the x-axis on the bottom and the vertical lines x and a, as shown in Figure 6.1.

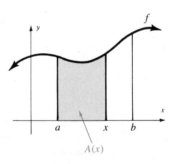

FIGURE 6.1
The area function

EXAMPLE 1 Let $a = 0$ and find $A(x)$ for each of the given functions.

a. $f(x) = 4$ **b.** $f(x) = 4x$ **c.** $f(x) = 4x + 3$

Solution **a.**

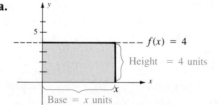

c. The figure formed is a rectangle, so

$$A(x) = bh$$
$$= x(4)$$
$$= 4x$$

b.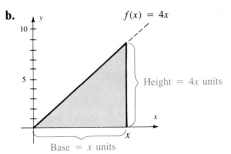

The figure formed is a triangle, so

$$A(x) = \frac{1}{2}bh$$
$$= \frac{1}{2}(x)(4x)$$
$$= 2x^2$$

c.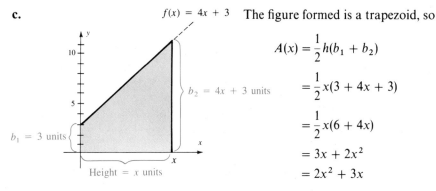

The figure formed is a trapezoid, so

$$A(x) = \frac{1}{2}h(b_1 + b_2)$$
$$= \frac{1}{2}x(3 + 4x + 3)$$
$$= \frac{1}{2}x(6 + 4x)$$
$$= 3x + 2x^2$$
$$= 2x^2 + 3x$$

The area function for each part of Example 1 is the antiderivative of f where $C = 0$. This is true whenever $a = 0$. That is,

Function	Area function	Antiderivative
$f(x) = 4$	$A(x) = 4x$	$F(x) = 4x + C$
$f(x) = 4x$	$A(x) = 2x^2$	$F(x) = 2x^2 + C$
$f(x) = 4x + 3$	$A(x) = 2x^2 + 3x$	$F(x) = 2x^2 + 3x + C$

Area Under a Curve

The fact that the area of the figures bounded by f in Example 1 turned out to be the antiderivatives of f is not a coincidence. In elementary mathematics, areas are defined in terms of the area of a square (i.e., square units). However, when we want to find the area of a region with a curved boundary, the situation is much more complicated. Even the formula for the area of a circle that you remember from elementary school is a result of the calculus. In order to find the area of regions bounded by curves (or functions in general), we make the following definition.

Area Under a Curve

If f is a continuous nonnegative function on $[a, b]$, then the **area** of the region bounded by the graph of f and the x-axis and the vertical lines $x = a$ and $x = b$ is given (exactly) by

$$A = \int_a^b f(x)\,dx$$

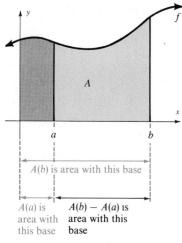

FIGURE 6.2
$A = A(b) - A(a)$

The area A in this definition is shown with the color screen in Figure 6.2. We see that the idea of area as defined by the integral in the box can be justified in terms of area functions (shown in Figure 6.1)—namely, $A(b) - A(a)$:

$$A = \int_a^b f(x)\,dx = A(b) - A(a)$$

Next, we wish to find a relationship between the function f and the area functions $A(b) - A(a)$. We begin with the area function $A(x)$ and shall show that $A'(x) = f(x)$, which means that $A(x)$ is an antiderivative of $f(x)$. From the definition of derivative,

$$A'(x) = \lim_{h \to 0} \frac{A(x+h) - A(x)}{h}$$

Consider $\dfrac{A(x+h) - A(x)}{h}$ in Figure 6.3. Notice that $A(x+h) - A(x)$ is the region shaded in color. This area is approximated by the area of the rectangle with base h and height $f(x)$. Thus

$$A(x+h) - A(x) \approx f(x) \cdot h$$
$$\frac{A(x+h) - A(x)}{h} \approx f(x)$$

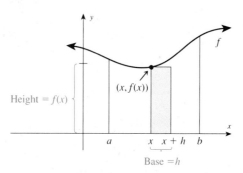

FIGURE 6.3
Area of shaded rectangle is $f(x)h$, which approximates $A(x+h) - A(x)$, as shown in color.

Now if we let $h \to 0$, then the left side has $A'(x)$ as a limit, which is equal to the right side. Thus,

$$A'(x) = f(x)$$

It is therefore reasonable to denote $A(x) = F(x)$, where F is an antiderivative of f. We can now restate the area definition:

$$A = \int_a^b f(x)\,dx = F(x)\Big|_a^b = F(b) - F(a)$$

6.4 AREA BETWEEN CURVES

We will now see by examples that this idea of area under a curve compares exactly with your previous knowledge of area from plane geometry.

EXAMPLE 2 Find the areas of the rectangle, triangle, and trapezoid from Example 1 for $a = 0$ and $b = 10$ using both the results of Example 1 and the area under a curve definition.

a. $f(x) = 4$ **b.** $f(x) = 4x$ **c.** $f(x) = 4x + 3$

Solution **a.** From Example 1a, $A(x) = 4x$, so $A(10) = 40$ and $A(0) = 0$. Using calculus,

$$\int_0^{10} 4\,dx = 4x \Big|_0^{10} = 4(10 - 0) = 40$$

b. From Example 1b, $A(x) = 2x^2$, so $A(10) = 2(10)^2 = 200$ and $A(0) = 0$. Using calculus,

$$\int_0^{10} 4x\,dx = 2x^2 \Big|_0^{10} = 2(10)^2 - 2(0)^2 = 200$$

c. From Example 1c, $A(x) = 2x^2 + 3x$, so $A(10) = 2(10)^2 + 3(10) = 230$ and $A(0) = 0$. Using calculus,

$$\int_0^{10} (4x + 3)\,dx = (2x^2 + 3x) \Big|_0^{10} = [2(10)^2 + 30] - (2 \cdot 0^2 + 0) = 230 \quad \blacksquare$$

EXAMPLE 3 Find the area under the curve $y = x^2$ on $[2, 4]$.

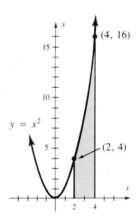

Solution $\int_2^4 x^2\,dx = \dfrac{x^3}{3}\Big|_2^4 = \dfrac{4^3}{3} - \dfrac{2^3}{3} = \dfrac{64}{3} - \dfrac{8}{3} = \dfrac{56}{3} = 18\dfrac{2}{3}$ $\blacksquare$

Area Between a Curve and the x-Axis

There is an important restriction on the conditions of the area under a curve definition. This condition is that the function f be nonnegative. Another way of looking at this is to say that the definite integral does not always give the area! You must, therefore, be careful about the way you state your conclusions. Consider the next example carefully.

EXAMPLE 4 Evaluate $\int_{-2}^{2} x^3 \, dx$.

Solution It would **not** be correct to say that this integral is the area bounded by the curve $f(x) = x^3$, the x-axis, and the lines $x = -2$ and $x = 2$. If you graph $f(x) = x^3$ (shown in the margin) we see that there is an area, but if we evaluate

$$I = \int_{-2}^{2} x^3 \, dx = \frac{x^4}{4} \Big|_{-2}^{2} = \frac{2^4}{4} - \frac{(-2)^4}{4} = 0$$

the definite integral is not the area. ∎

Compare the areas shown by the two graphs in Figure 6.4.

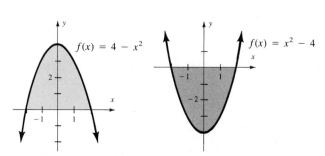

FIGURE 6.4
Areas bounded by two curves

For the curve bounded by $f(x) = 4 - x^2$, the integral **is** the area.

$$\int_{-2}^{2} (4 - x^2) \, dx = 4x - \frac{x^3}{3} \Big|_{-2}^{2}$$
$$= 8 - \frac{8}{3} - \left(-8 + \frac{8}{3}\right)$$

This is the area. $= 16 - \frac{16}{3} = \frac{32}{3}$

For the curve bounded by $f(x) = x^2 - 4$, the integral is **not** the area.

$$\int_{-2}^{2} (x^2 - 4) \, dx = \frac{x^3}{3} - 4x \Big|_{-2}^{2}$$
$$= \frac{8}{3} - 8 - \left(-\frac{8}{3} + 8\right)$$

This is not the area. $= \frac{16}{3} - 16 = -\frac{32}{3}$

It appears that if f is negative over $[a, b]$, then the area is the opposite of the given integral. Consider the function shown in Figure 6.5.

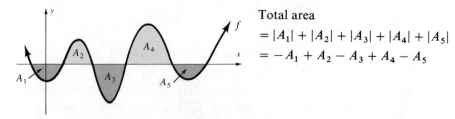

Total area
$= |A_1| + |A_2| + |A_3| + |A_4| + |A_5|$
$= -A_1 + A_2 - A_3 + A_4 - A_5$

FIGURE 6.5
Area defined by a function over $[a, b]$. A_2 and A_4 are positive numbers representing areas and A_1, A_3, and A_5 are opposites of the enclosed areas.

EXAMPLE 5 Find the area bounded by the curve $f(x) = x^3$, the x-axis, and the lines $x = -2$ and $x = 2$. See Example 4.

Solution We need to know which part of this curve is above the x-axis and which is below. We solve $f(x) = 0$, which tells us where the curve crosses the x-axis; in this example the curve crosses at $x = 0$. Let A be the desired area.

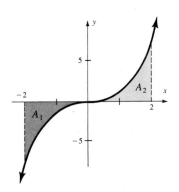

$$A = |A_1| + |A_2| \quad \text{where} \quad A_1 = \int_{-2}^{0} x^3 \, dx = \left.\frac{x^4}{4}\right|_{-2}^{0} = 0 - \frac{16}{4} - 0 = -4$$

$$\text{and} \quad A_2 = \int_{0}^{2} x^3 \, dx = \left.\frac{x^4}{4}\right|_{0}^{2} = 4$$

Thus, $A = -A_1 + A_2 = -(-4) + 4 = 8$. ∎

Area Between Curves

The difficulties encountered when dealing with functions like the one in Example 5 which have both positive and negative values (shown in Figure 6.5) can be simplified if we find the area *between* two curves.

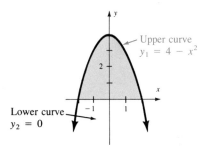

Difference between two curves:

$y_1 - y_2 = (4 - x^2) - 0 = 4 - x^2$

$$\text{Area} = \int_a^b (y_1 - y_2) \, dx$$

Upper curve ↑   ↓ Lower curve

Difference between two curves:

$y_2 - y_1 = 0 - (x^2 - 4) = 4 - x^2$

$$\text{Area} = \int_a^b (y_2 - y_1) \, dx$$

Upper curve ↑   ↓ Lower curve

Area Between Two Curves

Suppose that f and g are continuous functions on $[a, b]$ and that $f(x) \geq g(x)$ for $a \leq x \leq b$.* The area of the region bounded by f above, g below, $x = a$ on the left, and $x = b$ on the right is

$$A = \int_a^b [\underbrace{f(x)}_{\text{Upper curve}} - \underbrace{g(x)}_{\text{Lower curve}}] \, dx$$

EXAMPLE 6 Find the area between the curves $y = x^2$ and $y = 6 - x$ from $x = -3$ to $x = 2$.

Solution First sketch the curves. The top curve is $y = 6 - x$ and the bottom curve is $y = x^2$. Thus the area between the curves is given by the definite integral

$$\int_{-3}^{2} [(6 - x) - x^2] \, dx = \left(6x - \frac{x^2}{2} - \frac{x^3}{3} \right) \Bigg|_{-3}^{2}$$

$$= \left(12 - \frac{4}{2} - \frac{8}{3} \right) - \left(-18 - \frac{9}{2} + \frac{27}{3} \right)$$

$$= \frac{125}{6} \quad \text{or} \quad 20\frac{5}{6} \quad \blacksquare$$

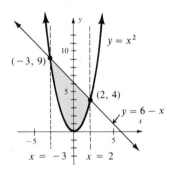

Area Between Curves That Intersect

Sometimes the limits of integration for problems like Example 6 are implied instead of given. In such cases you need to solve the system of equations formed by the boundaries in the problem:

$$\begin{cases} y = x^2 \\ y = 6 - x \end{cases}$$

By substitution,

$$6 - x = x^2$$
$$x^2 + x - 6 = 0$$
$$(x + 3)(x - 2) = 0$$
$$x = -3, 2$$

The points of intersection for the curves are now found:

If $x = -3$, then $y = 6 - (-3) = 9$. Point: $(-3, 9)$
If $x = 2$, then $y = 6 - 2 = 4$. Point: $(2, 4)$

You would use these values for the limits of integration as shown in Example 6.

A condition of the formula for the area between two curves is that you subtract the lower curve from the upper curve. This means you must know whether the curves intersect over the interval of integration. If they do intersect, then you must divide the problem into separate integrals as illustrated by Example 7.

* This means that the curve f is never below g anywhere on the interval from a to b.

6.4 AREA BETWEEN CURVES

EXAMPLE 7 **a.** Evaluate $\int_0^2 (x^3 - x)\,dx$.

b. Find the area of the region bounded by $y = x^3$ and $y = x$ on the interval $[0, 2]$ and compare with the answer you found in part **a**.

Solution **a.** Integral $= \int_0^2 (x^3 - x)\,dx = \left(\dfrac{x^4}{4} - \dfrac{x^2}{2}\right)\Big|_0^2 = \left(\dfrac{16}{4} - \dfrac{4}{2}\right) - \left(\dfrac{0}{4} - \dfrac{0}{2}\right) = 2$

b. When finding the area you must make sure that you identify the upper and lower curves on the interval. To see whether the curves cross we can solve the system

$$\begin{cases} y = x^3 \\ y = x \end{cases}$$

Use substitution to find

$$x^3 = x$$
$$x^3 - x = 0$$
$$x(x^2 - 1) = 0$$
$$x(x - 1)(x + 1) = 0$$
$$x = 0, 1, -1$$

The graph is shown in the margin. In order to find the area, we must break the integral into two parts in order to make sure that the equation for the lower curve is subtracted from the equation of the upper curve.

$$\text{Area} = \int_0^1 (x - x^3)\,dx + \int_1^2 (x^3 - x)\,dx$$

$$= \left(\dfrac{x^2}{2} - \dfrac{x^4}{4}\right)\Big|_0^1 + \left(\dfrac{x^4}{4} - \dfrac{x^2}{2}\right)\Big|_1^2$$

$$= \left(\dfrac{1}{2} - \dfrac{1}{4}\right) - 0 + \left[\left(\dfrac{16}{4} - \dfrac{4}{2}\right) - \left(\dfrac{1}{4} - \dfrac{1}{2}\right)\right]$$

$$= \dfrac{5}{2}$$

Notice that the area is not the same as the integral over the interval $[0, 2]$. ∎

EXAMPLE 8 Find the area between the curve $y = x^2 - 9$, the x-axis, and the lines $x = 1$ and $x = 9$.

Solution Check to see whether the curves $y = x^2 - 9$ and $y = 0$ cross by solving

$$x^2 - 9 = 0$$
$$(x - 3)(x + 3) = 0$$
$$x = 3, -3$$
$$\uparrow$$
Not an interval

The curves cross at $x = 3$.

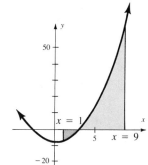

The area is found by dividing the interval into two different integrals:

$$\int_1^3 [0-(x^2-9)]\,dx + \int_3^9 [\underbrace{(x^2-9)}_{\text{Top curve}}-0]\,dx = \left(-\frac{x^3}{3}+9x\right)\bigg|_1^3 + \left(\frac{x^3}{3}-9x\right)\bigg|_3^9$$

(Top curve labels point to $-(x^2-9)$ and (x^2-9).)

$$= (-9+27) - \left(-\frac{1}{3}+9\right) + (243-81) - (9-27)$$

$$= \frac{568}{3} \quad \text{or} \quad 189\frac{1}{3} \quad \blacksquare$$

EXAMPLE 9 The 1986 crash of oil prices brought about a change in the worldwide consumption rate of petroleum products (*Wall Street Journal*, February 11, 1986). We can measure the effect of that change by finding the area between two curves. For example, suppose that

pre-1985 rate of consumption: $R_1(t) = 78e^{-.04t}$ $(0 \le t \le 5)$
post-1985 rate of consumption: $R_2(t) = 50.2e^{.048t}$ $(t \ge 5)$

in billions of barrels for a base year of 1980 ($t = 0$). How much extra oil would be consumed between 1985 and 1990?

Solution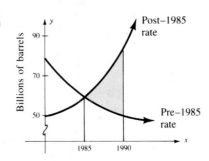

$$\int_5^{10} (50.2e^{.048t} - 78e^{-.04t})\,dt$$

$$= \frac{50.2}{.048}e^{.048t} - \frac{78}{-.04}e^{-.04t}\bigg|_5^{10}$$

$$\approx 2{,}997.27 - 2{,}926.04$$

$$\approx 71.2$$

The additional oil consumed would be about 71.2 billion barrels. ∎

6.4

Problem Set

Write integrals to express the shaded region in Problems 1–6. Both positive regions (color) and negative regions (gray) should be included.

1.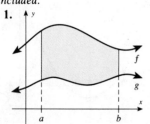

2.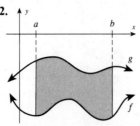

3.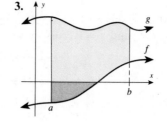

4.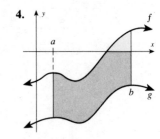

5.

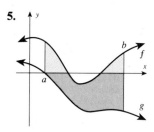

6.

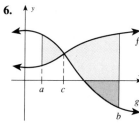

Find the area bounded by the conditions given in Problems 7–35.

7. $y = 6x^2$, the x-axis, and the lines $x = 1$, $x = 5$
8. $y = 4x^3$, the x-axis, and the lines $x = 1$, $x = 4$
9. $y = 3x^2 + 4$, $y = 0$, $x = -1$, and $x = 2$
10. $y = 12x^2 - 2x + 5$, $y = 0$, $x = 0$, $x = 2$
11. $y = 4x^3 - 3x^2 + 2x$, the x-axis on $[0, 2]$
12. $y = e^x$, the x-axis on $[-1, 2]$
13. $y = e^{-x}$, the x-axis on $[-2, 1]$
14. $y = x^2 - 4$ and the x-axis
15. $y = x^2 - 9$ and the x-axis
16. $y = x^2$ and $y = 6 - x$ on $[-3, 2]$
17. $y = x^2$ and $y = 6 - x$ on $[3, 5]$
18. $y = x^3$ and $y = x^2$ on $[0, 1]$
19. $y = x^3$ and $y = x^2$ on $[0, 2]$
20. $y = x^2$ and $y = 6x$ on $[0, 5]$
21. $y = x^2$ and $y = 2x + 4$ on $[0, 1]$
22. $y = x^2 + 4$ and $y = 2x + 4$ on $[-1, 1]$
23. $y = x^2 + 4$ and $y = 2x + 4$ on $[1, 5]$
24. $y = x^3$ and $y = x$
25. $y = x$ and $y = x^4$
26. $y = x^2 - x$ and $y = 6$
27. $y = x^2 - 5x$ and $y = 6$
28. $y = x^{-1}$ on $[.5, 1]$
29. $y = x^{-1}$ on $[.01, 1]$
30. $y = x^2$ and $y = \sqrt{x}$
31. $y = x$ and $y = \sqrt{x}$
32. $y = \sqrt{x}$ and $y = 4$
33. $y = \sqrt{x}$ and $y = 9$
34. $y = x^{-1}$ and $y = e^x$ on $[1, 2]$
35. $y = e^{.5x}$ and $y = x^{-1}$ on $[1, 2]$

APPLICATIONS

36. The Sorite Corporation has found that production costs are determined by the function

$$C(x) = \frac{100}{\sqrt{x + 1}} + 50$$

where $C(x)$ is the cost to produce the xth item. What is the approximate total cost of producing 2,000 items?

37. The EPA (Environmental Protection Agency) estimates that pollutants are being dumped into the Russian River according to the formula

$$P(t) = \frac{1{,}000}{\sqrt[3]{(t + 5)^2}}$$

where t is the year and $P(t)$ is pollutants in tons. What is the total amount (to the nearest ton) of pollutants that will be dumped into the river in the next 25 years?

38. The marginal revenue and marginal cost (in dollars) per day are given in terms of the tth month by the formulas

$$R'(t) = 500e^{.01t} \quad \text{and} \quad C'(t) = 50 - .1t$$

where $R(0) = C(0) = 0$, and t is the month. Find the total profit for the first year.

39. A chemical spill is known to kill fish according to the formula

$$N(t) = \frac{100{,}000}{\sqrt{t + 250}}$$

where N is the number of fish killed on the tth day since the spill. What is the approximate total number of fish killed in 30 days?

40. A culture is growing at a rate of

$$R'(x) = 200e^{.05t} \quad \text{for } 0 \leq t \leq 10$$

per hour. Find the area between the graph of this equation and the t-axis. What do you think this area represents?

41. The rate of change of the demand for a product with respect to time (in weeks) is given by the equation

$$D'(x) = 20 + .012t^2 \quad \text{for } 0 \leq t \leq 25$$

Find the area between the graph of this equation and the t-axis. What do you think this area represents?

42. Suppose the accumulated cost of a piece of equipment is $C(t)$ and the accumulated revenue is $R(t)$, where both of these are measured in thousands of dollars and t is the number of years after the piece of equipment was installed. If it is known that

$$C'(t) = 1 \quad \text{and} \quad R'(t) = 2e^{-.1t}$$

find the area (to the nearest unit) between the graphs of C' and R' (do not forget $t \geq 0$). What do you think this area represents?

43. Suppose the accumulated cost of a piece of equipment is $C(t)$ and the accumulated revenue is $R(t)$, where both of these are measured in thousands of dollars and t is the number of years after the piece of equipment was installed. If it is known that

$$C'(t) = 18 \quad \text{and} \quad R'(t) = 21e^{-.01t}$$

find the area (to the nearest unit) between the graphs of C' and R' (do not forget $t \geq 0$). What do you think this area represents?

6.5
The Fundamental Theorem of Calculus

In this section we will formulate a very important theorem in calculus—so important, in fact, that it carries the name *The Fundamental Theorem of Calculus*. In order to integrate a function f we need to find an antiderivative F. What if we cannot find such an antiderivative? For example,

$$\int_1^5 \frac{x}{x+1}\,dx$$

certainly should have a value, but with the techniques of integration considered thus far, we cannot find an antiderivative of

$$f(x) = \frac{x}{x+1}$$

We will evaluate this integral in Example 2, but we begin with a simpler example, one which we can evaluate.

EXAMPLE 1 Find the area bounded by $f(x) = x^2$, the x-axis, and the lines $x = 1$ and $x = 5$.

Solution The region is shown in Figure 6.6.

$$\int_1^5 x^2\,dx = \left.\frac{x^3}{3}\right|_1^5$$
$$= \frac{125}{3} - \frac{1}{3}$$
$$= \frac{124}{3}$$

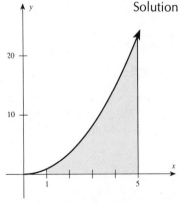

FIGURE 6.6

Suppose that we were not able to find the antiderivative of $f(x) = x^2$ in Example 1. We could approximate the area by dividing the area bounded by f, the x-axis, and the lines $x = 1$, $x = 5$ into one, two, and four rectangles, as shown in Figure 6.7 on page 256.

COMPUTER APPLICATION

These computer-generated drawings show the area under the curve $y = x^2$ as approximated by rectangles using left endpoints. The Riemann sum, defined later in this section, approximates the actual area.

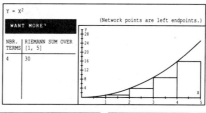
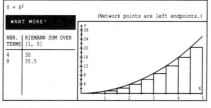
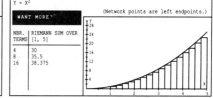
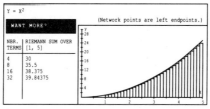
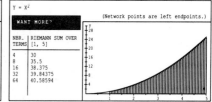
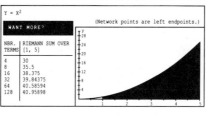
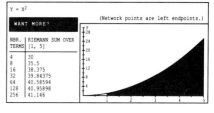

One rectangle

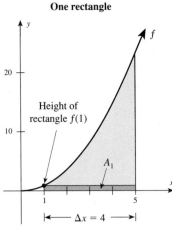

$$\int_1^5 x^2 \, dx \approx A_1$$
$$= \Delta x \cdot f(1)$$
$$= 4 \cdot 1^1$$
$$= 4$$

Two rectangles

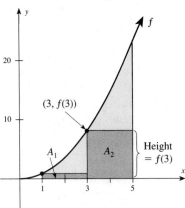

$$\int_1^5 x^2 \, dx \approx A_1 + A_2$$
$$= \Delta x \cdot f(1) + \Delta x \cdot f(3)$$
$$= 2 \cdot 1^2 + 2 \cdot 3^2$$
$$= 20$$

Four rectangles

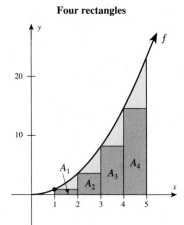

$$\int_1^5 x^2 \, dx \approx A_1 + A_2 + A_3 + A_4$$
$$= \Delta x \cdot f(1) + \Delta x \cdot f(2) + \Delta x \cdot f(3) + \Delta x \cdot f(4)$$
$$= 1 \cdot 1^2 + 1 \cdot 2^2 + 1 \cdot 3^2 + 1 \cdot 4^2$$
$$= 30$$

FIGURE 6.7

Notice that the approximations get better. We would expect the approximations to continue to improve as we use more and more rectangles. Also notice that we picked the left endpoint of each subinterval. We could have selected the right endpoint, the midpoint, or any point c_k in the kth subinterval. The following calculations show the results obtained by picking the midpoints for the subintervals shown in Figure 6.7.

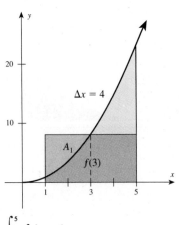

$$\int_1^5 x^2 \, dx \approx A_1$$
$$= \Delta x \cdot f(3)$$
$$= 4 \cdot 3^2$$
$$= 36$$

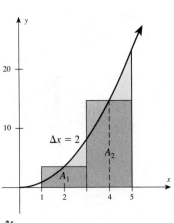

$$\int_1^5 x^2 \, dx \approx A_1 + A_2$$
$$= \Delta x \cdot f(2) + \Delta x \cdot f(4)$$
$$= 2 \cdot 2^2 + 2 \cdot 4^2$$
$$= 40$$

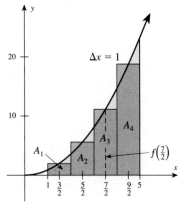

$$\int_1^5 x^2 \, dx \approx A_1 + A_2 + A_3 + A_4$$
$$= \Delta x \cdot f(\tfrac{3}{2}) + \Delta x \cdot f(\tfrac{5}{2}) + \Delta x \cdot f(\tfrac{7}{2}) + \Delta x \cdot f(\tfrac{9}{2})$$
$$= 1 \cdot (\tfrac{3}{2})^2 + 1 \cdot (\tfrac{5}{2})^2 + 1 \cdot (\tfrac{7}{2})^2 + 1 \cdot (\tfrac{9}{2})^2$$
$$= 41$$

As you can see, selecting the midpoint of each subinterval usually approximates the area with fewer approximations. However, if you take enough equal subintervals the rectangles are so small that it does not matter which point c_k you choose in the kth subinterval. This leads us to what we call the rectangular approximation to the integral, which is true for any continuous function f.

Rectangular Approximation

Divide the interval from $x = a$ to $x = b$ into n equal subintervals, each with length

$$\Delta x = \frac{b-a}{n}$$

Let c_k be *any* point in the kth subinterval. Then,

$$\int_a^b f(x)\,dx \approx \Delta x \cdot f(c_1) + \Delta x \cdot f(c_2) + \cdots + \Delta x \cdot f(c_n)$$

$$= \Delta x [f(c_1) + f(c_2) + \cdots + f(c_n)]$$

EXAMPLE 2 Evaluate $\int_1^5 \frac{x}{x+1}\,dx$. Approximate this integral (to two decimal places) by using the rectangular approximation with $n = 4$.

Solution We want the value of this integral. This is not an area problem; a rectangular approximation exists for any continuous function. If f is not positive on $[a, b]$, then neither the integral nor the sum represents an area. In this problem, we want the value of the integral.

In order to use the rectangular approximation for this integral, we must make sure that f is continuous over $[1, 5]$. The only suspicious point is $x = -1$ which is not in the interval, so f is continuous on $[1, 5]$.

The four subintervals are

$[a, a + \Delta x], [a + \Delta x, a + 2\Delta x], [a + 2\Delta x, a + 3\Delta x],$ and $[a + 3\Delta x, a + 4\Delta x]$
$[1, 2]$, $[2, 3]$, $[3, 4]$,   and $[4, 5]$

As a check, you should note that $a + 4\Delta x = b$, the right endpoint. In general,

$$a + n\Delta x = a + n\left(\frac{b-a}{n}\right)$$
$$= a + b - a$$
$$= b$$

See Appendix B for a discussion of spreadsheets.

```
What is a?     1
What is b?     5
What is n?     4           Delta x is 1

c values    f values     Sum of f values
  1.5         0.6
  2.5       0.71428571429
  3.5       0.77777777778
  4.5       0.81818181818   2.91024531
```

Choose the midpoints of each of these intervals:

| Intervals: | $[1, 2]$ | $[2, 3]$ | $[3, 4]$ | $[4, 5]$ |

Midpoints: $\quad c_1 = \dfrac{3}{2} \qquad c_2 = \dfrac{5}{2} \qquad c_3 = \dfrac{7}{2} \qquad c_4 = \dfrac{9}{2}$

Heights: $\quad f\left(\dfrac{3}{2}\right) = \dfrac{3}{5} \quad f\left(\dfrac{5}{2}\right) = \dfrac{5}{7} \quad f\left(\dfrac{7}{2}\right) = \dfrac{7}{9} \quad f\left(\dfrac{9}{2}\right) = \dfrac{9}{11}$

Bases: $\qquad\quad \Delta x = 1 \qquad \Delta x = 1 \qquad \Delta x = 1 \qquad \Delta x = 1$

Then, the rectangular approximation for $n = 4$ is

$$\int_1^5 \frac{x}{x+1} dx \approx \Delta x [f(c_1) + f(c_2) + \cdots + f(c_n)]$$

$$= (1)\left[\frac{3}{5} + \frac{5}{7} + \frac{7}{9} + \frac{9}{11}\right]$$

$$\approx 2.91$$

Areas with Tabular Functions

It is possible to find the area of an irregularly shaped region by using a rectangular approximation and a function defined by a table rather than by a formula. A function defined by tabular values is called a **tabular function**.

EXAMPLE 3 Find the surface area of a swimming pool whose dimensions are shown in the following figure.

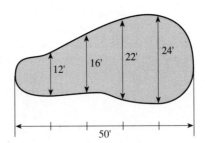

Solution We can represent the information in the sketch in tabular form by letting x be the horizontal distance (in feet) and f be the width of the pool at x (we are using the right-hand endpoints of each subinterval).

x	10	20	30	40	50
$f(x)$	12	16	22	24	0

$$\int_0^{50} f(x)\,dx \approx \Delta x[f(10) + f(20) + f(30) + f(40) + f(50)]$$
$$= 10(12 + 16 + 22 + 24 + 0) = 740$$

We would estimate the surface area of the swimming pool to be 740 ft². ∎

The Definite Integral as a Limit of a Sum

In Example 2 we found the area under the curve $y = f(x)$ over the closed interval $[a, b]$, by partitioning the interval into n subintervals of equal width $\Delta x = \dfrac{b-a}{n}$. Then we evaluated f at the midpoint of the kth subinterval for $k = 1, 2, 3, 4$. We then formed the approximating sum S_4 of the areas of the rectangles. Instead of taking 4 rectangles, consider n rectangles and the sum S_n. Since we expect the estimates S_n to improve as Δx decreases, we *define* the area A under the curve, above the x-axis, bounded by the lines $x = a$ and $x = b$ to be the limit of S_n as $\Delta x \to 0$.

This approach to the area problem contains the essentials of integration, but there is no compelling reason for the partition points to be evenly spaced or to insist on evaluating f at midpoints. These conventions are for convenience of computation, and to accommodate applications other than area, it is necessary to consider a more general type of approximating sum and to specify what is meant by the limit of such a sum. The approximating sums that occur in integration problems are called **Riemann sums**, and the following definition contains a step-by-step description of how such sums are formed.

Riemann Sum

Suppose a continuous function f is given, along with a closed interval $[a, b]$ on which f is defined. Then:

Step 1: Partition the interval $[a, b]$ into n subintervals by choosing points $\{x_0, x_1, \ldots, x_n\}$ arranged so that

$$a = x_0 < x_1 < x_2 < \cdots < x_{n-1} < x_n = b$$

Call this partition P. For $k = 1, 2, \ldots, n$, the kth subinterval width is $\Delta x_k = x_k - x_{k-1}$. The largest of these widths is called the **norm** of the partition P and is denoted by $\|P\|$; that is,

$$\|P\| = \max_{k=1,2,\ldots,n} \{\Delta x_k\}$$

Step 2: Choose a number arbitrarily from each subinterval. For $k = 1, 2, \ldots, n$, the number x_k^* chosen from the kth subinterval is called the kth *subinterval representative* of the partition P.

Step 3: Form the sum:

$$R_n = f(x_1^*)\Delta x_1 + f(x_2^*)\Delta x_2 + \cdots + f(x_n^*)\Delta x_n$$

This is the **Riemann sum** associated with f, the given partition P, and the chosen subinterval representatives $x_1^*, x_2^*, \ldots, x_n^*$.

EXAMPLE 4 Suppose the interval $[-2, 1]$ is partitioned into 6 subintervals with subdivision points

$$a = x_0 = -2$$
$$x_1 = -1.6$$
$$x_2 = -.93$$
$$x_3 = -.21$$
$$x_4 = .35$$
$$x_5 = .82$$
$$x_6 = 1 = b$$

Find the norm of this partition P and the Riemann sum associated with the function $f(x) = 2x$, the given partition, and the subinterval representatives $x_1^* = -1.81$, $x_2^* = -1.12$, $x_3^* = -.55$, $x_4^* = -.17$, $x_5^* = .43$, $x_6^* = .94$.

Solution Before we can find the norm of the partition or the required Riemann sum, we must compute the subinterval width Δx_k and evaluate f at each subinterval representative x_k^*. These values are shown in Figure 6.8 and the computations are shown at the top of page 261.

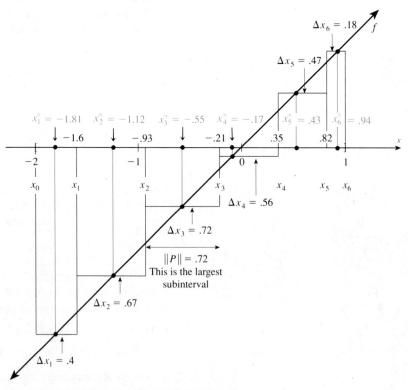

FIGURE 6.8
Graphical representation of a Riemann sum

k	$x_k - x_{k-1} = \Delta x_k$ Given	x_k^* Given	$f(x_k^*) = 2x_k^*$ Given
1	$-1.6 - (-2) = .40$	-1.81	$f(-1.81) = -3.62$
2	$-.93 - (-1.6) = .67$	-1.12	$f(-1.12) = -2.24$
3	$-.21 - (-.93) = .72$	$-.55$	$f(-.55) = -1.10$
4	$.35 - (-.21) = .56$	$-.17$	$f(-.17) = -.34$
5	$.82 - .35 = .47$	$.43$	$f(.43) = .86$
6	$1.00 - .82 = .18$	$.94$	$f(.94) = 1.88$

From the table, we see that the largest subinterval width is $\Delta x_3 = .72$, so the partition has norm $\|P\| = .72$. Finally, by using the definition, we find the Riemann sum:

$$R_6 = (-3.62)(.4) + (-2.24)(.67) + (-1.10)(.72) + (-.34)(.56) + (.86)(.47) + (1.88)(.18)$$
$$= -3.1886 \quad \blacksquare$$

WARNING Notice from Example 4 that the Riemann sum does not necessarily represent an area. The sum found is negative (and areas must be nonnegative). Also notice from Figure 6.8 that the nonnegativity requirement for the function in the definition of area is not met.

Definition of Definite Integral as a Riemann Sum

By comparing the formula for Riemann sum with that of area at the beginning of this section, we recognize that the sum S_n used to approximate area is actually a special kind of Riemann sum which has

$$\Delta x_k = \Delta x = \frac{b-a}{n} \quad \text{and} \quad x_k^* = a + k\Delta x$$

for $k = 1, 2, \ldots, n$. Since each subinterval in the partition P associated with S_n has width Δx, the norm of the partition is

$$\|P\| = \Delta x = \frac{b-a}{n}$$

This kind of partition is called a **regular partition**. When we express the area under the curve $y = f(x)$ as $A = \lim_{\Delta x \to 0} S_n$, we are actually saying that A can be estimated to any desired accuracy by finding a Riemann sum of the form S_n with norm

$$\|P\| = \frac{b-a}{n}$$

sufficiently small. We use this interpretation as a model for the following definition.

Definite Integral

> If f is defined on the closed interval $[a,b]$ we say f is **integrable** on $[a,b]$ if
>
> $$I = \lim_{||P|| \to 0} f(x_1)\Delta x_1 + f(x_2)\Delta_2 + \cdots + f(x_n)\Delta x_n$$
>
> exists. This limit is called the **definite integral** of f from a to b. The definite integral is denoted by
>
> $$I = \int_a^b f(x)\,dx$$

In other words, if the limit of the Riemann sum of f exists, then we say that f is *integrable*. In particular, it means that the number I can be approximated to any prescribed degree of accuracy by any Riemann sum with norm sufficiently small.

The last step in putting together this Riemann sum definition and the ideas developed earlier in this chapter, is a result that forms the basis for much of the calculus. It reformulates the informal definition of definite integral given in Section 6.3 as a Riemann sum, and is known as the **Fundamental Theorem of Calculus**.

Fundamental Theorem of Calculus

> If a function f is continuous on an interval $[a,b]$, then
>
> $$\int_a^b f(x)\,dx = F(b) - F(a)$$
>
> where F is any antiderivative of f.

The fundamental theorem of calculus is the most important theorem of calculus. It ties together the ideas of limits, derivatives, areas, and antiderivatives. What good is all of this? Just as the derivative was motivated by looking at a rate of change, but was then found to be useful in many settings, so we see that the limit

$$\lim_{||P|| \to 0} [\Delta x_1 \cdot f(c_1) + \Delta x_2 \cdot f(c_2) + \cdots + \Delta x_n \cdot f(c_n)]$$

occurs in many practical problems. Just as the derivative is not easy to find from the definition, so is the integral not easy to find from the definition. However, once we recognize a particular application as a derivative, or as an integral, we can apply the appropriate derivative or integral rules. The following application shows how we can do this for a Riemann sum.

Average Value of a Function

Suppose f is a continuous function over some interval $[a,b]$, and we wish to find the average value of f over this interval. For example, suppose a typist's speed in words

6.5 THE FUNDAMENTAL THEOREM OF CALCULUS

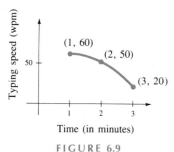

FIGURE 6.9
Typing speed as a function of the length of time of the typing test

per minute is a function of the length of a typing test which is administered, as shown in Figure 6.9. What is the typist's average speed for the three timed tests?

$$\text{Average} = \frac{60 + 50 + 20}{3} = \frac{130}{3} \text{ wpm}$$

However, this average does not accurately represent the situation because there are infinitely many values for t over the interval $[1, 3]$. That is, suppose that typing speed $w(t)$ is represented by the continuous function

$$w(t) = -10t^2 + 20t + 50 \quad \text{for } t \text{ in } [1, 3]$$

Suppose we divide the interval into n subintervals, and let c_k be any point in the kth subinterval. Then the average is given by

$$\text{Average} = \frac{1}{n}[w(c_1) + w(c_2) + \cdots + w(c_n)]$$

Now the limit as $n \to \infty$ is the average speed over the time interval $[1, 3]$. This looks like the formula for a Riemann sum except it does not have the Δx_k. For this example, we are looking for

$$\Delta t = \frac{b - a}{n}$$

Multiply the expression on the right by 1 written as $\frac{b - a}{b - a}$:

$$\text{Average} = \frac{1}{n} \cdot \frac{b - a}{b - a}[w(c_1) + w(c_2) + \cdots + w(c_n)]$$

$$= \frac{1}{b - a} \cdot \frac{b - a}{n}[w(c_1) + w(c_2) + \cdots + w(c_n)]$$

$$= \frac{1}{b - a} \Delta t[w(c_1) + w(c_2) + \cdots + w(c_n)]$$

Thus, over the interval $[a, b]$,

$$\text{Average} = \lim_{\|P\| \to 0} \frac{1}{b - a} \Delta t[w(c_1) + w(c_2) + \cdots + w(c_n)]$$

$$= \frac{1}{b - a} \lim_{n \to \infty} \Delta t[w(c_1) + w(c_2) + \cdots + w(c_n)]$$

We see that this is a Riemann sum, so by the definition of a definite integral we have

$$\text{Average value over the interval } [a, b] = \frac{1}{b - a} \int_a^b w(t) \, dt$$

The following definition of average value is motivated by the previous discussion.

Average Value

The **average value** of a continuous function f over $[a, b]$ is

$$\frac{1}{b - a} \int_a^b f(x) \, dx$$

EXAMPLE 5 If the typist's speed in the above discussion is given by the continuous function
$$w(t) = -10t^2 + 20t + 50 \quad \text{for } t \text{ in } [1, 3]$$
find the average typing speed in words per minute.

Solution The average is

$$\frac{1}{3-1}\int_1^3 (-10t^2 + 20t + 50)\,dt = \frac{1}{2}\left[\frac{-10t^3}{3} + \frac{20t^2}{2} + 50t\right]_1^3$$

$$= \frac{1}{2}\left[(-90 + 90 + 150) - \left(-\frac{10}{3} + 10 + 50\right)\right]$$

$$\approx 47 \text{ wpm} \blacksquare$$

EXAMPLE 6 Suppose the cost function for producing x items is
$$C(x) = 5{,}000 + 8x$$

a. What is the total cost of producing 100 items?
b. What is the average cost per unit if 100 items are produced?
c. What is the average value of the cost function over the interval $[0, 100]$?

Solution a. The total cost of producing 100 items is $C(100) = 5{,}000 + 8(100) = 5{,}800$.

b. The average cost per unit is $\dfrac{c(100)}{100} = \dfrac{5{,}800}{100} = 58$.

c. The average value of the cost function over $[0, 100]$ is

$$\frac{1}{100 - 0}\int_0^{100}(5{,}000 + 8x)\,dx = \frac{1}{100}(5{,}000x + 4x^2)\Big|_0^{100} = 5{,}400 \blacksquare$$

Notice that the average value of the cost function (part c of Example 6) is considerably different from the average cost of an item (part b). This is because the cost value gives the average value of producing 100 items.

EXAMPLE 7 Given the demand function
$$p(x) = 50e^{.03x}$$
find the average value of the price (in dollars) over the interval $[0, 100]$.

Solution Average value of the price $= \dfrac{1}{100 - 0}\int_0^{100} 50e^{.03x}\,dx$

$$= \frac{1}{2}\int_0^{100} e^{.03x}\,dx$$

$$= \frac{e^{.03x}}{.06}\Big|_0^{100}$$

$$= \frac{50}{3}(e^{.03(100)} - e^{.03(0)})$$

$$= \frac{50}{3}(e^3 - 1) \approx 318.09$$

The average value is approximately $318. $\blacksquare$

6.5 Problem Set

In Problems 1–6 find the area bounded by the x-axis, the curve $y = f(x)$, and the given vertical lines by

a. using an antiderivative.
b. sketching the curve, and then by doing a rectangular approximation (correct to two decimal places) by choosing $n = 4$, and c_k to be the left endpoint of each subinterval.

1. $y = x^2 + 1$, $x = -2$, $x = 2$
2. $y = x^2 + 2$, $x = -2$, $x = 2$
3. $y = 3x^2$, $x = 1$, $x = 4$
4. $y = 6x^2$, $x = 1$, $x = 4$
5. $y = 4 - x^2$, $x = 0$, $x = 2$
6. $y = 9 - x^2$, $x = -1$, $x = 3$

Evaluate the integrals in Problems 7–12 by using a rectangular approximation. Give your answer to two decimal places by using a rectangular approximation with $n = 4$ and with Δ_k the midpoint of each subinterval.

7. $\int_1^5 \frac{1}{(2x+1)^2} dx$
8. $\int_1^5 \frac{1}{2x+1} dx$
9. $\int_0^2 e^{.5x} dx$
10. $\int_1^5 \frac{\ln x}{x} dx$
11. $\int_0^{1/2} \frac{x}{\sqrt{1+3x^2}} dx$
12. $\int_0^{1/2} \frac{1}{\sqrt{1+3x^2}} dx$

Evaluate the integrals for the tabular functions in Problems 13–16.

13. $\int_0^6 f(x) dx$

x	1	3	5
f(x)	2.4	5.5	6.1

14. $\int_1^9 f(x) dx$

x	2	4	6	8
f(x)	4.9	5.1	4.3	6.2

15. $\int_5^{45} f(x) dx$

x	10	20	30	40
f(x)	130	180	110	150

16. $\int_0^{.6} f(x) dx$

x	.1	.2	.3	.4	.5
f(x)	.32	12.1	8.2	2.0	.55

Find the average value of each function in Problems 17–24 over the indicated interval.

17. $f(x) = 2x^2 + 1$, $[0, 4]$
18. $f(x) = 3x^2 + 2$, $[2, 5]$
19. $f(x) = 300 - 4x$, $[0, 10]$
20. $f(x) = 4x - 5x^3$, $[-1, 2]$
21. $f(x) = \sqrt[3]{x}$, $[1, 8]$
22. $f(x) = \sqrt{2x+3}$, $[3, 11]$
23. $f(x) = e^{-.5x}$, $[0, 10]$
24. $f(x) = 128e^{.04x}$, $[0, 10]$

In Problems 25–36 find $F'(x)$.

25. $F(x) = \int_0^x (t+5) dt$
26. $F(x) = \int_0^x t(t^2+1)^3 dt$
27. $F(x) = \int_{-3}^x (t+5)^2 dt$
28. $F(x) = \int_1^x (t^2+4) dt$
29. $F(x) = \int_4^x \sqrt{t} \, dt$
30. $F(x) = \int_1^x e^{3t} dt$
31. $F(x) = \int_0^x (t^3 - 3t^2 + 5t - 7) dt$
32. $F(x) = \int_{-3}^x (t^2 - 2t + 5) dt$
33. $F(x) = \int_8^x \sqrt[3]{t} \, dt$
34. $F(x) = \int_0^x \sqrt[4]{t+3} \, dt$
35. $F(x) = \int_0^x t^2(t+1)^3 dt$
36. $F(x) = \int_2^x t(3t^4-1)^2 dt$

APPLICATIONS

37. The graph below, from the March 5, 1986, issue of the *Wall Street Journal*, shows the rate of U.S. automobile sales. Estimate the total sales for July–December.

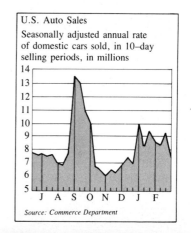

U.S. Auto Sales
Seasonally adjusted annual rate of domestic cars sold, in 10-day selling periods, in millions

Source: Commerce Department

38. Use the graph in Problem 37 to estimate the total automobile sales for the last quarter of the year (Oct. 1 to Dec. 31).

39. The graph below shows the rate of paper and paperboard production in the United States as reported by the American Paper Institute. Estimate the total production for 1985.

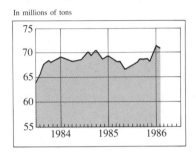
Paper and Paperboard

40. If the supply function is
$$p = S(x) = 10(e^{.03x} - 1)$$
what is the average price (in dollars) over the interval $[10, 50]$?

41. The number of rabbits in a limited geographical area (such as an island) is approximated by
$$P(t) = 500 + t - .25t^2$$
for the number t of years on the interval $[0, 5]$. What is the average number of rabbits in the area over the 5-year time period?

42. The world population (in billions) is given by
$$P(t) = 5e^{.03t} \qquad (t = 0 \text{ in } 1987)$$
What is the average population of the earth over the next 30 years?

43. The temperature (in degrees Fahrenheit) varies according to the formula
$$F(t) = 6.44t - .23t^2 + 30$$
Find the average daily temperature if t is the time of day (in hours) measured from midnight.

44. Repeat Problem 43 for the daylight hours—that is, for t on the interval $[6, 20]$.

45. Estimate the surface area of the following swimming pool.

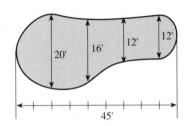

46. A surveyor made the following measurements on a parcel of land.

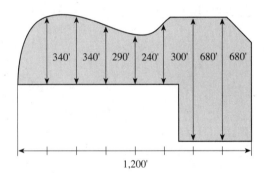

Estimate the area. (One acre is 43,560 ft^2.)

*6.6

Numerical Integration

Calculators or computers are being used more and more to evaluate definite integrals (see the Modeling Application at the beginning of Chapter 6). **Numerical integration** is the process of finding a numerical approximation for a definite integral *without actually carrying out the antidifferentiation process.* This means that you can easily write a computer or calculator program to evaluate definite integrals if you understand numerical integration. Numerical integration is also useful when it is difficult or impossible to find an antiderivative to carry out the integration process.

* Optional section.

Suppose a function f is continuous on $[a,b]$. Then there are three common approximations to the definite integral:

Approximations for the Definite Integral

$\int_a^b f(x)\,dx$ can be approximated by

RECTANGLES: $A_n = \Delta x[f(x_0) + f(x_1) + \cdots + f(x_{n-1})]$

TRAPEZOIDS: $T_n = \dfrac{\Delta x}{2}[f(x_0) + 2f(x_1) + 2f(x_2) + \cdots + 2f(x_{n-1}) + f(x_n)]$

PARABOLAS: $P_n = \dfrac{\Delta x}{3}[f(x_0) + 4f(x_1) + 2f(x_2) + \cdots + 4f(x_{n-1}) + f(x_n)]$

(true for n an even integer)

where $\Delta x = \dfrac{b-a}{n}$ for some positive integer n. (Δx is read "delta x.")

Even though there are many other approximation schemes, these will suffice for our needs. While we will not prove these approximations, it is instructive to see where these formulas come from.

Rectangular Approximation

We considered this method in the preceding section. Recall that we need to let f be a function that is continuous on a closed interval $[a,b]$. Suppose further that $a < b$, and that f is nonnegative on the interval, as shown in Figure 6.10.

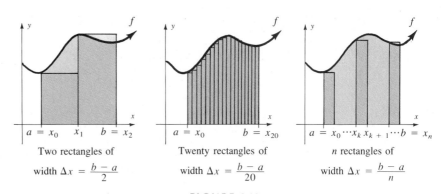

FIGURE 6.10
Rectangular approximation of the definite integral

Divide the interval into n subintervals, each of length $\Delta x = \dfrac{b-a}{n}$. The endpoints of these subintervals are

$$a, \quad a + \Delta x, \quad a + 2\Delta x, \quad a + 3\Delta x, \quad \ldots, \quad a + n\Delta x$$

which are, for convenience, labeled $x_0, x_1, x_2, \ldots, x_n$, respectively. The area of the kth rectangle is

$$f(x_{k-1}) \Delta x$$

If you add the areas of all the n rectangles, you clearly obtain A_n.

EXAMPLE 1 Let $f(x) = x^\pi + 1$. Find an approximation for

$$\int_1^3 (x^\pi + 1) \, dx$$

by using a rectangular approximation for $n = 1, 2, 4,$ and 8. (We pick $1, 2, 4, 8, \ldots$ only as a matter of convenience; *any* subdivision will suffice.)

Solution *First approximation, $n = 1$:*

$$\begin{aligned} A_1 &= \int_1^3 (x^\pi + 1) \, dx \approx f(x_0) \Delta x \\ &= f(1)(2) \\ &= (1^\pi + 1)(2) \\ &= 4 \end{aligned}$$

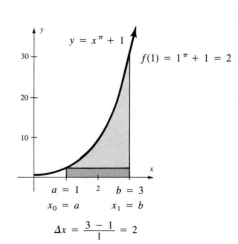

Second approximation, $n = 2$:

$$\begin{aligned} A_2 &= \int_1^3 (x^\pi + 1) \, dx \\ &\approx f(x_0) \Delta x + f(x_1) \Delta x \\ &= f(1)(1) + f(2)(1) \\ &= (1^\pi + 1) + (2^\pi + 1) \\ &\approx 2 + 9.82 \\ &= 11.82 \end{aligned}$$

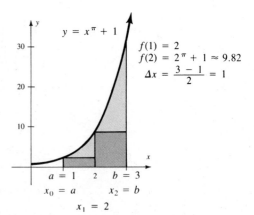

Third approximation, $n = 4$:

$$A_4 = \int_1^3 (x^\pi + 1)\,dx$$
$$\approx f(1)(.5) + f(\tfrac{3}{2})(.5) + f(2)(.5) + f(\tfrac{5}{2})(.5)$$
$$\approx 2(.5) + 4.57(.5) + 9.82(.5) + 18.79(.5)$$
$$\approx 17.59$$

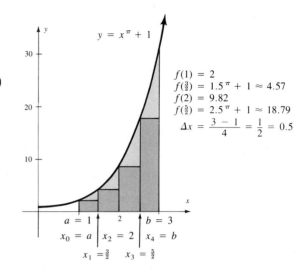

$f(1) = 2$
$f(\tfrac{3}{2}) = 1.5^\pi + 1 \approx 4.57$
$f(2) = 9.82$
$f(\tfrac{5}{2}) = 2.5^\pi + 1 \approx 18.79$
$\Delta x = \dfrac{3-1}{4} = \dfrac{1}{2} = 0.5$

Notice that as $n \to \infty$, the approximate area (gray) is more closely approximating the actual area (color).

Fourth approximation, $n = 8$:

$$A_8 = \int_1^3 (x^\pi + 1)\,dx \approx f(1)(.25) + f(1.25)(.25) + f(1.5)(.25) + \cdots + f(2.75)(.25)$$
$$\approx .25[f(1) + f(1.25) + f(1.5) + f(1.75) + f(2) + f(2.25) + f(2.5) + f(2.75)]$$
$$\approx .25(83.78)$$
$$\approx 20.95$$

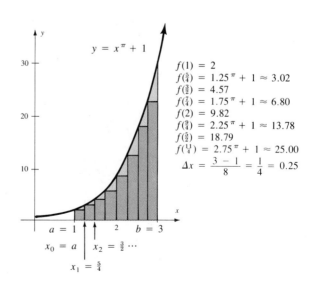

$f(1) = 2$
$f(\tfrac{5}{4}) = 1.25^\pi + 1 \approx 3.02$
$f(\tfrac{3}{2}) = 4.57$
$f(\tfrac{7}{4}) = 1.75^\pi + 1 \approx 6.80$
$f(2) = 9.82$
$f(\tfrac{9}{4}) = 2.25^\pi + 1 \approx 13.78$
$f(\tfrac{5}{2}) = 18.79$
$f(\tfrac{11}{4}) = 2.75^\pi + 1 \approx 25.00$
$\Delta x = \dfrac{3-1}{8} = \dfrac{1}{4} = 0.25$

COMPUTER APPLICATION

One of the most useful computer programs in integral calculus is called *Numerical integration*. This program will find rectangular approximations for left endpoints, right endpoints, midpoints, as well as trapezoidal and Simpson's approximations. In addition, look at Appendix B for a spreadsheet application of these approximation methods.

Trapezoidal Approximation

In order to carry out a trapezoidal approximation, you simply approximate the area represented by the definite integral using trapezoids instead of rectangles, as shown in Figure 6.11. To derive the trapezoidal approximation, use the formula for the area of a trapezoid:

$$A = \left(\frac{b_1 + b_2}{2}\right)h$$

for bases b_1 and b_2 and height h. From this formula you can find the area of the kth trapezoid:

$$\frac{f(x_{k-1}) + f(x_k)}{2} \Delta x \qquad \Delta x = h \text{ in the trapezoidal formula } b_1 = f(x_{k-1}) \text{ and } b_2 = f(x_k)$$

Therefore, the sum of the areas of all the trapezoids is

$$T_n = \left[\frac{f(x_0) + f(x_1)}{2} + \frac{f(x_1) + f(x_2)}{2} + \cdots + \frac{f(x_{n-1}) + f(x_n)}{2}\right]\Delta x$$

$$= \frac{\Delta x}{2}[f(x_0) + 2f(x_1) + 2f(x_2) + \cdots + 2f(x_{n-1}) + f(x_n)]$$

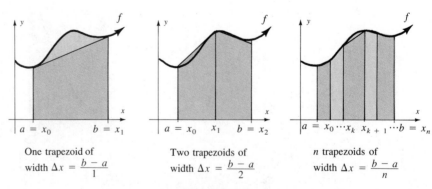

FIGURE 6.11
Trapezoidal approximation of the definite integral

EXAMPLE 2 Let $f(x) = x^\pi + 1$. Find an approximation for

$$\int_1^3 (x^\pi + 1)\,dx$$

by using a trapezoidal approximation for $n = 1$ and $n = 4$.

Solution *First approximation, $n = 1$:*

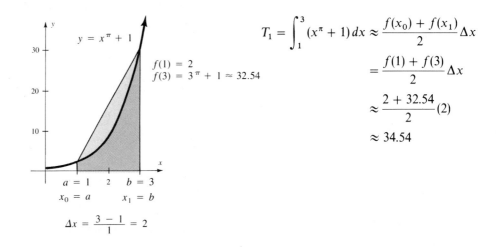

$$T_1 = \int_1^3 (x^\pi + 1)\,dx \approx \frac{f(x_0) + f(x_1)}{2}\Delta x$$

$$= \frac{f(1) + f(3)}{2}\Delta x$$

$$\approx \frac{2 + 32.54}{2}(2)$$

$$\approx 34.54$$

$f(1) = 2$
$f(3) = 3^\pi + 1 \approx 32.54$

$x_0 = a$, $x_1 = b$, $a = 1$, $b = 3$

$\Delta x = \dfrac{3 - 1}{1} = 2$

Second approximation, $n = 4$:

$$T_4 = \int_1^3 (x^\pi + 1)\,dx \approx \frac{.5}{2}[f(1) + 2f(1.5) + 2f(2) + 2f(2.5) + f(3)]$$

$$\approx .25(2 + 9.14 + 19.65 + 37.58 + 32.54)$$

$$\approx .25(100.92)$$

$$\approx 25.23$$

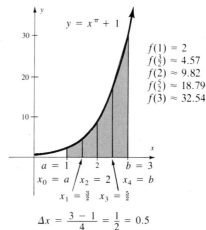

$f(1) = 2$
$f(\tfrac{3}{2}) \approx 4.57$
$f(2) \approx 9.82$
$f(\tfrac{5}{2}) \approx 18.79$
$f(3) \approx 32.54$

$\Delta x = \dfrac{3 - 1}{4} = \dfrac{1}{2} = 0.5$

∎

COMPUTER APPLICATION

In the *Numerical integration* program accompanying this book, you have the following options:
- Rectangular approximation: It uses midpoints.
- Trapezoidal approximation: Apply Trapezoidal rule.
- Parabola approximation: Apply Simpson's rule.

For Example 2 we obtain:

n	Trapezoid	Midpoint	Simpson
1	34.54428070	19.64995565	24.61473066
2	27.09711817	23.36399996	24.60837270
4	25.23055907	24.29665927	24.60795920
8	24.76360917	24.53009496	24.60793303
16	24.64685206	24.58847105	24.60793138
32	24.61766155	24.60306615	24.60793128

Parabolic Approximation

The parabolic approximation, also known as **Simpson's rule**, uses parabolas to approximate the area under the graph of a function, as shown in Figure 6.12.

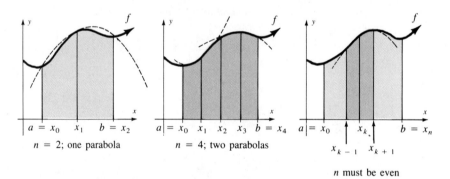

FIGURE 6.12
Parabolic approximation of the definite integral

We will not attempt to derive Simpson's rule, but instead illustrate it in Example 3. Note that to use Simpson's rule, n must be even.

EXAMPLE 3 Let $f(x) = x^\pi + 1$. Find an approximation for

$$\int_1^3 (x^\pi + 1)\,dx$$

by using Simpson's approximation for $n = 6$.

Solution   First, gather the necessary values,

$$\Delta x = \frac{3-1}{6} = \frac{1}{3}$$

and $f(1) = 2$, $f(\frac{4}{3}) \approx 3.47$, $f(\frac{5}{3}) \approx 5.98$, $f(2) \approx 9.82$, $f(\frac{7}{3}) \approx 15.32$, $f(\frac{8}{3}) \approx 22.79$, $f(3) \approx 32.54$. Thus the integral is approximated by

$$P_6 \approx \frac{\frac{1}{3}}{3}[2 + 4(3.47) + 2(5.98) + 4(9.82) + 2(15.32) + 4(22.79) + 32.54]$$

$$\approx \frac{1}{9}(221.46) \approx 24.61 \quad \blacksquare$$

If you compare Examples 1–3 with the exact value of the definite integral, you will note that each successive approximation is better than the previous one and that the larger the value for n the more accurate the results. For the definite integral in Examples 1–3 we see that

$$\int_1^3 (x^\pi + 1)\,dx = \left.\frac{x^{\pi+1}}{\pi+1} + x\right|_1^3 \approx 24.60793128$$

EXAMPLE 4   Use Simpson's rule with $n = 4$ to approximate the value of

$$\int_0^1 \frac{dx}{x+1}$$

to four decimal places. Check your answer by finding the exact value of the definite integral.

Solution $\Delta x = \dfrac{b-a}{n} = \dfrac{1-0}{4} = .25$

Therefore, since $f(x) = \dfrac{1}{x+1}$,

$$P_4 = \int_0^1 \frac{dx}{x+1} = \frac{.25}{3}[f(x_0) + 4f(x_1) + 2f(x_2) + 4f(x_3) + f(x_4)]$$

$$= \frac{1}{12}[f(0) + 4f(.25) + 2f(.5) + 4f(.75) + f(1)]$$

$$\approx \frac{1}{12}[1 + 4(.8) + 2(.6) + 4(.5714285714) + .5]$$

$$\approx \frac{1}{12}(8.319047619)$$

$$\approx .6932539683 \quad \text{or} \quad .6933 \quad \text{(to four decimal places)}$$

The exact value of the definite integral is found as follows:

$$\int_0^1 \frac{dx}{x+1} = \left.|\ln(x+1)|\right|_0^1 = \ln 2 - \ln 1 = \ln 2$$

(This is .6931 to four decimal places.) $\blacksquare$

6.6 Problem Set

Consider $\int_1^6 \sqrt{x+3}\, dx$. Then

$$\int_1^6 \sqrt{x+3}\, dx = \frac{2}{3}(x+3)^{3/2}\Big|_1^6 = \frac{2}{3}(27) - \frac{2}{3}(8) = \frac{38}{3}$$

Approximate this integral by finding the requested values in Problems 1–12.

1. A_1
2. A_2
3. A_3
4. A_4
5. T_1
6. T_2
7. T_3
8. T_4
9. P_2
10. P_4
11. P_6
12. T_8

Consider $\int_2^4 \frac{x\, dx}{(1+2x)^2}$. Then

$$\int_2^4 \frac{x\, dx}{(1+2x)^2} = \frac{1}{4}\left[\ln|1+2x| + \frac{1}{1+2x}\right]_2^4$$

$$= .25\left(\ln 9 + \frac{1}{9}\right) - .25\left(\ln 5 + \frac{1}{5}\right)$$

$$\approx .5770839221 - .4523594781$$

$$\approx .124724444$$

Approximate this integral by finding the requested values in Problems 13–24.

13. A_1
14. A_2
15. A_3
16. A_4
17. T_1
18. T_2
19. T_3
20. T_4
21. P_2
22. P_4
23. P_6
24. P_8

Approximate the integrals in Problems 25–28 by using a rectangular approximation (pick left endpoints). Round your answer to two decimal places using $n = 4$.

25. $\int_3^7 \frac{\sqrt{1+x}}{x^3}\, dx$
26. $\int_1^4 \ln x\, dx$
27. $\int_1^3 x^x\, dx$
28. $\int_0^2 e^{-x^2}\, dx$

Approximate the integrals in Problems 29–32 by using a trapezoidal approximation. Round your answer to two decimal places using $n = 4$.

29. $\int_3^7 \frac{\sqrt{1+x}}{x^3}\, dx$
30. $\int_1^4 \ln x\, dx$
31. $\int_1^3 x^x\, dx$
32. $\int_0^2 e^{-x^2}\, dx$

Approximate the integrals in Problems 33–36 by using Simpson's rule. Round your answer to two decimal places using $n = 4$.

33. $\int_3^7 \frac{\sqrt{1+x}}{x^3}\, dx$
34. $\int_1^4 \ln x\, dx$
35. $\int_1^3 x^x\, dx$
36. $\int_0^2 e^{-x^2}\, dx$

APPLICATIONS

37. A landscape architect needs to estimate the number of cubic yards of fill necessary to level the area shown below. Use Simpson's rule with $\Delta x = 5$ feet and y values equal to the distances measured in the figure to find the surface area. Assume that the depth averages 3 feet. (*Note:* 1 cubic yard equals 27 cubic feet.)

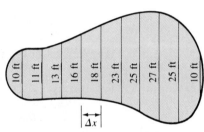

Each contour line is 5 feet from the others

38. The rate at which flashbulbs give off light varies during the flash. For some bulbs, the light output, measured in lumens, reaches a peak and fades quickly (as shown in part **a** of the figure below) and for others the light output stays level for a relatively longer period of time (shown in part **b**).

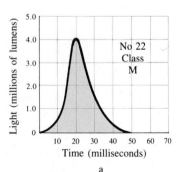

a

b

The amount A of light emitted by the flashbulb in the interval from 20 to 70 milliseconds after the button is pressed is given by the formula

$$A = \int_{20}^{70} L(t)\,dt \text{ lumens-milliseconds}$$

where $L(t)$ is the lumen output of the bulb as a function of time. Use the trapezoidal rule and the numerical data from the table below to determine which bulb gets more light to the film.*

Light Output Versus Time

Time after ignition (in milliseconds)	Light output (#22 bulb)	Light output (#31 bulb)
0	0	0
5	.2	.1
10	.5	.3
15	2.6	.7
20	4.2	1.0
25	3.0	1.2
30	1.7	1.0
35	.7	.9
40	.35	1.0
45	.2	1.1
50		1.3
55		1.4
60		1.3
65		1.0
70		.8
75		.6
80		.3
85		.2
90		0

---

* From W. U. Walton et al., *Integration* (Newton, MA: Project CALC, Education Development Center, 1975), p. 83. Data in the table from *Photographic Lamp and Equipment Guide*, P4–15P, General Electric Company, Cleveland, Ohio. The material is quoted from Thomas and Finney, *Calculus and Analytic Geometry*, 6th ed. (Reading, MA: Addison-Wesley Publishing Company, 1984), pp. 311–312.

*6.7
Review

The material of this chapter is reviewed in the following list of objectives. After each objective there are some practice questions. For a sample test select the first question of each set and check your answers. The second question for each objective has no answer given. If you are having trouble with a particular type of problem, look back at the indicated section in the text. When you are finished reviewing these objectives, a sample examination is given at the end of this section.

[6.1]
Objective 6.1: *Evaluate an indefinite integral using one of the three direct integration formulas.*

1. $\int x^6\,dx$

2. $\int \dfrac{dx}{x}$

3. $\int e^u\,du$

4. $\int u^{-5}\,du$

Objective 6.2: *Evaluate an indefinite integral using one of the integration formulas.*

5. $\int (6x^2 - 2x - 5)\,dx$

6. $\int \dfrac{du}{12}$

7. $\int \dfrac{(x-2)\,dx}{3x^2 - 5x - 2}$

8. $\int \left(e^{5x+1} + \dfrac{1}{e}\right)dx$

Objective 6.3: *Find an antiderivative that satisfies given initial conditions.*

9. $f(x) = e^x + 9$; $F(0) = 10$

10. $f(x) = 3x^2 - 9$; $F(10) = 0$

---

* Optional section.

11. $f(x) = \dfrac{1}{x};\quad F(1) = 0$

12. $f(x) = \dfrac{x^2 - 9}{x + 3};\quad F(10) = 100$

Objective 6.4: *Find the cost function, given the marginal cost function.*

13. $C'(x) = 9x^2 + 1;$ fixed cost, $25,000
14. $C'(x) = 250 - 8x;$ fixed cost, $8,000
15. $C'(x) = 10 - \dfrac{1}{\sqrt{x}};$ fixed cost, $1,500
16. $C'(x) = .005x^3;$ fixed cost, $35,000

[6.2]
Objective 6.5: *Find indefinite integrals by substitution.*

17. $\displaystyle\int \left(e^{-3x} - \dfrac{1}{e}\right) dx$

18. $\displaystyle\int (2 + 3x^2)^5 x\, dx$

19. $\displaystyle\int \dfrac{2x - 5}{\sqrt{(x^2 - 5x)^3}}\, dx$

20. $\displaystyle\int x^2 \sqrt{10 - x}\, dx$

[6.3]
Objective 6.6: *Evaluate definite integrals.*

21. $\displaystyle\int_1^2 \dfrac{x^3 + 2}{x^2}\, dx$

22. $\displaystyle\int_0^2 x\sqrt{2x^2 + 1}\, dx$

23. $\displaystyle\int_1^8 x^{1/3}(1 + x^{4/3})^3\, dx$

24. $\displaystyle\int_{\sqrt{17}}^5 x\sqrt{x^2 - 16}\, dx$

[6.4]
Objective 6.7: *Find the area bounded by the x-axis, the curve $y = f(x)$, and two given vertical lines.*

25. $y = \sqrt{5x + 1};\quad x = 3, x = \dfrac{8}{5}$

26. $y = 5x^4 - 20;\quad x = -1; x = 2$

27. $y = \dfrac{1}{(1 - 2x)^3};\quad x = 1, x = 5$

28. $y = \dfrac{\ln(3x + 4)}{3x + 4};\quad x = 1; x = 3$

Objective 6.8: *Write an integral to express a shaded region.*

29.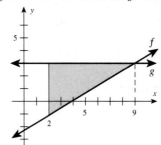

30.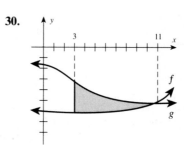

31.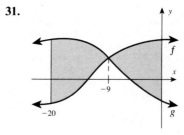

32.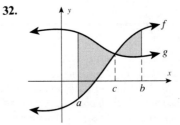

Objective 6.9: *Find the area between two given curves.*

33. $y = \tfrac{1}{2}x^2$ and $y = 4 - x$ from $x = -2$ to $x = 3$
34. $y = x^2 - 16$ and the x-axis
35. $y = x^2 - 3x$ and $x - 4y = 0$
36. $y = x^3$ and $y = x$

[6.5]
Objective 6.10: *Approximate integrals using a rectangular approximation. Use left endpoints to find*

$$\int_4^{25} \ln\sqrt{x}\, dx$$

correct to two decimal places where n is specified in Problems 37–40.

37. $n = 1$ **38.** $n = 2$
39. $n = 4$ **40.** $n = 8$

Objective 6.11: *Evaluate integrals for tabular functions.*

41. $\int_0^{10} f(x)\, dx$

x	1	3	5	7	9
f(x)	63	49	55	82	80

42. $\int_0^2 f(x)\, dx$

x	.5	1	1.5	2
f(x)	8.4	12.1	16.4	11.4

43. The graph shown here from the *Wall Street Journal* presents the rate of domestic cars sold from June 1 to February 28. Estimate the sales for the 4th quarter (September 1–December 31). (*Hint:* Imagine a rectangle 30 days wide with an area equal to the part under the curve for that month.)

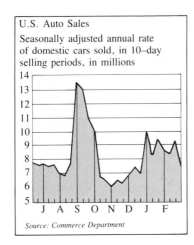

44. Using the graph in Problem 43, estimate the accumulated sales from July 1 to November 1.

Objective 6.12: *Find the average value for a given function.*
45. $f(x) = 6x^2 - 2x + 5$ on $[-2, 1]$
46. $f(x) = \dfrac{3x - 2}{x^2}$ on $[1, 4]$
47. $f(x) = xe^{x^2}$ on $[1, 3]$
48. $f(x) = x\sqrt{9 - 2x^2}$ on $[0, 2]$

[6.6]
Objective 6.13: *Approximate integrals using a trapezoidal approximation. Find*

$$\int_0^2 \frac{dx}{1 + x^2}$$

correct to two decimal places where n is specified in Problems 49–52.
49. $n = 1$ **50.** $n = 2$
51. $n = 4$ **52.** $n = 8$

Objective 6.14: *Approximate integrals using Simpson's rule. Find*

$$\int_0^2 \sqrt{4 - x^2}\, dx$$

correct to two decimal places where n is specified in Problems 53–56.
53. $n = 1$ **54.** $n = 2$
55. $n = 8$ **56.** $n = 4$

Objective 6.15: *Solve applied problems based on the preceding objectives.*

57. *Marginal profit.* The marginal profit of Campress Press-Ons is

$$P'(x) = 250x - 5{,}000$$

where x is the number of items. Find the profit function if there is a $1,500 loss when no items are produced.

58. *Marginal cost.* The marginal cost of Campress Press-Ons is

$$C'(x) = \frac{\sqrt{x}}{1{,}000} + .0001x$$

If the fixed costs are $1,500, what is the cost of producing 5,000 items (to the nearest thousand dollars)?

59. Industrial waste is pumped into a 1,000-gallon holding tank at the rate of 1 gal/min and the well-stirred mixture is removed at the same rate. If the concentration of waste in the tank is changing at a rate of

$$c'(t) = .001e^{-.001t}$$

after t minutes, find the concentration of waste in the tank after one hour.

60. The graph below (from the *Wall Street Journal*, February 26, 1986) shows the 10-year Treasury Bond rate for the years 1980–1985. Estimate the area under this curve using a rectangular approximation where $n = 6$.

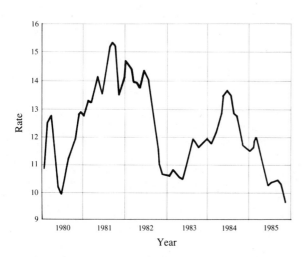

SAMPLE TEST

The following sample test (45 minutes) is intended to review the main ideas of this chapter.

Evaluate the integrals in Problems 1–12. Be sure to include the constant of integration when working with indefinite integrals.

1. $\int x^5 \, dx$

2. $\int \dfrac{dx}{x}$

3. $\int e^u \, du$

4. $\int u^{-4} \, du$

5. $\int (18x^2 - 6x - 3) \, dx$

6. $\int \dfrac{(x+1)}{2x^2 + 4x + 5} \, dx$

7. $\int \left(e^{3x+1} + \dfrac{1}{e}\right) dx$

8. $\int \dfrac{6x^2 - 3x + 2}{x} \, dx$

9. $\int_{-1}^{3} \dfrac{x^3 + 2}{x^2} \, dx$

10. $\int_{2}^{2\sqrt{3}} x\sqrt{2x^2 + 1} \, dx$

11. $\int_{1}^{8} [x^{1/3}(1 + x^{4/3})^3] \, dx$

12. $\int_{3\sqrt{2}}^{5} x\sqrt{x^2 - 9} \, dx$

13. Find the area bounded by the x-axis, the curve $y = \sqrt{3x + 4}$, and the lines $x = 4$ and $x = \tfrac{5}{3}$.

14. Find the area bounded by the curves $y = x^2$, $y = 6 - x$, and the lines $x = 0$ and $x = 4$.

15. Find the area bounded by the curves $y = x^3$ and $y = x$.

16. The marginal profit of Campress Press-Ons is

$$P'(x) = 250x - 5{,}000$$

where x is the number of items. Find the profit function if there is a $2,500 loss when no items are produced.

17. One of the fastest growing trees is the eucalyptus. In order to determine whether it is economically feasible to harvest it as a crop for firewood, it must grow at least 6 feet in 5 years ($t = 6$, since it is 1-year-old when planted). If the growth rate is

$$1.2 + 5t^{-4} \qquad t \geq 1$$

feet per year, where t is the time in years, is it economically feasible to plant these trees for commercial harvesting if they are 12 inches tall when planted (that is, when $t = 1$)?

18. In 1986 the growth rate of Chicago, Illinois, was

$$P'(t) = -.035e^{-.011t}$$

where t is measured in years after 1980 and $P(t)$ gives the population in millions. If the population in 1984 was 3,000,000, predict the population at the turn of the century using the 1986 growth rate.

19. The rate of consumption (in billions of barrels per year) for oil conforms to the formula

    $R(t) = 32.4e^{.048t}$

    for t years after 1985. If the total oil still left in the earth is estimated to be 670 billion barrels, estimate the length of time before all available oil is consumed if the rate does not change.

20. Find a decimal approximation for $\int_0^3 \sqrt{9 - x^2}\, dx$ correct to two decimal places. Use the left endpoints for a rectangular approximation where $n = 4$.

7
Applications and Integration

CHAPTER OVERVIEW
The calculus you learned in Chapter 6 is put to work in this chapter as we look at some important applications in order to show some of the power and versatility of integral calculus. In addition, our study of integration is expanded by the introduction of two new integration techniques: by parts, and by table. It might be noted that more and more emphasis today is given to integration by using tables. Although techniques of integration are important, new computer technology has made numerical integration and integration by table more important than ever before.

PREVIEW
We can now use the techniques of integration in a variety of interesting and important applications in mathematics and business. These applications include total value, money flow, consumers' and producers' surplus, lease payments, depreciation, optimum time to overhaul equipment, net investment flow, and capital formation. We learn the counterpart of the product rule for differentiation in a process called *integration by parts*. Probabilistic models are also considered (in an optional section), the most important of these being the normal curve, or normal density function.

PERSPECTIVE
This chapter completes your introduction to elementary calculus. The ideas of limit, the derivative, and the integral have been introduced and some of their applications discussed. If you have mastered these ideas, you have the skill to build and solve mathematical models that will help you analyze, understand, and forecast many real-life situations. The next chapter considers an additional useful topic: functions of several variables.

7.1 Business Models Using Integration
7.2 Integration by Parts
7.3 Using Tables of Integrals
7.4 Improper Integrals
7.5 Probability Density Functions
7.6 Chapter 7 Review
    Chapter Objectives
    Sample Test

MODELING APPLICATION 7

Modeling the Nervous System

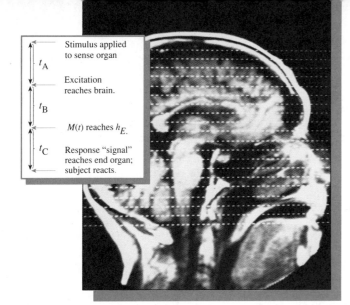

The nervous system has the ability to respond to a variety of stimuli by generating electrical impulses and transmitting them through nerve cells. Nerve cells vary in length from a few micrometers to more than a meter. Nerve cells that transmit impulses from sense organs to the spinal cord and brain are called *sensory* or *afferent neurons*. Those that transmit messages to the muscles and glands are called *motor* or *efferent neurons*. The central nervous system (CNS) consists of the brain and spinal cord; the peripheral nervous system (PNS) consists of the sensory and motor neurons; and the autonomic nervous system (ANS) consists of neurons that regulate processess not under conscious control.

Do some reading in the following sources* and write a paper discussing a model that describes reaction time in the central nervous system. For general guidelines about writing this essay, see the commentary for Modeling Application 1 on page 51.

*Horelick, Brindell, Sinan Koont, and Sheldon F. Gottleif. *Modeling the Nervous System: Reaction Time and the Central Nervous System.* © 1983 COMAP, Inc., 271 Lincoln Street, Suite 4, Lexington, MA 02173. COMAP is a nonprofit corporation engaged in research and development in mathematics education.

Erlanger, J., and H. S. Gasser. *Electrical Signs of Nervous Activity* (Philadelphia: University of Pennsylvania Press, 1937).

APPLICATIONS

Management (*Business, Economics, Finance, and Investments*)
Total value/total income
  (7.1, Problems 1–12)
Money flow (7.1, Problems 13–18)
Finding total sales (7.1, Problems 19–24)
Monthly lease payments
  (7.1, Problems 27–28)
Net excess profit (7.1, Problems 29–30)
Time period for a piece of equipment to pay for itself (7.1, Problems 33–34)
Consumer's/producer's surplus
  (7.1, Problems 35–38; 7.6, Problems 17–20)
Depreciation
  (7.1, Problems 39–40; 7.6, Problem 58; Test Problem 18)
Optimum time to overhaul equipment
  (7.1, Problem 41)
Net investment flow (7.1, Problem 42)

Management (*continued*)
Capital value
  (7.1, Problem 43; 7.4, Problems 23–25; 7.6, Problem 59; Test Problem 16)
Total profit (7.2, Problem 38)
Total cost (7.3, Problem 66)
Accumulated sales (7.3, Problem 67)
Endowment fund (7.4, Problem 31)

Life Sciences (*Biology, Ecology, Health, and Medicine*)
Reaction time of a drug
  (7.1, Problems 25–26)
Growth rate of bacteria (7.2, Problem 39)
Pollutants dumped into a river
  (7.3, Problem 65; 7.4, Problem 32)
Oil production (7.4, Problem 33)
Seed germination (7.5, Problem 37)
Rainfall in Ferndale, CA (7.5, Problem 40)
Spread of a disease (7.6, Problem 57)

Social Sciences (*Demography, Political science, Population, Psychology, Society, and Sociology*)
Rate of learning
  (7.5, Problem 36; 7.6, Test Problem 17)
Grading on a curve (7.6, Problem 61)

General Interest
Future value (7.1, Problems 31–32)
Maintenance costs (7.2, Problem 41)
Radioactive materials in the atmosphere
  (7.4, Problems 26–28)
Breaking strength (7.5, Problems 38–39)
Life of a light bulb
  (7.5, Problems 41–43; 7.6, Problem 60)
Telephone waiting time (7.5, Problem 44)
Battery life (7.5, Problem 45)

Modeling Application—
Modeling the Nervous System

7.1
Business Models Using Integration

Rates

Suppose Wayne Savick introduced a new batching process computer with an annual rate of sales t years after it was first introduced given by the function $S(t)$. He found that

$$S(t) = 5 + 15t^2$$

for $0 \leq t \leq 3$. What is the rate of sales at $t = 2$ years? Consider

$$S(2) = 5 + 15(2)^2 = 65$$

How should this be interpreted? Can we say 65 computers sold during the second year? Not really; notice that

$$S(1) = 5 + 15(1)^2 = 20$$

Since the rate of sales is 20 units when $t = 1$ and 65 units when $t = 2$, is seems reasonable that the number of units sold at different times during the second year is between 20 and 65. Remember, you cannot assume that the annual rate of sales is constant on a yearly basis. The graph of S is shown in Figure 7.1.

A better approximation for finding the total number of computers sold during the entire second year is found when we calculate

$$S\left(\frac{3}{2}\right) = 5 + 15(1.5)^2 = 38.75$$

Thus, during the first half of the second year, the number of units produced should be at least

$$\frac{1}{2}(20) = 10$$

and during the second half of the second year it should be

$$\frac{1}{2}(38.75) = 19.375$$

This approximation is shown in Figure 7.2.

The number of square units in the area of the rectangles in Figure 7.2 represents the total sales if sales during the first half of the year remain at a constant 20 units and during the second half at a constant 38.75 units. However, sales are not constant. So we now see (by using Riemann sums) that the desired answer is

$$\int_1^2 (5 + 15t^2)\,dt = 5t + \frac{15t^3}{3}\bigg|_1^2$$
$$= [10 + 5(2)^3] - [5 + 5(1)^3]$$
$$= 40$$

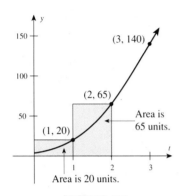

FIGURE 7.1
Graph of the sales of a new batching process computer

7.1 BUSINESS MODELS USING INTEGRATION

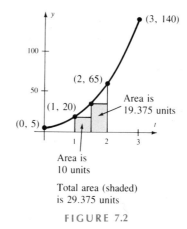

Area is 10 units

Total area (shaded) is 29.375 units

FIGURE 7.2

EXAMPLE 1 During the first year for which a new product is in the market, $S(t)$ is the rate of sales after t months and is given by the formula

$$S(t) = 90\sqrt{t} + 150$$

Find the total sales for the first 9 months.

Solution
$$\int_0^9 (90\sqrt{t} + 150)\,dt = \frac{90t^{3/2}}{\frac{3}{2}} + 150t \Big|_0^9$$

$$= 60t^{3/2} + 150t \Big|_0^9$$

$$= 60(9)^{3/2} + 150(9) = 2{,}970$$

This means that about 2,970 units are sold in the first 9 months. ∎

The definite integral we used in Example 1 is called the **total value** over a period of time.

Total Value of a Function over Time

The **total value** of an integrable function f over $[a, b]$ is given by

$$\int_a^b f(x)\,dx$$

A common variation of total value is to determine the length of time it will take for a piece of new equipment to pay for itself.

EXAMPLE 2 A solar heater is being considered as a replacement for an electric heater. The cost of the solar heater is bid at $38,500, and the rate of savings in operating costs by using a solar instead of an electric heater is given by

$$S(x) = 8{,}000x + 3{,}000$$

where $S(x)$ is in dollars per year and x is the number of years the solar heater will be in use. How long will it take the solar heater to pay for itself?

Solution Let t be the number of years the solar heater is used until the operating cost savings is equal to the initial replacement cost. Then

$$\int_0^t (8{,}000x + 3{,}000)\,dx = 38{,}500$$

$$\left. \frac{8{,}000x^2}{2} + 3{,}000x \right|_0^t = 38{,}500$$

$$\frac{8{,}000t^2}{2} + 3{,}000t = 38{,}500$$

$$4{,}000t^2 + 3{,}000t - 38{,}500 = 0$$

$$8t^2 + 6t - 77 = 0$$

$$(4t - 11)(2t + 7) = 0$$

$$t = \frac{11}{4},\ -\frac{7}{2}$$

Reject negative in this application.

Thus it would take about $\frac{11}{4}$ years or 2 years and 9 months for the solar heater to pay for itself. ∎

Continuous Income Stream

The growth rate of money with respect to time is stated as a compound interest formula or interest with continuous compounding. If this growth rate is denoted by $f'(t)$, then the integral of that rate of growth is the function $f(t)$ that gives the accumulated money. The value of this function at $t = k$ is the total amount at this time, and the value at $t = 0$ is the amount at the beginning of the time period. Thus, the total growth is $f(k) - f(0)$ or

$$\int_0^k f'(x)\,dx$$

This value denotes what is called a **continuous income stream**. For example, suppose money flows continuously into a computer video game and grows at a rate given by

$$f(t) = 10e^{.08t}$$

where t is in hours and $0 \leq t \leq 12$. Find the total amount of money that accumulates in the machine during the 12-hour period:

$$\int_0^{12} 10e^{.08t}\,dt = \left. 10\frac{e^{.08t}}{.08} \right|_0^{12} \approx 201.46$$

The function f is called the **rate of flow** and even though in reality the income would be collected at specific intervals (each month, each week, or each day), we assume that the income is actually received continuously. That is, we say that the income is received in a **continuous stream**, and the total value formula for this application is referred to as **total income**.

7.1 BUSINESS MODELS USING INTEGRATION

EXAMPLE 3 If the rate of flow for an investment is

$$f(x) = 25{,}000e^{.09t}$$

find the total income produced in one year and in ten years.

Solution The total income for each part is found with these respective integrals:

$$\int_0^1 25{,}000e^{.09t}\, dt = 25{,}000 \left. \frac{e^{.09t}}{.09} \right|_0^1 = \$26{,}159.52$$

$$\int_0^{10} 25{,}000e^{.09t}\, dt = 25{,}000 \left. \frac{e^{.09t}}{.09} \right|_0^{10} = \$405{,}445.31$$

The total income produced in one year is $26,159.52 and in 10 years is $405,445.31. ∎

You might recognize the formula used in Example 3 as the formula for the continuous compounding of $25,000 invested at 9% per year. If we calculate $A(t) = 25{,}000e^{.09t}$ for $t = 1$ we obtain $27,354.36. How does this relate to the number we found in Example 3? This represents the future value of a $25,000 deposit. In other words, of this $27,354.36, $25,000 is the initial investment and $2,354.36 is the interest. On the other hand, when we integrate as we did in Example 3, we are finding the *total income for the year*—namely, $26,159.52. These differences are even more remarkable if we let $t = 10$:

$$A(t) = 25{,}000e^{.09t} \text{ for } t = 10 \text{ gives } \$61{,}490.08.$$

This means that the future value for a $25,000 deposit at 9% in 10 years compounded continuously is $61,490.08, while the *total income* for the same period is given by the rate of flow formula, which yields $405,445.31.

If the rate of flow formula $f(x)$ in the total value of a function formula is the continuous interest formula, namely $f(t) = Pe^{rt}$, then we have still another name for the total value formula—**total money flow**.

Total Money Flow

The **total money flow** for P dollars over a period of T years at a rate r is given by the formula

$$F = \int_0^T Pe^{rt}\, dt$$

The total money flow is sometimes called **the amount of an annuity**. An *annuity* is a sequence of equal periodic payments into an account. If the interest is earned continuously at an annual rate of r, and a payment of m dollars is made n times a year for t years, then the above formula for N payments ($N = nt$) is

$$A = \int_0^N me^{it}\, dt$$

where $i = \dfrac{r}{n}$.

EXAMPLE 4 If a person places $50 a month into an account paying 9% compounded continuously, how much will be in the account after 5 years?

Solution $m = 50$, $n = 12$, $N = 5(12) = 60$, $r = .09$, and $i = \frac{.09}{12} = .0075$

$$A = \int_0^{60} 50 e^{.0075t}\, dt = 50 \frac{e^{.0075t}}{.0075}\bigg|_0^{60} = \frac{50}{.0075}(e^{.45} - 1) = 3{,}788.747903$$

There would be $3,788.75 in the account. ∎

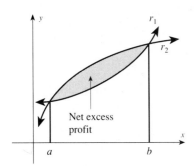

FIGURE 7.3

There are three business applications of integration that involve the area between two curves which we will now consider. These are *net excess profit, consumers' surplus,* and *producers' surplus.*

Net Excess Profit

Suppose you are interested in comparing the profit from two investments t years from now and you know that the investments are generating profits at the rates of $r_1(t)$ and $r_2(t)$ dollars per year. Suppose also that you know that for the next N years $r_2(t) > r_1(t)$. The *net excess profit* is defined to be the area between the curves representing these rates as shown in Figure 7.3.

Net Excess Profit If profit is generated by the rates $r_1(t)$ and $r_2(t)$ at time t over a closed interval of time $[0, N]$ where $r_2(t) > r_1(t)$, then the **net excess profit** is defined by the formula

$$\int_0^N [r_2(t) - r_1(t)]\, dt$$

EXAMPLE 5 Suppose that t years from now, two investment plans will be generating profits at the rates of $10 + t^2$ and $50 + 6t$ in thousands of dollars per year, respectively.

a. When will these investments be generating the same rate of profit?

b. Which is the better investment for the interval of time $[0, N]$ if N is the length of time until the investments generate the same rate of profit? See Figure 7.4.

c. What is the net excess profit?

Solution **a.** $10 + t^2 = 50 + 6t$
$t^2 - 6t - 40 = 0$
$(t - 10)(t + 4) = 0$
$t = 10, -4$

Since t is a length of time, reject $t = -4$, so $t = 10$; it will be 10 years before these investments are generating the same rate of profit; that is $N = 10$.

b. On the interval $[0, 10]$ we have $50 + 6t \geq 10 + t^2$, so $r_2(t) = 50 + 6t$ and $r_1(t) = 10 + t^2$.

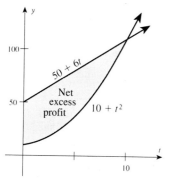

FIGURE 7.4

c. NET EXCESS PROFIT $= \int_0^N [r_2(t) - r_1(t)]\, dt$

$= \int_0^{10} [(50 + 6t) - (10 + t^2)]\, dt = \int_0^{10} (-t^2 + 6t + 4)\, dt$

$= \left(\dfrac{-t^3}{3} + 3t^2 + 4t \right) \Big|_0^{10}$

$= \dfrac{20}{3}$

The excess profit is $6,667. (This is $\tfrac{20}{3}$ thousands of dollars.) ∎

Consumers' and Producers' Surplus

An important business application of the area between curves involves the concepts of **consumers' surplus** and **producers' surplus**. The intersection of the demand and supply functions provides an *equilibrium price*, or point, as shown in Figure 7.5.

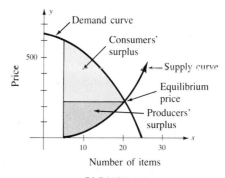

FIGURE 7.5
Consumers' and producers' surplus

The demand function, $D(x)$, gives the price per unit that consumers are willing to pay to get the xth item of the same commodity. For example, a company might be willing to spend $1,000 for a personal computer. However, once they have the first computer, they might be willing to pay only $800 for the second computer, and $500 for a third computer. The price that consumers are willing to pay to get one additional unit usually decreases as the number of units already bought increases. This demand function, which in economics is called the **marginal willingness to spend**, can be thought of as the rate of change with respect to x of the *total amount consumers are willing to spend for x units*. Thus, if $A(x)$ is the total amount (in dollars) that consumers are willing to spend to get x units of the commodity, and if $A(x)$ is differentiable, then

$$D(x) = \dfrac{dA}{dx}$$

Thus, the total amount consumers are willing to spend to get n units of the commodity is the definite integral

$$A(n) = A(n) - A(0) = \int_0^n \frac{dA}{dx} dx = \int_0^n D(x)\, dx$$

In a competitive economy, the total amount that consumers *actually spend* on a commodity is generally less than the total amount they would be *willing* to spend. The difference between these two amounts is known as the **consumers' surplus** because it can be thought of as a savings realized by the consumer.

CONSUMERS' SURPLUS

= TOTAL AMOUNT CONSUMERS WOULD BE WILLING TO SPEND

− ACTUAL CONSUMER EXPENDITURE

In order to understand this example, go back to the company that was willing to purchase a computer. Reconsider the fact that the company was willing to spend $1,000 on the first computer, $800 on the second computer, and $500 for the third. Also suppose that the marked price is $800. Then the company would buy two computers for a total of

2($800) = $1,600

This is less than the company would have been willing to spend to get two computers—namely, $1,000 + $800 = $1,800. The savings

$1,800 − $1,600 = $200

is the company's consumer surplus. This information is summarized in Figure 7.6.

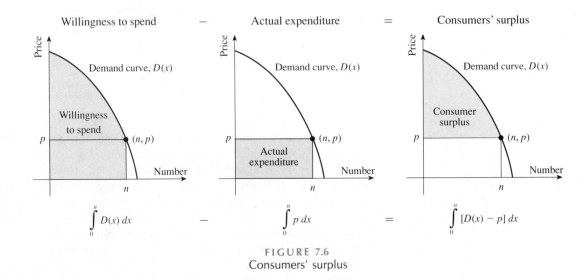

FIGURE 7.6
Consumers' surplus

On the other hand, suppose $S(x)$ is the price–supply function that represents the price at which a producer would be willing to sell the xth item of the same commodity. For example, a company might be willing to sell a single cut glass for

$100, but would be willing to sell a dozen of the same cut glasses for $75 each. The **producers' surplus** is the difference between the price (the actual consumer expenditure), p, and the total amount the producers would have been willing to accept for the sale of their product. These ideas are summarized in the following box.

Consumers' Surplus

If n items are supplied at a price of p dollars per item for a demand function $D(x)$, then

$$\text{CONSUMERS' SURPLUS} = \int_0^n [D(x) - p]\, dx$$

Producers' Surplus

If n items are produced at a price of p dollars per item for a supply function $S(x)$, then

$$\text{PRODUCERS' SURPLUS} = \int_0^n [p - S(x)]\, dx$$

The most common situation is for (n, p) to be the equilibrium point for the demand and supply functions.

EXAMPLE 6 The price, in dollars, for a product is

$$D(x) = 650 - x - x^2$$

where x is the number of items in the domain $[0, 25]$. The supply curve, in dollars, is given by

$$S(x) = x^2 - 9x + 10$$

where x is in the domain $[10, 25]$. Find the consumers' and producers' surplus.

Solution First, find the equilibrium price.

$$D(x) = 650 - x - x^2$$
$$S(x) = x^2 - 9x + 10$$

Find x so that $D(x) = S(x)$:

$$650 - x - x^2 = x^2 - 9x + 10$$
$$2x^2 - 8x - 640 = 0$$
$$x^2 - 4x - 320 = 0$$
$$(x - 20)(x + 16) = 0$$
$$x = 20, -16$$

↑ Reject the negative value

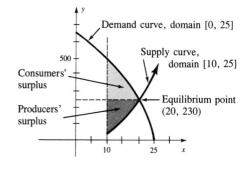

The domain for the surplus is the intersection of the domain for the supply and demand; in this example it is given as $[10, 25]$. For $x = 20$, the price is 230, so the equilibrium point is $(20, 230)$.

$$\text{Consumers' surplus} = \int_{10}^{20} [(650 - x - x^2) - 230] \, dx$$

$$= \int_{10}^{20} (420 - x - x^2) \, dx$$

$$= \left(420x - \frac{x^2}{2} - \frac{x^3}{3} \right) \Big|_{10}^{20}$$

$$= 8,400 - 200 - \frac{8,000}{3} - 4,200 + 50 + \frac{1,000}{3}$$

$$\approx 1,716.67$$

$$\text{Producers' surplus} = \int_{10}^{20} [230 - (x^2 - 9x + 10)] \, dx$$

$$= \int_{10}^{20} (220 + 9x - x^2) \, dx$$

$$= \left(220x + \frac{9x^2}{2} - \frac{x^3}{3} \right) \Big|_{10}^{20}$$

$$\approx 1,216.67$$

The consumers' surplus is $1,716.67, and the producers' surplus is $1,216.67. ∎

7.1

Problem Set

APPLICATIONS

Find the total value of the functions in Problems 1–6.
1. $f(x) = \sqrt{x} + 100$ on $[4, 9]$
2. $S(x) = 50\sqrt[3]{x} - 30x$ on $[1, 8]$
3. $m(x) = \dfrac{x - 1}{x + 1}$ on $[0, 2]$
4. $t(x) = \dfrac{2x + 1}{x - 2}$ on $[3, 5]$
5. $f(t) = 65e^{-5t}$ on $[0, 3]$
6. $f(t) = 1,000e^{.005t}$ on $[0, 2]$

Find the total income produced in the first 5 years for the rate of flow functions given in Problems 7–12.
7. $f(x) = 500$
8. $f(x) = 4,000$
9. $f(x) = 200e^{.08t}$
10. $f(x) = 500e^{.11t}$
11. $f(x) = 150e^{-.05t}$
12. $f(x) = 850e^{-.09t}$

Determine the total money flow for Problems 13–18.
13. $8,000 at 8% for 6 years
14. $83,000 at 10% for 2 years
15. $2,500 at 5% for 1 year
16. $125,000 at 11% for 15 years
17. $6,500 at 7% for 5 years
18. $4 million at 10% for 8 years
19. A new product has a rate of sales given by

$$S(t) = 60 + 24t^2 \quad \text{for } 0 \le t \le 3$$

where t is the number of years since the product was introduced. Find the number of items sold in the second year.

20. Repeat Problem 19 for the third year.
21. Repeat Problem 19 for the total 3-year period.
22. During the first year a new commodity has been on the market, y units per month were sold after x months elapsed, according to the formula

$$y = 300\sqrt[3]{x} + 50 \quad 0 \le x \le 12$$

Find the total sales during the first 6 months.

23. Repeat Problem 22 for the second 6 months.
24. Repeat Problem 22 for the entire first year.

25. The rate of reaction to a certain drug is measured in minutes after administration according to the formula
$$R(t) = x^{-.5}$$
Evaluate the total reaction in the first 10 minutes.

26. Repeat Problem 25 for the first hour.

27. An equipment leasing company determines that the rate of maintenance cost in dollars per year on a piece of new equipment is
$$M(x) = 150(1 + \sqrt[3]{x^2})$$
where x is the number of years the equipment has been leased. What should be the total amount for maintenance costs for a 4-year lease? What amount should be added to the monthly payment to pay for all the maintenance costs for a 4-year lease?

28. Repeat Problem 27 for a 3-year lease.

29. Find the net excess profit for the functions $r_1(x) = x^2 + 30$ and $r_2(x) = 15 + 8x$ on $[3, 5]$.

30. Find the net excess profit for the functions $r_1(x) = x^2 + 20$ and $r_2(x) = 24 + 3x$ on $[0, 4]$.

31. If a person deposits $2,250 each year into a tax-sheltered retirement account paying 9% compounded continuously, how much will be in the account in 20 years?

32. If a person deposits $25 per month into an account paying 7% compounded continuously, how much will be in the account in 8 years?

33. Suppose that a new piece of equipment costs $15,000. The rate of operating cost savings is $S(x)$ in dollars per year and is given by the formula
$$S(x) = 2{,}400x + 10{,}500$$
where x is the number of years the equipment has been used. How long will it take the piece of equipment to pay for itself?

34. If a piece of machinery has a rate of operating cost savings in dollars per year given by
$$S(x) = 4{,}800x + 22{,}500$$
how long it will take for the piece of machinery to pay for itself if x is the number of years the equipment has been used and the original cost was $5,100?

35. The price, in dollars, for a product is
$$D(x) = 2{,}000 - 15x - x^2$$
where x is the number of items in the domain $[10, 30]$. The supply function is given by
$$S(x) = x^2 - 3x + 560$$
Find the producers' surplus.

36. Find the consumers' surplus for the information in Problem 35.

37. Find the producers' surplus if the demand and supply functions are
$$D(x) = \frac{236 - 107x}{(x-3)^2} \quad \text{and} \quad S(x) = x + 20$$
for x on $[0, 3)$, where x is the number of items (in thousands).

38. Find the consumers' surplus for the functions given in Problem 37.

39. If the total depreciation at the end of t years is represented by $f(t)$, then the depreciation rate is $f'(t)$ over an interval $[0, t]$. Since $f'(t)$ is usually known, the total depreciation can be found by using the formula
$$f(t) = \int_0^t f'(x)\,dx$$
Suppose a $38,000 piece of equipment is depreciated over a 10-year period using *straight-line depreciation*. That is, each year the depreciation is
$$\frac{38{,}000}{10} = 3{,}800$$
So $f'(x) = 3{,}800$. Find the total depreciation for the first 3 years.

40. Many items do not depreciate at a constant rate (see Problem 39). Automobiles or computers, for example, depreciate much more quickly in the early years and more slowly toward the end of the time interval. Suppose an automobile depreciates according to the formula
$$f'(x) = 3{,}000\sqrt{5-x}$$
Find the total depreciation for the first 3 years.

41. A piece of machinery requires an overhaul after time t. If $E(t)$ represents the expense connected with the equipment, we see that
$$E(t) = C + \text{total depreciation}$$
where C is the cost of overhaul. Now, from Problem 39,
$$E(t) = C + \int_0^t f'(x)\,dx$$
(*Also note:* $E'(t) = f'(t)$ since the derivative of C is zero.) The average expense, $A(t)$, is found by dividing by the number of years t:
$$A(t) = \frac{E(t)}{t}$$

If no other factors are involved, the best time to overhaul the equipment is at the value of t for which E has a relative minimum:

$$A'(t) = \frac{tE'(t) - E(t)}{t^2} \quad \text{provided } E'(t) \text{ exists}$$

Set $A'(t) = 0$ and solve for $E'(t)$ to find the critical value:

$$\frac{tE'(t) - E(t)}{t^2} = 0$$

$$tE'(t) - E(t) = 0 \quad \text{Multiply both sides by } t^2$$

$$\underbrace{E'(t)}_{E'(t) = f'(x) \text{ is the rate of depreciation}} = \underbrace{\frac{E(t)}{t}}_{\text{Average expense}}$$

This critical value is a relative minimum (this is left for you to verify), so you see that the best time to overhaul occurs when the rate of depreciation equals the average expense. If a piece of equipment depreciates according to the formula

$$f'(x) = 1,000\sqrt{x} \quad \text{and} \quad C = 500$$

when should the equipment be overhauled to minimize the average expense?

42. If P dollars is invested at a rate of r percent per year compounded continuously for t years, then

$$A = Pe^{rt}$$

is the future amount. The rate of change of A with respect to time t is called the *net investment flow* and is found by

$$\frac{dA}{dt} = Pe^{rt}(r) = Pre^{rt}$$

Find the net investment flow for $P = \$50$ and $r = .09$.

43. The process by which a corporation increases its accumulated wealth is called *capital formation*. If the net investment flow (see Problem 42) is given by a function $f(x)$, then the increase in capital over the interval $[a, b]$ is

$$\int_a^b f(t)\, dt = F(b) - F(a)$$

Suppose Intel Corporation has a net investment flow approximated by the function $f(t) = \sqrt{t}$, where t is in years and f is in millions of dollars per year. What is the amount of capital formation over the next 5 years?

7.2
Integration by Parts

In Section 7.1 we were able to restate integration formulas corresponding to the power, sum, and difference differentiation rules. Now we develop an integration formula from the product rule for differentiation.

Suppose that u and v are differentiable functions of x. Then

$$\frac{d}{dx}(uv) = u\frac{dv}{dx} + v\frac{du}{dx}$$

or, using differentials,

$$d(uv) = u\, dv + v\, du$$

Integrate both sides to obtain

$$\int d(uv) = \int u\, dv + \int v\, du$$

$$uv = \int u\, dv + \int v\, du$$

Solve this equation for $\int u\,dv$ and obtain a formula called **integration by parts**:

Integration by Parts

$$\int u\,dv = uv - \int v\,du$$

EXAMPLE 1 Evaluate $\int xe^x\,dx$ using integration by parts.

Solution To use integration by parts, you must choose u and dv so that the new integral is easier to integrate than the original.

Integrate by parts:

Formula:
$$\int u\,dv = uv - \int v\,du$$
$$\int xe^x\,dx = xe^x - \int e^x\,dx$$
$$= xe^x - e^x + C$$

> Let $u = x$ and $dv = e^x\,dx$;
> then $du = dx$ $\quad v = \int e^x\,dx = e^*$
>
> *See note below

Integration by parts is often confusing the first time you try to do it because there is no absolute choice for u and dv. Experience will help you in deciding. In the

---

Note: When we wrote $v = \int e^x\,dx = e^x$ above, we should have written $v = \int e^x\,dx = e^x + K$. Then our work would have looked like

$$\int xe^x\,dx = x(e^x + K) - \int (e^x + K)\,dx$$
$$= xe^x + xK - \int e^x\,dx - \int K\,dx$$
$$= xe^x + xK - \int e^x\,dx - Kx + C_1$$
$$= xe^x - \int e^x\,dx + C_1$$
$$= xe^x - e^x + C_2 + C_1$$
$$= xe^x - e^x + C \quad \text{where} \quad C = C_1 + C_2$$

The intermediate constant K in integration by parts will always drop out, so we usually omit the constant when calculating v from dv and include only the constant C at the end.

previous example, you might have chosen

$$u = e^x \quad \text{and} \quad dv = x\, dx$$
$$du = e^x\, dx \quad \quad v = \int x\, dx = \frac{x^2}{2}$$

Then

$$\int \underbrace{e^x}_{u} \underbrace{x\, dx}_{dv} = \underbrace{e^x}_{u} \underbrace{\frac{x^2}{2}}_{v} - \int \underbrace{\frac{x^2}{2}}_{v} \underbrace{e^x\, dx}_{du}$$

$$= \frac{1}{2} x^2 e^x - \frac{1}{2} \int x^2 e^x\, dx$$

Note, however, that this choice of u and dv leads to a more complicated form than the original. Therefore, when you are integrating by parts, if you make a choice for u and dv that leads to a more complicated form than when you started, consider going back and making another choice for u and dv.

Sometimes you may have to integrate by parts more than once, as illustrated by Example 2.

EXAMPLE 2 Evaluate $\int 5x^2 e^{3x}\, dx$ using integration by parts.

Solution

$$\text{Let} \quad u = x^2 \quad \text{and} \quad dv = e^{3x}\, dx$$
$$du = 2x\, dx \quad \quad v = \int e^{3x}\, dx = \frac{1}{3} e^{3x}$$

Integrate by parts:

$$\int 5 \underbrace{x^2}_{u} \underbrace{e^{3x}\, dx}_{dv} = 5\left[\underbrace{x^2}_{u} \underbrace{\left(\frac{1}{3} e^{3x}\right)}_{v} - \int \underbrace{\frac{1}{3} e^{3x}}_{v} \underbrace{2x\, dx}_{du} \right]$$

$$= \frac{5}{3} x^2 e^{3x} - \frac{10}{3} \int x e^{3x}\, dx$$

$$= \frac{5}{3} x^2 e^{3x} - \frac{10}{3} \int x e^{3x}\, dx \quad \quad \text{Now use integration by parts again.}$$

$$u = x \quad \text{and} \quad dv = e^{3x}\, dx$$
$$du = dx \quad v = \int e^{3x}\, dx = \frac{1}{3}e^{3x}$$

$$= \frac{5}{3}x^2 e^{3x} - \frac{10}{3}\left[x\left(\frac{1}{3}e^{3x}\right) - \int \frac{1}{3}e^{3x}\, dx\right]$$
$$= \frac{5}{3}x^2 e^{3x} - \frac{10}{9}x e^{3x} + \frac{10}{9}\int e^{3x}\, dx$$
$$= \frac{5}{3}x^2 e^{3x} - \frac{10}{9}x e^{3x} + \frac{10}{9}\frac{e^{3x}}{3} + C$$
$$= \frac{5}{3}x^2 e^{3x} - \frac{10}{9}x e^{3x} + \frac{10}{27}e^{3x} + C$$

EXAMPLE 3 Evaluate $\int \ln x\, dx$ using integration by parts.

Solution

$$\text{Let} \quad u = \ln x \quad \text{and} \quad dv = dx$$
$$du = \frac{1}{x}\, dx \quad v = \int dx = x$$

$$\int \ln x\, dx = x \ln x - \int x\left(\frac{1}{x}\right) dx$$
$$= x \ln x - x + C$$

If you use integration by parts with the definite integral, be sure to evaluate the first part at both of the limits of integration, as illustrated by Example 4.

WARNING: Do not forget this evaluation.

EXAMPLE 4 $\int_0^1 xe^{2x}\, dx = \frac{1}{2}xe^{2x}\Big|_0^1 - \frac{1}{2}\int_0^1 e^{2x}\, dx$

$$u = x \quad \text{and} \quad dv = e^{2x}\, dx$$
$$du = dx \quad v = (\tfrac{1}{2})e^{2x}$$

It is usually easier to simplify the algebra and then do one evaluation here at the end of the problem.

$$= \left(\frac{1}{2}xe^{2x} - \frac{1}{4}e^{2x}\right)\Big|_0^1$$

$$= \left(\frac{1}{2}e^2 - \frac{1}{4}e^2\right) - \left(0 - \frac{1}{4}\right)$$

$$= \frac{1}{4}e^2 + \frac{1}{4}$$

$$\approx 2.097264 \qquad \blacksquare$$

Remember that integration by parts, as with other integration methods, is not a method that "always works." There are many functions whose integrals cannot be found by the methods described in this book. In these cases you can use approximate integration for definite integrals as discussed in Section 6.6.

COMPUTER APPLICATION

Do not forget that approximate (or computer) integration can work for definite integrals that might be difficult to integrate to a closed form. For Example 4, we found the exact value to be $.25e^2 + .25$, which can be approximated as 2.097264. We have shown the *Riemann sum* program for this same example.

```
Y = Xe^2X

How many rectangles? 512

a = 0     b = 1.00000

norm = 1.953E-4

Riemann sum = 2.09726
```

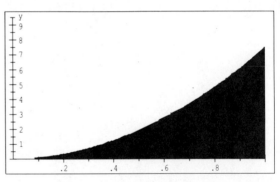

7.2
Problem Set

Evaluate the integrals in Problems 1–34.

1. $\displaystyle\int 12xe^{3x}\,dx$

2. $\displaystyle\int 60xe^{5x}\,dx$

5. $\displaystyle\int x\sqrt{1-x}\,dx$

6. $\displaystyle\int 3x\sqrt{1-2x}\,dx$

3. $\displaystyle\int_0^1 xe^{-x}\,dx$

4. $\displaystyle\int_0^{1/2} 9xe^{-2x}\,dx$

7. $\displaystyle\int x(x+2)^3\,dx$

8. $\displaystyle\int x(3x-1)^4\,dx$

9. $\int \ln(x+1)\,dx$
10. $\int \ln(2x-1)\,dx$
11. $\int \dfrac{x\,dx}{\sqrt{1-x}}$
12. $\int \dfrac{x\,dx}{\sqrt{2-3x}}$
13. $\int 6x^2 e^{2x}\,dx$
14. $\int 8x^2 e^{-2x}\,dx$
15. $\int x^2\sqrt{1-2x}\,dx$
16. $\int 5x^2\sqrt{1-x}\,dx$
17. $\int x^3 \sqrt[3]{1-x^2}\,dx$
18. $\int x \ln x\,dx$
19. $\int x^2 \ln x\,dx$
20. $\int x^2 \ln 3x\,dx$
21. $\int 6x^2 e^{4x}\,dx$
22. $\int 24x^2 e^{3x}\,dx$
23. $\int \dfrac{x^3\,dx}{\sqrt{1-x^2}}$
24. $\int \dfrac{x^3\,dx}{\sqrt{x^2+1}}$
25. $\int x^3 e^{x^2}\,dx$
26. $\int x^3 \sqrt{x^2+1}\,dx$
27. $\int_1^2 \ln 3x\,dx$
28. $\int_1^3 \ln 2x\,dx$
29. $\int_0^1 (x-4)e^x\,dx$
30. $\int_0^2 (2x-3)e^x\,dx$
31. $\int_e^{e^2} \dfrac{dx}{x \ln x}$
32. $\int_e^{e^2} \sqrt{x} \ln x\,dx$
33. $\int_1^e (\ln x)^2\,dx$
34. $\int_{1/3}^e 3(\ln 3x)^2\,dx$

35. Derive the formula
$$\int x^m e^x\,dx = x^m e^x - m \int x^{m-1} e^x\,dx$$

36. Derive the formula
$$\int (\ln x)^m\,dx = x(\ln x)^m - m \int (\ln x)^{m-1}\,dx$$

37. **a.** Evalute the following integral by parts:
$$\int \dfrac{x^3}{x^2+1}\,dx$$

b. Evaluate the integral in part **a** by first dividing the integrand.

APPLICATIONS

38. The marginal cost and revenue equations (in millions of dollars) for a product are given by
$$R'(x) = \dfrac{x}{\sqrt{x-1}} \quad \text{and} \quad C'(x) = .2x$$
where x is the time in years. The area between the graphs of the marginal functions for a time period where $R'(x) > C'(x)$ is the total accumulated profit bounded by two values of x. What is the total profit for years 2 through 5?

39. A contaminated water supply is treated, and the rate of harmful bacteria t days after the treatment is given by
$$N'(t) = te^{-.05t} \qquad 0 \le t \le 10$$
where $N(t)$ is the number of millions of bacteria per liter of water. The initial count is 1 million.
a. Find $N(t)$.
b. Find the total bacteria count after 7 days.

40. If the marginal profit is given by
$$P'(t) = \dfrac{2t}{(1+3t)^{2/3}}$$
where t is the time in years, find P if the profit at time 0 is $0.

41. Maintenance costs for a certain building can be approximated by
$$M'(x) = 100x^2 + \dfrac{5{,}000x}{x^2+1}$$
where x is the age of the building in years and $M(x)$ is the total accumulated cost of maintenance for x years. Find the total maintenance costs for the period from 5 to 10 years after the building was built.

7.3
Using Tables of Integrals

Professional mathematicians and scientists often integrate functions by using tables of integrals. One of the most common sources of integral tables is the Chemical Rubber Company's *Standard Mathematical Tables*, which is much more extensive than what we need for this course. We have, however, a shortened version on the endpapers (the inside covers) of this book.

To use a Brief Integral Table, first classify the integral by type. The types listed in our table include the following:

Forms containing $a + bx$
Forms containing $\sqrt{x^2 \pm a^2}$ or $\sqrt{a^2 - x^2}$
Logarithmic forms
Exponential forms

More extensive tables of integrals are available, but these four fundamental types will serve our purposes, and if you learn how to use this integral table you will be able to use more extensive tables.

After deciding which form applies, match the individual type with the problem at hand by making appropriate choices for the constants. More than one form may apply, but the results derived by using different formulas will be the same (except for the constant).

Take a few moments to look at the **Brief Integral Table**. Notice that this integral table has two basic types of **integration formulas**. The first gives a formula that is the antiderivative, while the second simply rewrites the integral in another form. Examples 1 and 2 illustrate each of these types.

EXAMPLE 1 Evaluate $\int x^2(3 - x)^5 \, dx$.

Solution This is an integral of the form $a + bx$; from the Brief Integral Table, we see that Formula 11 applies, where $a = 3$, $b = -1$, and $n = 5$:

$$\int x^2(3 - x)^5 \, dx = \frac{1}{(-1)^3}\left[\frac{(3 - x)^{5+3}}{5 + 3} - 2(3)\frac{(3 - x)^{5+2}}{5 + 2} + 3^2\frac{(3 - x)^{5+1}}{5 + 1}\right] + C$$

$$= -\left[\frac{(3 - x)^8}{8} - 6\frac{(3 - x)^7}{7} + 9\frac{(3 - x)^6}{6}\right] + C$$

$$= -\frac{1}{8}(3 - x)^8 + \frac{6}{7}(3 - x)^7 - \frac{3}{2}(3 - x)^6 + C$$

$$= -\frac{1}{112}(3 - x)^6[14(3 - x)^2 - 96(3 - x) + 168] + C$$

$$= -\frac{1}{112}(3 - x)^6[14(9 - 6x + x^2) - 288 + 96x + 168] + C$$

$$= -\frac{1}{112}(3 - x)^6(14x^2 + 12x + 6) + C$$

$$= -\frac{1}{56}(3 - x)^6(7x^2 + 6x + 3) + C$$

EXAMPLE 2 Evaluate $\int \ln^4 x \, dx$.

Solution This is an integral in logarithmic form; from the Brief Integral Table we see that Formula 34 applies. Notice that this form involves another integral.

$$\int \ln^4 x \, dx = x \ln^4 x - 4 \int \ln^{4-1} x \, dx$$

$$= x \ln^4 x - 4 \left(x \ln^3 x - 3 \int \ln^{3-1} x \, dx \right) \quad \text{Formula 34, again}$$

$$= x \ln^4 x - 4x \ln^3 x + 12 \int \ln^2 x \, dx$$

Now use Formula 33:

$$= x \ln^4 x - 4x \ln^3 x + 12(x \ln^2 x - 2x \ln x + 2x) + C$$
$$= x \ln^4 x - 4x \ln^3 x + 12x \ln^2 x - 24x \ln x + 24x + C \quad \blacksquare$$

It is often necessary to make substitutions before using one of the integration formulas, as shown in Example 3.

EXAMPLE 3 Evaluate $\int \dfrac{x \, dx}{\sqrt{8 - 5x^2}}$.

Solution This is an integral of the form $\sqrt{a^2 - x^2}$, but it does not exactly match any of the Formulas 15–29. Note, however, that, except for the coefficient of 5, it is like Formula 24. Let $u = \sqrt{5} x$ (so $u^2 = 5x^2$); then $du = \sqrt{5} \, dx$:

$$x = \frac{u}{\sqrt{5}}$$
$$dx = \frac{du}{\sqrt{5}}$$

$$\int \frac{x \, dx}{\sqrt{8 - 5x^2}} = \int \frac{\frac{u}{\sqrt{5}} \cdot \frac{du}{\sqrt{5}}}{\sqrt{8 - u^2}} = \frac{1}{5} \int \frac{u \, du}{\sqrt{8 - u^2}}$$

Now apply Formula 24, where $a^2 = 8$:

$$\frac{1}{5} \int \frac{u \, du}{\sqrt{8 - u^2}} = \frac{1}{5}(-\sqrt{8 - u^2}) + C$$

$$= -\frac{1}{5}\sqrt{8 - 5x^2} + C \quad \blacksquare$$

As you can see from Example 3, using an integral table is not a trivial task. In fact, other methods of integration are often preferable, if possible. For Example 3, you can let $u = 8 - 5x^2$ and integrate by substitution:

$$u = 8 - 5x^2 \quad \text{and} \quad du = -10x \, dx \quad \text{so} \quad dx = \frac{du}{-10x}$$

Substitute:

$$\int \frac{x\,dx}{\sqrt{8-5x^2}} = \int \frac{x \cdot \frac{du}{-10x}}{\sqrt{u}}$$

$$= \frac{-1}{10} \int u^{-1/2}\,du$$

$$= \frac{-1}{10} 2u^{1/2} + C$$

$$= \frac{-1}{5}\sqrt{8-5x^2} + C$$

This answer is the same, of course, as the one we obtained in Example 3. The point of this calculation is to emphasize that you should try simple methods of integration before turning to the table of integrals.

Also, since sometimes more than one formula can be used, it is to our advantage to pick the one that best simplifies the algebra, as shown by Example 4.

EXAMPLE 4 Evaluate $\int \sqrt{1-3x}\,dx$.

Solution *Method* I: Substitution (without integral tables). Let $u = 1 - 3x$, then $du = -3\,dx$, and

$$\int \sqrt{1-3x}\,dx = \int u^{1/2}\left(\frac{du}{-3}\right) = -\frac{1}{3}\frac{u^{3/2}}{\frac{3}{2}} + C$$

$$= -\frac{2}{9}u^{3/2} + C$$

$$= -\frac{2}{9}(1-3x)^{3/2} + C$$

Method II: Use Formula 9, with $a = 1$, $b = -3$, and $n = \frac{1}{2}$:

$$\int \sqrt{1-3x}\,dx = \frac{(1-3x)^{3/2}}{(\frac{1}{2}+1)(-3)} + C = -\frac{2}{9}(1-3x)^{3/2} + C$$

Method III: Use Formula 13, with $a = 1$, $b = -3$, and $m = 0$:

$$\int \sqrt{1-3x}\,dx = \frac{2}{-3(2 \cdot 0 + 3)}[x^0\sqrt{(1-3x)^3} - 0] + C$$

$$= -\frac{2}{9}\sqrt{(1-3x)^3} + C$$

$$= -\frac{2}{9}(1-3x)^{3/2} + C$$

Note: Even though this integral is 0, you must not forget the constant of integration, which was incorporated into the formula.

EXAMPLE 5 Evaluate $\int 5x^2\sqrt{3x^2+1}\,dx$.

Solution This is similar to Formula 25, but you must take care of the 5 (using procedural Formula 1) and the 3 (by making a substitution). Let $u = \sqrt{3}\,x$, then $du = \sqrt{3}\,dx$, and

$$\int 5x^2\sqrt{3x^2+1}\,dx = \int 5\left(\frac{u^2}{3}\right)\sqrt{u^2+1}\,\frac{du}{\sqrt{3}}$$

$dx = du/\sqrt{3}$
$3x^2 = u^2$
$x^2 = u^2/3$

$$= \frac{5}{3\sqrt{3}}\int u^2\sqrt{u^2+1}\,du$$

$$= \frac{5}{3\sqrt{3}}\left[\frac{u}{4}\sqrt{(u^2+1)^3} - \frac{1^2 u}{8}\sqrt{u^2+1} - \frac{1^4}{8}\ln|u+\sqrt{u^2+1}|\right] + C$$

Note the use of the $\mp$ symbol in conjunction with the $\pm$ symbol in Formula 25. If you use the top symbol in one place, then you must use the top symbol throughout.

$$= \frac{5u}{12\sqrt{3}}(u^2+1)^{3/2} - \frac{5u}{24\sqrt{3}}(u^2+1)^{1/2} - \frac{5}{24\sqrt{3}}\ln|u+\sqrt{u^2+1}| + C$$

$$= \frac{5\sqrt{3}\,x}{12\sqrt{3}}(3x^2+1)^{3/2} - \frac{5\sqrt{3}\,x}{24\sqrt{3}}(3x^2+1)^{1/2} - \frac{5}{24\sqrt{3}}\ln|\sqrt{3}\,x+\sqrt{3x^2+1}| + C$$

$$= \frac{5x}{12}(3x^2+1)^{3/2} - \frac{5x}{24}(3x^2+1)^{1/2} - \frac{5}{24\sqrt{3}}\ln|\sqrt{3}\,x+\sqrt{3x^2+1}| + C \blacksquare$$

7.3
Problem Set

Integrate the expressions in Problems 1–12 using the Brief Integral Table. Indicate the formula used.

1. $\int (1+bx)^{-1}\,dx$
2. $\int (a+bx)^5\,dx$
3. $\int \dfrac{x\,dx}{\sqrt{x^2+a^2}}$
4. $\int \dfrac{x\,dx}{\sqrt{a^2-x^2}}$
5. $\int \dfrac{dx}{x^2\sqrt{x^2-a^2}}$
6. $\int \dfrac{dx}{x^2\sqrt{a^2-x^2}}$
7. $\int x\ln x\,dx$
8. $\int x^2 \ln x\,dx$
9. $\int x^{-1}\ln x\,dx$
10. $\int x^5 \ln x\,dx$
11. $\int xe^{ax}\,dx$
12. $\int \dfrac{dx}{a+be^{2x}}$

Evaluate the integrals in Problems 13–64. If you use the Brief Integral Table, state the formula used.

13. $\int \dfrac{x^2\,dx}{\sqrt{x^2+1}}$
14. $\int \dfrac{x^2\,dx}{\sqrt{x^3+1}}$
15. $\int \dfrac{dx}{x^2\sqrt{x^2+16}}$
16. $\int \dfrac{dx}{x^2\sqrt{16x^2+1}}$
17. $\int \dfrac{x\,dx}{\sqrt{4x^2+1}}$
18. $\int \dfrac{x^3\,dx}{\sqrt{4x^4+1}}$
19. $\int \dfrac{dx}{x\sqrt{1-9x^2}}$
20. $\int \dfrac{\sqrt{4x^2+1}}{x}\,dx$
21. $\int \dfrac{x\,dx}{\sqrt{x^2+4}}$
22. $\int \dfrac{dx}{x\sqrt{x^2+4}}$

23. $\int (1+x)^3 \, dx$

24. $\int (1+5x)^3 \, dx$

25. $\int x(1+x)^3 \, dx$

26. $\int 4x(1-6x)^3 \, dx$

27. $\int x\sqrt{1+x} \, dx$

28. $\int x\sqrt{1+3x} \, dx$

29. $\int xe^{4x} \, dx$

30. $\int xe^{-5x} \, dx$

31. $\int x \ln 2x \, dx$

32. $\int x \ln 3x \, dx$

33. $\int x^2 \ln 5x \, dx$

34. $\int x^3 \ln x \, dx$

35. $\int \frac{dx}{3+5e^x}$

36. $\int \frac{dx}{1+e^{5x}}$

37. $\int \frac{dx}{1+e^{2x}}$

38. $\int \frac{e^{2x} \, dx}{1+e^{2x}}$

39. $\int \frac{dx}{\sqrt{1+x}}$

40. $\int \frac{x^3+2x+1}{x} \, dx$

41. $\int x^2(1+x)^3 \, dx$

42. $\int x^3(1+x)^3 \, dx$

43. $\int 5x^2(1-x)^3 \, dx$

44. $\int 3x^2(3-2x)^3 \, dx$

45. $\int 2x\sqrt{1-2x} \, dx$

46. $\int 3x\sqrt{4-3x} \, dx$

47. $\int x^2(2+3x)^3 \, dx$

48. $\int x^2\sqrt{4x^2+1} \, dx$

49. $\int x^2\sqrt{4x^3+1} \, dx$

50. $\int x^2 e^x \, dx$

51. $\int x^2 e^{3x} \, dx$

52. $\int \sqrt{2+9x^2} \, dx$

53. $\int \frac{\sqrt{2+9x}}{x^2} \, dx$

54. $\int \frac{\sqrt{2+9x}}{x} \, dx$

55. $\int \frac{\sqrt{2+9x^2}}{x} \, dx$

56. $\int 3\sqrt{5+4x^2} \, dx$

57. $\int \frac{1}{\sqrt{2+9x^2}} \, dx$

58. $\int x\sqrt{9x^2-5} \, dx$

59. $\int 5x\sqrt{9-16x^2} \, dx$

60. $\int \ln^5 x \, dx$

61. $\int \frac{\sqrt{x^2-1}}{x^2} \, dx$

62. $\int x^3(x^4+1)^8 \, dx$

63. $\int x^2\sqrt{3+10x} \, dx$

64. $\int (\ln 3x + e^{5x} + \sqrt{x^2+1}) \, dx$

APPLICATIONS

65. A local citizens' group estimates the rate at which pollutants are being dumped into the Eel River according to the formula

$$P(t) = \frac{5{,}000}{t\sqrt{100+t^2}} \qquad t \geq 1$$

where t is the month and $P(t)$ is the number of tons of pollutants per month. What is the total amount of pollutants that will be dumped into the river in the next 11 months (that is, from $t = 1$ to $t = 12$)?

66. Davis Industries has production costs for a product determined by the function

$$C(x) = \frac{100x}{\sqrt{x^2+100}} + 50$$

where $C(x)$ is the cost to produce the xth item. What is the approximate total cost of producing 1,000 items?

67. The sales of Simon's Simple Tool Kit are growing according to the function

$$S(t) = te^{.1t} \qquad t \geq 12$$

where $S(t)$ is the sales in thousands of dollars in the tth month. What are the accumulated sales for the second year (that is, for $t = 12$ to $t = 24$)?

7.4
Improper Integrals

The graph of the function $f(x) = e^{-x}$ is shown in Figure 7.7a on page 303. Note that the region between the graph of $f(x) = e^{-x}$ and the x-axis over the interval $[0, \infty]$ is unbounded. It is possible to define the area of this unbounded region. To do this, we consider a vertical line $x = b$, as shown in Figure 7.7b. We find the area

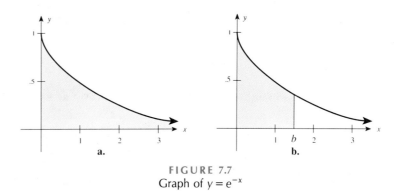

FIGURE 7.7
Graph of $y = e^{-x}$

of the shaded region:

$$A(b) = \int_0^b e^{-x}\,dx = -e^{-x}\Big|_0^b = (-e)^{-b} - (-e^0) = 1 - e^{-b}$$

Now we let the vertical line $x = b$ slide to the right; that is, suppose $b \to \infty$. The area changes according to

$$\lim_{b \to \infty}(1 - e^{-b}) = 1 \quad \text{since } \lim_{b \to \infty} e^{-b} = 0$$

Now

$$\lim_{b \to \infty} A(b) = \lim_{b \to \infty}(1 - e^{-b}) = 1$$

This limit is defined to be the area of the shaded region in Figure 7.7a. The

$$\lim_{b \to \infty} \int_0^b e^{-x}\,dx \quad \text{is denoted by} \quad \int_0^\infty e^{-x}\,dx$$

and is called an **improper integral**.

Improper Integral

If f is continuous over the indicated interval and the limit exists, then the following integrals are called **improper integrals**:

$$\int_a^\infty f(x)\,dx = \lim_{b \to \infty} \int_a^b f(x)\,dx \quad \text{if the limit exists}$$

$$\int_{-\infty}^b f(x)\,dx = \lim_{a \to -\infty} \int_a^b f(x)\,dx \quad \text{if the limit exists}$$

$$\int_{-\infty}^\infty f(x)\,dx = \int_{-\infty}^c f(x)\,dx + \int_c^\infty f(x)\,dx \quad \text{where } c \text{ is any point on } (-\infty, \infty) \text{ where both improper integrals on the right exist}$$

If the indicated limit exists, then the improper integral is said to **converge**; if the limit does not exist, then the improper integral is said to **diverge**.

EXAMPLE 1 Evaluate the following integrals.

a. $\int_1^x 2x^{-3}\, dx$　　b. $\int_{-\infty}^{-1} \dfrac{dx}{x}$　　c. $\int_{-\infty}^{\infty} xe^{-x^2}\, dx$

Solution

a. $\int_1^x 2x^{-3}\, dx = \lim_{b \to \infty} \int_1^b 2x^{-3}\, dx = \lim_{b \to \infty}(-x^{-2})\Big|_1^b = \lim_{b \to \infty}(-b^{-2} + 1) = 1$

b. $\int_{-\infty}^{-1} \dfrac{dx}{x} = \lim_{a \to -\infty} \int_a^{-1} x^{-1}\, dx = \lim_{a \to -\infty}(\ln|x|)\Big|_a^{-1} = \lim_{a \to -\infty}(0 - \ln|a|)$

This limit does not exist, so we say that this integral diverges.

c. $\int_{-\infty}^{\infty} xe^{-x^2}\, dx = \int_{-\infty}^{0} xe^{-x^2}\, dx + \int_{0}^{\infty} xe^{-x^2}\, dx$

$= \lim_{a \to -\infty} \int_a^0 xe^{-x^2}\, dx + \lim_{b \to \infty} \int_0^b xe^{-x^2}\, dx$

$= \lim_{a \to -\infty} \int_{x=a}^{x=0} \dfrac{-1}{2} e^u\, du + \lim_{b \to \infty} \int_{x=0}^{x=b} \dfrac{-1}{2} e^u\, du$　　Let $u = -x^2$　$du = -2x\, dx$

$= \lim_{a \to -\infty} (-.5e^{-x^2})\Big|_a^0 + \lim_{b \to \infty}(-.5e^{-x^2})\Big|_0^b$

$= \lim_{a \to -\infty}(-.5 + .5e^{-a^2}) + \lim_{b \to \infty}(-.5e^{-b^2} + .5)$

$= -.5 + 0 + 0 + .5 = 0$　　■

EXAMPLE 2 The **capital value**, V, of a property over T years is given by

$$V = \int_0^T Re^{-rt}\, dt$$

where R is the annual rent, or income, and r is the current interest rate. Suppose the current interest rate is 8% and you have an indeterminant lease (no date of termination) paying \$60,000 per year. What is the capital value?

Solution $V = \int_0^{\infty} 60{,}000 e^{-.08t}\, dt = \lim_{b \to \infty} \int_0^b 60{,}000 e^{-.08t}\, dt = 60{,}000 \lim_{b \to \infty} \int_0^b e^{-.08t}\, dt$

$= 60{,}000 \lim_{b \to \infty} \dfrac{e^{-.08t}}{-.08}\Big|_0^b = -750{,}000 \lim_{b \to \infty}(e^{-.08b} - 1)$

$= -750{,}000(0 - 1) = 750{,}000$

The capital value for the lease is \$750,000.　　■

7.4

Problem Set

Evaluate the integrals in Problems 1–22 that converge.

1. $\int_0^{\infty} e^{-x}\, dx$

2. $\int_{-\infty}^{0} e^{-x}\, dx$

3. $\int_0^{\infty} e^{-x/2}\, dx$

4. $\int_1^{\infty} \dfrac{dx}{x^4}$

5. $\int_1^{\infty} \dfrac{dx}{x^3}$

6. $\int_1^{\infty} \dfrac{dx}{\sqrt{x}}$

7. $\int_1^{\infty} \dfrac{dx}{\sqrt[3]{x}}$

8. $\int_1^{\infty} \dfrac{dx}{x^9}$

9. $\displaystyle\int_1^\infty \frac{dx}{x^{1.1}}$

10. $\displaystyle\int_1^\infty \frac{x\,dx}{\sqrt{x^2+2}}$

11. $\displaystyle\int_{-\infty}^0 \frac{2x\,dx}{x^2+1}$

12. $\displaystyle\int_1^\infty \frac{x\,dx}{(1+x^2)^2}$

13. $\displaystyle\int_{-\infty}^\infty x^2\,dx$

14. $\displaystyle\int_{-\infty}^\infty (x^2+1)^{-1/2}\,dx$

15. $\displaystyle\int_{-\infty}^\infty \frac{3x\,dx}{(3x^2+2)^3}$

16. $\displaystyle\int_e^\infty \frac{dx}{x\ln x}$

17. $\displaystyle\int_e^\infty \frac{dx}{x(\ln x)^2}$

18. $\displaystyle\int_1^\infty \ln x\,dx$

19. $\displaystyle\int_1^\infty \frac{e^{-\sqrt{x}}}{\sqrt{x}}\,dx$

20. $\displaystyle\int_1^\infty \frac{x^2}{\sqrt[3]{x^3+2}}\,dx$

21. $\displaystyle\int_1^\infty \frac{dx}{x^2\sqrt{x^2+4}}$

22. $\displaystyle\int_{-\infty}^1 \frac{dx}{x^2\sqrt{4-x^2}}$

APPLICATIONS

23. Suppose the current interest rate is 6% and you have an indeterminant lease paying $20,200 per year. What is the capital value?

24. Suppose the current interest rate is 11% and you have an indeterminant lease paying $3,600 per year. What is the capital value?

25. Show that the capital value of a rental property with an indeterminant lease paying an annual rent of R with a current interest rate of r percent is R/r.

26. The amount of radioactive material being released into the atmosphere annually—that is, the amount present at time T—is given by

$$A = \int_0^T Pe^{-rt}\,dt$$

If a recent United Nations publication estimates that $r = .002$ and $P = 200$ millirems, estimate the total future buildup of radioactive material in the atmosphere.

27. Suppose that $r = .05$ in Problem 26. Estimate the total future buildup of radioactive material in the atmosphere.

28. Show that the buildup of radioactive material in the atmosphere as reported in Problem 26 approaches a limiting value of P/r.

29. Find the area of the unbounded region between the x-axis and the curve $y = \dfrac{2}{(x-4)^3}$ for $x \geq 6$.

30. Find the area of the unbounded region between the x-axis and the curve $y = \dfrac{2}{(x-4)^3}$ for $x \leq 2$.

31. Suppose that an endowment produces a perpetual income with a rate of income (in dollars per year)

$$f(t) = 25{,}000e^{.08t}$$

The capital value formula for a variable rate of income is the same as the one in Example 2 with the constant R replaced by the rate of flow function. Find the capital value at 12% compounded continuously.

32. The rate at which a chemical is being released into a river at time t is given by $455e^{-.02t}$ tons per year. Find the total amount of chemical that will be released into the river in the indefinite future.

33. Suppose that an oil well produces $P(t)$ thousand barrels of crude oil per month according to the formula

$$P(t) = 100e^{-.02t} - 100e^{-.1t}$$

where t is the number of months that the well has been in production. What is the total amount of oil produced by the oil well?

*7.5

Probability Density Functions

The models we have considered so far have been deterministic models; now we turn to a **probabilistic model**, used for situations that are random in character and that attempts to predict the outcomes of these events with a certain stated or known degree of accuracy. For example, if we toss a coin, it is impossible to predict in advance whether the outcome will be a head or a tail. Our intuition tells us that the outcome is equally likely to be a head or a tail, and somehow we sense that if we repeat the experiment of tossing a coin a large number of times, heads will occur

---

* Optional section.

"about half the time." To check this out, I recently flipped a coin 1,000 times and obtained 460 heads and 540 tails. The percentage of heads is $\frac{460}{1,000} = .46 = 46\%$, which is called the **relative frequency**.

Relative Frequency of a Repeated Experiment

> If an experiment is repeated n times and an event occurs m times, then
>
> $$\frac{m}{n}$$
>
> is called the **relative frequency** of the event.

Our task is to create a model that will assign a number p, called the *probability of an event*, which will represent the relative frequency. This means that for a *sufficiently large number of repetitions* of an experiment

$$p \approx \frac{m}{n}$$

Probabilities can be obtained in one of three ways:

1. *Theoretical probabilities* (also called *a priori* models) are obtained by logical reasoning. For example, the probability of rolling a die and obtaining a 3 is $\frac{1}{6}$ because there are six possible outcomes, each with an equal chance of occurring, so a 3 should appear $\frac{1}{6}$ of the time.
2. *Empirical probabilities* (also called *a posteriori* models) are obtained from experimental data. For example, an assembly line producing brake assemblies for General Motors produces 1,500 brakes per day. The probability of a defective brake can be obtained by experimentation. Suppose the 1,500 brakes are selected at random and are tested; 3 are found to be defective. Then the relative frequency, or probability, is

    $$\frac{3}{1,500} = .002 \text{ or } .2\%$$

3. *Subjective probabilities* are obtained from experience and indicate a measure of "certainty." For example, a TV reporter studies the satellite maps and issues a prediction about tomorrow's weather based on experience under similar circumstances: "80% chance of rain tomorrow."

A probability measure must conform to these different ways of using the word *probability*. We begin by defining some terms. An **experiment** is the observation of any physical occurrence. A **sample space** of an experiment is the set of all possible outcomes. An **event** is a subset of the sample space. If an event is the empty set, it is called the **impossible event**; and if it has only one element, it is called a **simple event**.

EXAMPLE 1   List the sample space for each experiment, and then list one event for each of the sample spaces and state whether it is a simple event.

   a. Experiment 1: recording the results of the UCLA football team in a given season
   b. Experiment 2: simultaneously tossing a coin and rolling a die
   c. Experiment 3: installing a computer chip and recording the time to failure

Solution a. Experiment 1: $S = \{w, l, t\}$
where w, l, and t denote win, lose, and tie, respectively. An example of an event E is that "UCLA wins," which is $E = \{w\}$. This is a simple event. Another example is event F, "UCLA does not lose," which is $F = \{w, t\}$; this is not a simple event.

b. Experiment 2: $S = \{1H, 1T, 2H, 2T, 3H, 3T, 4H, 4T, 5H, 5T, 6H, 6T\}$
An example of an event is "obtaining an even number or a head," which is $E = \{1H, 2H, 2T, 3H, 4H, 4T, 5H, 6H, 6T\}$. E is not a simple event.

c. Experiment 3: $S = \{t \mid t \geq 0\}$
An example of an event is "the chip lasts more than 1,000 hours," which is $E = \{t \mid t > 1,000\}$. E is not a simple event. ∎

A **random variable** X is a function that has the following properties: the domain of X is contained in a set S, certain of whose subsets correspond to events for which there is an associated *probability function* (several of which will be defined in this section); the range of the function is a real number. A **discrete random variable** is one that has only a finite number of values, each of which is associated with a nonnegative probability, the sum of these probabilities being 1. If X can assume infinitely many values arranged in a sequence, it is called an **infinite discrete random variable**; and if X can assume any real value on an interval, then it is called a **continuous random variable**.

Suppose the heights of 60 students in a mathematics class are recorded (to the nearest inch), as shown in Table 7.1. The results in the table are often referred to as a **frequency distribution** because such results show the distribution of the numbers for each occurrence.

TABLE 7.1 Frequency Distribution of Student Heights

X = height (in inches)	61 or smaller	62	63	64	65	66	67	68	69	70	71	72	73 or taller
Number of occurrences (frequency)	2	2	2	3	5	9	14	10	5	3	2	1	2
Relative frequency	$\frac{2}{60}$	$\frac{2}{60}$	$\frac{2}{60}$	$\frac{3}{60}$	$\frac{5}{60}$	$\frac{9}{60}$	$\frac{14}{60}$	$\frac{10}{60}$	$\frac{5}{60}$	$\frac{3}{60}$	$\frac{2}{60}$	$\frac{1}{60}$	$\frac{2}{60}$

Probability Distribution

A **probability distribution** is the collection of all values that a random variable assumes along with the probabilities that correspond to these values. Furthermore,

1. $P(X = x_1) + P(X = x_2) + \cdots + P(X = x_n) = 1$
2. $0 \leq P(X = x_i) \leq 1$ for every $1 \leq i \leq n$

It is very useful to display the information in a probability distribution graphically in a bar graph called a **histogram**.

EXAMPLE 2 Represent the information in Table 7.1 in a histogram.

Solution Use the horizontal axis to delineate the values of the random variable (the heights) and the vertical axis to represent the relative frequencies. Draw each bar so that the width is 1 unit. The histogram is shown in Figure 7.8.

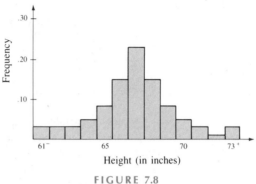

FIGURE 7.8
Histogram for relative frequencies

There is a very important relationship between the area of the rectangles in a histogram and the probabilities. Since the width of each bar is 1 unit, the **area of each bar is the probability of occurrence for that random variable**. Therefore the **sum of the areas of the rectangles is 1**. This property is very important in the study of probability distributions.

It is possible to record the heights of the students more accurately. In fact, theoretically, the height of a student could be any positive real number in some domain (say, 0 to 100 inches). We can therefore consider the graph of the relative frequencies as a continuous curve, as shown in Figure 7.9.

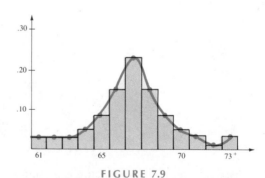

FIGURE 7.9
Relative frequencies as a continuous curve (this curve is drawn by connecting the points at the top of the bars and smoothing the resulting polygon into a curve)

If we now define the **probability** of an event as the relative frequency of occurrence of the event, and if the random variable is assumed to be continuous, then we can use the area under the curve between two values to find the probability that the random variable is between those values. For example,

$$P(65 \leq X \leq 71)$$

is the probability that a student's height is between 65 and 71 inches and is shown by the shaded region in Figure 7.10.

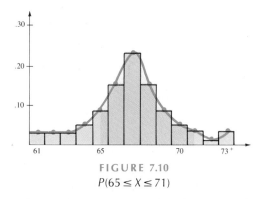

FIGURE 7.10
$P(65 \leq X \leq 71)$

Since the definite integral can be used to find the area under the graph of $f(x)$ from $x = a$ to $x = b$, if we can find a function f to describe a relative frequency curve, then the probability that a continuous random variable X associated with f will be between a and b is

$$P(a \leq X \leq b) = \int_a^b f(x)\,dx$$

A function f that can be used to describe a relative frequency curve is called a **probability density function**. Such a function must satisfy certain conditions, as summarized below:

Probability Density Function

If X is a continuous random variable associated with a function f, then f is a **probability density function** if

1. $f(x) \geq 0$ for all x in the domain $(-\infty, \infty)$
2. $\int_{-\infty}^{\infty} f(x)\,dx = 1$
3. If $[a, b]$ is a subinterval of $(-\infty, \infty)$, then

$$P[a \leq X \leq b] = \int_a^b f(x)\,dx$$

EXAMPLE 3 Show that $f(x) = \dfrac{x}{6} - \dfrac{1}{12}$ is a probability density function on $[1, 4]$.

Solution
1. $\displaystyle\int_1^4 \dfrac{x}{6} - \dfrac{1}{12}\,dx = \dfrac{x^2}{12} - \dfrac{x}{12}\bigg|_1^4 = \dfrac{16}{12} - \dfrac{4}{12} - \dfrac{1}{12} + \dfrac{1}{12} = \dfrac{12}{12} = 1$
2. If x is between 1 and 4, then f is between $\dfrac{1}{12}$ and $\dfrac{7}{12}$, so $f(x) \geq 0$ for all X on $[a, b]$.

Thus, f is a probability density function.

EXAMPLE 4 Suppose that the life of a computer chip is described by the probability density function

$$f(x) = \frac{72}{35x^3}$$

where x is the number of months it will function properly. The domain for x is the interval $[1, 6]$. What is the probability that the chip will last longer than 4 months?

Solution The probability that the chip will last longer than 4 months is

$$P(4 \leq X \leq 6) = \int_4^6 \frac{72}{35x^3} dx = \frac{72}{35} \frac{x^{-2}}{-2}\Big|_4^6 = -\frac{36}{35x^2}\Big|_4^6$$

$$= -\frac{1}{35} + \frac{9}{140} = .036 \qquad \blacksquare$$

Sometimes it is necessary to construct a probability density function. For example, suppose $f(x) = x^2$ on $[1, 4]$. Then

$$\int_1^4 x^2 = \frac{x^3}{3}\Big|_1^4 = \frac{64}{3} - \frac{1}{3} = \frac{63}{3} = 21$$

If you want to construct a probability density function using f, simply multiply by the reciprocal of the value of the integral, in this case, $\frac{1}{21}$, so that the value of the integral will be 1:

$$f(x) = \frac{1}{21} x^2$$

is a probability density function.

Uniform Probability Distribution

If a density function is a constant (the graph is a horizontal line), then the appropriate model is the **uniform probability distribution**. Suppose the given interval is $[1, 5]$. Then the length of the interval is $5 - 1 = 4$, so if the probability density function is a constant it must be $\frac{1}{4}$ (so that the area is 1), as shown in Figure 7.11 on page 311. Notice that if the interval is $[4, 10]$, then $f(x) = \frac{1}{6}$; if the interval is $[2, 4]$, then $f(x) = \frac{1}{2}$. This observation leads us to the following definition of the uniform probability distribution:

Uniform Distribution

A continuous random variable X is said to be **uniformly distributed** over an interval $[a, b]$ if it has a probability density function

$$f(x) = \frac{1}{b - a} \quad \text{for } a \leq x \leq b$$

The graph is shown in Figure 7.11.

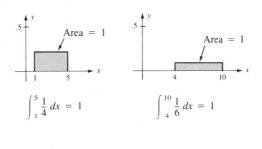

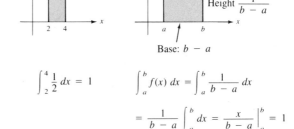

FIGURE 7.11
Some uniform distributions

EXAMPLE 5 Cholesterol levels are artificially introduced into blood samples so that the resulting mixtures are uniformly distributed over the interval [200, 300]. What is the probability that the cholesterol level is between 240 and 265?

Solution The probability density function for x is given by

$$f(x) = \frac{1}{300 - 200} = \frac{1}{100} = .01 \quad \text{for } 200 \leq x \leq 300$$

Then the desired probability is

$$\int_{240}^{265} .01 \, dx = .01x \Big|_{240}^{265}$$
$$= .01(265 - 240)$$
$$= .25 \qquad \blacksquare$$

Exponential Probability Distribution

A probability density function will often follow an exponential curve. We find this curve in response to the question, "How long do you need to wait if you are observing a sequence of events occurring in time in order to observe the first occurrence of the event?"

Exponential Distribution

A continuous random variable X is said to be **exponentially distributed** over an interval $[0, \infty)$ if it has a probability density function

$$f(x) = ke^{-kx} \quad \text{for } k > 0$$

The graph is shown in Figure 7.12.

EXAMPLE 6 The useful life of a machine part is given by

$$f(x) = .015e^{-.015t} \quad 0 \le t < \infty$$

where t is the number of months to failure. What is the probability that the part will fail in the first year?

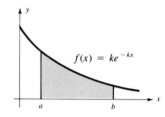

FIGURE 7.12
Exponential distribution

Solution
$$\int_0^{12} .015e^{-.015t}\, dt = \left.\frac{.015e^{-.015t}}{-.015}\right|_0^{12}$$
$$= -e^{-.015(12)} + e^{-.015(0)}$$
$$\approx .16$$

Normal Probability Distribution

A very common probability distribution is the one associated with the so-called bell-shaped curve. Suppose we survey the results of 100,000 IQ scores and obtain the frequency distribution shown in Figure 7.13.

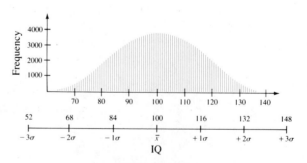

FIGURE 7.13
Frequencies of IQ scores

If we connect the endpoints of the bars in Figure 7.13 by drawing a smooth curve, we obtain a curve very close to a curve called the *normal distribution curve*, or simply the **normal curve**, as shown in Figure 7.14.

7.5 PROBABILITY DENSITY FUNCTIONS

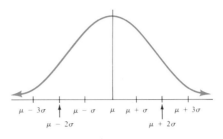

$$f(x) = \frac{e^{-(x-\mu)^2/(2\sigma^2)}}{\sigma\sqrt{2\pi}}$$

FIGURE 7.14
Normal distribution curve

Normal Distribution

A continuous random variable X is said to be **normally distributed** over an interval $(-\infty, \infty)$ if it has a probability density function

$$f(x) = \frac{e^{-(x-\mu)^2/(2\sigma^2)}}{\sigma\sqrt{2\pi}}$$

where μ and σ are real numbers with $\sigma \geq 0$. The number μ is called the **mean** and the number σ is called the **standard deviation**. The graph is shown in Figure 7.14.

Since the normal distribution is a probability distribution, we know the area under this curve is 1. Therefore we can relate the area to probabilities as follows for a random variable X (see Figure 7.15).

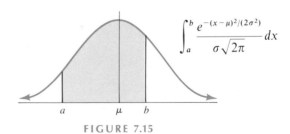

FIGURE 7.15

$P(a \leq X \leq b)$ is the area under the associated normal curve between a and b.
$P(X > \mu) = P(X < \mu) = \frac{1}{2}$; that is, the curve is symmetric about the mean.
$P(X = x) = 0$ for any real number x. (Since there are infinitely many possibilities, the probability of a particular value is 0.)
$P(X < x) = P(X \leq x)$ for any real number x.
$P(X > x) = 1 - P(X \leq x)$ for any real number x.
If $\mu = 0$ and $\sigma = 1$, then the curve is called the **standard normal curve**.

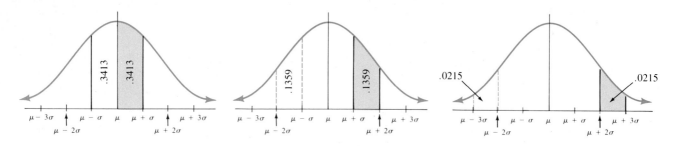

Since the normal distribution is so important for so many applications, and since the calculation of the integral for the probability density function is so complicated, tables of approximate values of the definite integral of the standard normal curve have been calculated. Table 7.2 is such a table. It contains values of

$$P(X \leq b) = \int_{-\infty}^{b} \frac{1}{\sqrt{2\pi}} e^{-x^2/2} \, dx$$

EXAMPLE 7 Find $P(X \leq .57)$, $P(X > -.13)$, and $P(-.05 < X < .93)$.

Solution
$P(X \leq .57) = .7157$ From Table 7.2
$P(X > -.13) = 1 - P(X \leq -.13)$ From Table 7.2
$ = 1 - .4483$
$ = .5517$

$P(-.05 < X < .93) = .8238 - .4801$ From Table 7.2
$\phantom{P(-.05 < X < .93)} = .3437$ ∎

What if you are not working with a standard normal curve? That is, suppose the curve is a normal curve but does not have a mean of 0. You can then use the information in Figure 7.16.

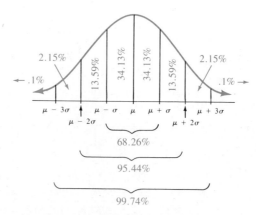

FIGURE 7.16
Areas under a normal curve. Notice that approximately 34% (.3413) lies between μ and $\mu + \sigma$; 14% (.1359) lies between $\mu + \sigma$ and $\mu + 2\sigma$; and 2% (.0215) lies between $\mu + 2\sigma$ and $\mu + 3\sigma$.

TABLE 7.2 Standard Normal Cumulative Distribution

z	0.00	0.01	0.02	0.03	0.04	0.05	0.06	0.07	0.08	0.09
−3.4	0.0003	0.0003	0.0003	0.0003	0.0003	0.0003	0.0003	0.0003	0.0003	0.0002
−3.3	0.0005	0.0005	0.0005	0.0004	0.0004	0.0004	0.0004	0.0004	0.0004	0.0003
−3.2	0.0007	0.0007	0.0006	0.0006	0.0006	0.0006	0.0006	0.0005	0.0005	0.0005
−3.1	0.0010	0.0009	0.0009	0.0009	0.0008	0.0008	0.0008	0.0008	0.0007	0.0007
−3.0	0.0013	0.0013	0.0013	0.0012	0.0012	0.0011	0.0011	0.0011	0.0010	0.0010
−2.9	0.0019	0.0018	0.0017	0.0017	0.0016	0.0016	0.0015	0.0015	0.0014	0.0014
−2.8	0.0026	0.0025	0.0024	0.0023	0.0023	0.0022	0.0021	0.0021	0.0020	0.0019
−2.7	0.0035	0.0034	0.0033	0.0032	0.0031	0.0030	0.0029	0.0028	0.0027	0.0026
−2.6	0.0047	0.0045	0.0044	0.0043	0.0041	0.0040	0.0039	0.0038	0.0037	0.0036
−2.5	0.0062	0.0060	0.0059	0.0057	0.0055	0.0054	0.0052	0.0051	0.0049	0.0048
−2.4	0.0082	0.0080	0.0078	0.0075	0.0073	0.0071	0.0069	0.0068	0.0066	0.0064
−2.3	0.0107	0.0104	0.0102	0.0099	0.0096	0.0094	0.0091	0.0089	0.0087	0.0084
−2.2	0.0139	0.0136	0.0132	0.0129	0.0125	0.0122	0.0119	0.0116	0.0113	0.0110
−2.1	0.0179	0.0174	0.0170	0.0166	0.0162	0.0158	0.0154	0.0150	0.0146	0.0143
−2.0	0.0228	0.0222	0.0217	0.0212	0.0207	0.0202	0.0197	0.0192	0.0188	0.0183
−1.9	0.0287	0.0281	0.0274	0.0268	0.0262	0.0256	0.0250	0.0244	0.0239	0.0233
−1.8	0.0359	0.0352	0.0344	0.0336	0.0329	0.0322	0.0314	0.0307	0.0301	0.0294
−1.7	0.0446	0.0436	0.0427	0.0418	0.0409	0.0401	0.0392	0.0384	0.0375	0.0367
−1.6	0.0548	0.0537	0.0526	0.0516	0.0505	0.0495	0.0485	0.0475	0.0465	0.0455
−1.5	0.0668	0.0655	0.0643	0.0630	0.0618	0.0606	0.0594	0.0582	0.0571	0.0559
−1.4	0.0808	0.0793	0.0778	0.0764	0.0749	0.0735	0.0722	0.0708	0.0694	0.0681
−1.3	0.0968	0.0951	0.0934	0.0918	0.0901	0.0885	0.0869	0.0853	0.0838	0.0823
−1.2	0.1151	0.1131	0.1112	0.1093	0.1075	0.1056	0.1038	0.1020	0.1003	0.0985
−1.1	0.1357	0.1335	0.1314	0.1292	0.1271	0.1251	0.1230	0.1210	0.1190	0.1170
−1.0	0.1587	0.1562	0.1539	0.1515	0.1492	0.1469	0.1446	0.1423	0.1401	0.1379
−0.9	0.1841	0.1814	0.1788	0.1762	0.1736	0.1711	0.1685	0.1660	0.1635	0.1611
−0.8	0.2119	0.2090	0.2061	0.2033	0.2005	0.1977	0.1949	0.1922	0.1894	0.1867
−0.7	0.2420	0.2389	0.2358	0.2327	0.2296	0.2266	0.2236	0.2206	0.2177	0.2148
−0.6	0.2743	0.2709	0.2676	0.2643	0.2611	0.2578	0.2546	0.2514	0.2483	0.2451
−0.5	0.3085	0.3050	0.3015	0.2981	0.2946	0.2912	0.2877	0.2843	0.2810	0.2776
−0.4	0.3446	0.3409	0.3372	0.3336	0.3300	0.3264	0.3228	0.3192	0.3156	0.3121
−0.3	0.3821	0.3783	0.3745	0.3707	0.3669	0.3632	0.3594	0.3557	0.3520	0.3483
−0.2	0.4207	0.4168	0.4129	0.4090	0.4052	0.4013	0.3974	0.3936	0.3897	0.3859
−0.1	0.4602	0.4562	0.4522	0.4483	0.4443	0.4404	0.4364	0.4325	0.4286	0.4247
−0.0	0.5000	0.4960	0.4920	0.4880	0.4840	0.4801	0.4761	0.4721	0.4681	0.4641
0.0	0.5000	0.5040	0.5080	0.5120	0.5160	0.5199	0.5239	0.5279	0.5319	0.5359
0.1	0.5398	0.5438	0.5478	0.5517	0.5557	0.5596	0.5636	0.5675	0.5714	0.5753
0.2	0.5793	0.5832	0.5871	0.5910	0.5948	0.5987	0.6026	0.6064	0.6103	0.6141
0.3	0.6179	0.6217	0.6255	0.6293	0.6331	0.6368	0.6406	0.6443	0.6480	0.6517
0.4	0.6554	0.6591	0.6628	0.6664	0.6700	0.6736	0.6772	0.6808	0.6844	0.6879
0.5	0.6915	0.6950	0.6985	0.7019	0.7054	0.7088	0.7123	0.7157	0.7190	0.7224
0.6	0.7257	0.7291	0.7324	0.7357	0.7389	0.7422	0.7454	0.7486	0.7517	0.7549
0.7	0.7580	0.7611	0.7642	0.7673	0.7704	0.7734	0.7764	0.7794	0.7823	0.7852
0.8	0.7881	0.7910	0.7939	0.7967	0.7995	0.8023	0.8051	0.8078	0.8106	0.8133
0.9	0.8159	0.8186	0.8212	0.8238	0.8264	0.8289	0.8315	0.8340	0.8365	0.8389
1.0	0.8413	0.8438	0.8461	0.8485	0.8508	0.8531	0.8554	0.8577	0.8599	0.8621
1.1	0.8643	0.8665	0.8686	0.8708	0.8729	0.8749	0.8770	0.8790	0.8810	0.8830
1.2	0.8849	0.8869	0.8888	0.8907	0.8925	0.8944	0.8962	0.8980	0.8997	0.9015
1.3	0.9032	0.9049	0.9066	0.9082	0.9099	0.9115	0.9131	0.9147	0.9162	0.9177
1.4	0.9192	0.9207	0.9222	0.9236	0.9251	0.9265	0.9278	0.9292	0.9306	0.9319
1.5	0.9332	0.9345	0.9357	0.9370	0.9382	0.9394	0.9406	0.9418	0.9429	0.9441
1.6	0.9452	0.9463	0.9474	0.9484	0.9495	0.9505	0.9515	0.9525	0.9535	0.9545
1.7	0.9554	0.9564	0.9573	0.9582	0.9591	0.9599	0.9608	0.9616	0.9625	0.9633
1.8	0.9641	0.9649	0.9656	0.9664	0.9671	0.9678	0.9686	0.9693	0.9699	0.9706
1.9	0.9713	0.9719	0.9726	0.9732	0.9738	0.9744	0.9750	0.9756	0.9761	0.9767
2.0	0.9772	0.9778	0.9783	0.9788	0.9793	0.9798	0.9803	0.9808	0.9812	0.9817
2.1	0.9821	0.9826	0.9830	0.9834	0.9838	0.9842	0.9846	0.9850	0.9854	0.9857
2.2	0.9861	0.9864	0.9868	0.9871	0.9875	0.9878	0.9881	0.9884	0.9887	0.9890
2.3	0.9893	0.9896	0.9898	0.9901	0.9904	0.9906	0.9909	0.9911	0.9913	0.9916
2.4	0.9918	0.9920	0.9922	0.9925	0.9927	0.9929	0.9931	0.9932	0.9934	0.9936
2.5	0.9938	0.9940	0.9941	0.9943	0.9945	0.9946	0.9948	0.9949	0.9951	0.9952
2.6	0.9953	0.9955	0.9956	0.9957	0.9959	0.9960	0.9961	0.9962	0.9963	0.9964
2.7	0.9965	0.9966	0.9967	0.9968	0.9969	0.9970	0.9971	0.9972	0.9973	0.9974
2.8	0.9974	0.9975	0.9976	0.9977	0.9977	0.9978	0.9979	0.9979	0.9980	0.9981
2.9	0.9981	0.9982	0.9982	0.9983	0.9984	0.9984	0.9985	0.9985	0.9986	0.9986
3.0	0.9987	0.9987	0.9987	0.9988	0.9988	0.9989	0.9989	0.9989	0.9990	0.9990
3.1	0.9990	0.9991	0.9991	0.9991	0.9992	0.9992	0.9992	0.9992	0.9993	0.9993
3.2	0.9993	0.9993	0.9994	0.9994	0.9994	0.9994	0.9994	0.9995	0.9995	0.9995
3.3	0.9995	0.9995	0.9995	0.9996	0.9996	0.9996	0.9996	0.9996	0.9996	0.9997
3.4	0.9997	0.9997	0.9997	0.9997	0.9997	0.9997	0.9997	0.9997	0.9997	0.9998

EXAMPLE 8 A teacher claims to grade "on a curve." That is, the teacher believes that the scores on a given test are normally distributed. If 200 students take the exam, with mean 73 and standard deviation 9, how would the teacher grade the students?

Solution First, draw a normal curve with a mean 73 and standard deviation 9, as shown in Figure 7.17.

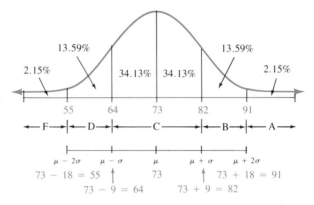

FIGURE 7.17

The interval $[73, 82]$ $(73 + 1\sigma = 82)$ contains about 34% of the class, and the interval $[82, 91]$ $(73 + 2\sigma = 91)$ contains about 14%. Finally, about 2% of the class will score above 91 or below 55:

Score	Letter grade	Number	Percent
92–100	A	4	2%
83–91	B	28	14%
64–82	C	136	68%
55–63	D	28	14%
0–54	F	4	2%

EXAMPLE 9 The Ridgemont Light Bulb Company tests a new line of light bulbs and finds their lifetimes to be normally distributed, with a mean life of 98 hours and a standard deviation of 13.

a. What percentage of bulbs will last less than 72 hours?

b. What is the probability that a bulb selected at random will last more than 111 hours?

Solution Draw a normal curve with mean 98 and standard deviation 13. (See the figure at the top of page 317.)

a. $P(X < 72) = .02$

b. $P(X > 111) = .1359 + .0215 \approx .1574$

7.5 PROBABILITY DENSITY FUNCTIONS

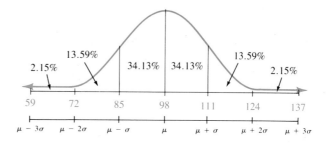

z-Scores

If a division of standard deviations finer than 1, 2, or 3 is needed, a **z-score** is used. The z-score translates any normal curve into a standard normal curve by using the simple calculation given below:

z-Score

The area to the left of a value x under a normal curve with mean μ and standard deviation σ is the same as the area under a standard normal curve to the left of the following z-value:

$$z = \frac{x - \mu}{\sigma}$$

Area between $z = 1$ and $z = 2$ is shaded

EXAMPLE 10 Find the probability that one of the light bulbs described in Example 9 will last between 110 and 120 hours.

Solution From Example 9, $\mu = 98$ and $\sigma = 13$. Let $x = 110$; then

$$z = \frac{110 - 98}{13} \approx .92$$

Now, look up $z = .92$ in Table 7.2 and find

$$P(X < 110) = P(z < .92) = .8212$$

For $x = 120$,

$$z = \frac{120 - 98}{13} \approx 1.69$$

Again, from Table 7.2,
$$P(X < 120) = P(z < 1.69) = .9545$$
Thus
$$P(110 \leq X \leq 120) = .9545 - .8212 = .1333$$

7.5

Problem Set

APPLICATIONS

Describe the sample space for the experiments in Problems 1–6. Define a random variable and characterize it as a discrete or continuous random variable.

1. Ten people are asked if they graduated from college, and the number of people responding yes is recorded.
2. An Everready Long-Life® battery is tested, and the life of the battery is recorded.
3. The number of words in which an error is made on a typing test consisting of 500 words is recorded.
4. A person is randomly selected for a public opinion poll and asked two questions: Are you male (m) or female (f)? Do you consider yourself a Democrat (d), Republican (r), Independent (i), or are you not registered (n) to vote?
5. A psychologist is studying sibling relationships in families with three children. Each family is asked about the sex of the children in the family and the results are recorded. If a family has a girl, then a boy, and finally another boy, this would be recorded as {gbb}.
6. A sample of five radios is selected from an assembly line and tested. The number of defective radios is recorded.
7. The heights of 30 students are (numbers are rounded to the nearest inch): 66, 68, 64, 70, 67, 67, 68, 64, 65, 66, 64, 70, 72, 71, 69, 64, 63, 70, 71, 63, 68, 67, 65, 69, 65, 67, 66, 69, 69, 67. Give the frequency distribution, define the random variable, and draw a histogram.
8. The wages of employees of a small accounting firm are (in thousands of dollars): 10, 15, 15, 15, 20, 20, 25, 40, 60, 8, 8, 8, 8, 8, 6, 4. Give the frequency distribution, define the random variable, and draw a histogram.
9. The times that a bank teller spent on each transaction were recorded as follows (rounded to the nearest minute): 8, 7, 3, 6, 5, 7, 3, 4, 5, 5, 4, 2, 1, 5, 2, 1, 3, 3, 4, 6, 4, 4, 2, 5, 4, 6, 4, 5, 4, 5. Give the frequency distribution, define the random variable, and draw a histogram.

Determine whether the functions in Problems 10–17 are probability density functions.

10. $f(x) = 6x$ on $[1, 5]$
11. $f(x) = \dfrac{x^2}{72}$ on $[1, 5]$
12. $f(x) = 6x$ on $[0, \sqrt{3}/3]$
13. $f(x) = \tfrac{1}{4}$ on $[1, 5]$
14. $f(x) = \tfrac{1}{4}$ on $[3, 6]$
15. $f(x) = 4$ on $[1, 4]$
16. $f(x) = \tfrac{3}{125} x^2$ on $[0, 5]$
17. $f(x) = \tfrac{3}{64} x^2$ on $[1, 4]$

Find k such that each function in Problems 18–27 is a probability density function over the given interval.

18. $f(x) = kx$ on $[2, 6]$
19. $f(x) = kx$ on $[-2, 3]$
20. $f(x) = kx^2$ on $[-1, 3]$
21. $f(x) = kx^2$ on $[0, 5]$
22. $f(x) = 3k$ on $[1, 4]$
23. $f(x) = \dfrac{k}{4}$ on $[1, 7]$
24. $f(x) = kx^{2/3}$ on $[0, 8]$
25. $f(x) = kx^{1/2}$ on $[1, 4]$
26. $f(x) = k/x$ on $[1, 6]$
27. $f(x) = k/x^2$ on $[1, 4]$

Find the area under the standard normal curve satisfying the conditions in Problems 28–33.

28. $X < -2$
29. $X < 1.23$
30. $X > 1$
31. $X > 1.69$
32. $.5 < X < 1.61$
33. $-2.8 \leq X \leq -.46$

34. The price of an item in dollars is a continuous random variable with a probability density function of $f(x) = 2$ for $1.25 \leq x \leq 1.75$. What is the probability that the price is more than $1.50?
35. A number is selected at random from the interval $[0, 100]$. The probability density function is $f(x) = .01$. Find the probability that the number selected is between 35 and 75.
36. The number of minutes to learn a certain task is a random variable with a probability density function of
$$f(t) = .1 e^{-.1t}$$
Find the probability that the task is learned in less than 10 minutes.
37. The seeds of many plants are dispersed by the wind, and the distance x (in feet) that the seed travels is given by a probability density function
$$d(x) = .5 e^{-.5x}$$
Find the probability that the seeds will be dispersed from 5 to 10 feet away.
38. The breaking strength of a rope (in pounds) is normally distributed, with a mean of 100 pounds and a standard deviation of 16. What is the probability that the rope will break will a force of 132 pounds or less?

39. The diameter of an electric cable is normally distributed, with a mean of .9 inch and a standard deviation of .01. What is the probability that the diameter will exceed .91 inch?

40. The annual rainfall in Ferndale, California, is known to be normally distributed, with a mean of 35.5 inches and a standard deviation of 2.5. What is the probability that the rainfall will exceed 32 inches?

41. If the life of a light bulb is normally distributed with a mean of 250 hours and a standard deviation of 25 hours, what is the probability that the bulb with burn for less than 210 hours?

42. What is the probability that the bulb described in Problem 41 will burn for more than 270 hours?

43. What is the probability that the bulb described in Problem 41 will burn between 240 and 280 hours?

44. At Caltex Corporation the probability density function for the length of time (in minutes) that a randomly selected telephone customer must wait to be helped is given by

$$f(t) = .25e^{-t/4}$$

What is the probability that a customer is kept waiting more than 3 minutes?

45. For a particular type of battery, the probability density function for the life (in hours) of a battery selected at random is given by

$$f(t) = .01e^{-.01t}$$

What is the probability that the battery will last at least 50 hours?

46. The probability density function for the time (in months) after it is purchased that a new telephone will need servicing is given by

$$f(x) = .05e^{-x/20}$$

If the phone is guaranteed for a year, what is the probability that a customer selected at random will have a phone that is no longer covered by the guarantee but that needs servicing?

*7.6
Review

The material of this chapter is reviewed in the following list of objectives. After each objective there are some practice questions. For a sample test select the first question of each set and check your answers. The second question for each objective has no answer given. If you are having trouble with a particular type of problem, look back at the indicated section in the text. When you are finished reviewing these objectives, a sample examination is given at the end of this section.

[7.1]
Objective 7.1: *Find the total value of a given function.*
1. $f(x) = \sqrt[3]{2x + 5}$ on $[1, 40]$
2. $f(x) = .25e^{-.12x}$ on $[0, 4]$
3. $f(x) = \dfrac{x}{2x - 3}$ on $[2, 4]$
4. $f(x) = \ln x$ on $[1, 10]$

Objective 7.2: *Find net excess profit for the given functions over the given intervals.*
5. $r_1(x) = 10 + x^2$; $r_2(x) = 34 + 5x$ on $[0, 8]$
6. $r_1(x) = 15 + x^2$; $r_2(x) = 25 + 3x$ on $[0, 5]$
7. $r_1(x) = 40 + x^2$; $r_2(x) = 32 + 6x$ on $[2, 4]$
8. $r_1(x) = 30 + x^2$; $r_2(x) = 12 + 9x$ on $[3, 6]$

Objective 7.3: *Find the total income for a given rate of flow of a continuous income stream over a given interval $[a, b]$.*
9. $f(t) = 1{,}000$, $[0, 5]$
10. $f(t) = 100e^{.09t}$, $[0, 3]$
11. $f(t) = 250e^{-.11t}$, $[0, 4]$
12. $f(t) = 100 + t^2$, $[0, 2]$

Objective 7.4: *Find the total money flow for P dollars over a period of T years at a given rate.*
13. $3,900 at 8% for 30 years
14. $1{,}000\, e^{rt}$ for a rate of 8% and time 4 years
15. Find the amount of an annuity with an annual deposit of $2,500 for 25 years into an account paying 7.5% compounded continuously.
16. If a person places $125 a month into an account paying 8% compounded continuously, how much will be in the account in 15 years?

Objective 7.5: *Find the consumers' or producers' surplus.*
17. The price, in dollars, for a product is given by the demand function

$$D(x) = 250 + 10x - x^2$$

where x is the number of items (in thousands) in the domain $[10, 20]$. The supply function is given by

$$S(x) = x^2 - 20x + 250$$

Find the consumers' surplus.

*Optional section.

18. The price, in dollars, for a product is given by the demand function

$$D(x) = 4(25 - x^2)$$

where x is the number of items (in thousands) in the domain $[0, 10]$. The supply function is given by

$$S(x) = x^2 + 5x + 40$$

Find the consumers' surplus.

19. Using the information in Problem 17, find the producers' surplus.

20. Using the information in Problem 18, find the producers' surplus.

[7.2]
Objective 7.6: *Evaluate indefinite integrals by parts.*

21. $\int \dfrac{x}{e^{2x}} \, dx$

22. $\int \dfrac{x^5}{(x^3 + 1)^2} \, dx$

23. $\int \ln \sqrt{2x} \, dx$

24. $\int x^2 \sqrt{10 - x} \, dx$

[7.3]
Objective 7.7: *Evaluate indefinite integrals by using the Brief Integral Table.*

25. $\int x^2 \sqrt{x^2 - 16} \, dx$

26. $\int (3 - 4x)^6 \, dx$

27. $\int \ln^3 3x \, dx$

28. $\int x^2 \sqrt{10 - x} \, dx$

Objective 7.8: *Evaluate indefinite integrals using any appropriate method.*

29. $\int x^2 e^{-2x} \, dx$

30. $\int \dfrac{3 \, dx}{1 - e^{3x}}$

31. $\int \dfrac{8x + 5}{4x^2 + 5x - 3} \, dx$

32. $\int \dfrac{5 \, dx}{2 - e^{-x}}$

[7.4]
Objective 7.9: *Evaluate improper integrals that converge.*

33. $\displaystyle\int_0^\infty xe^{-x^2} \, dx$

34. $\displaystyle\int_{-\infty}^1 \dfrac{x \, dx}{x^2 + 1}$

35. $\displaystyle\int_0^\infty .5e^{-.5x} \, dx$

36. $\displaystyle\int_e^\infty \ln x \, dx$

*[7.5]
Objective 7.10: *Describe the sample space for a given experiment. Define a random variable for the experiment and be able to tell if it is a discrete or a continuous random variable.*

37. A pair of dice is rolled and the sum of the top faces is recorded.

* Optional section.

38. A coin is tossed three times and the sequence of heads and tails is recorded.

39. The number of minutes it takes a rat to make its way through a maze is recorded; compare with Problem 40.

40. The number of rats that can make their way through a maze is recorded; compare with Problem 39.

Objective 7.11: *Given a set of data, prepare a frequency distribution and draw a histogram.*

41. A pair of dice is rolled and the sum of the spots on the tops of the dice is recorded as follows: 3, 2, 6, 5, 3, 8, 8, 7, 10, 9, 7, 5, 12, 9, 6, 8, 11, 11, 8, 7, 7, 7, 10, 7, 9, 7, 9, 6, 6, 9, 4, 4, 6, 3, 4, 10, 6, 9, 6, 11.

42. Blane, Inc. a consulting firm was employed to perform an efficiency study at National City Bank. As part of the study, they found the number of times per day that people were waiting in line to be the following: 2 were waiting 20 times; 3 were waiting 15 times; 4, 7 times; 5, 5 times; 6, 2 times; 7, 1 time; and there were never more than 7 people in line at any time during the day.

43. Blane, Inc., the consulting firm described in Problem 42, also noted the transaction times (rounded to the nearest minute) for customers at National City Bank, as follows: 1 minute, 10 times; 2 minutes, 12 times; 3 minutes, 18 times; 4 minutes, 25 times; 5 minutes, 16 times; 6 minutes, 10 times; 7 minutes, 6 times; 8 minutes, 1 time; 9 minutes, none; and 10 minutes, 2 times. No transaction took more than 10 minutes.

44. Three coins are tossed onto a table and the following frequencies are noted: 0 heads, 17 times; 1 head, 59 times; 2 heads, 56 times; and 3 heads, 18 times.

Objective 7.12: *Determine whether a given function is a probability density function.*

45. $f(x) = \dfrac{3}{63} x^2$ on $[1, 4]$

46. $f(x) = \dfrac{x^2}{3}$ on $[-1, 1]$

47. $f(x) = \dfrac{2}{3}$ on $[2, 5]$

48. $f(x) = .1$ on $[0, 10]$

Objective 7.13: *Find a constant k in order to define a probability density function over a given interval.*

49. $f(x) = kx^{1/4}$ on $[0, 16]$

50. $f(x) = k$ on $[-4, 0]$

51. $f(x) = 5kx$ on $[1, 10]$

52. $f(x) = k\sqrt{2x}$ on $[2, 8]$

Objective 7.14: *Find the area under the standard normal curve.*

53. $x < 0$

54. $-1.03 < x < 1.59$

55. $x > -.11$

56. $-.5 \le x \le 1.5$

Objective 7.15: *Solve applied problems based on the preceding objectives.*

57. *Spread of a disease.* If a disease is spreading at a rate of $40t - 6t^2$ cases per day where t is the number of days measured from the first outbreak, give the number of

people affected on the seventh day. Assume that there were 4 cases recorded on the first day.

58. *Depreciation.* If the total depreciation at the end of t years is given by $f(t)$ and the depreciation rate is $f'(t) = 200\sqrt{10-x}$, find the total depreciation for the first five years.

59. *Capital value.* Suppose that the current interest rate is 12% and the British Embassy in Washington, D. C., has an indeterminant lease paying $576,000 per year. What is the capital value of this lease?

*60. If the life of a light bulb is normally distributed with a mean of 250 hours and a standard deviation of 25 hours, find the following probabilities:
   a. $P(X > 250)$
   b. $P(X < 220)$
   c. $P(200 < X < 300)$
   d. $P(220 \leq X \leq 320)$

*61. *Grading on a curve.* Suppose that for a certain exam a teacher grades on a curve. It is known that the mean is 50 and the standard deviation is 5. There are 45 students in the class.
   a. How many students should receive a C?
   b. How many students should receive an A?
   c. What score would be necessary to obtain an A?
   d. If an exam paper is selected at random, what is the probability that it will be a failing paper?

SAMPLE TEST

The following sample test (75 minutes) is intended to review the main ideas of this chapter.

* Optional section.

Evaluate the integrals in Problems 1–9. Use substitution, integration by parts, or the Brief Integral Table.

1. $\int (2x-1)^3 \, dx$

2. $\int x(2x-1)^3 \, dx$

3. $\int x^2(2x-1)^3 \, dx$

4. $\int x(2x^2-1)^3 \, dx$

5. $\int \frac{8x+5}{4x^2+5x-3} \, dx$

6. $\int \frac{4x^2+5x-3}{x} \, dx$

7. $\int \frac{3}{1-e^{3x}} \, dx$

8. $\int \ln^2 x \, dx$

9. $\int \ln^3 x \, dx$

State the formula you would use from the Brief Integral Table to approximate the integrals in Problems 10–12. You do not need to actually carry out the integration.

10. $\int x^2(20-5x)^4 \, dx$

11. $\int \frac{\sqrt{100-x}}{x^5} \, dx$

12. $\int_{.5}^{2} t^4 e^{-t} \, dt$

Evaluate the improper integrals in Problems 13–14 that converge.

13. $\int_1^\infty x^2 e^{-x^3} \, dx$

14. $\int_{-\infty}^{1} \frac{x^2 \, dx}{x^3+1}$

15. Determine the total money flow for $35,000 at 12% for 6 years.

16. Suppose that the current interest rate is 10% and the French Embassy in Washington, D.C., has a 50-year lease paying $682,000 per year. What is the capital value of this lease?

CUMULATIVE REVIEW
for chapters 5–7

1. Graph $y = e^x$.
2. Graph $y = \ln\sqrt{x}$.

Solve the equations in Problems 3–6.

3. $\log\left(\dfrac{x-5}{6}\right) = 2$
4. $3\ln\dfrac{e}{\sqrt[3]{5}} = 3 - \ln x$
5. $10^{-x} = .5$
6. $e^{1-x} = 105$

7. Fill in the blanks to complete the properties of the definite integral.

   a. $\displaystyle\int_a^a f(x)\,dx = $ _____

   b. $\displaystyle\int_a^b dx = $ _____

   c. $\displaystyle\int_a^b f(x)\,dx = $ _____ where $b < a$

   d. $\displaystyle\int_a^c f(x)\,dx + \int_c^b f(x)\,dx = $ _____ where $a < c < b$

   e. $\displaystyle\int_a^b kf(x)\,dx = $ _____

   f. $\displaystyle\int_a^b [f(x) \pm g(x)]\,dx = $ _____

   g. $\displaystyle\int_{x=a}^{x=b} u\,dv = $ _____

Evaluate the integrals in Problems 8–15.

8. a. $\displaystyle\int \dfrac{du}{u}$ b. $\displaystyle\int du$

9. a. $\displaystyle\int (5x^4 + 3x^2 + 5)\,dx$ b. $\displaystyle\int e^{2x}\,dx$

10. $\displaystyle\int x^2(5 - 2x^3)^4\,dx$

11. $\displaystyle\int \ln^4 5x\,dx$

12. $\displaystyle\int \dfrac{100\,dx}{5 - e^{-x}}$

13. $\displaystyle\int \dfrac{8x - 3}{4x^2 - 3x + 2}\,dx$

14. $\displaystyle\int_1^\infty x^{-3/2}\,dx$

15. $\displaystyle\int_{-\infty}^\infty \dfrac{x\,dx}{(x^2 + 1)^2}$

16. Write a formula for the area bounded above by the curve $y = 1/\sqrt{x^2 + 4}$, below by the x-axis, on the left by the y-axis, and on the right by the line $x = t$.

17. Find the area bounded by $y = x^2$ and $y = 32 - x^2$.

18. Find $\displaystyle\int_1^2 \ln x^2\,dx$ correct to two decimal places using one (or all) of the following approximations for $n = 4$.
    a. Rectangular approximation
    b. Trapezoidal approximation
    c. Simpson's rule

APPLICATIONS

19. A piece of replacement equipment costs \$48,000. The rate of operating cost savings is $S(x)$ in dollars and is given by the formula
$$S(x) = 5{,}000x + 2{,}000$$
where x is the number of years the piece of equipment will be used. How long will it take the piece of equipment to pay for itself?

20. In the course of any year, the number y of cases of a disease is reduced by 10% according to the growth formula
$$y = P_0 e^{-.1t}$$
for t years with P_0 cases reported today. If there are 100,000 cases today, how long will it take to reduce the number of cases to less than 10,000?

21. Use the formula $A = P(1 + i)^N$ to find out how long after depositing \$1,000 at 8% compounded daily (365-day year) you must wait in order to have \$5,000.

22. In 1986 the growth rate of Houston, Texas, was given by the formula
$$P'(x) = 1.6e^{.025t}$$
where t is measured in years since 1980 and $P(t)$ is the population in millions. If the 1986 population was 1,860,000, predict the population at the turn of the century. Comment on this result and growth rate.

23. A businessperson receives a shipment of Christmas trees on November 20. The sales pattern is such that the inventory moves slowly at the beginning but as Christmas approaches, the demand increases so that x days after

November 20 the inventory is y trees, where
$$y = 2{,}450 - 2x^2 \quad \text{for } 0 \leq x \leq 35$$
What is the average inventory for the first 30 days?

24. Let X be a normally distributed random variable with mean 55 and standard deviation 10. Find the following probabilities.
   a. $P(X \geq 55)$
   b. $P(X > 60)$
   c. $P(40 \leq X \leq 50)$

25. The wait time (in seconds) for a response at a particular terminal in a time-share network is a continuous random variable (X) with a probability density function
$$f(x) = .01e^{-x/100}$$
What is the probability that a user must wait more than 1 minute for a response?

8 Functions of Several Variables

CHAPTER OVERVIEW

In order to build realistic mathematical models for real-life situations it is necessary to consider functions of more than one variable. This chapter defines and discusses such functions.

PREVIEW

The important concept of this chapter is that of a partial derivative. We use it in maximum–minimum applications and the method of least squares. Another method of maximization, Lagrange multipliers, is introduced in Section 8.4. Multiple integration applications, including finding volumes, are developed in Section 8.5.

PERSPECTIVE

Even though this chapter may be considered optional because of time constraints, it is very important in serious model building. The concept of a function of two, three, or more variables is an easy one; unfortunately, the graphical representation of these functions is not. If you do not have time to consider the ideas of this chapter in class, it is useful to study this chapter on your own after you have completed this course.

8.1 Three-Dimensional Coordinate System
8.2 Partial Derivatives
8.3 Maximum–Minimum Applications
8.4 Lagrange Multipliers
8.5 Multiple Integrals
8.6 Correlation and Least Squares Applications
8.7 Chapter 8 Review
   Chapter Objectives
   Sample Test

MODELING
APPLICATION 8

The Cobb–Douglas Production Function

Karlin Corporation manufactures only one product, which is sold at the price P_0. The firm employs a labor force L, which must be paid an average wage p_1. The firm also requires capital K in terms of tools, buildings, and so forth. The cost of using one unit of capital is p_2. Karlin wishes to maximize its profit. Determine what data need to be collected and construct a model that will accomplish Karlin's goal of maximizing profit.

Developing a mathematical model is no easy task. As you have realized by now, there is usually a great deal of work involved. Every good model should include listing the assumptions, translating the assumptions into mathematical notation, building the model, and then checking the model against tabulated data. Read "The Cobb–Douglas Production Function" by Robert Geitz (The UMAP Journal, Unit 509. © 1981 Education Development Center, Inc.) This paper is a perfect illustration of the modeling process. After reading it, develop a model to accomplish Karlin Corporation's goal.

APPLICATIONS

Management (*Business, Economics, Finance, and Investments*)
Cost function
  (8.1, Problems 43, 45–46; 8.2, Problem 50; 8.7, Problem 54, Test Problem 15)
Revenue function
  (8.1, Problem 44; 8.2, Problem 51)
Profit (8.2, Problem 49)
Rate of change of an investment
  (8.2, Problems 55–56)
Rate of change of an annuity
  (8.2, Problem 57)
Maximize profit (8.3, Problems 19, 21)
Minimize labor cost for a function of two variables (8.3, Problem 20)
Minimize cost of shipping container
  (8.3, Problem 22)
Least material to construct a shipping container (8.3, Problem 23)
Maximum yield on farm production
  (8.4, Problem 15)

Management (*continued*)
Minimize cost relative to supply
  (8.4, Problem 16)
Maximum area for a fenced enclosure, given fixed costs (8.4, Problem 17)
Minimum surface area for a standard size Coke can (8.4, Problem 18)
Marginal cost and revenue of a function of two variables
  (8.4, Problem 20; 8.7, Problems 55–57)
Average value of a function
  (8.5, Problems 55–56)
Cobb–Douglas production function
  (8.5, Problem 57)

Life Sciences (*Biology, Ecology, Health, and Medicine*)
Amount of blood flow as a function of blood vessel size (8.1, Problem 47)
Poiseuille's law for blood flow
  (8.2, Problem 52)

Life Sciences (*continued*)
Surface area of a human body
  (8.2, Problems 53–54)
Supplying a 1,000 calorie diet while minimizing the cost (8.4, Problem 19)

Social Sciences (*Demography, Political science, Population, Psychology, Society, and Sociology*)
Intelligence quotient (IQ) (8.1, Problem 42)

General Interest
Cost function for finishing a room
  (8.1, Problems 45–46)
Largest volume that can be mailed in the U.S.
  (8.3, Problem 24)

Modeling Application—
The Cobb–Douglas Production Function

8.1

Three-Dimensional Coordinate System

Many real-life models involve more than one variable. Suppose, for example, we consider one of the most fundamental applications, that of the total cost of producing an item. If Ballad Corporation produces a single record with fixed costs of $2,000 and a unit cost of $.35, then

$$C(x) = 2{,}000 + .35x$$

for x records produced. However, if a second record is produced with additional fixed costs of $500 and a unit cost of $.30, then the total cost of producing x records of the first type and y records of the second type requires what we call a **function of two independent variables** x and y:

$$C(x, y) = 2{,}500 + .35x + .30y$$

Function of Two or More Variables

Suppose D is a collection of ordered n-tuples of real numbers $(x_1, x_2, \ldots, x_n)$. Then a function f with **domain** D is a rule that assigns a number

$$z = f(x_1, x_2, \ldots, x_n)$$

to each n-tuple in D. The function's **range** is the set of z values the function assumes. The symbol z is called the **dependent variable** of f, and f is said to be a **function of the n independent variables** $x_1, x_2, \ldots, x_n$.

You have already considered many examples of functions of several variables, as Example 1 shows.

EXAMPLE 1 Area of a rectangle: $K(l, w) = lw$
Volume of a box: $V(l, w, h) = lwh$
Simple interest: $I(P, r, t) = P(1 + rt)$
Compound interest: $A(P, r, t, n) = P(1 + \frac{r}{n})^{nt}$

Find each of the requested values and interpret your results in terms of what you know about each of these formulas.

a. $K(25, 15)$
b. $V(5, 20, 30)$
c. $I(100000, .08, 15)$
d. $A(450000, .09, 30, 12)$

Solution
a. $K(25, 15) = 25(15) = 375$; the area of a 25 by 15 rectangle is 375 square units.
b. $V(5, 20, 30) = 5(20)(30) = 3{,}000$; the volume of a 5 by 20 by 30 box is 3,000 cubic units.
c. $I(100000, .08, 15) = 100{,}000[1 + .08(15)] = 100{,}000[2.2] = 222{,}000$; future value of a $100,000 investment at 8% simple interest for 15 years is $220,000.
d. $A(450000, .09, 30, 12) = 450{,}000(1 + \frac{.09}{12})^{30(12)} = 450{,}000(1.0075)^{360} \approx 6{,}628{,}759.26$; the future value of a $450,000 investment at 9% compounded monthly is approximately $6,628,759.26. ∎

Even though a function of several variables has been defined for the general case and Example 1 shows functions of several variables, this chapter focuses primarily on functions of two variables. That is, $z = f(x, y)$ is the notation used for z, a function of two independent variables x and y. In order to graph such a function we need to consider **ordered triplets** (x, y, z) and a **three-dimensional coordinate system**, just as we have already considered ordered pairs (x, y) and a two-dimensional coordinate system. We draw a coordinate system with three mutually perpendicular axes, as shown in Figure 8.1.

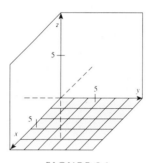

FIGURE 8.1
Three-dimensional coordinate system

Think of the x-axis and the y-axis as the floor and the z-axis as a line perpendicular to the floor. All of the graphs we have done up to now in this book would now be drawn on the "floor." Example 2 shows how to plot points in three dimensions.

EXAMPLE 2 Graph the following ordered triplets.

Solution
a. $(10, 20, 10)$ **b.** $(-12, 6, 12)$
c. $(-12, -18, 6)$ **d.** $(20, -10, 18)$

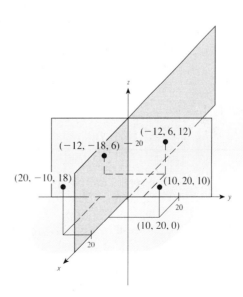

If you orient yourself in a room (your classroom, for example), as shown in Figure 8.2, you will notice certain important planes:

FIGURE 8.2
A typical classroom; assume the dimensions are 25 by 30 feet with an 8 foot ceiling.

Floor: **xy-plane**
  Equation is $z = 0$.
Ceiling: a plane parallel to the xy-plane
  Equation is $z = 8$.
Front wall: **yz-plane**
  Equation is $x = 0$.
Back wall: plane parallel to the yz-plane
  Equation is $x = 30$.
Left-side wall: **xz-plane**
  Equation is $y = 0$.
Right-side wall: plane parallel to the xz-plane
  Equation is $y = 25$.

The xy-, xz-, and yz-planes are called the **coordinate planes**. Name the coordinates of several objects in the figure.

Just as points in the plane are associated with ordered pairs satisfying an equation in two variables, points in space are associated with ordered triplets satisfying an equation. The graph of any function of the form $z = f(x, y)$ is called a **surface**. It is beyond the scope of this course to have you spend a great deal of time graphing three-dimensional surfaces, but you should be aware that computer programs have simplified the task of graphing surfaces, as shown in Figure 8.3.

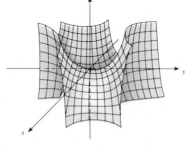

a. $z = x^3 - 3xy^2$

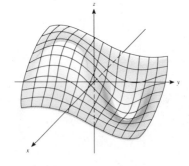

b. $z = \dfrac{-1}{x^2 + y^2 + 1}$

FIGURE 8.3
Graphs of surfaces in three dimensions

Planes

The graph of $ax + by + cz = d$ is a **plane** if $a, b, c,$ and d are real numbers ($a, b,$ and c not all zero).

EXAMPLE 3 Graph the planes defined by the given equations.

a. $x + 3y + 2z = 6$
b. $y + z = 5$
c. $x = 4$

Solution It is customary to show only the portion of the graph that lies in the **first octant** (that is, where $x, y,$ and z are all positive). To graph a plane, find some ordered triplets satisfying the equation. The best ones to use are often those on one of the coordinate axes.

a. Let $x = 0$ and $y = 0$; then $z = 3$; point is $(0, 0, 3)$.
   Let $x = 0$ and $z = 0$; then $y = 2$; point is $(0, 2, 0)$.
   Let $y = 0$ and $z = 0$; then $x = 6$; point is $(6, 0, 0)$.

   Plot these points as shown in Figure 8.4a and use them to draw the plane.

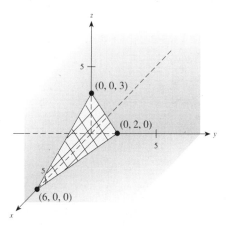

FIGURE 8.4a
Graphing planes

b. If exactly two of the coefficients a, b, c are not zero (i.e., one of the variables is missing from the equation of a plane), then that plane is parallel to the axis corresponding to the missing variable; in this case it is parallel to the x-axis. Draw the line $y + z = 5$ on the yz-plane, and then complete the plane as shown in Figure 8.4b on page 330.

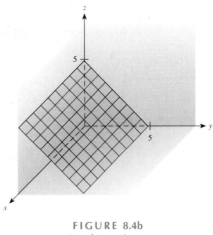

FIGURE 8.4b
Graphing planes

c. If exactly one of a, b, c is nonzero (i.e., two variables are missing), then the plane is parallel to the plane of the two variables missing in the equation, as shown in Figure 8.4c.

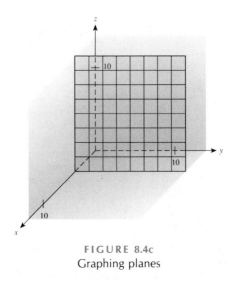

FIGURE 8.4c
Graphing planes

Quadric Surfaces

The graph of the equation

$$Ax^2 + By^2 + Cz^2 + Dxy + Exz + Fyz + Gy + Hy + Iz + J = 0$$

is called a **quadric surface**. The **trace** of a curve is found by setting one of the variables equal to a constant and then graphing the resulting curve. If $x = k$ (k a constant), then the resulting curve is drawn in the plane $x = k$, which is parallel to the yz-plane; similarly, if $y = k$, then the curve is drawn in the plane $y = k$, which is parallel to the xz-plane; and if $z = k$, then the curve is drawn in the plane $z = k$, which is parallel to the xy-plane. Table 8.1 shows the quadric surfaces.

TABLE 8.1 Quadric Surfaces

Surface	Description	Surface	Description	Surface	Description
Elliptic cone	The trace in the xy-plane is a point; in planes parallel to the xy-plane it is an ellipse. Traces in the xz- and yz-planes are intersecting lines; in planes parallel to these they are hyperbolas $$z^2 = \frac{x^2}{a^2} + \frac{y^2}{b^2}$$	Elliptic paraboloid	The trace in the xy-plane is a point; in planes parallel to the xy-plane it is an ellipse. Traces in the xz- and yz-planes are parabolas $$z = \frac{x^2}{a^2} + \frac{y^2}{b^2}$$	Ellipsoid or sphere	The traces in the coordinate planes are ellipses. $$\frac{x^2}{a^2} + \frac{y^2}{b^2} + \frac{z^2}{c^2} = 1$$ If $a^2 = b^2 = c^2 = r^2$, then the graph is a sphere $$x^2 + y^2 + z^2 = r^2$$
Hyperboloid of one sheet	The trace in the xy-plane is an ellipse; in the xz- and yz-planes the traces are hyperbolas $$\frac{x^2}{a^2} + \frac{y^2}{b^2} - \frac{z^2}{c^2} = 1$$	Hyperboloid of two sheets	There is no trace in the xy-plane. In planes parallel to the xy-plane, which intersect the surface, the traces are ellipses. Traces in the xz- and yz-planes are the hyperbolas $$\frac{x^2}{a^2} + \frac{y^2}{b^2} - \frac{z^2}{c^2} = -1$$		
Hyperbolic paraboloid	The trace in the xy-plane is two intersecting lines; in planes parallel to the xy-plane the traces are hyperbolas. Traces in the xz- and yz-planes are parabolas $$z = \frac{y^2}{b^2} - \frac{x^2}{a^2}$$				

Circular Cylinders

The graphs of

$$y^2 + z^2 = r^2 \qquad x^2 + z^2 = r^2 \qquad \text{and} \qquad x^2 + y^2 = r^2$$

are **right circular cylinders** of radius r, parallel to the x-axis, y-axis, and z-axis, respectively.

EXAMPLE 4 Graph the following equations.

a. $x^2 + y^2 = 9$ **b.** $y^2 + z^2 = 16$ **c.** $x^2 + z^2 = 25$

Solution **a.** This is a cylinder parallel to the z-axis (the z variable is missing), as shown in Figure 8.5a.
b. This is a cylinder parallel to the x-axis, as shown in Figure 8.5b.
c. This is a cylinder parallel to the y-axis, as shown in Figure 8.5c.

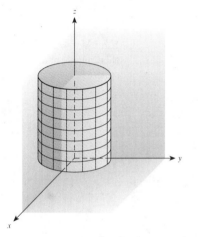

a. Graph of
$x^2 + y^2 = 9$

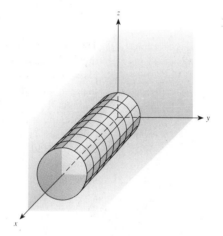

b. Graph of
$y^2 + z^2 = 16$

c. Graph of
$x^2 + z^2 = 25$

FIGURE 8.5
Graphs of right circular cylinders

8.1

Problem Set

Evaluate the functions K, V, I, and A from Example 1 in Problems 1–10, and interpret your results.

1. $K(15, 35)$
2. $K(45, 90)$
3. $V(3, 5, 8)$
4. $V(15, 25, 8)$
5. $I(500, .05, 3)$
6. $I(1250, .08, 12)$
7. $A(2500, .12, 6, 4)$
8. $A(5500, .18, 5, 6)$
9. $A(110000, .09, 30, 12)$
10. $A(250000, .15, 15, 12)$

Evaluate $f(x, y) = x^2 - 2xy + y^2$ for the values in Problems 11–14.

11. $f(2, 3)$
12. $f(-1, 4)$
13. $f(-2, 5)$
14. $f(0, 6)$

Evaluate $g(x, y) = \dfrac{2x - 4y}{x^2 + y^2}$ for the values in Problems 15–18.

15. $g(2, 1)$
16. $g(-3, 2)$
17. $g(5, -3)$
18. $g(-3, -4)$

Evaluate $h(x, y) = \dfrac{e^{xy}}{\sqrt{x^2 + y^2}}$ for the values in Problems 19–22.
Round answers to the nearest hundredth.

19. $h(0, 5)$
20. $h(-2, 3)$
21. $h(-2, -3)$
22. $h(1, 1)$

Graph the ordered triplets in Problems 23–25.

23. **a.** $(1, 2, 3)$ **b.** $(-3, 2, 4)$
  c. $(1, -4, 3)$ **d.** $(-5, -9, -8)$
24. **a.** $(2, 4, 3)$ **b.** $(-3, 2, 4)$
  c. $(10, -20, -5)$ **d.** $(-1, -2, -3)$
25. **a.** $(10, 5, 20)$ **b.** $(5, -15, -5)$
  c. $(3, 2, -4)$ **d.** $(-5, -1, 3)$

Graph the surfaces in Problems 26–41.

26. $2x + y + 3z = 6$
27. $x + 2y + 5z = 10$
28. $x + y + z = 1$
29. $3x - 2y - z = 12$
30. $z^2 = \dfrac{x^2}{4} + \dfrac{y^2}{9}$
31. $z = \dfrac{x^2}{4} + \dfrac{y^2}{9}$
32. $\dfrac{x^2}{1} + \dfrac{y^2}{4} + \dfrac{z^2}{9} = 1$
33. $\dfrac{x^2}{9} + \dfrac{y^2}{4} + \dfrac{z^2}{25} = 1$
34. $x^2 + y^2 + z^2 = 9$
35. $z = x^2 + y^2$
36. $\dfrac{x^2}{9} - \dfrac{y^2}{1} + \dfrac{z^2}{4} = 1$
37. $\dfrac{x^2}{9} + \dfrac{y^2}{1} - \dfrac{z^2}{4} = -1$
38. $y^2 + z^2 = 25$
39. $x^2 + y^2 = 36$
40. $x^2 + z^2 = 4$
41. $y^2 + z^2 = 20$

APPLICATIONS

42. The intelligence quotient (IQ) is defined as $Q(x, y) = \dfrac{100x}{y}$ where Q is the IQ, x is a person's mental age as measured on a standardized test, and y is a person's chronological age measured in years. Find (to the nearest unit) and interpret
  a. $Q(15, 13)$ **b.** $Q(6, 9)$
  c. $Q(15, 15)$ **d.** $Q(10.5, 9.8)$

43. A company manufactures two types of golf carts. The first has a fixed cost of $2,500, a variable cost of $800, and x are produced. The second has a fixed cost of $1,200, a variable cost of $550, and y are produced. Write a cost function $C(x, y)$ and find
  a. $C(10, 15)$ **b.** $C(5, 25)$
  c. $C(15, 10)$ **d.** $C(0, 30)$

44. If the revenue function for the golf carts in Problem 43 is $R(x, y) = 1,500x + 900y$ find
  a. $R(10, 15)$ **b.** $R(5, 25)$
  c. $R(15, 10)$ **d.** $R(0, 30)$

45. If the dimensions of the room in Figure 8.2 are x feet wide, y feet long, and z feet high, and if ceiling material is $2 per square foot, wall material is $.75 per square foot, and floor material is $1.25 per square foot, write a cost function for the ceiling, floor, and wall (assuming no doors or windows).

46. If $C(x, y, z)$ is the cost function for Problem 45, find
  a. $C(25, 30, 8)$ **b.** $C(12, 14, 8)$
  c. $C(15, 20, 10)$

47. The amount of blood flowing in a blood vessel measured in milliliters is given by $F(l, r) = .002l/r^4$ where l is the length of the blood vessel and r is the radius. Find
  a. $F(3.1, .002)$ **b.** $F(15.3, .001)$
  c. $F(6, .005)$

8.2
Partial Derivatives

One of the most important and useful concepts in mathematics is that of a derivative. In this section we consider the derivative of a function of several variables. We begin with a geometric interpretation and then apply this interpretation to some particular examples.

Consider a surface $z = f(x, y)$, as shown in Figure 8.6a on page 334. Hold one of the variables constant, say, $y = b$. This gives a curve $z = f(x, b)$, which is the intersection of the plane $y = b$ and the surface. The slope of this curve is called the **partial derivative** of f with respect to x and is denoted by f_x or $\partial z/\partial x$ (see Figure 8.6b). Similarly, if we let $x = a$ (a constant), then the curve that is the intersection of this plane and the surface has slope f_y or $\partial z/\partial y$, which is called the **partial derivative** of f with respect to y (see Figure 8.6c).

334 CHAPTER EIGHT FUNCTIONS OF SEVERAL VARIABLES

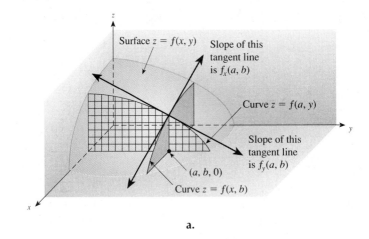

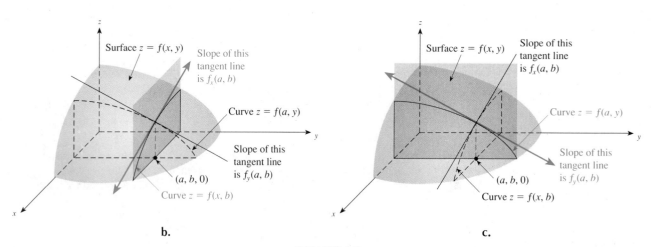

FIGURE 8.6
Geometric interpretation of a partial derivative

Partial Derivative

If $z = f(x, y)$, then the **partial derivative of f with respect to x**, denoted by $\partial z/\partial x$, f_x, or $f_x(x, y)$, is defined by

$$\frac{\partial z}{\partial x} = \lim_{h \to 0} \frac{f(x + h, y) - f(x, y)}{h} \qquad \text{Hold y constant.}$$

Also, the **partial derivative of f with respect to y**, denoted by $\partial z/\partial y$, f_y, or $f_y(x, y)$, is defined by

$$\frac{\partial z}{\partial y} = \lim_{h \to 0} \frac{f(x, y + h) - f(x, y)}{h} \qquad \text{Hold x constant.}$$

provided these limits exist.

This definition says that these are the usual derivatives found by holding y or x, respectively, constant. Similar definitions could be formulated for functions with three or more independent variables.

EXAMPLE 1 Let $f(x, y) = 3xy^2 - 2x^3y + 5x$ and $g(x, y) = (3x - 5y)^4$. Find the requested partial derivatives.

a. f_x **b.** f_y **c.** g_x **d.** g_y

Solution **a.** Treat y as a constant and find the derivative with respect to x:

$$f_x = \frac{\partial}{\partial x} f(x, y) = 3y^2 - 6x^2y + 5$$

b. Treat x as a constant and find the derivative with respect to y:

$$f_y = \frac{\partial}{\partial y} f(x, y) = 6xy - 2x^3$$

c. Treat y as a constant:

$$g_x = \frac{\partial}{\partial x} g(x, y) = 4(3x - 5y)^3(3) = 12(3x - 5y)^3$$

d. Treat x as a constant:

$$g_y = \frac{\partial}{\partial y} g(x, y) = 4(3x - 5y)^3(-5) = -20(3x - 5y)^3 \quad \blacksquare$$

EXAMPLE 2 Suppose that x is the inventory (in thousands of dollars) and y is the number of employees of a dress store, where $80 \leq x \leq 250$ and $3 \leq y \leq 8$. Also, suppose the weekly profit function (in dollars) is

$$P(x, y) = 5{,}000 + 10x - 15xy + 10x^2 - 3(y - 8)^2$$

At the present time the inventory is \$105,000 ($x = 105$) and there are four employees ($y = 4$). Approximate the rate of change of P per unit change in x if y remains fixed at 4.

Solution This is the partial derivative of P with respect to x at $(105, 4)$:

$$P_x = (10 - 15y + 20x)\Big|_{(105, 4)} = 10 - 15(4) + 20(105) = 2{,}050$$

This says that the instantaneous rate of change of the profit is changing \$2,050 per \$1,000 change in x when y remains fixed at four employees. $\blacksquare$

It is possible to find the partial derivative of a partial derivative. The idea is very straightforward. Study the following definition of **higher-order partial derivatives**.

Higher-Order Partial Derivatives

If $z = f(x, y)$, then

$$\frac{\partial^2 z}{\partial x^2} = \frac{\partial}{\partial x}\left(\frac{\partial z}{\partial x}\right) = f_{xx}(x, y) = f_{xx}$$

$$\frac{\partial^2 z}{\partial y^2} = \frac{\partial}{\partial y}\left(\frac{\partial z}{\partial y}\right) = f_{yy}(x, y) = f_{yy}$$

$$\frac{\partial^2 z}{\partial x\, \partial y} = \frac{\partial}{\partial x}\left(\frac{\partial z}{\partial y}\right) = f_{yx}(x, y) = f_{yx}$$

$$\frac{\partial^2 z}{\partial y\, \partial x} = \frac{\partial}{\partial y}\left(\frac{\partial z}{\partial x}\right) = f_{xy}(x, y) = f_{xy}$$

For the mixed partial derivative, $\dfrac{\partial^2 z}{\partial x\, \partial y} = f_{yx}$, start with z and first differentiate with respect to y (keep x constant) and then with respect to x. For $\dfrac{\partial^2 z}{\partial y\, \partial z} = f_{xy}$, start with z and first differentiate with respect to x (keep y constant) and then with respect to y. That is, read f_{yx} from left to right; y first, then x.

EXAMPLE 3 For $z = f(x, y) = 5x^2 - 2xy + 3y^3$, find the requested higher-order partial derivatives. Pay particular attention to the notation—part of what this example is illustrating is the variety in notation that can be used for higher-order partial derivatives.

a. $\dfrac{\partial^2 z}{\partial x\, \partial y}$ **b.** $\dfrac{\partial^2 z}{\partial y\, \partial x}$ **c.** $\dfrac{\partial^2 z}{\partial x^2}$ **d.** $f_{xy}(3, 2)$

e. $\dfrac{\partial^2}{\partial x\, \partial y}(2x^4 - 3x^2 y^2 + 5y^3 + 25)\Big|_{(-2, 5)}$

Solution **a.** First differentiate with respect to y:

$$\frac{\partial z}{\partial y} = -2x + 9y^2$$

Then differentiate with respect to x:

$$\frac{\partial^2 z}{\partial x\, \partial y} = \frac{\partial}{\partial x}\left(\frac{\partial z}{\partial y}\right)$$

$$= \frac{\partial}{\partial x}(-2x + 9y^2)$$

$$= -2$$

b. First differentiate with respect to x, then with respect to y:

$$\frac{\partial z}{\partial x} = 10x - 2y \quad \text{and} \quad \frac{\partial^2 z}{\partial y\, \partial x} = \frac{\partial}{\partial y}(10x - 2y) = -2$$

c. Differentiate with respect to x twice:

$$\frac{\partial z}{\partial x} = 10x - 2y \quad \text{and} \quad \frac{\partial^2 z}{\partial x^2} = \frac{\partial}{\partial x}(10x - 2y) = 10$$

d. Differentiate first with respect to x and then with respect to y; finally, evaluate at $(3, 2)$:

$$f_x(x, y) = 10x - 2y \quad \text{and} \quad f_{xy}(x, y) = -2$$

At $(3, 2)$ the value is -2. Notice that since the value is a constant, it is -2 at all points.

e. First differentiate with respect to y, then with respect to x; finally, evaluate at $(-2, 5)$:

$$\frac{\partial^2}{\partial x \, \partial y}(2x^4 - 3x^2 y^2 + 5y^3 + 25) = \frac{\partial}{\partial x}(-6x^2 y + 15y^2) = -12xy$$

At the point $(-2, 5)$:

$$\frac{\partial^2}{\partial x \, \partial y}(2x^4 - 3x^2 y^2 + 5y^3 + 25)\bigg|_{(-2,5)} = -12xy \bigg|_{(-2,5)} = -12(-2)(5) = 120 \quad \blacksquare$$

Notice from parts **a** and **b** that

$$\frac{\partial^2 z}{\partial x \, \partial y} = \frac{\partial^2 z}{\partial y \, \partial x}$$

but in general *this is not true*. However, for all of the functions in this book it will be true.

8.2
Problem Set

Let

$$z = f(x, y) = 5x^2 - 3x^3 y^4 + 2y^3 - 15 \quad \text{and}$$
$$w = g(x, y) = (4x - 3y)^5$$

Find the derivatives in Problems 1–20.
1. f_x
2. g_x
3. g_y
4. f_y
5. $f_x(1, 2)$
6. $g_x(2, -1)$
7. $g_y(3, -1)$
8. $f_y(-2, 3)$
9. $\dfrac{\partial^2 z}{\partial x \, \partial y}$
10. $\dfrac{\partial^2 z}{\partial y \, \partial x}$
11. $\dfrac{\partial^2 w}{\partial y \, \partial x}$
12. $\dfrac{\partial^2 w}{\partial x \, \partial y}$
13. $f_{xx}(0, 2)$
14. $f_{xy}(1, 2)$
15. $f_{yx}(-1, 0)$
16. $f_{yy}(2, -1)$
17. $g_{xx}(0, 2)$
18. $g_{xy}(1, 2)$
19. $g_{yx}(-1, 0)$
20. $g_{yy}(2, -1)$

Let

$$z = f(x, y) = e^{3x + 2y} \quad \text{and}$$
$$w = g(x, y) = \sqrt{x^2 - 3y^2}$$

Find the derivatives in Problems 21–40.
21. f_x
22. g_x
23. g_y
24. f_y
25. $f_x(1, 2)$
26. $g_x(-1, 2)$
27. $g_y(3, -2)$
28. $f_y(-2, 3)$

29. $\dfrac{\partial^2 z}{\partial x\, \partial y}$

30. $\dfrac{\partial^2 z}{\partial y\, \partial x}$

31. $\dfrac{\partial^2 w}{\partial y\, \partial x}$

32. $\dfrac{\partial^2 w}{\partial x\, \partial y}$

33. $f_{xx}(0, 2)$

34. $f_{xy}(1, 2)$

35. $f_{yx}(-1, 0)$

36. $f_{yy}(2, -1)$

37. $g_{xx}(2, 0)$

38. $g_{xy}(2, 1)$

39. $g_{yx}(-1, 0)$

40. $g_{yy}(2, -1)$

Find f_x, f_y, and f_λ for the functions in Problems 41–44.

41. $f(x, y, \lambda) = x + 2xy + \lambda(xy - 10)$

42. $f(x, y, \lambda) = 2x + 2y + \lambda xy$

43. $f(x, y, \lambda) = x^2 + y^2 - \lambda(3x + 2y - 6)$

44. $f(x, y, \lambda) = x^2 - y^2 - \lambda(5x - 3y + 10)$

Find $\partial f/\partial b$ and $\partial f/\partial m$ in Problems 45–48.

45. $f(b, m) = (10m + 5b)^2 + (2m + b)$

46. $f(b, m) = (m + b - 4)^2 + (2m + 2b - 8)^2$

47. $f(b, m) = (m + b + 1)^2 + (2m + 2b + 2)^2 + (3m + 3b + 3)^2$

48. $f(b, m) = (2m - b - 3)^3 + (2m - b - 3)^2 + (2m - b - 3)$

APPLICATIONS

49. Hartwell Corporation sells microwave ovens and finds that its profit is a function of the price, p, of the oven as well as the amount spent on advertising, a, according to the function

$$P(p, a) = 2ap + 50p - 10p^2 - .1a^2 p - 100$$

a. $\dfrac{\partial P}{\partial a}$ is the rate of change of profit as a function of the change in advertising spending. Find $\dfrac{\partial P}{\partial a}$.

b. $\dfrac{\partial P}{\partial p}$ is the rate of change of profit as a function of a change in price. Find $\dfrac{\partial P}{\partial p}$.

50. Ritetex is producing x units of one item and y units of another. The cost function is

$$C(x, y) = 3{,}700 + 2{,}500x + 550y$$

Find and interpret $C_x(x, y)$ and $C_y(x, y)$.

51. The revenue function for the items produced by Ritetex in Problem 50 is

$$R(x, y) = 1{,}500x + 900y$$

If P is the profit function, find $P_x(5, 10)$ and $P_y(5, 10)$ and interpret what these numbers mean.

52. The amount of blood flowing in a blood vessel (in milliliters) is given by

$$F(l, r) = .002 l / r^4$$

where l is the length of the blood vessel and r is the radius. (This formula is called *Poiseuille's law*.) Find $\partial F/\partial r$ and $\partial F/\partial l$ and interpret.

53. The number of square inches of a person's body area is a function of the person's height and weight:

$$A(w, h) = 40.5 w^{.425} h^{.725}$$

where A is the surface area in square inches, w is the weight in pounds, and h is the height in feet. Find your own surface area.

54. Find $A_w(180, 6)$ and $A_h(180, 6)$ for the function A of Problem 53 and interpret the results.

55. If \$100 is invested at an annual rate of r for t years, the future value, A, is given by the formula

$$A = 100(1 + r)^t$$

What is the instantaneous rate of change of A per unit change in r if t remains fixed at 5?

56. If \$25 is deposited monthly to an account paying an annual rate of r compounded monthly for 5 years, the future value, A, is given by the formula

$$A = 25 \left[\dfrac{(1 + i)^{60} - 1}{i} \right]$$

where $i = \dfrac{r}{12}$. What is the instantaneous rate of change of A per unit change of i?

57. If P is the present value of an ordinary annuity of equal payments of \$25 per month for 5 years at an annual interest rate of r, then

$$P = 25 \left[\dfrac{1 - (1 + i)^{-60}}{i} \right]$$

where $i = \dfrac{r}{12}$. What is the instantaneous rate of change of P per unit change in r if t remains fixed at 5?

8.3
Maximum–Minimum Applications

We have seen how important optimization is for functions of a single variable. In this section we consider relative maximums and minimums of functions of two variables—that is, of the type $z = f(x, y)$. In this discussion we assume that all second-order partial derivatives exist. Geometrically, the existence of all second-order partial derivatives guarantees that the surface has no tears, ruptures, sharp points, edges, or corners, as shown in Figure 8.7. We will be looking for bulges or dents, which are called **relative maximums** and **minimums**.

Relative Maximum and Relative Minimum

Let a function be defined by $f(x, y)$ for each point in some region of the xy-plane containing the x- and y-axes. Let (a, b) be some point in this region. If there exists a circular region with center at (a, b) such that for all (x, y) in that region,

$f(x, y) \leq f(a, b)$ then $f(a, b)$ is a **relative maximum**

$f(x, y) \geq f(a, b)$ then $f(a, b)$ is a **relative minimum**

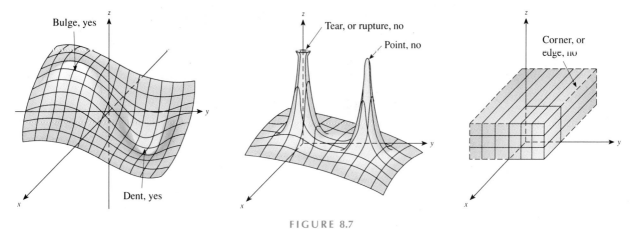

FIGURE 8.7
Surfaces that have or do not have partial derivatives existing at all points

The top of the bulge in Figure 8.7 is a high point which gives a relative maximum for the function and the bottom of the dent is a low point which gives a relative minimum. In this section we are not concerned with boundary points or absolute maximum–minimum theory but, nevertheless, we will be able to consider a great many maximum and minimum problems.

Remember when you did maximums and minimums for a function of a single variable? You first found critical values $x = c$ (values of x where $f'(x) = 0$ or $f'(x)$

does not exist); $f(c)$ was not necessarily a maximum or a minimum but just a possibility. You needed to test further to determine if $f(c)$ was a maximum, minimum, or neither by using the first- or second-derivative tests. We do the same for functions of two variables. We find **critical points** (a, b). These are points for which the partial derivatives f_x and f_y are equal to zero.

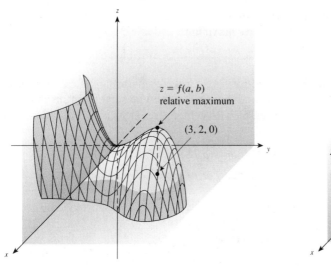

a. Relative maximum at $z = f(a, b)$ at the point $(3, 2, f(3, 2))$

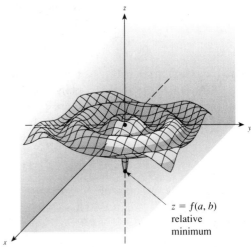

b. Relative minimum $z = f(a, b)$ at $(0, 0, f(0, 0))$

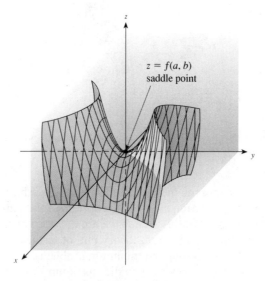

c. Saddle point $(0, 0, 0)$

FIGURE 8.8
Critical points

Critical Points

Let $z = f(x, y)$ have a relative maximum or a relative minimum at the point (a, b). If both f_x and f_y exist at (a, b), then

$$f_x(a, b) = 0 \quad \text{and} \quad f_y(a, b) = 0$$

On the other hand, if (a, b) is a point for which

$$f_x(a, b) = 0 \quad \text{and} \quad f_y(a, b) = 0$$

then (a, b) is called a **critical point** for the function f. The value $f(a, b)$ is *not necessarily* a relative maximum or a minimum, but is a candidate.

If you find a point (a, b) that is a critical point, then there are three possibilities, as shown in Figure 8.8. Notice that a saddle point is a point for which the two first partial derivatives are zero but is not itself a relative maximum or a relative minimum. For example, $z = x^2 - y^2$ has both $f_x(0, 0) = 0$ and $f_y(0, 0) = 0$ and $(0, 0, f(0, 0))$ is a saddle point. A **saddle point** gives neither a maximum nor a minimum value for the function.

EXAMPLE 1 Find critical points for the function

$$f(x, y) = 3x^2 + 4y^2 - 2xy + 22x + 30$$

Solution First find the partial derivatives:

$$f_x(x, y) = 6x - 2y + 22 \quad \text{and} \quad f_y(x, y) = 8y - 2x$$

Next, set these partials equal to zero:

$$6x - 2y + 22 = 0 \quad \text{and} \quad 8y - 2x = 0$$

Finally, solve the system

$$3 \begin{cases} 6x - 2y = -22 & \leftarrow \text{This is } 6x - 2y + 22 = 0. \\ -2x + 8y = 0 & \leftarrow \text{This is } 8y - 2x = 0. \end{cases}$$

$$+ \begin{cases} 6x - 2y = -22 \\ -6x + 24y = 0 \end{cases}$$

$$22y = -22$$

$$y = -1$$

To find x, substitute into either of the equations:

$$8y - 2x = 0 \quad \text{so} \quad 8(-1) - 2x = 0$$

$$x = -4$$

A critical point is thus $(-4, -1)$. This says that if f has any maximums or minimums they must occur at $(x, y) = (-4, -1)$. ∎

EXAMPLE 2 Find the critical points for the surface defined by

$$f(x, y) = 6xy - 4x^3 - 4y^3 - 10$$

Solution $f_x = 6y - 12x^2$ and $f_y = 6x - 12y^2$

Set these equal to zero and solve the system:
$$\begin{cases} 6y - 12x^2 = 0 \\ 6x - 12y^2 = 0 \end{cases}$$

Solve by substitution; from the first equation,
$$6y = 12x^2$$
$$y = 2x^2$$

Substitute into the second equation:
$$6x - 12(2x^2)^2 = 0$$
$$6x - 48x^4 = 0$$
$$x - 8x^4 = 0$$
$$x(1 - 8x^3) = 0$$
$$x(1 - 2x)(1 + 2x + 4x^2) = 0$$

Solve by setting each factor equal to zero:

$x = 0$ $1 - 2x = 0$ $1 + 2x + 4x^2 = 0$

$\qquad\qquad x = \dfrac{1}{2}$ No real solution (the discriminant is negative)

Finally, find y:

$y = 2x^2$ so if $x = 0,$ $y = 0$

$\qquad\qquad$ if $x = \dfrac{1}{2},$ $y = 2\left(\dfrac{1}{2}\right)^2 = \dfrac{1}{2}$

The critical points are $(0, 0)$ and $(\tfrac{1}{2}, \tfrac{1}{2})$. This means that if f has any maximums or minimums they must occur at $(x, y) = (0, 0)$ or $(x, y) = (\tfrac{1}{2}, \tfrac{1}{2})$. ∎

Since there is no guarantee that $f(a, b)$ is a maximum or a minimum at the critical point (a, b), we need a further test, called a **second-derivative test**, to find the relative maximums or minimums.

Second-Derivative Test for Functions of Two Variables

If *all* of the following conditions hold:
1. $z = f(x, y)$
2. All second-order partial derivatives exist in some circular region containing (a, b) as the center.
3. (a, b) is a critical point [that is, $f_x(a, b) = 0$ and $f_y(a, b) = 0$].
4. $A = f_{xx}(a, b)$, $B = f_{xy}(a, b)$, $C = f_{yy}(a, b)$, $D = B^2 - AC$

Also if:

$D < 0$ and $A < 0$, then $f(a, b)$ is a **relative maximum**.
$D < 0$ and $A > 0$, then $f(a, b)$ is a **relative minimum**.
$D > 0$, then f has a **saddle point** at (a, b) and $f(a, b)$ is neither a relative maximum nor a relative minimum.
$D = 0$, then the *test fails*.

EXAMPLE 3 Apply the second-derivative test for the critical values found in Examples 1 and 2.

Solution **a.** For Example 1, $f(x, y) = 3x^2 + 4y^2 - 2xy + 22x + 30$ has critical point $(-4, -1)$. $A = f_{xx} = 6$, $B = f_{xy} = -2$, and $C = f_{yy} = 8$, so $D = B^2 - AC = 4 - 6(8) < 0$. Thus, $D < 0$ and $A > 0$, so $f(-4, -1)$ is a relative minimum. Even though you would not be expected to graph this surface, it is shown in Figure 8.9a.

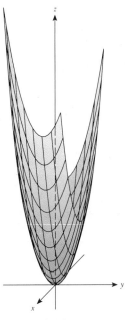
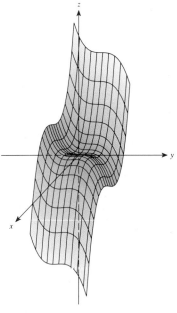

a. Graph of
$f(x, y) = 3x^2 + 4y^2 - 2xy + 22x + 30$

b. Graph of
$f(x, y) = 6xy - 4x^3 - 4y^3 - 10$

FIGURE 8.9
Graphs of surfaces from Examples 1 and 2

b. For Example 2, $f(x, y) = 6xy - 4x^3 - 4y^3 - 10$ has critical points $(0, 0)$ and $(\frac{1}{2}, \frac{1}{2})$. For point $(0, 0)$:

$$f_{xx}(x, y) = -24x \quad \text{so} \quad A = f_{xx}(0, 0) = 0$$
$$f_{xy}(x, y) = 6 \quad \text{so} \quad B = f_{xy}(0, 0) = 6$$
$$f_{yy}(x, y) = -24y \quad \text{so} \quad C = f_{yy}(0, 0) = 0$$
$$D = B^2 - AC = 36 - 0 > 0$$

Thus $D > 0$, so there is a saddle point at $(0, 0)$ and $f(0, 0)$ is neither a relative maximum nor a relative minimum. This means that when $(x, y) = (0, 0)$,

$$z = f(0, 0) = 6(0)(0) - 4(0)^3 - (0)^3 - 10$$
$$= -10$$

Thus the saddle point is $(0, 0, -10)$.

For point $(\frac{1}{2}, \frac{1}{2})$:

$$A = f_{xx}\left(\frac{1}{2}, \frac{1}{2}\right) = -12$$

$$B = f_{xy}\left(\frac{1}{2}, \frac{1}{2}\right) = 6$$

$$C = f_{yy}\left(\frac{1}{2}, \frac{1}{2}\right) = -12$$

$$D = B^2 - AC = 36 - 144 < 0$$

$$f\left(\frac{1}{2}, \frac{1}{2}\right) = 6\left(\frac{1}{2}\right)\left(\frac{1}{2}\right) - 4\left(\frac{1}{2}\right)^3 - 4\left(\frac{1}{2}\right)^3 - 10$$

$$= -\frac{19}{2}$$

Thus $D < 0$ and $A < 0$, so there is a relative maximum of -9.5 at $(\frac{1}{2}, \frac{1}{2})$. Even though you are not expected to graph this surface, it is shown in Figure 8.9b so that you can see the high point and the saddle point. ∎

EXAMPLE 4   A manufacturer makes two models of widgets, standard and deluxe. Market research shows the following profit equations:

Standard = $[44{,}000 + 500(2y - 5x)]x$
Deluxe   = $[500(x - y)]y$

where x is the price of the standard model and y the price of the deluxe model. How should the widgets be priced to maximize the profit?

Solution   Let P = total profit, then

$P(x, y)$ = Profit from standard widget + profit from deluxe widget
$= [44{,}000 + 500(2y - 5x)]x + [500(x - y)]y$
$= 44{,}000x + 1{,}000xy - 2{,}500x^2 + 500xy - 500y^2$
$= 44{,}000x + 1{,}500xy - 2{,}500x^2 - 500y^2$

$P_x(x, y) = 44{,}000 + 1{,}500y - 5{,}000x$ and $P_y(x, y) = 1{,}500x - 1{,}000y$

Set these equal to zero and solve:

$$y = \frac{3}{2}x \quad \text{From } P_y(x, y) = 0$$

Substitute this into the equation for $P_x(x, y) = 0$:

$$44{,}000 + 1{,}500\left(\frac{3}{2}x\right) - 5{,}000x = 0$$

$$-2{,}750x = -44{,}000$$

$$x = 16$$

If $x = 16$, then $y = (\frac{3}{2})16 = 24$; a critical point is thus $(16, 24)$. Finally, apply the second-derivative test:

$$A = P_{xx} = -5{,}000$$
$$B = P_{xy} = 1{,}500$$
$$C = P_{yy} = -1{,}000$$
$$D = B^2 - AC = (1{,}500)^2 - (-5{,}000)(-1{,}000) < 0$$

Thus, $D < 0$ and $A < 0$, so $(16, 24)$ is a relative maximum. Also, since this is the *only* critical value, we conclude this must also be the absolute maximum. This is analogous to the second-derivative test in two dimensions (see Section 4.4). This means that the profit is maximized if the standard model is priced at \$16 and the deluxe model is priced at \$24. ∎

8.3 Problem Set

Find the relative maximums, relative minimums, and saddle points in Problems 1–18. If the second-derivative test fails, simply say so and do not continue with further analysis.

1. $f(x, y) = 3x^2 + 5xy + y^2$
2. $f(x, y) = x^2 + xy - 3x + 2y + 5$
3. $f(x, y) = x^2 + xy + y^2 - 3y$
4. $f(x, y) = x^2 + xy + 2x - 3y + 1$
5. $f(x, y) = x^3 - 3xy - y^3$
6. $f(x, y) = x^3 + y^3 - 3x - 3y$
7. $f(x, y) = 3xy - 5x^2 - y^2 + 3x - 5y - 4$
8. $f(x, y) = 4xy - 6x^2 - y^2 + 2x - 6y - 4$
9. $f(x, y) = xy + 2x - 3y - 4$
10. $f(x, y) = xy - x^2 - 2y^2 + x - y - 5$
11. $f(x, y) = x^2 - y + e^x$
12. $f(x, y) = x^2 + y - e^y$
13. $f(x, y) = xe^y$
14. $f(x, y) = ye^x$
15. $f(x, y) = 4xy - x^4 - y^4$
16. $f(x, y) = xy - x^4 - y^2$
17. $f(x, y) = x^3 + 3xy + y^3$
18. $f(x, y) = x^3 - 3xy + y^3 + 15$

APPLICATIONS

19. Miltex Corporation finds that its profit (in thousands of dollars) is given by the function

$$P(a, n) = -3a^2 - 5n^2 + 34a - 2n + 2an + 40$$

where a is the amount spent on advertising (in thousands of dollars), and n is the number of items (in thousands). Find the maximum value of P and the values of a and n that yield this maximum.

20. The labor cost (per item) for Miltex Corporation is approximated by the function

$$L(x, y) = .5x^2 + 5y^2 - 8x - 26y + 3xy + 55$$

where x is the number of hours of machine time, and y is the number of hours of finishing time. Fine the minimum labor cost and the values of x and y that yield this minimum.

21. A manufacturer makes two models of class rings, standard and deluxe. Market research has shown the following profit equation for x standard rings and y deluxe rings:

$$P(x, y) = 3{,}950x + 150xy - 150x^2 - 50y^2 - 110y$$

How many of each ring should be manufactured in order to maximize the profit?

22. A closed rectangular box with a volume of 8 cubic feet is made from two kinds of material. The top and bottom are made of material costing \$.25 per square foot and the sides require special reinforcing that raises the cost to \$.50 per square foot. What are the cost and dimensions of the box (to the nearest inch) so that the cost of materials is minimized?

23. What are the dimensions of a rectangular box, open at the top, having a volume of 64 cubic feet, and using the least amount of material for its construction?

24. The U.S. Postal Service states that a package cannot have a combined length and girth (distance around) exceeding 120 inches. What are the dimensions of the largest (in volume) box that can be mailed?

25. The second-derivative test does not mention the case where $D < 0$ and $A = 0$. Show that this case is not possible.

8.4

Lagrange Multipliers

In the last section we found the relative maximums and relative minimums of functions of two variables. Sometimes the function to be maximized or minimized has one or more secondary conditions. These conditions are called **constraints**, and the function to be maximized or minimized is called the **objective function**.

The method we will present in this section was first presented by Joseph Lagrange (1736–1813) in a paper he wrote when he was 19, and is called the **method of Lagrange multipliers**.

Method of Lagrange Multipliers

The relative maximums and minimums of the function $z = f(x, y)$ subject to the constraint $g(x, y) = 0$ will be attained at those points (x_0, y_0) which can be found as follows (provided all the partial derivatives exist):

1. Formulate the problem in the form of the objective and constraint functions.
2. Write the equation $F(x, y, \lambda) = f(x, y) + \lambda g(x, y)$. The variable λ is called the *Lagrange multiplier*.
3. Find the partial derivatives F_x, F_y, and F_λ and solve the system

$$\begin{cases} F_x(x, y, \lambda) = 0 \\ F_y(x, y, \lambda) = 0 \\ F_\lambda(x, y, \lambda) = 0 \end{cases}$$

Solutions (x_0, y_0, λ_0) of this system are called **critical points** for F.

4. Evaluate $z = f(x, y)$ at each critical point. The relative maximums and minimums will be among this list of values.

In addition, if the endpoints of the constraint curve (if any) are also included in the list found in step 3, then the largest value yields the maximum of z subject to the constraint $g(x, y) = 0$ and the smallest value yields the minimum of z subject to the constraint $g(x, y) = 0$.

Keep in mind that this is an alternate method to the second-derivative test introduced in the last section. We begin by considering an example.

EXAMPLE 1 A university extension agricultural service concludes that on a particular farm the yield of wheat per acre (measured in bushels) is a function of the number of acre-feet of water, x, applied and the pounds of fertilizer, y, applied during the growing season according to the formula

$$f(x, y) = 140 - x^2 - 2y^2$$

where f is the yield function. Suppose that water costs $20 per acre-foot, fertilizer costs $12 per pound, and the farmer will invest $236 per acre for water and fertilizer. How much water and fertilizer should the farmer buy to maximize the yield?

Solution In this problem the objective function is f and the constraint is
$$20x + 12y = 236$$

(Note that a more realistic constraint would be that the farmer would spend no more than \$236, in which case this constraint is $20x + 12y \leq 236$.)

Step 1: Formulate the problem in the form of objective and constraint functions.

Maximize: $f(x, y) = 140 - x^2 - 2y^2$
Subject to: $g(x, y) = 20x + 12y - 236 = 0$

Step 2: Write the function $F(x, y, \lambda)$ by introducing the Lagrange multiplier λ.

$$\begin{aligned} F(x, y, \lambda) &= f(x, y) + \lambda g(x, y) \\ &= 140 - x^2 - 2y^2 + \lambda(20x + 12y - 236) \\ &= 140 - x^2 - 2y^2 + 20\lambda x + 12\lambda y - 236\lambda \end{aligned}$$

Step 3: Find the partial derivatives F_x, F_y, and F_λ and solve the system $F_x = 0$, $F_y = 0$, $F_\lambda = 0$. Solutions to this system are called *critical points for F*.

$$\begin{aligned} F_x &= -2x + 20\lambda \\ F_y &= -4y + 12\lambda \\ F_\lambda &= 20x + 12y - 236 \end{aligned}$$

These partial derivatives lead to the system of equations

$$\begin{cases} -2x + 20\lambda = 0 \\ -4y + 12\lambda = 0 \\ 20x + 12y - 236 = 0 \end{cases} \quad \text{or} \quad \begin{cases} x - 10\lambda = 0 \\ y - 3\lambda = 0 \\ 5x + 3y = 59 \end{cases}$$

Begin by solving the first pair of equations simultaneously:

$$\begin{array}{r} 3 \\ -10 \end{array} \begin{cases} x - 10\lambda = 0 \\ y - 3\lambda = 0 \end{cases} + \begin{cases} 3x - 30\lambda = 0 \\ \underline{-10y + 30\lambda = 0} \\ 3x - 10y = 0 \end{cases}$$

Use this result along with the third equation of the original system to complete the solution:

$$\begin{array}{r} -5 \\ 3 \end{array} \begin{cases} 3x - 10y = 0 \\ 5x + 3y = 59 \end{cases} + \begin{cases} -15x + 50y = 0 \\ \underline{15x + 9y = 177} \\ 59y = 177 \\ y = 3 \end{cases}$$

If $y = 3$, then $3x - 10(3) = 0$ implies that $x = 10$. Also $y - 3\lambda = 0$. This means that $3 - 3\lambda = 0$ or $\lambda = 1$. This system has only one critical point: $(x, y, \lambda) = (10, 3, 1)$.

Step 4: Evaluate $z = f(x, y)$ at each critical point. The relative maximum or relative minimum values of $f(x, y)$ will be among these values in the problem. In this example there is only one critical point, and since there is no minimum, it follows that this point must provide an absolute maximum. Since $f(x, y) = 140 - x^2 - 2y^2$, we see

$$f(10, 3) = 140 - 10^2 - 2(3)^2 = 22$$

This says that the maximum yield per acre is 22 bushels. ∎

EXAMPLE 2 Minimize $f(x, y) = x^2 + y^2$ subject to $x + y - 1 = 0$.

Solution   *Step 1:* This was done for us in the statement of the problem:

Minimize: $f(x, y) = x^2 + y^2$
Subject to: $g(x, y) = x + y - 1 = 0$

Step 2: $F(x, y, \lambda) = f(x, y) + \lambda g(x, y) = x^2 + y^2 + \lambda x + \lambda y - \lambda$

Step 3: $F_x = 2x + \lambda$
$F_y = 2y + \lambda$
$F_\lambda = x + y - 1$

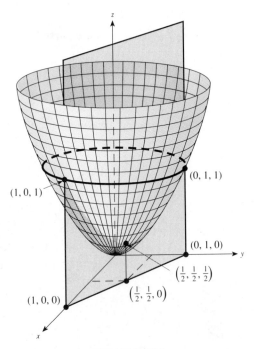

FIGURE 8.10
Graph of $f(x, y) = x^2 + y^2$ showing that the minimum value on this surface satisfying $x + y = 1$ is .5.

Solve the system
$$\begin{cases} 2x + \lambda = 0 \\ 2y + \lambda = 0 \\ x + y - 1 = 0 \end{cases}$$
to find $x = .5$, $y = .5$, $\lambda = -1$, or the critical point: $(.5, .5, -1)$.

Step 4: $\quad f(.5, .5) = .5^2 + .5^2 = .5$

By checking other points near $(.5, .5)$, you can see that this point is a minimum. This situation is illustrated quite nicely in Figure 8.10. ∎

EXAMPLE 3 Find two numbers whose sum is 100 and whose product is a maximum.

Solution *Step 1:* Let x and y be the two numbers. Then:

| Maximize: | $f(x, y) = xy$ | That is, maximize the product. |
| Subject to: | $g(x, y) = x + y - 100 = 0$ | This is the constraint; the sum must be 100. |

Step 2:
$$\begin{aligned} F(x, y, \lambda) &= f(x, y) + \lambda g(x, y) \\ &= xy + \lambda(x + y - 100) \\ &= xy + \lambda x + \lambda y - 100\lambda \end{aligned}$$

Step 3:
$$F_x = y + \lambda$$
$$F_y = x + \lambda$$
$$F_\lambda = x + y - 100$$

By solving the system
$$\begin{cases} y + \lambda = 0 \\ x + \lambda = 0 \\ x + y = 100 \end{cases}$$
simultaneously, we find that $x = 50$, $y = 50$, and $\lambda = -50$.

Step 4: Since $f(x, y) = xy$, we see that $f(50, 50) = 50(50) = 2{,}500$. Nearby values of x and y yield smaller products, so we conclude that the maximum product is 2,500, which occurs for the numbers 50 and 50. ∎

The method of Lagrange multipliers can easily be extended to functions of more than two variables, as illustrated by Example 4.

EXAMPLE 4 Suppose the temperature T at any point (x, y, z) in a region of space is given by the formula $T = 5{,}000 - (xy + xz + yz)$. Find the lowest temperature on the plane $x + y + z = 100$.

Solution *Step 1:* Minimize: $\quad f(x, y, z) = 5{,}000 - xy - xz - yz$
Subject to: $\quad g(x, y, z) = x + y + z - 100 = 0$

Step 2:
$$F(x, y, z, \lambda) = f(x, y, z) + \lambda g(x, y, z)$$
$$= 5{,}000 - xy - xz - yz + \lambda(x + y + z - 100)$$
$$= 5{,}000 - xy - xz - yz + \lambda x + \lambda y + \lambda z - 100\lambda$$

Step 3:
$$F_x = -y - z + \lambda$$
$$F_y = -x - z + \lambda$$
$$F_z = -x - y + \lambda$$
$$F_\lambda = x + y + z - 100$$

Solve the system
$$\begin{cases} y + z - \lambda = 0 \\ x + z - \lambda = 0 \\ x + y - \lambda = 0 \\ x + y + z = 100 \end{cases}$$

From the first two equations: $x - y = 0$
From the first and third equations: $x - z = 0$
From the fourth equation: $z = 100 - x - y$; substitute this result into the equation $x - z = 0$ to find
$$x - (100 - x - y) = 0$$
$$2x + y = 100$$

Finally,
$$\begin{cases} 2x + y = 100 \\ x - y = 0 \end{cases}$$
$$3x = 100$$
$$x = \frac{100}{3}$$

By substitution, $y = \frac{100}{3}$, $z = \frac{100}{3}$, and $\lambda = \frac{200}{3}$.

Step 4: $f(x, y, z) = 5{,}000 - xy - xz - yz$

So
$$f\left(\frac{100}{3}, \frac{100}{3}, \frac{100}{3}\right) = 5{,}000 - \left(\frac{100}{3}\right)\left(\frac{100}{3}\right) - \left(\frac{100}{3}\right)\left(\frac{100}{3}\right) - \left(\frac{100}{3}\right)\left(\frac{100}{3}\right)$$
$$= 5{,}000 - 3\left(\frac{100}{3}\right)^2 = 5{,}000 - \frac{10{,}000}{3}$$
$$= \frac{5{,}000}{3} \approx 1{,}667$$

By checking other nearby points on the plane, you see that this is a minimum. Thus the minimum temperature is about 1,667 degrees. ■

8.4
Problem Set

Find the relative maximums for $f(x, y)$ in Problems 1–6.
1. $f(x, y) = xy$ subject to $x + y = 20$
2. $f(x, y) = 2xy - 5$ subject to $x + y = 12$
3. $f(x, y) = -2x^2 - 3y^2$ subject to $x + 2y = 24$
4. $f(x, y) = 16 - x^2 - y^2$ subject to $x + 2y = 6$
5. $f(x, y) = x^2 y$ subject to $x + 2y = 14$
6. $f(x, y) = 4xy^2$ subject to $x - 4y = 16$

Find the relative minimums for $f(x, y)$ in Problems 7–12.
7. $f(x, y) = x^2 + y^2$ subject to $x + y = 24$
8. $f(x, y) = x^2 + y^2$ subject to $x + y = 8$
9. $f(x, y) = x^2 + y^2 - xy - 4$ subject to $x + y = 6$
10. $f(x, y) = x^2 + y^2$ subject to $x + y = 9$
11. $f(x, y) = x^2 + y^2$ subject to $x + y = 16$
12. $f(x, y) = x^2 + y^2$ subject to $2x + y = 20$
13. Find two numbers whose sum is 10 and whose product is a maximum.
14. Find two numbers whose sum is 120 and whose product is a maximum.

APPLICATIONS

15. How would the farmer of Example 1 maximize the yield if the amount spent is $100 instead of $236?
16. A wholesaler supplies two types of radio-controlled airplanes, models A and B. Suppose x units of model A and y units of model B can be supplied at a cost of

    $$C(x, y) = 6x^2 + 18y^2$$

    If the supplier is limited to shipping no more than 100 models, how much of each item should be supplied in order to minimize the cost? (Assume that the number shipped is exactly 100 models.)

17. A rancher needs to build a rectangular fenced enclosure. Because of a difference in terrain, one width costs $3 per foot and the other costs $9 per foot. The length costs $6 per foot. What is the maximum area that can be enclosed if the rancher has $4,000 to spend?

18. A can of Classic Coke® holds about 25 cubic inches. Find the minimum surface area of a can with a volume of 25 cubic inches. This gives the appropriate minimum amount of material required. Then measure a can of Coke to compare your answer with the size of the actual can.

19. A patient is put on a 1,000 calorie diet. Let us suppose (for purposes of this problem) that these calories will be supplied by two food types, meat and vegetables. Let x be the number of ounces of meat costing $.15 per ounce and y the number of ounces of vegetables costing $.03 per ounce. If the two foods produce

    $$C(x, y) = 20xy$$

    calories, determine what mixture of meat and vegetables should be supplied to minimize the cost.

20. A company buys x items of new equipment and uses y hours of labor at a cost of $C(x, y)$ dollars and a revenue of $R(x, y)$ dollars. The partial derivatives C_x and R_y are the marginal cost and marginal revenue with respect to x, and the partial derivatives C_y and R_y are the marginal cost and marginal revenue with respect to y. Show that if a Lagrange multiplier is used to find the maximum of $R(x, y)$ subject to the constraint $C(x, y) =$ constant, then the Lagrange multiplier is the ratio of the marginal revenue and the marginal cost with respect to each variable at the maximum.

Problems 21–24 involve functions of three variables.

21. Find the maximum value of

    $$f(x, y, z) = x - y + z$$

    on the sphere

    $$x^2 + y^2 + z^2 = 100$$

22. Find the minimum value of

    $$f(x, y, z) = x - y + z$$

    on the sphere

    $$x^2 + y^2 + z^2 = 100$$

23. The temperature T at point (x, y, z) in a region of space is given by the formula

    $$T = 100 - xy - xz - yz$$

    Find the lowest temperature on the plane

    $$x + y + z = 10$$

24. Find the largest product of numbers x, y, and z such that their sum is 24.

*8.5
Multiple Integrals

In this chapter we have defined functions with two or more independent variables, defined the derivatives of such functions, and then looked at some applications of the derivatives. It seems reasonable to now ask how we integrate functions of two or more variables.

Integration was originally motivated by looking at antiderivatives as the reverse of the process of differentiation. With functions of several variables the question becomes one of reversing the process of partial differentiation. We write $\int f(x, y)\, dx$ to indicate that we are to find the antiderivative with respect to x while holding y fixed, and $\int f(x, y)\, dy$ to indicate that we are to antidifferentiate $f(x, y)$ with respect to y, holding x fixed.

EXAMPLE 1 **a.** Evaluate $\int (12x^2 y^3 + 2x - 6y^2)\, dx$.

b. Evaluate $\int (12x^2 y^3 + 2x - 6y^2)\, dy$.

Check your results by using partial derivatives.

Solution **a.** $\int (12x^2 y^3 + 2x - 6y^2)\, dx = \dfrac{12 y^3 x^3}{3} + \dfrac{2x^2}{2} - 6y^2 x + C(y)$

$= 4x^3 y^3 + x^2 - 6xy^2 + C(y)$

Integrate with respect to x; note that the constant of integration is a function of y since the derivative of a function of y with respect to x is 0 (remember to think of y as a constant in this context).

Check by taking the partial derivative with respect to x:

$\dfrac{\partial}{\partial x}[4x^3 y^3 + x^2 - 6xy^2 + C(y)]$

$= 4(3)x^2 y^3 + 2x - 6y^2 + 0$

$= 12x^2 y^3 + 2x - 6y^2$

b. $\int (12x^2 y^3 + 2x - 6y^2)\, dy = \dfrac{12x^2 y^4}{4} + 2xy - \dfrac{6y^3}{3} + C(x)$

$= 3x^2 y^4 + 2xy - 2y^3 + C(x)$

This time integrate with respect to y (and treat x as a constant).

Check: $\dfrac{\partial}{\partial y}[3x^2 y^4 + 2xy - 2y^3 + C(x)] = 12x^2 y^3 + 2x - 6y^2$ ∎

* Optional section.

This indefinite integration of a function of two variables extends quite easily to definite integration, as shown in Example 2.

EXAMPLE 2 Evaluate

$$\int_{-1}^{4} (x^2 - xy + y^2)\,dx = \frac{x^3}{3} - \frac{x^2 y}{2} + xy^2 \bigg|_{x=-1}^{x=4}$$

$$= \left(\frac{64}{3} - 8y + 4y^2\right) - \left[\frac{-1}{3} - \frac{1}{2}y + (-1)y^2\right]$$

$$= \frac{65}{3} - \frac{15}{2}y + 5y^2$$

Notice in Example 2 that integrating and evaluating a definite integral, which is a function $f(x, y)$ with respect to x, produces a function of y alone (including constants). This integral, in turn, can be integrated with respect to y:

$$\int_{1}^{3} \left(\frac{65}{3} - \frac{15}{2}y + 5y^2\right) dy = \frac{65}{3}y - \frac{15}{4}y^2 + \frac{5y^3}{3}\bigg|_{y=1}^{y=3}$$

$$= \left(65 - \frac{135}{4} + 45\right) - \left(\frac{65}{3} - \frac{15}{4} + \frac{5}{3}\right)$$

$$= \frac{170}{3}$$

What we have done here is a process called **double integration** that can be summarized using the notation

$$\iint_R f(x, y)\,dA$$

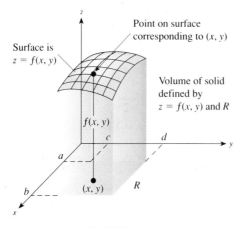

FIGURE 8.11
Geometric interpretation of a double integral

where, in this example, $f(x, y) = x^2 - xy + y^2$, dA indicates that this is an integral over a two-dimensional region, and R is a region on which the function is defined—in this example,

$$-1 \leq x \leq 4$$
$$1 \leq y \leq 3$$

Geometrically, this process can be viewed as the volume of a solid under the surface $f(x, y)$ (where $f(x, y) \geq 0$) and directly over the rectangle R, as shown in Figure 8.11.

$$V = \iint_R (x^2 - xy + y^2)\,dA$$

Double Integral

The **double integral** of a function $f(x, y)$ over a rectangle R defined by the boundaries

$$a \leq x \leq b \quad \text{and} \quad c \leq y \leq d$$

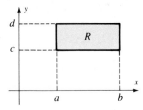

is

$$\iint_R f(x, y)\, dA = \int_a^b \int_c^d f(x, y)\, dy\, dx = \int_c^d \int_a^b f(x, y)\, dx\, dy$$

The function $f(x, y)$ is called the **integrand** and R is called the **region of integration**. The expression dA indicates that it is an integral over a two-dimensional region (i.e., an area). The integrals at the right are called **iterated integrals** and the order in which dx and dy are written indicates the order of integration as follows:

$$\int_c^d \int_a^b f(x, y)\, dx\, dy = \int_c^d \left[\int_a^b f(x, y)\, dx \right] dy \quad \text{Integrate first with respect to } x, \text{ then with respect to } y.$$

$$\int_a^b \int_c^d f(x, y)\, dy\, dx = \int_a^b \left[\int_c^d f(x, y)\, dy \right] dx \quad \text{Integrate first with respect to } y, \text{ then with respect to } x.$$

EXAMPLE 3 Evaluate

a. $\displaystyle\int_0^3 \int_0^1 (12x^2 y^3 - 4xy)\, dy\, dx$ **b.** $\displaystyle\int_0^1 \int_0^3 (12x^2 y^3 - 4xy)\, dx\, dy$

Solution **a.** $\displaystyle\int_0^3 \int_0^1 (12x^2 y^3 - 4xy)\, dy\, dx = \int_0^3 \left(\frac{12x^2 y^4}{4} - \frac{4xy^2}{2} \right) \bigg|_0^1 dx$ Integrate first with respect to y.

$$= \int_0^3 (3x^2 - 2x)\, dx$$

$$= \frac{3x^3}{3} - \frac{2x^2}{2} \bigg|_0^3$$

$$= 27 - 9 = 18$$

b. $\displaystyle\int_0^1 \int_0^3 (12x^2 y^3 - 4xy)\, dx\, dy = \int_0^1 \left(\frac{12x^3 y^3}{3} - \frac{4x^2 y}{2} \right) \bigg|_0^3 dy$ Integrate first with respect to x.

$$= \int_0^1 (108 y^3 - 18y)\, dy$$

$$= \frac{108 y^4}{4} - \frac{18 y^2}{2} \bigg|_0^1$$

$$= 27 - 9 = 18$$

Notice that the integrals in parts **a** and **b** of Example 3 are the same except for the order of integration. The fact that the answers are the same is no coincidence. We will be reversing the order of integration later in this section.

Double Integrals over General Regions

The limits of integration in Examples 1–3 are constant, which corresponds to integration over a rectangular region. We now consider variable limits of integration. There are four types of regions with variable limits:

I. Variable limits of integration for y: $\iint_R f(x, y)\, dA = \int_a^b \int_{g_1(x)}^{g_2(x)} f(x, y)\, dy\, dx$

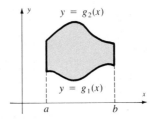

II. Variable limits of integration for x: $\iint_R f(x, y)\, dA = \int_c^d \int_{h_1(y)}^{h_2(y)} f(x, y)\, dx\, dy$

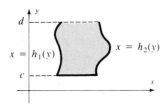

III. Variable limits of integration for either x or y:

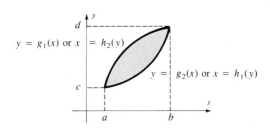

$$\iint_R f(x, y)\, dA = \int_a^b \int_{g_2(x)}^{g_1(x)} f(x, y)\, dy\, dx = \int_c^d \int_{h_2(y)}^{h_1(y)} f(x, y)\, dx\, dy$$

IV. A region that does not have boundaries specified as functions. A region like this, which does not have a boundary that can easily be expressed as a function of x or a function of y, can be broken up into smaller regions that fit one of the previous three types. However, regions such as this are beyond the scope of this book.

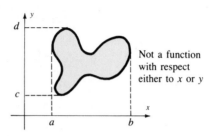

Double Integrals over Variable Regions

Let $z = f(x, y)$ be a function of two variables. If R is a region defined by $g_1(x) \le y \le g_2(x)$, $a \le x \le b$ (type I), then

$$\iint_R f(x, y)\, dA = \int_a^b \int_{g_1(x)}^{g_2(x)} f(x, y)\, dy\, dx$$

If R is a region defined by $h_1(y) \le x \le h_2(y)$, $c \le y \le d$ (type II), then

$$\iint_R f(x, y)\, dA = \int_c^d \int_{h_1(y)}^{h_2(y)} f(x, y)\, dx\, dy$$

WARNING When setting up these double integrals, remember that the variable limits of integration are on the inner integral and the constant limits of integration are on the outer integral.

EXAMPLE 4 Evaluate the double integral of $f(x, y) = 6xy$ over the region shown in the figure.

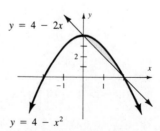

Solution   This is a type I region with variable limits for y, so we can set up the following integral:

$$\iint_R f(x,y)\,dA = \int_0^2 \int_{4-2x}^{4-x^2} 6xy\,dy\,dx$$

$$= \int_0^2 3xy^2 \Big|_{4-2x}^{4-x^2} dx$$

$$= \int_0^2 3x(4-x^2)^2 - 3x(4-2x)^2\,dx$$

$$= \int_0^2 3x(16 - 8x^2 + x^4) - 3x(16 - 16x + 4x^2)\,dx$$

$$= \int_0^2 (3x^5 - 36x^3 + 48x^2)\,dx$$

$$= \left(\frac{3x^6}{6} - \frac{36x^4}{4} + \frac{48x^3}{3}\right)\Big|_0^2$$

$$= 32 - 144 + 128 - 0 = 16 \quad \blacksquare$$

EXAMPLE 5   Integrate $f(x,y) = x - y$ over the region shown in the figure.

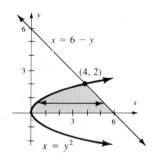

Solution   This is a type II region with variable limits for x:

$$\int_0^2 \int_{y^2}^{6-y} (x-y)\,dx\,dy = \int_0^2 \left(\frac{x^2}{2} - xy\right)\Big|_{y^2}^{6-y} dy$$

$$= \int_0^2 \left[\frac{1}{2}(6-y)^2 - (6-y)y - \frac{1}{2}(y^2)^2 + y^3\right] dy$$

$$= \int_0^2 \left(-\frac{1}{2}y^4 + y^3 + \frac{3}{2}y^2 - 12y + 18\right) dy$$

$$= \left(-\frac{y^5}{10} + \frac{y^4}{4} + \frac{3y^3}{6} - \frac{12y^2}{2} + 18y\right)\Big|_0^2$$

$$= -\frac{32}{10} + \frac{16}{4} + \frac{24}{6} - \frac{48}{2} + 36 - 0$$

$$= 16\frac{4}{5} \quad \blacksquare$$

One of the hardest parts in evaluating a double integral is deciding which variable to integrate with respect to first, and which second. After that, the next step is deciding the limits of integration. These steps are illustrated for the regions we used in Examples 4 and 5.

Problem: Evaluate $\iint_R f(x, y)\, dA$.

Example 4

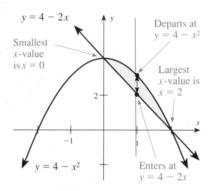

Type I: vertical line; integrate with respect to y first:

$$\int_0^2 \int_{4-2x}^{4-x^2} f(x, y)\, dy\, dx$$

Example 5

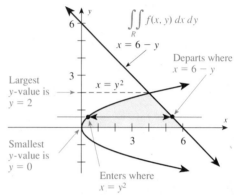

Type II: horizontal line; integrate with respect to x first:

$$\int_0^2 \int_{y^2}^{6-y} f(x, y)\, dx\, dy$$

Type I. Integrate first with respect to y, then with respect to x:

1. Imagine a vertical line L cutting through R in the direction of increasing y.
2. Integrate from the y value where L enters R to the y value where L departs R (these should be functions of x; that is, solve for y).
3. Choose the x limits that include all the vertical lines that pass through R.

Type II. Integrate first with respect to x, then with respect to y:

1. Imagine a horizontal line L cutting through R in the direction of increasing x.
2. Integrate from the x value where L enters R to the x value where L departs R (these should be functions of y; that is, solve for x).
3. Choose the y limits that include all the horizontal lines that pass through R.

EXAMPLE 6   Evaluate $f(x, y) = x + y$ two different ways over the region shown in the figure.

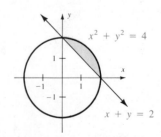

Solution This is a type III region with variable limits.

I. Integrate *first with respect to y*, then *x*; consider a **vertical line**, as shown in the figure at the right:

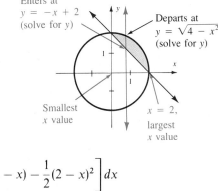

Enters at $y = -x + 2$ (solve for y)

Departs at $y = \sqrt{4-x^2}$ (solve for y)

Smallest x value

$x = 2$, largest x value

$$\int_0^2 \int_{2-x}^{\sqrt{4-x^2}} (x+y)\,dy\,dx$$

$$= \int_0^2 \left(xy + \frac{1}{2}y^2\right)\Big|_{2-x}^{\sqrt{4-x^2}} dx$$

$$= \int_0^2 \left[x\sqrt{4-x^2} + \frac{1}{2}(4-x^2) - x(2-x) - \frac{1}{2}(2-x)^2\right] dx$$

$$= \int_0^2 x\sqrt{4-x^2}\,dx = -\frac{1}{2}\int_4^0 u^{1/2}\,du$$

Let $u = 4 - x^2$ If $x = 0$, then $u = 4$
$du = -2x\,dx$ If $x = 2$, then $u = 0$

$$= -\frac{1}{2}\left(\frac{2}{3}\right)u^{3/2}\Big|_4^0 = \frac{8}{3}$$

II. Integrate *first with respect to x*, then *y*; consider a **horizontal line**, as shown in the figure at the right:

$$\int_0^2 \int_{2-y}^{\sqrt{4-y^2}} (x+y)\,dx\,dy$$

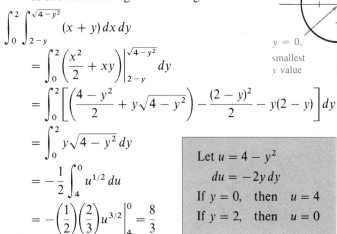

$y = 2$, largest y value

Departs at $x = \sqrt{4-y^2}$ (solve for x)

$y = 0$, smallest y value

Enters at $x = 2 - y$ (solve for x)

$$= \int_0^2 \left(\frac{x^2}{2} + xy\right)\Big|_{2-y}^{\sqrt{4-y^2}} dy$$

$$= \int_0^2 \left[\left(\frac{4-y^2}{2} + y\sqrt{4-y^2}\right) - \frac{(2-y)^2}{2} - y(2-y)\right] dy$$

$$= \int_0^2 y\sqrt{4-y^2}\,dy$$

$$= -\frac{1}{2}\int_4^0 u^{1/2}\,du$$

Let $u = 4 - y^2$
$du = -2y\,dy$
If $y = 0$, then $u = 4$
If $y = 2$, then $u = 0$

$$= -\left(\frac{1}{2}\right)\left(\frac{2}{3}\right)u^{3/2}\Big|_4^0 = \frac{8}{3}$$

Reversing the Order of Integration

As Examples 4–6 show, sometimes we integrate first with respect to *y* (using a vertical line), sometimes with respect to *x* (using a horizontal line), and other times it is possible to do it either way. If the integration can be done either way, we can change the order of integration. Over a rectangular region, it is simply a matter of interchanging the numerical limits of integration, but when interchanging the order of integration with variable limits, we need to look at the region R more carefully, as Example 7 shows.

EXAMPLE 7 Reverse the order of integration for

$$\int_2^{11} \int_{\sqrt{y-2}}^3 f(x, y)\, dx\, dy$$

Solution You can see that this integral is integrated first with respect to x (so a horizontal line should be used). This means that to find R, you write

$$x = 3 \quad \text{and} \quad x = \sqrt{y-2}$$

and graph these curves as shown in the margin.

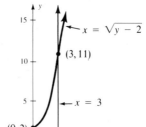

Next, install the limits of integration for y, namely,

$$y = 2 \quad \text{and} \quad y = 11$$

and graph these as shown at the right. Shade R and reverse the order of integration. Then solve each equation for y and draw a vertical line to find the limits of integration.

$$x = \sqrt{y-2}$$
$$x^2 = y - 2$$
$$y = x^2 + 2$$

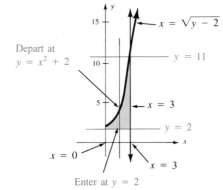

It enters at $y = 2$ and departs at $y = x^2 + 2$; the least value for x is 0 and the largest value for x is 3, so the limits can now be reversed:

$$\int_0^3 \int_2^{x^2+2} f(x, y)\, dy\, dx \qquad \blacksquare$$

Volume

As stated earlier, a geometrical interpretation of double integration is that of a **volume** under a surface. That idea is now formalized:

Volume Let $z = f(x, y)$ be a nonnegative surface over a region R, as illustrated by Figure 8.12. The volume of the solid under the surface of f and over the region R is

$$\iint_R f(x, y)\, dA$$

8.5 MULTIPLE INTEGRALS **361**

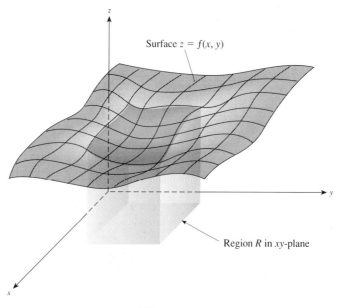

FIGURE 8.12

EXAMPLE 8 Find the volume of the solid bounded on the bottom by the region bounded by the curve with equation $y = x^2$ and the line $2x - y = 0$ and on the top by the plane $z = 10$.

Solution First draw the region R in two dimensions. (The three-dimensional figure is for reinforcement only; you do not need to draw it for your work.)

We will integrate with respect to y first (vertical line).

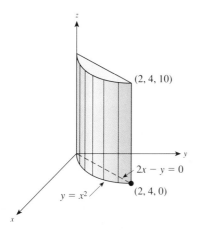

$$\iint_R z \, dA = \int_0^2 \int_{x^2}^{2x} 10 \, dy \, dx$$

$$= \int_0^2 10y \Big|_{x^2}^{2x} dx$$

$$= \int_0^2 10(2x - x^2) \, dx$$

$$= 10x^2 - \frac{10x^3}{3} \Big|_0^2$$

$$= \frac{40}{3}$$

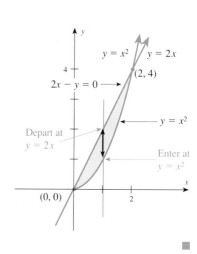

EXAMPLE 9 Find the volume of the function $z = x + y$ over the region R, where $0 \le x \le 2$ and $1 \le y \le 3$.

Solution We can choose either order of iteration. As a check, we will evaluate both ways.

$$\iint_R (x+y)\,dA$$

Method I:

$$\int_1^3 \int_0^2 (x+y)\,dx\,dy$$

$$= \int_1^3 \left(\frac{x^2}{2} + xy\right)\Big|_0^2 dy$$

$$= \int_1^3 (2 + 2y)\,dy$$

$$= 2y + y^2 \Big|_1^3$$

$$= 6 + 9 - 2 - 1 = 12$$

Method II:

$$\int_0^2 \int_1^3 (x+y)\,dy\,dx$$

$$= \int_0^2 \left(xy + \frac{y^2}{2}\right)\Big|_1^3 dx$$

$$= \int_0^2 \left(3x + \frac{9}{2} - x - \frac{1}{2}\right) dx$$

$$= \int_0^2 (2x + 4)\,dx = x^2 + 4x \Big|_0^2$$

$$= 4 + 8 - 0 - 0 = 12 \quad \blacksquare$$

8.5
Problem Set

Evaluate the integrals in Problems 1–12.

1. $\int_0^2 (x^2 y + y^2)\,dy$
2. $\int_0^2 (x^2 y + y^2)\,dx$
3. $\int_1^4 (xy^2 - x)\,dx$
4. $\int_1^4 (xy^2 - x)\,dy$
5. $\int_0^4 x\sqrt{x^2 + 2y}\,dy$
6. $\int_0^4 x\sqrt{x^2 + 2y}\,dx$
7. $\int_1^2 3e^{x+2y}\,dx$
8. $\int_1^2 3e^{x+2y}\,dy$
9. $\int_1^5 xe^{x^2+3y}\,dx$
10. $\int_1^5 xe^{x^2+3y}\,dy$
11. $\int_1^{e^x} \frac{x}{y}\,dy$
12. $\int_1^{e^x} \frac{x}{y}\,dx$

Evaluate the iterated integrals in Problems 13–24, and draw the region R defined by the double integral.

13. $\int_1^4 \int_0^3 dy\,dx$
14. $\int_{-2}^3 \int_0^4 dy\,dx$
15. $\int_{-3}^2 \int_2^4 dx\,dy$
16. $\int_2^5 \int_{-1}^3 dx\,dy$
17. $\int_1^3 \int_0^y (x + 3y)\,dx\,dy$
18. $\int_0^2 \int_{y^2}^{\frac{1}{2}y+1} xy\,dx\,dy$
19. $\int_0^2 \int_0^{y^2} xy\,dx\,dy$
20. $\int_0^1 \int_{y^2}^{(y+3)^{1/2}} 2xy\,dx\,dy$
21. $\int_0^5 \int_0^x (2x + y)\,dy\,dx$
22. $\int_1^3 \int_1^{x^2} 3xy\,dy\,dx$
23. $\int_0^1 \int_{x^3}^{x^2} x^2 y\,dy\,dx$
24. $\int_{-.5}^2 \int_{x^2}^{\frac{3}{2}x+1} 4x\,dy\,dx$

Evaluate the integrals in Problems 25–30 by first reversing the order of integration. (Note that these are the odd-numbered Problems 13–23.)

25. $\int_1^4 \int_0^3 dy\,dx$
26. $\int_{-3}^2 \int_2^4 dx\,dy$
27. $\int_1^3 \int_0^y (x + 3y)\,dx\,dy$
28. $\int_0^2 \int_0^{y^2} xy\,dx\,dy$
29. $\int_0^5 \int_0^x (2x + y)\,dy\,dx$
30. $\int_0^1 \int_{x^3}^{x^2} x^2 y\,dy\,dx$

Evaluate each double integral over the region R in Problems 31–38.

31. $\iint_R (x^2 + y)\,dx\,dy \quad \begin{array}{l} 0 \le x \le 2 \\ 0 \le y \le 3 \end{array}$

32. $\iint_R (x+2y)\,dy\,dx \quad \begin{array}{l}-1 \le x \le 2\\ 1 \le y \le 3\end{array}$

33. $\iint_R \dfrac{dy\,dx}{x} \quad \begin{array}{l}1 \le x \le 3\\ 0 \le y \le 1-x\end{array}$

34. $\iint_R (3-3x^2)\,dy\,dx \quad \begin{array}{l}0 \le x \le 2\\ 0 \le y \le -\tfrac{9}{2}x\end{array}$

35. $\iint_R y^3 e^{xy}\,dx\,dy \quad \begin{array}{l}0 \le x \le y^2\\ 1 \le y \le 2\end{array}$

36. $\iint_R e^{x+y}\,dx\,dy \quad \begin{array}{l}0 \le x \le 2y\\ 0 \le y \le 1\end{array}$

37. $\iint_R xy\sqrt{x^2+y^2}\,dx\,dy \quad \begin{array}{l}1 \le x \le 3\\ 2 \le y \le 4\end{array}$

38. $\iint_R xy\sqrt{x^2+y^2}\,dy\,dx \quad \begin{array}{l}0 \le x \le 4\\ 1 \le y \le 3\end{array}$

Set up the integrals in Problems 39–44 two ways.

39. $\iint_R (x+y)\,dA$, R bounded by $x=y^2$ and $y=x^2$

40. $\iint_R xy\,dA$, R bounded by $y=x^2$ and $y=2x$

41. $\iint_R 2xy\,dA$, R bounded by $x+y=5$, $x=0$, and $y=0$

42. $\iint_R (2x+y)\,dA$, R bounded by $y=x$, $y=2-x$, and the x-axis

43. $\iint_R dA$, R bounded by $y=x^2$, $y=8-x^2$, and the y-axis

44. $\iint_R 3xy^2\,dA$, R bounded by $y=2x$, $y=3-x$, and $y=0$

Find the volume of the solid bounded by the surface $z=f(x,y)$ about the region R as specified in Problems 45–54.

45. $z=5$; $\quad R: 1 \le x \le 5,\ -3 \le y \le 2$
46. $z=x+y$; $\quad R: -1 \le x \le 2,\ 0 \le y \le 4$
47. $z=xy\sqrt{x^2+y^2}$; $\quad R: 0 \le x \le 1,\ 0 \le y \le 1$
48. $z=x^2y$; $\quad R: 0 \le x \le 2,\ 0 \le y \le 4$
49. $z=5xy$; $\quad R$ bounded by $x=y^2$ and $y=x^2$
50. $z=2xy$; $\quad R$ bounded by $y=x^2$ and $y=x$
51. $z=6$; $\quad R$ bounded by $y=x^2$ and $y=3x$
52. $z=5$; $\quad R$ bounded by $x=y^2$ and $x=2y$
53. $z=xy^2$; $\quad R$ bounded by $y=x^2$ and $y=x$
54. $z=\sqrt{\tfrac{x}{y}}$; $\quad R$ bounded by $y=x^2$ and $y=x$

55. The *average value* of the function $z=f(x,y)$ over the rectangle $R: a \le x \le b,\ c \le y \le d$ is
$$\dfrac{1}{(b-a)(d-c)} \iint_R f(x,y)\,dA$$
Find the average value of the surface $z=f(x,y)=x-2y$ over the rectangle $0 \le x \le 2,\ 0 \le y \le 3$.

56. Find the average value of $f(x,y)=4-x-y$ over the rectangle $0 \le x \le 2,\ 0 \le y \le 1$. (See Problem 55 for the definition of average value.)

APPLICATION

57. If an industry invests L thousand labor-hours ($0 \le L \le 5$) and K million dollars ($0 \le K \le 2$) in the production of P thousand units of a commodity, then P is given by
$$P(L,K) = L^{.75}K^{.25}$$
This is called a *Cobb–Douglas production function*.* Find the average number of units produced for the indicated ranges of x and y. (See Problem 55 for a definition of average value.)

---

* You are asked to do some research on this model in the modeling application at the beginning of this chapter.

*8.6
Correlation and Least Squares Applications

We now consider functions of two variables and attempt to establish whether the variables are related to one another or whether they are indeed independent variables. If the variables are related, we say there is a **correlation**. If there is a

---

* Optional section.

correlation, we need to identify the nature of the relation. This is called **regression analysis**. If the relation between the variables is linear, then we find a *best-fitting line* by a technique called the **least squares method** (an application of the second-derivative test).

Our first consideration is one of correlation. We want to know whether two variables are related. Let us call one variable x and the other y. These variables can be represented by graphing (x, y) in a graph called a **scatter diagram**.

EXAMPLE 1 A survey of 20 students compared the grade received on an examination with the length of time the student studied. Draw a scatter diagram to represent the data in the table.

Student	Length of study time (nearest 5 minutes)	Grade (100 possible)
1	30	72
2	40	85
3	30	75
4	35	89
5	45	89
6	15	58
7	15	71
8	50	94
9	30	78
10	0	10
11	20	75
12	10	43
13	25	68
14	25	60
15	25	70
16	30	68
17	40	82
18	35	75
19	20	65
20	15	62

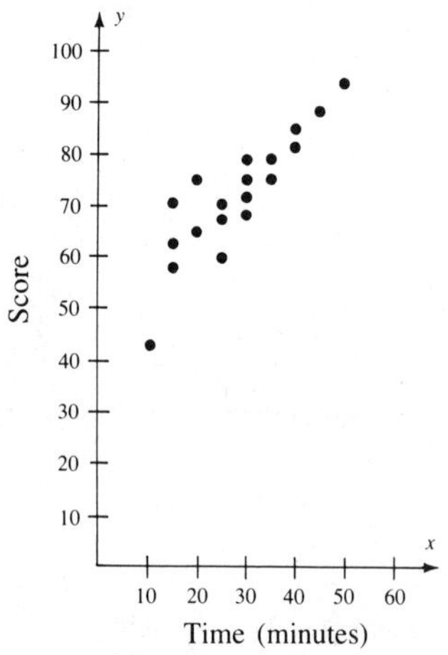

FIGURE 8.13
Scatter diagram

Solution Let x be the study time (in minutes) and let y be the grade (in points). The graph is shown in Figure 8.13. ■

Correlation is a measure of whether there is a relationship between two variables, as Figure 8.14 shows.

The **linear correlation coefficient**, r, which measures the correlation, is defined to have the following properties:

1. r measures the correlation between x and y.
2. r is between -1 and 1.
3. If r is close to 0, there is little correlation.
4. If r is close to 1, there is a strong positive correlation.
5. If r is close to -1, there is a strong negative correlation.

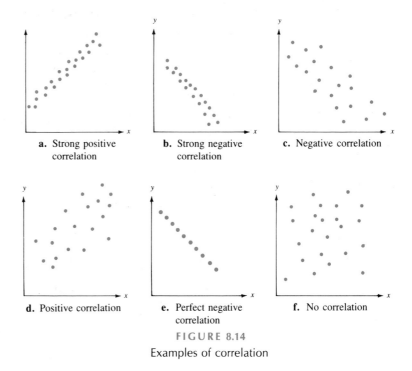

a. Strong positive correlation
b. Strong negative correlation
c. Negative correlation
d. Positive correlation
e. Perfect negative correlation
f. No correlation

FIGURE 8.14
Examples of correlation

To write a formula for r, we let n denote the number of pairs of data present. So we need a notation that means to add a set of values. This notation is called **summation notation**. We will use the following variations of summation notation:

Σx denotes the sum of the x values

Σx^2 means square the x values and then sum

$(\Sigma x)^2$ means sum the x values and then square

Σxy means multiply each x value by the corresponding y value and then sum

$n\Sigma xy$ means multiply n by Σxy

$(\Sigma x)(\Sigma y)$ means multiply Σx by Σy

We can now define the **correlation coefficient**:

Correlation Coefficient

The linear correlation coefficient r is defined as

$$r = \frac{n\Sigma xy - (\Sigma x)(\Sigma y)}{\sqrt{n(\Sigma x^2) - (\Sigma x)^2}\sqrt{n(\Sigma y^2) - (\Sigma y)^2}}$$

EXAMPLE 2   Find the correlation coefficient for the data given in Example 1.

Solution

Study time x	Score y	xy	x^2	y^2
30	72	2,160	900	5,184
40	85	3,400	1,600	7,225
30	75	2,250	900	5,625
35	78	2,730	1,225	6,084
45	89	4,005	2,025	7,921
15	58	870	225	3,364
15	71	1,065	225	5,041
50	94	4,700	2,500	8,836
30	78	2,340	900	6,084
0	10	0	0	100
20	75	1,500	400	5,625
10	43	430	100	1,849
25	68	1,700	625	4,624
25	60	1,500	625	3,600
25	70	1,750	625	4,900
30	68	2,040	900	4,624
40	82	3,280	1,600	6,724
35	75	2,625	1,225	5,625
20	65	1,300	400	4,225
15	62	930	225	3,844
535	1,378	40,575	17,225	101,104
Σx	Σy	Σxy	Σx^2	Σy^2

$n = 20$

TABLE 8.2 Correlation Coefficient r

n	$\alpha = .05$	$\alpha = .01$
4	.950	.999
5	.878	.959
6	.811	.917
7	.754	.875
8	.707	.834
9	.666	.798
10	.632	.765
11	.602	.735
12	.576	.708
13	.553	.684
14	.532	.661
15	.514	.641
16	.497	.623
17	.482	.606
18	.468	.590
19	.456	.575
20	.444	.561
25	.396	.505
30	.361	.463
35	.335	.430
40	.312	.402
45	.294	.378
50	.279	.361
60	.254	.330
70	.236	.305
80	.220	.286
90	.207	.269
100	.196	.256

NOTE: The derivation of this table is beyond the scope of this course. The table shows critical values of the *Pearson correlation coefficient*.

$$r = \frac{n\Sigma xy - (\Sigma x)(\Sigma y)}{\sqrt{n(\Sigma x^2) - (\Sigma x)^2}\sqrt{n(\Sigma y^2) - (\Sigma y)^2}}$$

$$= \frac{20(40,575) - (535)(1,378)}{\sqrt{20(17,225) - (535)^2}\sqrt{20(101,104) - (1,378)^2}} = \frac{74,270}{\sqrt{58,275}\sqrt{123,196}}$$

$$\approx .8765 \quad \blacksquare$$

Example 2 shows a very strong positive correlation. But if r for Example 2 had been .46, would we still be able to assume that there is a strong correlation? This question is a topic of major concern in statistics. The term **significance level** is used to denote the cutoff between results attributed to chance and results attributed to an inherent relationship between the variables. Table 8.2 gives *critical values* for determining whether two variables are correlated. If the magnitude of r is greater than the given table value, then you may assume that a correlation exists between the variables. If you use the column labeled $\alpha = .05$, you find that the significance level is 5%. This means that the probability is .05 that you will say the variables are correlated when, in fact, the results are attributed to chance. This is also true for a significance level of 1% ($\alpha = .01$). For Example 2, since $n = 20$, we see in Table 8.2 that $r = .46$ shows a linear correlation at a 5% significance level, but not at a 1% level.

EXAMPLE 3   Find the critical value of the linear correlation coefficient for 10 pairs of data and a significance level of .05.

Solution   From Table 8.2, for $n = 10$, the critical value is $r = .632$. Thus any value greater than $r = .632$ or less than $r = -.632$ indicates that the two variables are linearly correlated. ∎

EXAMPLE 4   If $r = -.85$ and $n = 10$, are the variables correlated at a significance level of 1%?

Solution   For $n = 10$ and $\alpha = .01$, the Table 8.2 entry is .765. Since r is negative and since $|r| > .765$, we see that there is a negative linear correlation. ∎

EXAMPLE 5   The following table shows a sample of some past annual mean salaries for teachers in elementary and secondary schools, along with the annual per capita beer consumption (in gallons) for Americans. Find the correlation coefficient.*

Year	1960	1965	1970	1972	1973	1983
Mean teacher salary (x dollars)	5,000	6,200	8,600	9,700	10,200	16,400
Per capita beer consumption (y gallons)	24.02	25.46	28.55	29.43	29.68	35.2

Solution $n = 6$;

$\Sigma x = 56,100$ $\Sigma y = 172.34$

$\Sigma x^2 = 604,490,000$ $\Sigma y^2 = 5,026.3418$

$(\Sigma x)^2 = 3,147,210,000$ $(\Sigma y)^2 = 29,701.0756$ $\Sigma xy = 1,688,969$

$$r = \frac{6(1,688,969) - (56,100)(172.34)}{\sqrt{6(604,490,000) - (56,100)^2}\sqrt{6(5,026.3418) - (172.34)^2}} \approx .994$$

The number $r \approx .994$ is so close to 1 that you hardly need to consult Table 8.2 to know that there is a strong positive linear correlation between teachers' salaries and per capita beer drinking. ∎

The correlation obtained in Example 5 certainly implies that teachers are using their raises to buy more beer, right? Wrong. Perhaps the salary increases led to higher taxes, which in turn caused all taxpayers to try to forget their financial difficulties by drinking more beer. Or perhaps higher teachers' salaries and greater beer consumption are both manifestations of some other factor, such as a general improvement in the standard of living. In any event, the techniques in this section can be used only to establish a *statistical* linear relationship. *We cannot establish the existence or absence of any inherent cause-and-effect relationship.*

* Example 5 and the paragraph following are from Mario F. Triola, *Elementary Statistics*, 2nd ed. (Menlo Park, Ca: Benjamin/Cummings). © 1983 by The Benjamin/Cummings Publishing Company, Inc. Reprinted by permission.

The next step in our discussion is to find the equation of the best-fitting line. That is, we want to find a line

$$y = mx + b$$

so that the sum of the distances of the data points from this line will be a minimum (as small as possible). Since some of these distances may be positive and some negative, and since we do not want large opposites to "cancel each other out," we minimize the sum of the *squares* of these distances. (The regression line is thus sometimes called the **least squares line**.) The derivation of the formula for the least squares line is lengthy (and is also optional), so it is presented in the form of an example that you can skip if you wish.

EXAMPLE 6 Find the best-fitting line through the points

$$(x_1, y_1), (x_2, y_2), (x_3, y_3), \ldots, (x_n, y_n)$$

Solution Let $y = mx + b$ be the equation of the line. We want to minimize the sum of the squares of the vertical distances from the points to the line, as shown in Figure 8.15.

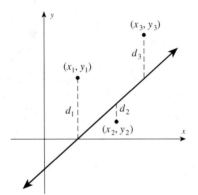

FIGURE 8.15
Best-fitting line

The distance of the first point from the line is (d_1): $y_1 - (mx_1 + b)$
The distance of the second point from the line is (d_2): $y_2 - (mx_2 + b)$
The distance of the third point from the line is (d_3): $y_3 - (mx_3 + b)$
$$\vdots$$

The sum of these distances is denoted by:

$$\Sigma[y - (mx + b)]$$

We wish to minimize the sum of the squares of these distances:

$$f(m, b) = \Sigma[y - (mx + b)]^2$$
$$= (y_1 - mx_1 - b)^2 + (y_2 - mx_2 - b)^2 + \cdots + (y_n - mx_n - b)^2$$

To find the minimum value of this function, we find the partial derivatives with respect to m and b and set each equal to 0:

$$f_m = -2x_1(y_1 - mx_1 - b) - 2x_2(y_2 - mx_2 - b) - \cdots - 2x_n(y_n - mx_n - b) = 0$$
$$f_b = -2(y_1 - mx_1 - b) - 2(y_2 - mx_2 - b) - \cdots - 2(y_n - mx_n - b) = 0$$

Clearing the parentheses on each of the equations and then combining similar terms gives the following system of equations:

$$\begin{cases} (x_1^2 + x_2^2 + \cdots + x_n^2)m + (x_1 + x_2 + \cdots + x_n)b = x_1y_1 + x_2y_2 + \cdots + x_ny_n \\ (x_1 + x_2 + \cdots + x_n)m + nb = y_1 + y_2 + \cdots + y_n \end{cases}$$

In summation notation this system is written as

$$\begin{cases} (\Sigma x^2)m + (\Sigma x)b = \Sigma xy \\ (\Sigma x)m + nb = \Sigma y \end{cases}$$

To solve this system, we multiply the first equation on both sides by $-n$ and the second on both sides by Σx; this leads to an equation we can use for m:

$$m = \frac{n(\Sigma xy) - (\Sigma x)(\Sigma y)}{n(\Sigma x^2) - (\Sigma x)^2}$$

The second equation gives an equation we can solve for b:

$$b = \frac{\Sigma y - m(\Sigma x)}{n}$$

Finally, the second-derivative test can be used to show that this value is a minimum.

Least Squares Line

The **least squares line**, also called the **regression line**, is the line $y = mx + b$ where

$$m = \frac{n(\Sigma xy) - (\Sigma x)(\Sigma y)}{n(\Sigma x^2) - (\Sigma x)^2} \quad \text{and} \quad b = \frac{\Sigma y - m(\Sigma x)}{n}$$

EXAMPLE 7 Find the best-fitting line for the data in Example 1.

Solution Many of the calculations we need for the best-fitting line were done in Example 2. Thus

$$n = 20 \quad \Sigma x = 535 \quad \Sigma y = 1{,}378 \quad \Sigma xy = 40{,}575$$
$$\Sigma x^2 = 17{,}225$$
$$(\Sigma x)^2 = (535)^2 = 286{,}225$$

Since m is part of the formula for b, we must first find m:

$$m = \frac{20(40{,}575) - (535)(1{,}378)}{20(17{,}225) - (535)^2}$$
$$= \frac{74{,}270}{58{,}275}$$
$$= 1.27447$$

We use this result to find b:

$$b = \frac{1{,}378 - 1.27447(535)}{20}$$
$$\approx 34.8078$$

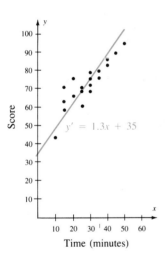

FIGURE 8.16
Regression line for the data in Example 1

Thus we may approximate the least squares line as $y = 1.3x + 35$. This line is shown in Figure 8.16.

EXAMPLE 8 Use the regression line of Example 7 to predict the score of a person who studied $\frac{1}{2}$ hour.

Solution $x = 30$ minutes, so $y = 1.3(30) + 35 = 74$

A final word of caution: Use the regression line only if r indicates that there is a significant linear correlation, as given in Table 8.2.

8.6
Problem Set

In Problems 1–12 a sample of paired data gives a linear correlation coefficient r. In each case use Table 8.2 to determine whether there is a significant linear correlation.

1. $n = 10$, $r = .7$, significance level 5%
2. $n = 10$, $r = .7$, significance level 1%
3. $n = 30$, $r = .4$, significance level 1%
4. $n = 30$, $r = .4$, significance level 5%
5. $n = 15$, $r = -.732$, significance level 5%
6. $n = 35$, $r = -.4127$, significance level 1%
7. $n = 50$, $r = -.3416$, significance level 1%
8. $n = 100$, $r = -.41096$, significance level 5%
9. $n = 23$, $r = .501$, significance level 1%
10. $n = 38$, $r = .416$, significance level 5%
11. $n = 28$, $r = -.214$, significance level 5%
12. $n = 55$, $r = -.14613$, significance level 1%

Draw a scatter diagram and find r for the data in each table in Problems 13–18 and determine whether there is a linear correlation at either the 5% or 1% level.

13.
x	1	2	3	4
y	1	5	8	13

14.
x	4	5	10	10	10
y	0	-10	-10		-20

15.
x	0	1	2	3	4
y	25	19	16	12	10

16.
x	1	3	3	5	8
y	30	22	19	15	10

17.
x	10	20	30	30	50	60
y	50	48	60	58	70	75

18.
x	85	90	100	102	105	110
y	80	40	30	28	25	15

In Problems 19–24 find the regression line for the indicated problem.

19. Problem 13
20. Problem 15
21. Problem 17
22. Problem 14
23. Problem 16
24. Problem 18

APPLICATIONS

25. A new computer circuit was tested and the times (in nanoseconds) required to carry out different subroutines were recorded as follows:

Difficulty Level	1	2	2	3	4	5	5	5
Time	10	11	13	8	15	18	21	19

Find r and determine whether it is statistically significant at the 1% level.

26. Ten people are given a standard IQ test. Their scores were then compared with their high school grades:

IQ	Grade (GPA)
117	3.1
105	2.8
111	2.5
96	2.8
135	3.4
81	1.9
103	2.1
99	3.2
107	2.9
109	2.3

Find r and determine whether it is statistically significant at the 1% level.

27. Find the regression line for the data in Problem 25.
28. Find the regression line for the data in Problem 26.

29. The following data are the number of years of full-time education (x) and the annual salary in thousands of dollars (y) for 15 persons. Is there a correlation between these variables, and if so, is it at the 5% or the 1% significance level?

Education:	20	27	28	18	13	18	9	16
Salary:	35.2	24.6	23.7	33.3	24.4	33.4	11.2	32.3
Education:	16	12	12	19	16	14	13	
Salary:	25.1	22.1	18.9	37.8	25.9	28.4	29.6	

30. The following data are measurements of temperature (°F) and chirping frequency (in chirps per second) for the striped ground cricket. Is there a correlation between these variables, and if so, is it at the 5% or the 1% significance level?

Temperature:	31.4	22.0	34.1	29.1	27.0	24.0	20.9	27.8
Frequency:	20.0	16.0	19.8	18.4	17.1	15.5	14.7	17.1
Temperature:	20.8	28.5	26.4	28.1	27.0	28.6	24.6	
Frequency:	15.4	16.2	15.0	17.2	16.0	17.0	14.4	

31. A bank records the number of mortgage applications and its own prevailing interest rate (at the first of the month) for each of 16 consecutive months. Is there a correlation between these variables, and if so, it is at the 5% or the 1% significance level?

Interest:	9.5	9.9	10.0	10.5	11.0	11.5	11.0	12.0
Number:	27	29	25	25	19	20	17	13
Interest:	12.0	12.5	13.0	13.5	13.0	12.5	11.5	11.5
Number:	15	10	10	6	5	5	11	14

32. A researcher chooses and interviews a group of 15 male workers in an automobile plant. The researcher then gives a score ranging from 1 to 20 based on a scale of patriotism—the higher the score, the more patriotic the person appeared to be. Each person is then given a written test and is scored on their patriotism. Is there a correlation between the researcher score and the test score, and if so, is it at the 5% or the 1% significance level?

Researcher:	10	14	15	17	17	18	18	19
Test:	15	12	19	8	9	16	17	6
Researcher:	16	18	20	12	14	9	17	
Test:	11	14	12	11	10	12	6	

33. Find the best-fitting line for the data in Problem 29.

34. Find the best-fitting line for the data in Problem 30.

35. Find the best-fitting line for the data in Problem 31.

36. Find the best-fitting line for the data in Problem 32.

*8.7
Review

The material of this chapter is reviewed in the following list of objectives. After each objective there are some practice questions. For a sample test select the first question of each set and check your answers. The second question for each objective has no answer given. If you are having trouble with a particular type of problem, look back at the indicated section in the text. When you are finished reviewing these objectives, a sample examination is given at the end of this section.

[8.1]

Objective 8.1: Evaluate functions of several variables.
Evaluate the functions in Problems 1–4 for the point $(5, -12)$.

1. $f(x, y) = 3x^2 - 2xy + y^2$
2. $g(x, y) = e^{x/y}$
3. $K(l, w) = lw$
4. $P(L, K) = L^{.5}K^{.3}$

Objective 8.2: Plot points in three dimensions.

5. **a.** $(5, 0, 0)$ **b.** $(3, -5, 5)$
6. **a.** $(0, 1, 0)$ **b.** $(-1, 2, 1)$
7. **a.** $(0, 0, 10)$ **b.** $(5, 10, -10)$
8. **a.** $(2, 5, 7)$ **b.** $(-2, -5, -4)$

Objective 8.3: Graph surfaces in space.

9. $x^2 + y^2 = 9$
10. $z = x^2 + y^2$
11. $x + y + 2z = 10$
12. $x^2 + y^2 - z^2 = 0$

[8.2]

Objective 8.4: Find partial derivatives.

13. Find f_x for $f(x, y) = (2x - 5y)^{12}$.
14. Find $g_y(1, 3)$ for $g(x, y) = e^{5xy}$.
15. If $z = T(x, y)$ find $\dfrac{\partial z}{\partial y}$ for $T(x, y) = \dfrac{1{,}000}{\sqrt{x^2 + y^2}}$.
16. If $f(x, y, \lambda) = 2xy + \lambda(2x + 3y - 100)$ find f_λ.

* Optional section.

Objective 8.5: Find higher-order partial derivatives.
Let $z = f(x, y) = 3x^5 - 5x^4y^3 + 2x^2 - 150$ for Problems 17–20.

17. $f_{xx}(x, y)$
18. $f_{yx}(1, -1)$
19. $\dfrac{\partial^2 z}{\partial y \, \partial x}$
20. $\dfrac{\partial^2 z}{\partial y^2}$

[8.3]
Objective 8.6: Find the relative maximums, relative minimums, and saddle points of a function of two variables.

21. $g(x, y) = \dfrac{y}{x}$
22. $f(x, y) = xy - 4x + 3y + 120$
23. $f(x, y) = 2x^2 - 3xy + y^2$
24. $g(x, y) = e^{xy}$

[8.4]
Objective 8.7: Find relative maximums or minimums using the method of Lagrange multipliers.

25. Find the relative maximum of $f(x, y) = 12 - x^2 - y^2$ subject to $x + y = 10$.
26. Find the relative minimum of $g(x, y) = x^2 + 4y^2$ subject to $x + 2y = 12$.
27. Find two numbers whose sum is 250 and whose product is a maximum.
28. Find the smallest value for a product of two numbers if their difference must be 10.

[8.5]
Objective 8.8: Evaluate integrals that are functions of two variables.

29. $\displaystyle\int_1^3 x^2 y^3 \, dx$
30. $\displaystyle\int_{-1}^4 5e^{x+2y} \, dy$
31. $\displaystyle\int_0^1 x^2 y^2 \sqrt{x^3 + 8} \, dx$
32. $\displaystyle\int_1^{e^\pi} \dfrac{x^2}{y} \, dy$

Objective 8.9: Evaluate iterated integrals.

33. $\displaystyle\int_0^9 \int_{\sqrt{x}}^{\sqrt{x}+2} y \, dy \, dx$
34. $\displaystyle\int_0^{\ln 3} \int_0^x e^y \, dy \, dx$
35. $\displaystyle\int_1^2 \int_0^{1/y} y^3 \, dx \, dy$
36. $\displaystyle\int_0^3 \int_0^{y\sqrt{9-y^2}} dx \, dy$

Objective 8.10: Evaluate integrals by reversing the order of integration.

37. $\displaystyle\int_1^8 \int_{2/y}^{\sqrt{2y}} dx \, dy$ (correct to the nearest tenth)
38. $\displaystyle\int_0^4 \int_{2y}^8 dx \, dy$
39. $\displaystyle\int_0^2 \int_{x^2}^4 dy \, dx$
40. $\displaystyle\int_1^{e^2} \int_{\ln y}^2 dx \, dy$

Objective 8.11: Find a volume between a given surface and a given region by evaluating a double integral.

41. $\displaystyle\iint_R (x + 2y) \, dA$, R bounded by $y = x^2$ and $y = 5x$
42. $\displaystyle\iint_R y^{-1} \, dy \, dx$, $\begin{array}{l} 1 \le x \le 2 \\ 1 \le y \le e^3 \end{array}$
43. $\displaystyle\iint_R x^2 y^3 \, dx \, dy$, $\begin{array}{l} 0 \le x \le 2 \\ 0 \le y \le 1 \end{array}$
44. $\displaystyle\iint_R xy^2 \, dA$, R bounded by $x = y^2$ and $y = x^2$

[8.6]
Objective 8.12: Draw a scatter diagram for a set of data.

45. A study was conducted to test the relationship between speed (mph) and fuel consumption (mpg). The following information was obtained:

Speed (mph)	20	30	40	50	60	70
Fuel Consumption (mpg)	24	28	29	27	23	19

46. Compare IQ with productivity on an assembly line.

IQ	105	87	110	101	85
Productivity	83	110	81	90	103

IQ	90	92	109	95	97
Productivity	109	111	80	105	103

47. Compare age with blood pressure.

Age	20	25	30	40	50	35	68	55
Blood Pressure	85	91	84	93	100	86	94	92

Objective 8.13: Find the linear correlation coefficient and determine whether the variables are significantly correlated at either the 1% or the 5% level.

48. Use the data in Problem 45.
49. Use the data in Problem 46.
50. Use the data in Problem 47.

Objective 8.14: *Find the regression line for a set of data.*
51. Use the data in Problem 45.
52. Use the data in Problem 46.
53. Use the data in Problem 47.

Objective 8.15: *Solve applied problems based on the preceding objectives.*
54. A company produces two types of skateboards, standard and competition models. The weekly demand and cost equations are given where p is the price (in dollars) of the standard skateboard and q is the price (in dollars) of the competition model:

$$p = 50 - .05x + .001y$$
$$q = 130 + .01x - .04y$$

for x the weekly demand (in hundreds) for standard skateboards and y the weekly demand (in hundreds) for competition skateboards. If the cost function is

$$C(x, y) = 90 + 20x + 90y$$

find $C_x(5, 8)$ and $C_y(5, 8)$ and interpret each of these.

55. Karlin Enterprises employs between 100 and 500 employees and has a capital investment of between 3 and 5 million dollars. A research company has determined that Karlin's productivity (units per employee per day) is approximated by the formula

$$z = P(x, y) = 4xy - 3x^2 - y^3$$

where x is the size of the labor force (in hundreds) and y is the capital investment (in millions of dollars). Find the marginal productivity of labor when $x = 2$ and $y = 4$ and interpret your answer.

56. Find the marginal productivity of capital in Problem 55 when $x = 5$ and $y = 2$ and interpret your answer.

57. Find the maximum productivity for Karlin Enterprises (see Problem 56) in terms of labor force and capital investment.

SAMPLE TEST

The following sample test (45 minutes) is intended to review the main ideas of this chapter.

1. Evaluate $P(L, K) = L^4 K^{-2}$ at the point $(10, 50)$.
2. Plot the following points in three dimensions.
   a. $(0, 3, 0)$ **b.** $(4, -2, 5)$
   c. $(2, 6, 3)$
3. Graph the surface $2x + y + z = 8$.
4. Graph the surface $y^2 + z^2 - x^2 = 0$.

Find the requested derivatives or integrals in Problems 5–10.
5. f_x for $f(x, y) = (x - 3y)^{10}$
6. f_λ for $f(x, y, \lambda) = 4xy + \lambda(2x + 3y - 50)$
7. f_{xx} for $f(x, y) = 4x^4 - 3x^3 y^2 + 2x^2 - 250$
8. $\displaystyle\int_0^1 x^2 y^2 \sqrt{x^3 + 8}\, dx$
9. $\displaystyle\int_0^9 \int_{\sqrt{x}}^{\sqrt{x}+2} dy\, dx$
10. $\displaystyle\int_0^3 \int_0^{y\sqrt{9-y^2}} dx\, dy$
11. Find the relative maximums, relative minimums, and saddle points of $f(x, y) = 2x^2 - 3xy + y^2$.
12. Use the method of Lagrange multipliers to find two numbers whose sum is 250 and whose product is a maximum.
13. Evaluate

$$\int_1^{e^2} \int_{\ln y}^{2} dx\, dy$$

by reversing the order of integration.

14. Find the volume of the region R bounded by $y = x^2$ and $y = 5x$.

$$\iint_R (x + 2y)\, dA$$

15. A company produces two types of skateboards, standard and competition. The weekly demand and cost equations are:

$$p = 50 - .05x + .001y$$
$$q = 130 + .01x - .04y$$

where p is the price of the standard skateboard and q is the price of the competition model for x, the weekly demand (in hundreds) for standard skateboards, and y, the weekly demand (in hundreds) for competition skateboards. If the cost function is

$$C(x, y) = 90 + 20x + 90y$$

find the revenue function, R, and evaluate $C(5, 8)$ and $R(5, 8)$.

9
Differential Equations

CHAPTER OVERVIEW
This concluding chapter discusses differential equations. The ideas here are an extension of those in Chapters 6–7, but here some new techniques and terminology are introduced.

PREVIEW
Differential equations are defined and different types of solutions are discussed. Then the main solution technique, separation of variables, is introduced. This idea is used for some interesting applications in Section 9.2. The chapter concludes with a discussion of some special types of second-order differential equations.

PERSPECTIVE
As you look back over this book, you will see that you have come a long way. Your mathematical confidence should have grown as you progressed through the book, for you should now have a fundamental understanding of mathematical models and the concepts of functions, limits, derivatives, and integrals. You also should be able to apply these concepts in a variety of contexts and in many different applications.

9.1 First-Order Differential Equations
9.2 Applications—Growth Models
9.3 Special Second-Order Differential Equations
9.4 Chapter 9 Review
    Chapter Objectives
    Sample Test

MODELING
APPLICATION 9

The Battle of Trafalgar

The French Revolution, beginning in 1789, paved the way to power for Napoleon Bonaparte, who then overran much of Europe. Britain's security depended on its control of the seas. In 1805, combined French and Spanish fleets were met off Cape Trafalgar by the British fleet under Admiral Nelson. In a prebattle memorandum, Nelson adopted a plan that can be mathematically verified.

After you have finished this chapter, write a paper based on this modeling application to verify Nelson's plan.

This Modeling Application is continued on page 397.

APPLICATIONS

Management (*Business, Economics, Finance, and Investments*)
Domar's capital expansion model (9.1, Problem 30)
Estimating sales from an increasing growth rate (9.2, Problem 1)
Value of a dollar based on inflation rate (9.2, Problem 2)
Finding increased production based on marginal productivity and increased capitalization (9.2, Problem 5)
Cost of producing rototillers given the marginal cost (9.2, Problem 6)
Predicting the GNP (9.2, Problems 7–8)
Sales related to advertising (9.2, Problem 17)
Finding the public debt relative to the growth rate of income (9.3, Problems 29–30)
Sales growth predictions (9.4, Problem 21; 9.4, Test Problem 18)
Growth rate of a stock (9.4, Problem 23; 9.4, Test Problem 20)

Life Sciences (*Biology, Ecology, Health, and Medicine*)
Growth rate of bacteria in milk (9.1, Problem 31)
Carbon-14 dating of artifacts (9.2, Problems 3–4)
Bacterial growth in a culture (9.2, Problems 14–15)
Estimating the population of animals on Catalina Island (9.2, Problem 16)
Contamination of the atmosphere due to the Soviet nuclear disaster at Chernobyl (9.2, Problem 18)

Social Sciences (*Demography, Political science, Population, Psychology, Society, and Sociology*)
Estimating the crime rate (9.1, Problem 29)
Population based on growth rate (9.2, Problem 9)

Social Sciences (*continued*)
Estimating the number of black and Hispanic people between census dates (9.2, Problems 10–11)
Predicting the number of divorces or marriages from divorce and marriage rates (9.2, Problems 12–13)
Hullian model of learning (9.2, Problem 19)
Size of a population with a limited food supply (9.2, Problem 20)
Spread of information about a new mathematical theorem (9.4, Problem 22; 9.4, Test Problem 19)
Inflation rate in Italy (9.4, Problem 24)

Modeling Application—
The Battle of Trafalgar

9.1
First-Order Differential Equations

An equation that involves only x, y (a function of x), and the derivative dy/dx is called a **first-order differential equation**. That is, given a function for three variables, the equation

$$F\left(x, y, \frac{dy}{dx}\right) = 0$$

is a first-order differential equation.*

Differential Equations with No y Terms

The simplest type of differential equation is one in which the coefficient of y is 0 (that is, the y term is missing). These differential equations can be solved by integration. This type of equation introduces the terminology we will be using in this chapter. Next, differential equations with both x and y terms in which the variables can be separated will be considered. Then, in the next section we will consider applications of growth, both uninhibited and inhibited. Finally, in the last section of this chapter we consider second-order differential equations.

The first type of differential equation we consider is solved by antidifferentiation, which is much of what we have been doing since Chapter 6.

Differential Equation Theorem 1

A differential equation of the type

$$\frac{dy}{dx} = f(x)$$

has the solution

$$y = \int f(x)\,dx + C$$

where C is an arbitrary constant.

This result is easily derived by integrating, as shown in Example 1.

EXAMPLE 1 Solve the differential equation $y' = 6x$.

Solution
$\dfrac{dy}{dx} = 6x$ First set up the problem by writing y' as $\dfrac{dy}{dx}$.

$dy = 6x\,dx$ Separate variables (multiply both sides by dx).

$\displaystyle\int dy = \int 6x\,dx$ Integrate both sides.

---

* Even though the notation of Chapter 8 is used here, it is not necessary to have studied that chapter in order to study this one.

$$y + C_1 = \frac{6x^2}{2} + C_2 \qquad \text{Evaluate the integrals.}$$
$$y = 3x^2 + (C_2 - C_1) \qquad \text{Combine constants.}$$
$$= 3x^2 + C$$

The last form shown in Example 1 is called the **general solution**, which combines all the constants into one arbitrary constant. In practice, include just one constant at the end of your work; in fact, your work for Example 1 should look like:

$$\frac{dy}{dx} = 6x$$
$$y = \int 6x\, dx + C$$
$$= 3x^2 + C$$

If we take different values for C, we obtain **particular solutions**. For example,

$$3x^2 \qquad 3x^2 + 5 \qquad \text{and} \qquad 3x^2 - 10$$

are all particular solutions. Thus, if we know the value of a function at a particular point, we can solve for C, as shown in Example 2.

EXAMPLE 2 Solve $y' = 4x - e^{-x} - \sqrt{x}$ if you know that $y = 50$ when $x = 0$.

Solution
$$y = \int (4x - e^{-x} - \sqrt{x})\, dx$$
$$= \frac{4x^2}{2} - \frac{e^{-x}}{-1} - \frac{x^{(3/2)}}{\frac{3}{2}} + C$$
$$= 2x^2 + e^{-x} - \tfrac{2}{3}x\sqrt{x} + C$$

Now, if $x = 0$, then $y = 50$, so use substitution to obtain

$$50 = 2(0)^2 + e^{-0} - \tfrac{2}{3} \cdot 0\sqrt{0} + C$$
$$50 = 1 + C$$
$$49 = C$$

Thus the particular solution is

$$y = 2x^2 + e^{-x} - \frac{2}{3}x\sqrt{x} + 49$$

If the known value for the independent variable is 0, then each of the values $x = 0$ and $y = 50$ is called an **initial condition** or a **boundary condition**.

You can verify that a function is a solution of a differential equation by differentiation. Check the results of Example 2 in this fashion.

EXAMPLE 3 Derive the formula for an initial amount of P dollars invested at an annual rate of r compounded continuously for t years.

Solution Let A be the amount in the account at any time t. Then dA/dt is the rate of growth of A with respect to time t; this means

$$\frac{dA}{dt} = rA \quad \text{with initial conditions } A(0) = P, A > 0, \text{ and } P > 0$$

We want to find a function $A = A(t)$ that satisfies these conditions. Solve the equation for r and integrate both sides with respect to t:

$$\frac{dA}{dt} = rA$$

$$\frac{1}{A}\frac{dA}{dt} = r$$

$$\int \frac{1}{A}\frac{dA}{dt}\,dt = \int r\,dt$$

$$\int A^{-1}\,dA = \int r\,dt$$

$$\ln|A| = rt + C$$

$$A = e^{rt+C}$$

$$= e^C e^{rt}$$

Since $A(0) = P$, we evaluate $A(t) = e^C e^{rt}$ at $t = 0$ to find

$$A(0) = e^C e^{r(0)}$$

$$P = e^C$$

Thus, the desired formula is

$$A = Pe^{rt}$$

Separation of Variables

If the differential equation involves two variables, say x and y, it is sometimes possible to separate those variables into the product of two separate functions, say s and t, as follows:

$$\frac{dy}{dx} = s(x) \cdot t(y)$$

$$\frac{1}{t(y)}\,dy = s(x)\,dx \quad t(y) \neq 0$$

$$[t(y)]^{-1}\,dy = s(x)\,dx$$

$$[t(y)]^{-1}\,dy - s(x)\,dx = 0$$

We say this equation is of the *form*

$$g(y)\,dy + f(x)\,dx = 0$$

If we integrate both sides of this equation, we have the following result.

Differential Equation Theorem 2

If a differential equation can be written in the form
$$g(y)\,dy + f(x)\,dx = 0$$
then the general solution is given by
$$\int g(y)\,dy + \int f(x)\,dx = C$$
where C is an arbitrary constant.

EXAMPLE 4   Find the general solution of
$$\frac{dy}{dx} = x^2 y$$
and check by differentiating.

Solution   Algebraically **separate the variables** by dividing each side by y and multiplying by dx:
$$y^{-1}\,dy = x^2\,dx$$
$$y^{-1}\,dy - x^2\,dx = 0$$
$$\int y^{-1}\,dy - \int x^2\,dx = C$$
$$\ln|y| - \frac{x^3}{3} = C$$
$$\ln|y| = \tfrac{1}{3}x^3 + C$$

Check: Take the derivative of both sides with respect to x:
$$\frac{1}{y}y' - \frac{3x^2}{3} = 0$$
Solve for $y' = dy/dx$:
$$\frac{y'}{y} = x^2$$
$$y' = x^2 y$$
Thus it checks.   ■

Note in Example 4 that we did not solve for y. Sometimes, however, we will want to solve for y. For Example 4,
$$\ln|y| = \tfrac{1}{3}x^2 + C$$
$$|y| = e^{(1/3)x^2 + C}$$
$$= e^{(1/3)x^2} e^C$$
$$y = \pm e^C e^{(1/3)x^2}$$
$$= M e^{(1/3)x^2}$$

where the constant $\pm e^C$ is written as the constant M. You can also check by finding the derivative of this result to confirm that it gives the original differential equation.

EXAMPLE 5 Find the general solution of $dy/dx = ky$, where k is a constant.

Solution   Separate the variables:

$$y^{-1}\,dy - k\,dx = 0$$

$$\int y^{-1}\,dy - \int k\,dx = C$$

$$\ln|y| - kx = C$$

$$\ln|y| = kx + C$$

$$|y| = e^{kx+C} = e^{kx}e^C$$

$$y = Me^{kx}$$

where the constant $\pm e^C$ is written as M. ∎

Examples 4 and 5 are so common that the result of these examples is stated as a theorem:

Differential Equation Theorem 3

If the rate of change of a variable y with respect to x is linear, then we say that $dy/dx = ky$ for some constant k. That is, the rate of change of y is directly proportional to the quantity itself. The solution of this differential equation is

$$y = Me^{kx}$$

EXAMPLE 6 If the divorce rate has been a constant 5% (1975–1985) and if the number of divorces in 1987 was 1,157,000, estimate the number of divorces in 1992.

Solution   Since the rate is a constant, we can write

$$\frac{dy}{dt} = .05y$$

where y is the number of divorces and t is the time. From Theorem 3,

$$y = Me^{.05t}$$

The initial condition is $y = 1{,}157{,}000$ for $t = 0$ (1987):

$$1{,}157{,}000 = Me^{.05(0)}$$

$$1{,}157{,}000 = M$$

Thus

$$y = 1{,}157{,}000e^{.05t}$$

Now, for 1992, $t = 5$, so

$$y = 1{,}157{,}000e^{.05(5)}$$

$$\approx 1{,}486{,}000 \qquad \text{(Calculator: } 1485617.4\text{)} \qquad ∎$$

9.1

Problem Set

Solve the differential equations in Problems 1–20.

1. $\dfrac{dy}{dx} = x^2$
2. $\dfrac{dy}{dx} = 5$
3. $\dfrac{dy}{dx} = 8x - 10$
4. $\dfrac{dy}{dx} = 5 - x$
5. $y' = 4x^3 - 3x^2 - 5$
6. $y' = 10 - 9x^2 + 2x$
7. $5\dfrac{dy}{dx} = e^x$
8. $8\dfrac{dy}{dx} = \sqrt{x}$
9. $12\dfrac{dy}{dx} = \sqrt{5x+1}$
10. $\dfrac{3}{2}\dfrac{dy}{dx} = e^{4x-3}$
11. $yy' = x$
12. $yy' = 5x^2$
13. $y^2\dfrac{dy}{dx} = x^3 - 3$
14. $y\dfrac{dy}{dx} = \sqrt{x} + e^{3x} + 4$
15. $\dfrac{dy}{dx} - 2xy$
16. $\dfrac{dy}{dx} - 4x^2y - 3xy + y$
17. $\dfrac{dP}{dt} = .02P$
18. $\dfrac{dP}{dt} = 1{,}250P$
19. $\dfrac{dN}{dt} = .001N$
20. $\dfrac{dN}{dt} = .15N$

Find a particular solution for the differential equations in Problems 21–28.

21. $\dfrac{dy}{dx} = x^2 y^{-2}$ for $y = 5$ when $x = 0$
22. $x^2 \dfrac{dy}{dx} = y$ for $y = 1$ when $x = 2$
23. $\dfrac{dP}{dt} = .02P$ for $P = e^3$ when $t = 0$
24. $\dfrac{dN}{dt} = .12N$ for $N = 2.3$ million when $t = 0$ (estimate to two places)
25. $x\dfrac{dy}{dx} - y\sqrt{x} = 0$ for $y = 1$ when $x = 1$
26. $5xy - 3y = \dfrac{dy}{dx}$ for $y = e^2$ when $x = 2$
27. $\dfrac{dy}{dx} = \dfrac{xy}{1+x^2}$ for $y = 2$ when $x = -1$
28. $\dfrac{dy}{dx} = \dfrac{1+x^2}{xy}$ for $y = 4$ when $x = 2$

APPLICATIONS

29. You read a newspaper story that says that the crime rate is increasing at a constant rate of 3% per year. If 3,500 major crimes were reported in 1986, estimate the number of crimes in 1996.

30. Domar's capital expansion model is based on the present value of an investment, P, the investment productivity (a constant h), the marginal productivity to consume (a constant k), and time, t:

$$\dfrac{dP}{dt} = hkP$$

Solve this equation for P where the initial investment is $150,000.

31. Suppose that the growth rate of bacteria in milk is a constant 10% per hour and that the maximum number (in millions) permitted is 1,000. How long will the milk be acceptable if we assume that initially there are 12 million bacteria?

9.2
Applications—Growth Models

As you work through the problems below, be aware that we have discussed only the simplest techniques for solving differential equations. You are still not able to solve most differential equations, so the problem set comprises a very select set of differential equations for you to solve.

Earlier in our work we discussed an important application of exponential equations—namely, the formula used as a model for human population growth:

$$P = P_0 e^{rt}$$

where P_0 is the size of the initial population, r is the growth rate, t is the length of time, and P is the size of the population after time t. This model is called an example of **uninhibited** or **unlimited growth**.

Two additional models for growth take into consideration the possibility of a limiting value L that might be due to space limitations, limitations of food, or limitations of other resources. The first of these models assumes that the growth rate decreases as the population size increases, and, in fact, the growth rate approaches zero as P approaches L. This model is an example of **inhibited** or **logistic growth**. The last growth model also assumes a limited population, but has a growth rate that is directly proportional *only* to its remaining possible room for growth. This model is called a **limited growth** model.

In these applications we consider a population of size P and growth rate r. Then dP/dt is the rate of change of the population with respect to time t. We now develop formulas for the following growth models:

Uninhibited growth:

$$\frac{dP}{dt} = rP$$

The rate of growth is directly proportional to the size of the population.

Inhibited growth:

$$\frac{dP}{dt} = rP(L - P)$$

The rate of growth is proportional to both the size of the population and the remaining room for growth.

Limited growth:

$$\frac{dP}{dt} = r(L - P)$$

The rate of growth is proportional only to the remaining possible room for growth.

Uninhibited Growth

We begin by deriving the formula first stated in Chapter 5. For this model we assume a constant growth rate r over a period of time t. That is, the *growth rate is directly proportional to the population itself*. Let $P(t)$ represent the population at time t and P_0 be the initial population (that is, the population at time $t=0$). Then the model is

$$\frac{dP}{dt} = rP$$

since the rate at which P is changing with respect to time is r. In other words, the rate of change of P with respect to time is linear. From Differential Equation Theorem 3, the solution is

$$P = Me^{rt}$$

If $t = 0$, $P = P_0$, so

$$P_0 = Me^{r(0)} = M$$

Thus the uninhibited growth model (constant rate r) is

$$P = P_0 e^{rt}$$

Uninhibited Growth Model

> The model for uninhibited growth (constant rate r) in which the rate of change is directly proportional to the population is
>
> $$\frac{dP}{dt} = rP$$
>
> and has the solution
>
> $$P(t) = P_0 e^{rt}$$
>
> where P_0 is the initial value when $t = 0$.

This model can be used not only for human population growth, but also for interest compounded continuously, growth of a bacteria culture, or (if r is negative) radioactive decay, or, in fact, any model that assumes a linear growth (or decay) rate.

EXAMPLE 1 A new spit-roasted chicken franchise estimates growth at the phenomenal rate of 22.5% per year. If there are 60 outlets at the present time, how many franchises will there be in 10 years?

Solution Our assumptions for this model are a constant growth rate of 22.5%, $P_0 = 10$, and an uninhibited growth model. Thus, for $t = 10$,

$$P = 60e^{.225(10)} \approx 569$$ ∎

EXAMPLE 2 Some bone artifacts found at the Lindenmeier site in northeastern Colorado were tested for their carbon-14 content. If 25% of the original carbon-14 was still present, what is the probable age of the artifacts?

Solution To answer this question, we need to know that carbon-14 has a half-life of 5,750 years. This means that in that length of time 50% of the carbon-14 would remain. That is,

$$P = P_0 e^{rt}$$

$$\frac{P}{P_0} = e^{rt}$$

$$.5 = e^{r(5,750)} \qquad \text{Since 50\% is present in 5,750 years}$$

$$5{,}750r = \ln .5$$

$$r \approx -.0001205473$$

We now use this value for r to date the artifacts:

$$\frac{P}{P_0} = e^{(-.0001205473)t}$$

$$-.0001205473t = \ln\frac{P}{P_0} \quad \text{or } \ln .25 \quad \text{Since 25\% is present}$$

$$t \approx 11{,}500$$

The artifacts are about 11,500 years old. ∎

Inhibited Growth

The model of inhibited growth supposes that there is some limiting value, L, for the population. This assumption can be mathematically stated by saying that the growth rate is proportional to *both* the size of the population, P, and the remaining room for growth, $L - P$. In symbols,

$$\frac{dP}{dt} = rP(L - P)$$

where $\lim_{t \to \infty} P = L$ and $r > 0$.

We need to solve this differential equation by separating the variables:

$$dP = rP(L - P)\,dt$$

$$\frac{dP}{P(L - P)} = r\,dt$$

We integrate both sides:

$$\int \frac{dP}{P(L - P)} = \int r\,dt + C$$

The integral on the right causes no difficulty, but the integral on the left is not easy to integrate in its present form. Consider the following algebraic manipulation, which allows us to rewrite the troublesome integral in a useful form for our purposes in this application:

$$\frac{1}{L}\left[\frac{1}{P} + \frac{1}{L - P}\right] = \frac{1}{L}\frac{L - P + P}{P(L - P)} = \frac{L}{LP(L - P)} = \frac{1}{P(L - P)}$$

We now substitute the form on the left into the equation and complete the integration:

$$\int \frac{1}{L}\left[\frac{1}{P} + \frac{1}{L - P}\right]dP = \int r\,dt + C$$

$$\frac{1}{L}\left[\int \frac{1}{P}\,dP + \int \frac{1}{L - P}\,dP\right] = r\int dt + C$$

$$\frac{1}{L}[\ln P + (-1)\ln(L - P)] = rt + C \qquad \text{Note that } |P| = P \text{ and } |L - P| = L - P \text{ since both } P > 0 \text{ and } (L - P) > 0.$$

$$\ln P - \ln(L - P) = Lrt + LC$$

$$\ln \frac{P}{L - P} = Lrt + LC$$

Now we solve for P:

$$\frac{P}{L-P} = e^{Lrt + LC}$$
$$= e^{Lrt}e^{LC}$$
$$= C_1 e^{Lrt} \qquad \text{Where } C_1 = e^{LC}, \text{ a constant}$$
$$= C e^{Lrt} \qquad \text{Since } C_1 \text{ is any constant,}$$
$$\qquad\qquad\qquad\qquad\quad \text{you can simply write } C.$$
$$P = (L-P)Ce^{Lrt}$$
$$= CLe^{Lrt} - PCe^{Lrt}$$
$$P + PCe^{Lrt} = CLe^{Lrt}$$
$$P(1 + Ce^{Lrt}) = CLe^{Lrt}$$
$$P = \frac{CLe^{Lrt}}{1 + Ce^{Lrt}}$$

Finally, to find the initial values we substitute $P = P_0$ for $t = 0$:

$$P_0 = \frac{CLe^{Lr(0)}}{1 + Ce^{Lr(0)}} = \frac{CL}{1 + C}$$

We solve for C:

$$C = \frac{P_0}{L - P_0}$$

and substitute into the formula for P:

$$P = \frac{\dfrac{P_0}{L - P_0} Le^{Lrt}}{1 + \dfrac{P_0}{L - P_0} e^{Lrt}}$$

$$= \frac{P_0 L e^{Lrt}}{L - P_0 + P_0 e^{Lrt}}$$

We algebraically simplify this result, and so obtain the following model for inhibited growth:

Inhibited Growth Model

The model for inhibited growth (rate r decreasing so that the population P has a limiting value L) in which the rate of change is directly proportional to both the population and the remaining room for growth is

$$\frac{dP}{dt} = rP(L - P)$$

and has the solution

$$P(t) = \frac{P_0 L}{P_0 + (L - P_0)e^{-Lrt}}$$

for an initial population of P_0 when $t = 0$.

EXAMPLE 3 Suppose that a small, fairly isolated town of Ferndale, CA has a population of 1,800 people. Also suppose that a flu epidemic is spreading through the town at a rate of .05% per week with 5 people initially affected. How many people will be affected in 10 weeks, and how long will it take for half the population to be affected?

Solution This is an example of an inhibited growth model where $L = 1{,}800$, $r = .0005$, and $P_0 = 5$:

$$P(t) = \frac{P_0 L}{P_0 + (L - P_0)e^{-Lrt}}$$

$$= \frac{5(1{,}800)}{5 + (1{,}800 - 5)e^{-1{,}800(.0005)t}}$$

$$= \frac{5(1{,}800)}{5 + 1{,}795 e^{-1{,}800(.0005)t}}$$

$$= \frac{1{,}800}{1 + 359 e^{-1{,}800(.0005)t}}$$

For the first part, we find $P(10)$:

$$P(10) = \frac{1{,}800}{1 + 359 e^{-1{,}800(.0005)(10)}}$$

$$\approx 1{,}724 \text{ people affected after 10 weeks}$$

To find the time for half the population to be affected, solve for t:

$$900 = \frac{1{,}800}{1 + 359 e^{-1{,}800(.0005)t}}$$

$$1 + 359 e^{-.9t} = 2$$

$$359 e^{-.9t} = 1$$

$$e^{-.9t} = \frac{1}{359}$$

$$-.9t = \ln \frac{1}{359}$$

$$t = \frac{-10}{9} \ln \frac{1}{359}$$

$$t \approx 6.537$$

Half the population will be affected in about $6\frac{1}{2}$ weeks. ∎

Limited Growth Model

The model for limited growth supposes that there is some limiting value L for the population. This assumption can be mathematically stated by saying that the growth rate is proportional only to the remaining possible room for growth. Symbolically,

$$\frac{dP}{dt} = r(L - P)$$

To solve this differential equation, we separate the variables

$$\frac{dP}{L-P} = r\,dt$$

and integrate both sides:

$$\int \frac{dP}{L-P} = \int r\,dt$$

$$-\ln(L-P) = rt + C_1$$
$$\ln(L-P) = -rt + C$$
$$L - P = e^{-rt+C}$$
$$= e^{-rt} \cdot e^C$$
$$= Me^{-rt} \quad \text{Where } M \text{ is the constant } e^C$$

If $P = 0$ when $t = 0$, then $0 = L - M$, so $L = M$. Thus

$$P = L - Le^{-rt} = L(1 - e^{-rt})$$

Limited Growth Model

> The model for limited growth (with growth rate r) in which the rate of growth is proportional only to the remaining possible room for growth is
>
> $$\frac{dP}{dt} = r(L - P)$$
>
> and has the solution
>
> $$P(t) = L(1 - e^{-rt})$$

EXAMPLE 4 Psychologists have found that in many learning situations a person's rate of learning is rapid at first and then slows down, so that the learning curve "fits" the limited growth model. Graph the learning curve

$$P(t) = 80(1 - e^{-.2t})$$

for learning the touch system of typing, where t is measured in months and P is the number of words typed per minute. ∎

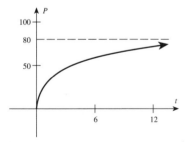

In Example 4 we graphed a limited growth model. For comparison, graphs of the other growth models are shown in Table 9.1 on page 388.

Uninhibited growth is called a J-shaped curve, uninhibited decay an L-shaped curve, and inhibited growth an S-shaped curve. Note that the concavity of the uninhibited growth models does not change, but that of the inhibited growth model does. The point at which the curve changes concavity is the **inflection point**. It can be shown (with a considerable amount of work and a little intuition) that this inflection point is at a value of t for which

$$P(t) = .5L$$

This is the point we found in Example 3.

TABLE 9.1 Summary of Growth Models

	Growth model	Equation and solution	Sample applications
P = Population size; r = Growth rate; t = Time; L = Limiting value of population; M = Size of initial population			
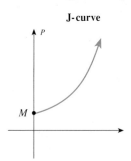 J-curve	**Uninhibited growth** Rate is proportional to the amount present. ($r > 0$)	$\dfrac{dP}{dt} = rP$ Solution: $P = Me^{rt}$	*Exponential growth;* short-term population growth; interest compounded continuously; inflation; price–supply curves
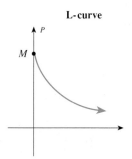 L-curve	**Uninhibited decay** Rate is proportional to the amount present. ($k > 0$)	$\dfrac{dP}{dt} = -kP$ Solution: $P = Me^{-kt}$	Radioactive decay; depletion of natural resources; price–demand curves
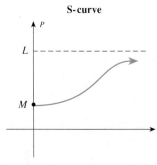 S-curve	**Inhibited growth or Logistic growth** Rate is proportional to the amount present and to the difference between the amount present and a fixed amount. ($r > 0$)	$\dfrac{dP}{dt} = rP(L - P)$ Solution: $P = \dfrac{ML}{M + (L - M)e^{-Lrt}}$	Long-term population growth (with a limiting value); spread of a disease in a population; sales fads (for example, singing flowers); growth of a business
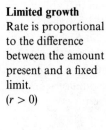 C-curve	**Limited growth** Rate is proportional to the difference between the amount present and a fixed limit. ($r > 0$)	$\dfrac{dP}{dt} = r(L - P)$ Solution: $P = L(1 - e^{-rt})$	Learning curve; diffusion of information by mass media; intravenous infusion of a medication; Newton's law of cooling; depreciation; sales of new products; growth of a business

9.2
Problem Set

APPLICATIONS

1. Sales of a new product are growing at a rate of 12% per year. If 25,000 products are presently being sold, what are the expected sales in 5 years, if the growth rate is directly proportional to the sales?

2. If the average inflation over the last 10 years has been 8% per year, how long will it take for $1 to be worth $.60?

3. Tests of an artifact discovered at the Debert site in Nova Scotia show that 28% of the original carbon-14 is still present. What is the probable age of the artifact?

4. An artifact is found and tested for its carbon-14 content. If 12% of the original carbon-14 is still present, what is the artifact's probable age?

5. The marginal productivity of a business is given by

$$P'(x) = 6x^2 - 10$$

where x is the amount of capitalization in millions of dollars. The operation presently produces 1,000 items per month with a capitalization of $3 million. How much would production increase if the capitalization were increased to $5 million?

6. The marginal cost of producing x rototillers is given by

$$y' = \frac{300}{\sqrt{x}}$$

If $y = 85,000$ when $x = 0$, find the cost of producing 100 rototillers.

7. In 1989 the gross national product (GNP) was $5,200.8 billion. If the growth rate is 7.15%, predict the GNP in 2000.

8. In 1980 the gross national product (GNP) in constant 1972 dollars was $1.481 billion. If the growth rate from 1980 to 1984 was 2.5%, predict the GNP in 1990.

9. Assume that the U.S. population increases at a rate proportional to the population. If the 1990 census recorded the population at 249,632,692, what would you expect the population to be in 2000 if the 1980 population was 226,545,805?

10. If the 1980 census recorded 26,496,000 blacks and if growth is proportional to the population, how many blacks should have been recorded in 1990 if the 1977 population was 25,199,000?

11. If the 1980 census recorded 14,609,000 Hispanics and if growth is proportional to the population, how many Hispanics should have been recorded in 1990 if the 1984 population was 15,575,000?

12. According to the Department of Health and Human Services, the divorce rate in 1990 was .47% and there were 1,177,000 divorces that year. How many divorces will there be in 2000 if the divorce rate is constant?

13. According to the Department of Health and Human Services, the marriage rate in 1990 was .97% and there were 2,423,000 marriages that year. How many marriages will there be in 2000 if the marriage rate is constant?

14. Bacteria usually grow at a rate proportional to the size of the culture. If there are 200 bacteria present initially, and if there are 20,000 present 10 hours later, how many will there be at time $t = 168$ (1 week later)?

15. Suppose a researcher adds a chemical to the bacteria culture described in Problem 14 so that the growth rate is modified according to the equation

$$\frac{dy}{dt} = .7 - .001y$$

where y is the size of the culture at time t (in hours). What is the size of the culture in 1 week?

16. A population of animals on Catalina Island is limited by the amount of food available. If there were 1,800 animals present in 1980 and 2,000 in 1986, and it is estimated that the population cannot exceed 5,000, estimate the population in the year 2000.

17. Market research for a certain product has shown that no matter how much it is advertised the percentage of people who would buy it will never reach or exceed 50% ($L = .5$). If the item is advertised 10 times per week in a certain market, then 20% [$P(10) = .2$] of the population will buy it. How many times should the product be advertised so that 30% of the population will purchase it?

18. In 1986 the Chernobyl, Ukraine, nuclear disaster in the former Soviet Union contaminated the atmosphere. The buildup of radioactive material in the atmosphere satisfies the differential equation

$$\frac{dM}{dt} = r\left(\frac{k}{r} - M\right) \quad M = 0 \quad \text{when} \quad t = 0$$

where M = mass of radioactive material in the atmosphere after time t (in years)
k = rate at which the radioactive material is introduced into the atmosphere
r = annual decay rate of the radioactive material

Find the solution, $M(t)$, of this differential equation in terms of k and r.

19. The Hullian model of learning is stated in terms of the differential equation

$$\frac{dP}{dt} = k(L - P)$$

where P = probability of mastery of a certain concept after t learning trials
$L = 1$, the limiting value of mastery of the learning
k = constant

Find the solution, $P(t)$, of this differential equation in terms of L and k.

20. An isolated population is limited by its food supply to 5,000 individuals. The present population is 1,500, and it was 1,400 5 years ago. Write an equation for the population, P, in terms of time, t. What will the population be in 5 years?

9.3
Special Second-Order Differential Equations

Even though it is beyond the scope of this course to consider the solution of all second-order differential equations, we can look at some of the more common types. An equation that contains only x, y, y', and y'' is called a differential equation of the **second order**. We will list which of the variables x, y, y' or y'' are missing in order to categorize different methods of solution.

Both y and y' Are Missing

If both y and y' are missing, the differential equation has the form

$$y'' = f \qquad \text{where } f \text{ is a function of } x$$

The solution of this type of differential equation can be found by integrating twice. The general solution will contain two constants of integration, and if you know two initial conditions, you can find a particular solution, as illustrated by Example 1.

EXAMPLE 1 Find the particular solution of

$$\frac{d^2y}{dx^2} = 6x + 4$$

if it is known that $y = -7$ if $x = 0$ and $y = 1{,}243$ if $x = 10$.

Solution Integrate both sides with respect to x:

$$y' = \int \frac{d^2y}{dx^2} dx = \int (6x + 4) dx$$

$$\frac{dy}{dx} = \frac{6x^2}{2} + 4x + C_1 \qquad \text{where } C_1 \text{ is some constant}$$

Integrate again with respect to x:

$$\int dy = \int (3x^2 + 4x + C_1)\, dx$$

$$y = \frac{3x^3}{3} + \frac{4x^2}{2} + C_1 x + C_2 \quad \text{where } C_2 \text{ is a constant}$$

$$= x^3 + 2x^2 + C_1 x + C_2$$

This is a general solution; for a particular solution substitute the known values into this equation:

$(0, -7)$: $\quad -7 = 0^3 + 2(0)^2 + C_1(0) + C_2$

$\quad\quad\quad\quad -7 = C_2$

We substitute this value into the equation and then evaluate for the other known point:

$(10, 1243)$: $\quad 1{,}243 = (10)^3 + 2(10)^2 + C_1(10) - 7$

$\quad\quad\quad\quad\quad 1{,}243 = 1{,}000 + 200 + 10C_1 - 7$

$\quad\quad\quad\quad\quad\quad 50 = 10C_1$

$\quad\quad\quad\quad\quad\quad\quad 5 = C_1$

Thus the particular solution is $y = x^3 + 2x^2 + 5x - 7$. To check this solution, differentiate two times. ∎

The Variable y Is Missing

Let $dy/dx = p$ so that $d^2y/dx^2 = dp/dx$, which reduces the equation to a first-order differential equation, as illustrated by Example 2.

EXAMPLE 2 Find the solution of

$$xy'' = y'$$

if it is known that $y = -10$ if $x = 0$ and that $y = 15$ if $x = 5$.

Solution Begin by letting $y' = p$, so that $y'' = dp/dx$:

$xy'' = y'$ $\quad$ Given

$x \dfrac{dp}{dx} = p$ $\quad$ Substitution, $y'' = \dfrac{dp}{dx}$ and $y' = p$

We solve this by using separation of variables (we divide both sides by px and then multiply both sides by dx):

$$\frac{1}{p} dp = \frac{1}{x} dx$$

$$\int p^{-1}\, dp = \int x^{-1}\, dx$$

$$\ln|p| = \ln|x| + k$$

We let $k = \ln C_1$, so that we can write

$$\ln|p| = \ln|x| + \ln C_1$$
$$= \ln|C_1 x|$$

Thus $|p| = |C_1 x|$. Then $p = \pm C_1 x = C_2 x$. But $p = dy/dx$, so that

$$\frac{dy}{dx} = C_2 x$$

$$\int dy = \int C_2 x \, dx$$

$$y = \frac{C_2 x^2}{2} + C_3$$

We find a particular solution by substitution:

$(0, -10)$: $\quad -10 = \dfrac{C_2 (0)^2}{2} + C_3$

$\quad -10 = C_3$

$(5, 15)$: $\quad 15 = \dfrac{C_2 (5)^2}{2} - 10$

$\quad 30 = 25 C_2 - 20$
$\quad 50 = 25 C_2$
$\quad 2 = C_2$

Thus the particular solution is $y = x^2 - 10$, which can be checked by differentiation:

$$y' = 2x \quad \text{and} \quad y'' = 2, \quad \text{so}$$

$xy'' = y'$ yields (by substitution) $x(2) = 2x$, which checks. ∎

The Variable x Is Missing

In this case, we let $p = dy/dx$, so that

$$\frac{d^2 y}{dx^2} = \frac{dp}{dx} = \frac{dp}{dy}\frac{dy}{dx} = \frac{dp}{dy} p$$

By making these substitutions you should be able to form a first-order differential equation that can be solved by integration. Notice in Example 3 that $A, B, C, \ldots$ are used as arbitrary constants. Since we do not have boundary conditions, you must find a general solution.

EXAMPLE 3 Find a general solution of $y'' - y = 0$.

Solution We let $y' = p$, so that $y'' = \dfrac{dy'}{dx} = \dfrac{dp}{dx} = \dfrac{dp}{dy} \cdot \dfrac{dy}{dx} = \dfrac{dp}{dy} p$. By making this substitution, we find

$$\frac{dp}{dy} p - y = 0$$

We separate the variables:

$$p\,dp = y\,dy$$

$$\int p\,dp = \int y\,dy$$

$$\frac{p^2}{2} = \frac{y^2}{2} + A$$

$$p^2 = y^2 + B \qquad \text{Multiply both sides by 2, and let } B = 2A.$$

$$p = \sqrt{y^2 + B}$$

$$\frac{dy}{dx} = \sqrt{y^2 + B}$$

Note: Technically $|p| = \sqrt{y^2 + B}$. In this derivation we assume $|p| = p$. In Problem 31 you are asked to assume $|p| = -p$ and to show that the same result is obtained.

$$(y^2 + B)^{-1/2}\,dy = dx$$

$$\ln|y + \sqrt{y^2 + B}| = x + C \qquad \text{Brief Integral Tables, Formula 16}$$

$$e^{x+C} = y + \sqrt{y^2 + B} \qquad \text{Definition of natural log}$$

$$De^x = y + \sqrt{y^2 + B} \qquad e^{x+C} = e^x e^C = De^x \text{ where } D = e^C$$

$$De^x - y = \sqrt{y^2 + B}$$

$$D^2 e^{2x} - 2yDe^x + y^2 = y^2 + B \qquad \text{Square both sides.}$$

$$2yDe^x = D^2 e^{2x} - B$$

$$y = \frac{D^2 e^{2x}}{2De^x} - \frac{B}{2De^x}$$

$$= \frac{D}{2} e^x - \frac{B}{2D} e^{-x}$$

$$= Ee^x + Fe^{-x} \qquad \text{Let } E = \frac{D}{2} \text{ and } F = \frac{-B}{2D}. \quad \blacksquare$$

Example 3 involves some difficult algebra that becomes more involved if the coefficients are not 1. However, general theorems have been developed to help us solve problems of this type.

Differential Equation Theorem 4

Suppose $ay'' + by' + cy = 0$ is a second-order differential equation with a, b, and c real constants ($a \neq 0$). The general solution is

$$y = Ae^{m_1 x} + Be^{m_2 x}$$

where m_1 and m_2 are solutions of an **auxiliary equation**

$$am^2 + bm + c = 0$$

EXAMPLE 4 Solve the differential equation of Example 3, $y'' - y = 0$, using Theorem 4.

Solution The auxiliary equation is

$$m^2 - 1 = 0 \quad \text{since } a = 1, b = 0, \text{ and } c = -1.$$

$$m^2 = 1$$

so $m = 1, -1$, and the solution to the differential equation is

$$y = Ae^x + Be^{-x}$$

EXAMPLE 5 Solve the differential equation

$$5y'' + 2y = 11y' \quad \text{or} \quad 5y'' - 11y' + 2y = 0$$

Solution Note that $a = 5$, $b = -11$, and $c = 2$; thus the auxiliary equation is

$$5m^2 - 11m + 2 = 0$$
$$(m - 2)(5m - 1) = 0$$
$$m = 2, \tfrac{1}{5}$$

so the general solution to the differential equation is

$$y = Ae^{2x} + Be^{(1/5)x} \quad \text{or} \quad y = Ae^{2x} + Be^{.2x}$$

WARNING Most differential equations are unsolvable by elementary techniques, so the differential equations that you are now able to solve are very limited. Entire courses on differential equations are offered if you want to expand your knowledge and techniques.

9.3

Problem Set

Find the particular solution for the differential equations in Problems 1–12.

1. $\dfrac{d^2y}{dx^2} = -24x$
   $y = 5$ when $x = 0$
   $y = 29$ when $x = 2$

2. $\dfrac{d^2y}{dx^2} = 8x$
   $y = 8$ when $x = 0$
   $y = -4$ when $x = 3$

3. $\dfrac{d^2y}{dx^2} = 10 - 6x$
   $y = 8$ when $x = 0$
   $y = 52$ when $x = 4$

4. $\dfrac{d^2y}{dx^2} = 5 - 9x$
   $y = -10$ when $x = 0$
   $y = 100$ when $x = 6$

5. $\dfrac{d^2y}{dx^2} = 2x^2 - 3x + 5$
   $y = 3$ when $x = 0$
   $y = 12$ when $x = 2$

6. $\dfrac{d^2y}{dx^2} = 6x^2 - 5x - 11$
   $x = 5$ when $x = 0$
   $y = 4$ when $x = 1$

7. $3x - 4 + y'' = 0$
   $y = 4{,}500$ when $x = 0$
   $y = 50$ when $x = 10$

8. $5 - 3x - y'' = 0$
   $y = 100$ when $x = 0$
   $y = 10$ when $x = 10$

9. $xy'' = 2y'$
   $y = -4$ when $x = 0$
   $y = 8$ when $x = 3$

10. $xy'' = 3y'$
    $y = 2$ when $x = 0$
    $y = 20$ when $x = 4$

11. $2y'' - y' = y$
    $y = 0$ when $x = 0$
    $y = 1$ when $x = 4$

12. $3y'' - 11y' = 4y$
    $y = 0$ when $x = 0$
    $y = 2$ when $x = 3$

Find a general solution for the differential equations in Problems 13–28.

13. $\dfrac{d^2y}{dx^2} = e^x$

14. $\dfrac{d^2y}{dx^2} = 12e^x$

15. $\dfrac{d^2y}{dx^2} = 1 - e^{-x}$

16. $\dfrac{d^2y}{dx^2} = x^2 - e^{-x}$

17. $5x^2 - y'' = 2$

18. $y'' + 6x^2 = 1$

19. $x + y'' = 2$

20. $x^{-1}y'' = 12x$

21. $6y'' + 23y' - 4y = 0$

22. $8y'' - 25y' + 3y = 0$

23. $y'' - 9y = 0$

24. $y'' = 16y$

25. $xy'' + 2y' = 0$ 26. $2xy'' - y' = 0$
27. $2y'' + x^3 y'' - 3x^2 y' = 0$ 28. $y'' + x^2 y'' - 2xy' = 0$

APPLICATIONS

29. The public debt, $D(t)$, can be modeled by the differential equation

$$D''(t) = MD(t)$$

where M is the constant relative growth rate of income. Find the general solution of this equation.

30. Suppose that the constant relative growth rate of income is 2.5% and also suppose that $D(0) = 1$ (in billions) and $D(12) = 150$, where D is the public debt function described in Problem 29. Find the particular solution. Round your answer to three decimal places.

31. In Example 3, it was assumed $|p| = p$. Suppose now, $|p| = -p$ and carry out the derivation to show that the same result is obtained.

*9.4

Review

The material of this chapter is reviewed in the following list of objectives. After each objective there are some practice questions. For a sample test select the first question of each set and check your answers. The second question for each objective has no answer given. If you are having trouble with a particular type of problem, look back at the indicated section in the text. When you are finished reviewing these objectives, a sample examination is given at the end of this section.

[9.1]
Objective 9.1: Solve first-order differential equations using Differential Equation Theorem 1.

1. $\dfrac{dy}{dx} = 8x^3 - 2x^2 + 1$ 2. $\dfrac{dy}{dx} = \sqrt{x+5}$
3. $3y' = 2x - 5$ 4. $6y' - 5 = 11x$

Objective 9.2: Solve first-order differential equations using Differential Equation Theorem 2.

5. $4yy' = x + 5$ 6. $yy' = e^{2x+1} - x$
7. $\dfrac{dy}{dx} = 5xy$
8. $\dfrac{dy}{dx} = 3x^3 y^2 - 2xy^2 - y^2$

[9.2]
Objective 9.3: Solve first-order differential equations for a particular solution.

9. $\dfrac{dy}{dx} = x^3 y^{-2}$ for $y = -5$ when $x = 0$
10. $y'x^2 = y$ for $y = 1$ when $x = 2$
11. $2xy = y'$ for $y = e^2$ when $x = 1$
12. $\dfrac{dy}{dx} = \dfrac{xy}{5 - y^2}$ for $y = 1$ when $x = 0$

[9.3]
Objective 9.4: Solve second-order differential equations where both the variables y and y' are missing.

13. $\dfrac{d^2 y}{dx^2} = 15 - 3x^2$ where $y = 3$ if $x = 0$ and $y = 9$ if $x = 2$
14. $y'' = 2x^2 + 5x - 3$ where $y = -1$ if $x = 0$ and $y = 5$ if $x = 1$
15. $2x^3 - y'' = 4$ 16. $x^{-2} y'' = 6x$

Objective 9.5: Solve second-order differential equations where either the variable x or the variable y is missing.

17. $y'' - 16y = 0$ 18. $\dfrac{8 d^2 y}{dx^2} = \dfrac{17 dy}{dx} - 2y$
19. $y'' = y$ where $y = 0$ when $x = 0$ and $y = \dfrac{2e}{1 - e^2}$ when $x = 1$
20. $y'' = y' + 6y$ where $y = 0$ when $x = 0$ and $y = 5$ when $x = 1$

[9.1–9.3]
Objective 9.6: Solve applied problems based on the preceding objectives.

21. Suppose that Melville's Scooter Sales are increasing at a constant rate of 5% per year. If the present sales are 1,250 scooters per year, how many scooters could they expect to sell in 3 years?

22. In 1989 a new mathematical theorem was proved for which a diffusion of information model is

$$\dfrac{dP}{dt} = k(1 - P)$$

P is the percentage of mathematicians who are aware of the new proof after t months.
   a. Find the constant k if it takes 6 months for one-fourth of the mathematicians to hear of the theorem.
   b. How long will it take for 90% of the mathematicians to hear of the new result?

* Optional section.

23. Pertec stock has enjoyed a growth rate according to the model

$$\frac{dV}{dt} = k(100 - V)$$

where V is the value of the stock (per share) after t months.
   a. If the stock's issue price was \$10 and if the value was \$40 after 18 months, find the value of k.
   b. What is the limiting value for this stock?

24. If inflation in Italy over the past 5 years has averaged 20.1%, how long will it take 10,000 lire to be worth what 100 lire is worth today?

SAMPLE TEST

The following sample test (45 minutes) is intended to review the main ideas of this chapter.

Solve the differential equations in Problems 1–17.

1. $\dfrac{dy}{dx} = 12x^3 - 6x^2 + 5$
2. $\dfrac{dy}{dx} = \sqrt{x+3}$
3. $9y' = 14x - 18$
4. $4y' - 3 = 21x$
5. $8yy' = x + 3$
6. $yy' = e^{3x+1} - 2x$
7. $\dfrac{dy}{dx} = 10xy$
8. $\dfrac{dy}{dx} = 3x^3y^2 - 2xy^2 - 5y^2$
9. $\dfrac{dy}{dx} = x^4 y^{-2}$
   for $y = -1$ when $x = 0$
10. $y'x^2 = y$
    for $y = 1$ when $x = 2$
11. $\dfrac{dy}{dx} = \dfrac{xy}{4 - y^2}$
    for $y = 1$ when $x = 1$
12. $\dfrac{d^2y}{dx^2} = 24 - 3x^2$
    where $y = 3$ when $x = 0$
    and $y = 9$ when $x = 2$
13. $x^3 - y'' = 8$
14. $y'' - 20y = 0$
15. $\dfrac{8d^2y}{dx^2} = \dfrac{17dy}{dx} - 2y$
16. $y'' = y$ where $y = 0$ when $x = 0$ and $y = \dfrac{1 - e^2}{2e}$ when $x = 1$
17. $y'' - y' - 6y = 0$ where $y = 0$ when $x = 0$ and $y = 4$ when $x = 1$
18. Suppose that sales of Melville scooters are increasing at a constant rate of 8% per year. If present sales are 1,520 scooters per year, how many scooters would be sold in 5 years?
19. In 1987 a new mathematical theorem was proved. A model for the diffusion of information is

$$\frac{dP}{dt} = k(1 - P)$$

where P is the percentage of mathematicians who are aware of the new proof after t months.
   a. Find the constant k if it takes 6 months for 25% of the mathematicians to hear of the theorem.
   b. How long will it take for 75% of the mathematicians to hear of the theorem?

20. Mastec stock has enjoyed a growth rate according to the model

$$\frac{dV}{dt} = k(100 - V)$$

where V is the value of the stock (per share) after t months.
   a. If the stock's issue price was \$1 and if the value was \$15 after 18 months, find the value of k.
   b. What is the limiting value for this stock if the rate of change is directly proportional to both the value and the remaining room for growth?
   c. What is the value of this stock after 3 years?

MODELING
APPLICATION 9

The Battle of Trafalgar*

The French Revolution, beginning in 1789, paved the way to power for Napoleon Bonaparte, who then overran much of Europe. Britain's security depended on its control of the seas. In 1805 combined French and Spanish fleets were met off Cape Trafalgar by the British fleet under Admiral Nelson. In a prebattle memorandum (shown in the figure), Nelson adopted a plan that can be mathematically verified.

Nelson decided to break the enemy line in two by concentrating 32 of his ships on the 23 ships at the rear of the enemy's line and using his remaining 8 ships to fight a delaying action against the remaining 23 enemy ships. At Trafalgar, Nelson lost his life but won the battle, a victory that gave Britain dominance of the seas in the nineteenth century.

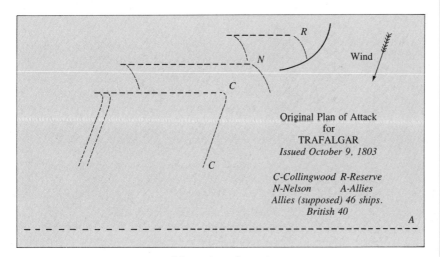

Trafalgar plan of attack

Develop a mathematical model to verify Nelson's plan. You will need to use the N-square law of fighting strength developed by the British engineer Frederick Lanchester. This law states that after an engagement of M units of the red army against N units of the blue army, where $M > N$,

The number of red units left = $\sqrt{M^2 - 23^2}$

The number of blue units left = 0

---

*This model is adapted from David H. Nash, "Differential Equations and the Battle of Trafalgar," *College Mathematics Journal*, March 1985, pp. 98–102. Nelson's attack plan is from A. T. Mahan, *The Life of Nelson*, Vol. II (London: Samson Low, Marston & Company, 1897) p. 345.

CUMULATIVE REVIEW

1. In your own words, discuss the meaning of limit. Use examples and graphs.
2. In your own words, discuss the definition of derivative. Include in your discussion reasons why the derivative is important, as well as some of the principal applications of derivative.
3. In your own words, discuss the definition of integral. Include in your discussion reasons why the integral is important, as well as some of the principal applications of the integral.
4. State the fundamental theorem of integral calculus. Discuss why you think this theorem is fundamental.

Find the derivatives of the functions in Problems 5–7.

5. $y = 5x^{-1} + 3\sqrt{x}$
6. $y = \dfrac{-3}{1 - 2x}$
7. $y = \ln|1 - 11x|$

Evaluate the integrals in Problems 8–10.

8. $\displaystyle\int (3x^2 - 2x)\,dx$
9. $\displaystyle\int_0^2 4x\,dx$
10. $\displaystyle\int_1^3 5x^{-1}\,dx$

11. Evaluate $\displaystyle\int_0^1 \int_{e^x}^{e} \dfrac{1}{\ln y}\,dy\,dx$ by reversing the order of integration.

APPLICATIONS

12. Changes in oxygen pressure (P_{O_2}) have been recorded on the graph below for time (in seconds) on the interval $[0, 20]$.

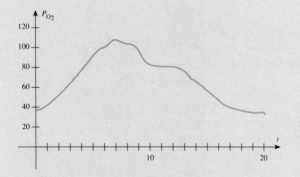

Is P_{O_2} increasing, decreasing, or constant for
a. $t < 3$?
b. $10 \le t \le 13$?
c. $t \ge 15$?

13. The cost of producing x units of a product is $C(x)$, where
$$C(x) = 5x + 8{,}000$$
The product sells for \$130 per unit.
a. What is the break-even point?
b. What revenue will the company receive if it sells just that number of units?

14. A company finds its sales are related to the amount spent on training programs by
$$T(x) = \dfrac{50 + 25x}{x + 5}$$
where T is the sales in hundreds of thousands of dollars when x thousand dollars are spent on training. Find the rate of change of sales when $x = 5$.

15. Suppose the profit in thousands of dollars from an item is $P(x) = x^3 - 13x^2 + 40x$, where x is the price in dollars. Find the maximum possible profit on $[0, 7]$.

16. The rate of sales of a brand of razor blades, in thousands, is given by $S(x) = 3x^2 + 4x$, where x is the time in months that the new product has been on the market. Find the total sales after 10 months.

17. Find the area of the region enclosed by the curves
$$y = 2x - 4 \quad \text{and} \quad y^2 = 4x \quad \text{for } x \ge 1$$
(*Note:* $y^2 = 4x$ can be broken into two functions, $y = 2\sqrt{x}$ and $y = -2\sqrt{x}$.)

18. A manufacturer finds it costs her $x^2 + 5x + 7$ dollars to produce x tons of dulconite. At production levels above 3 tons, she must hire additional workers, and her costs increase by $3x - 9$ dollars on the total production. If the price she receives is \$13 per ton regardless of how much she manufactures, and if her plant capacity is 10 tons, what level of output maximizes profits?

19. Graph $f(x) = \tfrac{1}{4}x^4 - \tfrac{3}{2}x^2$ showing the relative maximums, relative minimums, and points of inflection.

20. The functional life of a timing device selected at random is determined by the probability density function
$$f(t) = .005e^{-.005t}$$
where t is the number of months it has been in operation. What is the probability that the device will operate for more than 5 years ($t = 60$ months)?

398

A Review of Algebra

APPENDIX OVERVIEW

The appendix begins with an algebra pretest. Part I of the pretest reviews general algebraic concepts. If you have difficulty with these questions, take an algebra course before attempting the material in this book. Part II of the pretest reviews more advanced algebraic topics that are discussed in this appendix. If you have difficulty with these questions, review the appropriate section of this appendix.

PERSPECTIVE

The prerequisite for the material in this book is a course in basic algebra. Since it has probably been a while since you studied algebra, this appendix includes a review of the algebraic topics you need to understand the mathematical models presented in this book.

A.1 Algebra Pretests
A.2 Real Numbers
A.3 Algebraic Expressions
A.4 Factoring
A.5 Linear Equations and Inequalities
A.6 Quadratic Equations and Inequalities
A.7 Rational Expressions
A.8 Appendix A Review
    Appendix A Objectives
    Sample Test

A.1
Algebra Pretests

Part I
Basic Algebraic Concepts

Choose the best answer in Problems 1–15.

1. In $5x^2y$, the 5 is
   A. a term
   B. a binomial
   C. an exponent
   D. a literal factor
   E. a numerical coefficient

2. In $6x^2y + 3z$, the 6 and 3 are
   A. terms
   B. exponents
   C. binomials
   D. coefficients
   E. literal factors

3. If $(-2)(-2)(-2) = x$, the value of x is
   A. -8
   B. -6
   C. 6
   D. 8
   E. none of these

4. $21 - (-5)$ equals
   A. -26
   B. -16
   C. 16
   D. 26
   E. none of these

5. If $a = -3$ and $b = 5$, then $a^2 - 2ab + b^2$ equals
   A. -4
   B. 2
   C. 64
   D. -2
   E. none of these

6. The expression $8 + 2 \cdot 3 - 8 \div 2$ equals
   A. 11
   B. 10
   C. 3
   D. -3
   E. none of these

7. $3x - 5x$ equals
   A. $15x^2$
   B. $8x$
   C. $15x$
   D. $-2x$
   E. none of these

8. -5^2 equals
   A. -10
   B. 25
   C. -25
   D. 10
   E. none of these

9. $(-3y^2)^3$ equals
   A. $27y^6$
   B. $-27y^5$
   C. $-3y^6$
   D. $-27y^6$
   E. none of these

10. $\dfrac{a+b}{2}$ means
    A. $a + b \cdot \tfrac{1}{2}$
    B. $a + b \div 2$
    C. $a + (b \div 2)$
    D. $(a + b) \cdot \tfrac{1}{2}$
    E. all of these

11. If $3x + 12 = 6$, x equals
    A. 6
    B. 2
    C. -2
    D. -6
    E. none of these

12. If $6 + 3x = x - 4$, x equals
    A. 5
    B. $2\tfrac{1}{2}$
    C. 1
    D. 2
    E. none of these

13. If $2x - 16 = 3x - 9$, x equals
    A. $-\tfrac{7}{5}$
    B. -7
    C. 5
    D. 25
    E. none of these

14. $(32x^8) \div (-2x)$ equals
    A. $30x^7$
    B. $16x^8$
    C. $34x^8$
    D. $-30x^7$
    E. none of these

15. Simplify $3(x + 2) - (x - 4y) + (x + y)$.
    A. $3x + 11y$
    B. $3x - 3y + 6$
    C. $x + 5y + 6$
    D. $3x + 5y + 6$
    E. none of these

Indicate whether each of the statements in Problems 16–30 is true or false.

16. If $x \neq 0$, then $-x$ always indicates a number that is less than zero.

17. $2(xy) = (2x)(2y)$

18. The domain of a variable is a set from which values of the variable are chosen.

19. x^2 is always positive.

20. $(a + b)^2 = a^2 + b^2$

21. $\dfrac{A + C}{B + C} = \dfrac{A}{B}$

22. $\dfrac{2 + x}{6} = \dfrac{x}{3}$

23. $\dfrac{0}{5}$ is not defined.

24. $\sqrt{-4}$ is a real number.

25. $\dfrac{A}{B} + \dfrac{C}{B} = \dfrac{1}{B}(A + C)$

26. $\dfrac{-x}{-y} = -\dfrac{x}{y}$

27. $(x - y)^2 = (y - x)^2$

28. $\dfrac{x}{y} + \dfrac{y}{x} = 1$

29. $\dfrac{1}{x}y\dfrac{1}{z} = \dfrac{y}{xz}$

30. $\sqrt{x^2 + y^2} = x + y$

A.2
Real Numbers

In mathematics, we need to be clear about the types of numbers under consideration. For example, if someone asked you to "pick a number," you probably wouldn't choose $\sqrt{5}$ or $\pi/2$, but in this course we want to include such choices—the set of numbers called **real numbers**. The various **sets** of numbers include:

Natural numbers: $N = \{1, 2, 3, 4, \ldots\}$
Whole numbers: $W = \{0, 1, 2, 3, 4, \ldots\}$
Integers: $I = \{\ldots, -3, -2, -1, 0, 1, 2, 3, \ldots\}$
Rationals: $Q = \{$all Quotients p/q where p is an integer and q is a nonzero integer$\}$

Rationals are numbers whose decimal representations either terminate or eventually repeat, such as $\frac{1}{3}$, .5, $\frac{1}{8}$, $.\overline{1}$, $\frac{83}{74}$, or $-\frac{147}{44}$, but the set also includes integers such as 2, 6, −19,... since 2 = 2.0, 6 = 6.0, and so on.

Irrationals: $Q' = \{$numbers whose decimal representation does not terminate or repeat$\}$

For example, π, $\sqrt{5}$, $\sqrt[3]{2}$

Real numbers: $R = \{$numbers in Q or $Q'\}$*

A capital letter is usually used to name a set. Thus the set of natural numbers (also called counting numbers) is often referred to by the letter N, the set of integers by I, and the set of rationals by Q (for quotients). One method of designating a set is to enclose the list of its members, or elements, in braces, { }, and to use three dots, if needed, to indicate that the numbers continue in the shown pattern. The set is said to be **finite** if the number of elements in it is a natural number or less than some natural number. A set that is not finite is said to be **infinite**. A set with no members is called the **empty set**, or **null set**, and is labeled { } or ∅.

This course is focused on the set of real numbers, which is easily visualized by using a **coordinate system** called a **number line**, as shown in Figure A.1. A **one-to-one correspondence** exists between all real numbers and all points on such a number line:

1. Every point on the line corresponds to precisely one real number.
2. For each real number, there corresponds one and only one point.

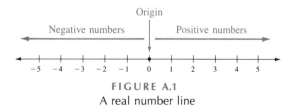

FIGURE A.1
A real number line

A point associated with a particular number is called the **graph** of that number, and the number is called the **coordinate** of the point. Numbers associated with points to the right of the origin are called **positive real numbers**, and those associated with

---
* There is another set of numbers used in mathematics, the set of **complex numbers**. In this book, however, we limit ourselves to the set of real numbers.

Part II
Algebra Pretest

Choose the best answer for each of the following problems.

[A.2]
1. $|2\pi - 7| =$
   A. $2\pi - 7$ B. $7 - 2\pi$
   C. $-.7168$ D. $.7168$
   E. The expression may have more than one value.

2. $6.23\overline{4}$ is an example of
   A. a natural number
   B. a real number
   C. an irrational number
   D. all the answers in parts **A**, **B**, and **C**
   E. none of these

3. The distance between the points (5) and $(\sqrt{10})$ on a real number line is
   A. -1.834 B. 1.834
   C. $5 - \sqrt{10}$ D. $\sqrt{10} - 5$
   E. none of these

[A.3]
4. If $a = 2$ and $b = -3$, then $2a(a + 2b)$ equals
   A. 32 B. 16
   C. -16 D. -48
   E. none of these

5. $(x + y)^2$ equals
   A. $x^2 + y^2$ B. $x^2 + xy + x$
   C. $x^2 + 2xy + y^2$ D. $x^2 y^2$
   E. none of these

6. The result of simplifying $42x - [10 - 2(3x - 4) - 2]$ is
   A. $48x + 16$ B. $48x - 16$
   C. $36x - 20$ D. $36x - 16$
   E. none of these

[A.4]
7. The greatest common factor of $10x^3$, $5x^2$, and $25x$ is
   A. $5x$ B. x
   C. $5x^2$ D. $10x^2$
   E. none of these

8. One of the factors of $x^2 - 18x + 80$ is
   A. $x + 4$ B. $x + 5$
   C. $x - 10$ D. $x - 16$
   E. none of these

9. The complete factorization of $15x^3 - 15xy^2$ is
   A. $x(15x^2 - 15y^2)$ B. $15x(x + y)(x - y)$
   C. $x(15x + y)(x - y)$ D. $3xy(5x^2 - 5xy)$
   E. none of these

[A.5]
10. Solve $\dfrac{(x + 3)(x - 1)}{x - 1} = 4$.

    A. 1 B. -1
    C. 3 D. -3
    E. none of these

11. The solution for $-2x \leq 8$ is
    A. $x \leq -4$
    B. $\{-4, -3, -2, -1, 0, 1, 2, 3, 4, \ldots\}$
    C. $x > -4$
    D. $x \geq -4$
    E. none of these

12. If $3 \leq x + 4 \leq 5$, then
    A. $-1 \geq x$ or $x \geq 1$ B. $-1 \leq x$ or $x \leq 1$
    C. $-1 \leq x \leq 1$ D. $7 \leq x \leq 9$
    E. none of these

[A.6]
13. Solve $x^2 - 7x + 12 = 0$.
    A. $\{-3, -4\}$ B. $\{-3, 4\}$
    C. $\{3, 4\}$ D. $\{2, -6\}$
    E. none of these

14. Solve $(x + 1)(2x - 3) = 25$.
    A. $\{-1, 3\}$ B. $\{4, 8\}$
    C. $\{4, -\frac{7}{2}\}$ D. $\{6, 2\}$
    E. none of these

15. If $(x + 1)(2 - x) < 0$, then
    A. $x < -1$ or $x > 2$ B. $x < -2$ or $x > 1$
    C. $x < -2$ or $x < 1$ D. $-1 < x < 2$
    E. none of these

16. Solve $x^2 - 2x - 2 = 0$.
    A. $-2 \pm \sqrt{3}$ B. $4 \pm \sqrt{3}$
    C. $2 \pm \sqrt{3}$ D. $4 \pm 4\sqrt{3}$
    E. none of these

[A.7]
17. Simplify $\dfrac{x^2 - 5x + 4}{x + 3} \cdot \dfrac{x^2 + 2x - 3}{x - 4}$.
    A. 1 B. $x - 1$
    C. $x^2 - 1$ D. $(x - 1)^2$
    E. none of these

18. Simplify $\dfrac{x}{x^2 - 4} - \dfrac{2}{4 - x^2}$.
    A. $x - 2$ B. $\dfrac{1}{x - 2}$
    C. $\dfrac{1}{2 - x}$ D. $\dfrac{x}{x - 2}$
    E. none of these

points to the left are called **negative real numbers**. Thus a number line is also a convenient way of ordering any two real numbers. If a point whose coordinate is a lies to the right of a point whose coordinate is b on a number line, then **a is greater than b**, written $a > b$. For example, $6 > 3$ and $-4 > -10$. If a point's coordinate b is to the left of a point's coordinate a, then we say that **b is less than a**, or $b < a$. For example, $3 < 6$ and $-10 < -4$. The other symbols of comparison are:

$a = b$ a is equal to b
$a \neq b$ a is not equal to b
$a > b$ a is greater than b
$a \geq b$ a is greater than or equal to b
$a < b$ a is less than b
$a \leq b$ a is less than or equal to b

The symbols $>$, $\geq$, $<$, and $\leq$ are referred to as the **inequality symbols**.

A concept called **absolute value** is also associated with points on a number line. The symbol $|a|$ is read "the absolute value of a" and means the distance between the graph of a and the origin. For example:

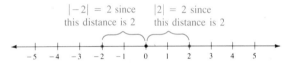

EXAMPLE 1
a. $|5| = 5$
b. $|-5| = 5$
c. $|-\tfrac{1}{2}| = \tfrac{1}{2}$
d. $|\sqrt{5}| = \sqrt{5}$
e. $-|-5| = -5$
f. $-|\pi| = -\pi$

The opposite of the absolute value of -5

The opposite of the absolute value of π ∎

Since we will use the notion of absolute value in a variety of contexts, we need a more formal definition:

Absolute Value

The absolute value of a real number a is defined by

$$|a| = \begin{cases} a & \text{if } a \geq 0 \\ -a & \text{if } a < 0 \end{cases}$$

EXAMPLE 2
a. $|5| = 5$ since $5 \geq 0$
b. $|-5| = -(-5)$ since $-5 < 0$
$ = 5$
c. $-|5| = -5$ since $5 \geq 0$
d. $|-\pi| = -(-\pi)$ since $-\pi < 0$
$ = \pi$
e. $|\pi - 3| = \pi - 3$ since $\pi - 3 \geq 0$
f. $|3 - \pi| = -(3 - \pi)$ since $3 - \pi < 0$
$ = -3 + \pi$
$ = \pi - 3$

g. $|\sqrt{30} - 5| = \sqrt{30} - 5$ since $\sqrt{30} - 5 \geq 0$

h. $|\sqrt{30} - 6| = -(\sqrt{30} - 6)$ since $\sqrt{30} - 6 < 0$
$= 6 - \sqrt{30}$

In algebra, it is necessary to assume certain properties of equality and inequality. The first of these is called the **trichotomy property**, or **property of comparison**:

Property of Comparison

Given any two real numbers a and b, exactly one of the following holds:
1. $a = b$ 2. $a < b$ 3. $a > b$

This property tells us that if we are given *any* two real numbers, either they are equal or one of them is greater than the other. It establishes order on the number line. We need this property to derive a formula for the distance between two points on a number line. For example, let P_1 and P_2 be any points on a number line with coordinates x_1 and x_2, usually denoted as $P_1(x_1)$ and $P_2(x_2)$. Then, by the property of comparison, we know that $x_1 = x_2$, $x_1 < x_2$, or $x_1 > x_2$. Consider these possibilities one at a time.

$x_1 = x_2$ That is, the distance between x_1 and x_2 is 0.

$x_1 < x_2$ $x_1 > x_2$

The distance is $x_2 - x_1$.   The distance is $x_1 - x_2$.

We can combine these different possibilities into one distance formula by using the idea of absolute value.

Distance on a Number Line

The distance d between points $P_1(x_1)$ and $P_2(x_2)$ is

$$d = |x_2 - x_1|$$

EXAMPLE 3 Let $A, B, C, D,$ and E be points with coordinates as shown on the number line below:

a. The distance from D to E is: $|5 - 3| = |2| = 2$.
b. The distance from E to D is: $|3 - 5| = |-2| = 2$.
c. The distance from A to D is: $|3 - (-5)| = |8| = 8$.
d. The distance from E to B is: $|-3 - 5| = |-8| = 8$.

A.2
Problem Set

Classify each example in Problems 1–10 as a natural number (N), whole number (W), integer (I), rational number (Q), irrational number (Q'), or real number (R). Examples may be in more than one set.

1. **a.** -9 **b.** $\sqrt{30}$
2. **a.** $\sqrt{49}$ **b.** $\frac{17}{2}$
3. **a.** 5 **b.** $\sqrt{4}$
4. **a.** $\frac{0}{5}$ **b.** $\frac{5}{0}$
5. **a.** $.\overline{4}$ **b.** $.5252\ldots$
6. **a.** 3.1416 **b.** π
7. **a.** $.381$ **b.** $\pi/6$
8. **a.** $\sqrt{121}$ **b.** $\sqrt{125}$
9. **a.** $16/0$ **b.** $0/16$
10. **a.** $\pi/3$ **b.** $\sqrt{4/25}$

Certain relationships among these sets are assumed from your previous courses and are illustrated in Figure A.2. For Problems 11–40 let

$P = \{\text{primes}\}$
$ = \{2, 3, 5, 7, 11, 13, \ldots\}$

Primes are natural numbers with exactly two divisors (itself and 1).

$C = \{\text{composites}\}$
$ = \{4, 6, 8, 9, 10, \ldots\}$

Composites are natural numbers greater than 1 that are *not* primes.

$E = \{\text{evens}\}$
$ = \{0, 2, 4, 6, 8, 10, \ldots\}$

Even numbers are whole numbers divisible by 2.

$D = \{\text{odds}\}$
$ = \{1, 3, 5, 7, 9, 11, \ldots\}$

Odd numbers are natural numbers that can be written as an even number plus 1.

11. Draw Figure A.2 and show the set P.
12. Draw Figure A.2 and show the set C.
13. Draw Figure A.2 and show the set E.
14. Draw Figure A.2 and show the set D.

Indicate whether the statements in Problems 15–40 are true or false. Test your understanding of these sets, both from this section and from previous courses.

15. 0 is a natural number.
16. $\sqrt{3}$ is a real number.
17. $.333\ldots$ is an irrational number.
18. $-\frac{1}{4}$ is an integer.
19. 2 is a real number.
20. 1 is a prime number.
21. $|5|$ is a prime number.
22. The sum of two odd numbers is an odd number.
23. The sum of two even numbers is an even number.
24. The sum of two prime numbers is a prime number.
25. The sum of two composite numbers is a composite number.
26. The product of two even numbers is an even number.
27. The product of two odd numbers is an odd number.
28. The product of two primes is a prime.

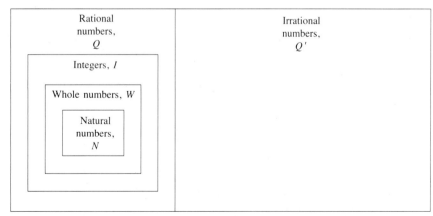

FIGURE A.2
Relationships within the set of real numbers

29. The product of two rational numbers is a rational number.
30. The product of two irrational numbers is an irrational number.
31. All natural numbers are integers.
32. All irrational numbers are real.
33. All repeating decimals are positive.
34. No even integers are primes.
35. All odd integers are primes.
36. All numbers are either even or odd.
37. Some negative numbers are primes.
38. a/b is a rational number for all numbers a and b.
39. Some real numbers are integers.
40. Some irrational numbers are integers.

Rewrite Problems 41–61 without the absolute value symbols.
41. $|9|$
42. $-|19|$
43. $|-19|$
44. $-|-8|$
45. $-|\pi|$
46. $|\pi - 2|$
47. $|2 - \pi|$
48. $|\pi - 5|$
49. $|\pi - 10|$
50. $|\sqrt{20} - 5|$
51. $|\sqrt{20} - 4|$
52. $|\sqrt{50} - 7|$
53. $|\sqrt{50} - 8|$
54. $|2\pi - 5|$
55. $|2\pi - 7|$
56. $|x + 3|$ if $x \geq 3$
57. $|x + 3|$ if $x \leq -3$
58. $|y - 5|$ if $y < 5$
59. $|y - 5|$ if $y \geq 5$
60. $|5 - 2s|$ if $s > 10$
61. $|4 + 3t|$ if $t < -10$

Find the distance between the pairs of points whose coordinates are given in Problems 62–70.
62. $A(5)$ and $B(17)$
63. $C(-5)$ and $D(-15)$
64. $E(22)$ and $F(-8)$
65. $G(-2)$ and $H(9)$
66. $I(\pi)$ and $J(5)$
67. $K(\pi)$ and $L(2)$
68. $M(\sqrt{3})$ and $N(1)$
69. $P(\sqrt{3})$ and $Q(2)$
70. $R(6)$ and $S(2\pi)$

A.3
Algebraic Expressions

An **algebraic expression** is a grouping of constants and variables obtained by applying a finite number of operations (such as addition, subtraction, multiplication, nonzero division, extraction of roots). **Constants** are symbols with just one possible value (such as 5, 11, 0, .5, or π), and **variables** are symbols (such as x, y, or z) used to represent unspecified numbers selected from some set called the **domain**. In this book, if a domain for a variable is not specified, the domain is the set of real numbers.

Numbers or expressions that are added are called **terms** while those that are multiplied are called **factors**. If a number and a variable are multiplied together, then the factor consisting of the number alone is called the **numerical coefficient**. It is very common to have repeated factors, which are indicated using **exponents**:

Exponent

If b is any real number and n is any natural number, then
$$b^n = \underbrace{b \cdot b \cdot b \cdots b}_{n \text{ factors}}$$

Furthermore, if $b \neq 0$, then
$$b^0 = 1 \quad \text{and} \quad b^{-n} = \frac{1}{b^n}$$

The number b is called the **base**, n is called the **exponent**, and b^n is called a **power**.

If an algebraic expression is a finite sum of terms with variables containing whole-number exponents, it is called a **polynomial**. For example,

$$3x, \quad 5, \quad 2x^2 - 3x + 5, \quad \frac{2}{3}x, \quad 0, \quad 5.5, \quad \frac{x-2}{4}$$

are polynomials, but

$$\sqrt{x}, \quad \frac{2}{x}, \quad \frac{x}{0}, \quad x^{1/3}$$

are not. The general form of a polynomial with a single variable is:

General Form of an nth-Degree Polynomial in x

$$a_n x^n + a_{n-1} x^{n-1} + \cdots + a_2 x^2 + a_1 x + a_0 \qquad a_n \neq 0$$

Polynomials are frequently classified according to the number of terms they contain. A **monomial** is a polynomial with one term, a **binomial** has two terms, and a **trinomial** has three. The **degree** of a monomial is the number of variable factors in that monomial. For example, the degree of $3x^2$ is two because $3x^2 = 3xx$ (two variable factors); $3^2 x$ is of degree one because there is one variable factor; and $3x^2 y^3$ is of degree five because there are five variable factors ($xxyyy$). The expressions 5, 6^2, and π are all of degree zero since none contains a variable factor. A special case is the monomial 0, to which no degree is assigned. The degree of a simplified polynomial is the degree of the highest-degree term.

From beginning algebra, recall the proper order of operations given in the following box.

Order of Operations Agreement

When simplifying an algebraic expression:

1. Carry out all operations within parentheses (begin with the innermost parentheses).
2. Do exponents next.
3. Complete multiplications and divisions, working from left to right.
4. Finally, do additions and subtractions, working from left to right.

EXAMPLE 1 $2 + 3 \cdot 4 = 2 + 12 = 14$ ∎

EXAMPLE 2 $9 + 12 \div 3 + 4 \div 2 + 1 \cdot 2 = 9 + 4 + 2 + 2 = 17$ ∎

In algebra we rarely use the symbol $\div$ but, instead, write the expression in Example 2 as

$$9 + \frac{12}{3} + \frac{4}{2} + 1 \cdot 2$$

The fractional bar used for division is also used as a grouping symbol:

$$6 + 4 \div 2 \quad \text{is written as} \quad 6 + \frac{4}{2}$$

whereas

$$(6 + 4) \div 2 \quad \text{is written as} \quad \frac{6+4}{2}$$

EXAMPLE 3 In algebra, $6 + 4 \div 2 + (6 + 4) \div 2 - (6 + 4 \div 2)$ is written as

$$6 + \frac{4}{2} + \frac{6+4}{2} - \left(6 + \frac{4}{2}\right) = 6 + \frac{4}{2} + \frac{10}{2} - (6 + 2)$$

$$= 6 + \frac{4}{2} + \frac{10}{2} - 8$$

$$= 6 + 2 + 5 - 8$$

$$= 5 \qquad ■$$

EXAMPLE 4 Simplify: $2 + 4[1 - 3(3 - 5) - \frac{12}{4}]$

Solution Do the innermost parentheses first:

$$2 + 4\left[1 - 3(3 - 5) - \frac{12}{4}\right] = 2 + 4\left[1 - 3(-2) - \frac{12}{4}\right] \qquad \text{Multiply and divide inside the brackets.}$$

$$= 2 + 4[1 + 6 - 3] \qquad \text{Add and subtract inside the grouping symbols.}$$

$$= 2 + 4[4] \qquad \text{Multiply before adding.}$$

$$= 2 + 16$$

$$= 18 \qquad ■$$

To **evaluate** an expression means to replace the variables with a given numerical value and then to carry out the order of operations to simplify the resulting numerical expression. This process is illustrated in Example 5.

EXAMPLE 5 Find $-(x + y)[-(s - t)]$ where $x = -2$, $y = -3$, $s = 4$, and $t = 7$.

Solution $-[(-2) + (-3)][-(4 - 7)] = -[-5][-(-3)]$

$$= -[-5][3]$$

$$= -[-15] = 15 \qquad ■$$

Addition and Subtraction of Polynomials

Terms that are identical, except for their numerical coefficients, are called **similar terms**. That is, similar terms contain the same exponent (or exponents) on the variables. Since the variable parts are identical, they can be simplified as follows:

$$5x + 4x = (5 + 4)x = 9x$$

If you add five apples to four apples, you obtain nine apples; so, too, if you add 5x's to 4x's, you get 9x's. This is a statement of an algebraic property called the **distributive property**:

Distributive Property

For real numbers a, b, and c,
$$a(b + c) = ab + ac$$

You will, of course, use this property and combine similar terms mentally as shown by the following more complicated examples.

EXAMPLE 6

$$(5x^2 + 2x + 1) + (3x^3 - 4x^2 + 3x - 2) = 3x^3 + \underbrace{5x^2 + (-4)x^2}_{\text{Similar terms}} + \underbrace{2x + 3x}_{\text{Similar terms}} + \underbrace{1 + (-2)}_{\text{Similar terms}}$$

Third-degree term; Second-degree terms; First-degree terms; Zero-degree terms

$$= 3x^3 + x^2 + 5x - 1$$ ∎

EXAMPLE 7 $(4x - 5) + (5x^2 + 2x + 1) = 5x^2 + (4x + 2x) + (-5 + 1) = 5x^2 + 6x - 4$ ∎

EXAMPLE 8 $(4x - 5) - (5x^2 + 2x + 1) = 4x - 5 - 5x^2 - 2x - 1 = -5x^2 + 2x - 6$ ∎

Note that in Example 7 we added similar terms. In Example 8, notice that the procedure for subtracting a polynomial is to subtract *each* term.

EXAMPLE 9 $(5x^2 + 2x + 1) - (3x^3 - 4x^2 + 3x - 2) = 5x^2 + 2x + 1 - 3x^3 + 4x^2 - 3x + 2$
$$= -3x^3 + 9x^2 - x + 3$$ ∎

Multiplication of Polynomials

To understand multiplication of polynomials, it is necessary to first understand multiplication of monomials.

EXAMPLE 10
$x^2(x^3) = xx(xxx) = x^5$ There are five x's used as a factor.
$(x^2)^3 = (x^2)(x^2)(x^2) = (xx)(xx)(xx) = x^6$ There are six x's.
$(x^2y^3)^4 = (x^2y^3)(x^2y^3)(x^2y^3)(x^2y^3) = (x^2x^2x^2x^2)(y^3y^3y^3y^3) = x^8y^{12}$ ∎

Example 10 illustrates three laws of exponents that are used to simplify algebraic expressions.

Laws of Exponents

Let a and b be any real numbers, m and n be any integers, and assume that each expression is defined.

FIRST LAW: $b^m \cdot b^n = b^{m+n}$
SECOND LAW: $(b^n)^m = b^{mn}$
THIRD LAW: $(ab)^m = a^m b^m$

These laws of exponents are used along with the distributive property when multiplying polynomials.

EXAMPLE 11 $(4x - 5)(5x^2 + 2x + 1) = (4x - 5)5x^2 + (4x - 5)2x + (4x - 5)1$
$= 20x^3 - 25x^2 + 8x^2 - 10x + 4x - 5$
$= 20x^3 - 17x^2 - 6x - 5$ ∎

EXAMPLE 12 $(2x - 3)(x + 4) = (2x - 3)x + (2x - 3)4$
$= 2x^2 - 3x + 8x - 12$
$= 2x^2 + 5x - 12$ ∎

Notice that one term, $P = 4x - 5$ in Example 11 and $P = 2x - 3$ in Example 12, is simply distributed to each term in the other factor. We call this **distributive multiplication** to remind us to use the distributive property to do multiplication of polynomials.

Distributive Multiplication

$$P(A_1 + A_2 + \cdots + A_n) = PA_1 + PA_2 + \cdots + PA_n$$

EXAMPLE 13 $(2x + 3y + 1)(5x - 2y - 3)$
$= (2x + 3y + 1)5x + (2x + 3y + 1)(-2y) + (2x + 3y + 1)(-3)$
$= 10x^2 + 15xy + 5x - 4xy - 6y^2 - 2y - 6x - 9y - 3$
$= 10x^2 + 11xy - 6y^2 - x - 11y - 3$ ∎

It is frequently necessary to multiply binomials, and even though we use distributive multiplication, we want to be able to carry out the process quickly and efficiently in our heads. Consider Example 14.

EXAMPLE 14 **a.** $(2x + 3)(4x - 5) = (2x + 3)(4x) + (2x + 3)(-5)$
$= \mathbf{8x^2} + \underbrace{\mathbf{12x + (-10x)}} + 3(-5)$
   ↑   ↑   ↑
Product of   Sum of inner and   Product
first terms   outer terms   of last terms
$= 8x^2 + 2x - 15$

b. $(5x - 3)(2x + 3) = 10x^2 + \underbrace{(15x - 6x)}_{} + (-9)$

Mentally: Product of first terms ↑, Sum of outer and inner terms ↑, Product of last terms ↑

$= 10x^2 + 9x - 9$

c. $(4x - 3)(3x - 2) = 12x^2 - 17x + 6$

Product of first terms ↑, Sum of outer and inner terms ↑, Product of last terms ↑

It is frequently necessary to multiply binomials such as those shown in Example 12; however the process illustrated in that example is too lengthy, so a shortened version, called **FOIL**, is often used instead. Four pairs of terms are multiplied.

Foil

$(ax + b)(cx + d) = acx^2 + \underbrace{(ad + bc)}_{②+③}x + bd$

① First terms
② Outer terms
③ Inner terms
④ Last terms

Do this mentally.

EXAMPLE 15 $(2x - 3)(x + 3) = 2x^2 + 3x - 9$

① ④

② + ③: $6x + (-3x)$

$(x + 3)(3x - 4) = 3x^2 + 5x - 12$
$(5x - 2)(3x + 4) = 15x^2 + 14x - 8$

A.3
Problem Set

Simplify the expressions in Problems 1–14.

1. $3 + 2 \cdot 5$
2. $4 + 2 \cdot 3$
3. $5 + 3 \cdot 2$
4. $(-5)^2$
5. $(-6)^2$
6. $(-7)^2$
7. -5^2
8. -6^2
9. -7^2
10. $10 + 6(3 - 5)$
11. $8 - 5(2 - 7)$
12. $3 \cdot 5 - (-2)(-4)$
13. $\dfrac{(-2)6 + (-4)}{-4} + \dfrac{(-3)(-4)}{6}$
14. $(-2) + \dfrac{6}{-3} - 4 + \dfrac{-8}{-4}$

Let $x = -3$, $y = 2$, $z = -1$, and $w = -4$. Evaluate the expressions in Problems 15–27.

15. x^2
16. $-x^2$
17. y^2
18. $-y^2$
19. z^2
20. $-z^2$
21. $x + y - z$
22. $x - (y - z)$
23. $(xy)^2 + xy^2 + x^2y$
24. $5x - (4x + 3w)$
25. $x^2 - w^2(y^2 + z^2)$
26. $\dfrac{x - w^2}{z}$
27. $\dfrac{x - z}{y}$

Mentally multiply the expressions in Problems 28–41.

28. **a.** $(x + 3)(x + 2)$ **b.** $(x + 1)(x + 5)$
29. **a.** $(x - 2)(x + 6)$ **b.** $(x + 5)(x - 4)$
30. **a.** $(x + 1)(x - 2)$ **b.** $(x - 3)(x + 2)$
31. **a.** $(x - 5)(x - 3)$ **b.** $(x + 3)(x - 4)$
32. **a.** $(y + 1)(y - 7)$ **b.** $(y - 3)(y + 5)$
33. **a.** $(2y + 1)(y - 1)$ **b.** $(2y - 3)(y - 1)$
34. **a.** $(y + 1)(3y + 1)$ **b.** $(y + 1)(3y + 2)$
35. **a.** $(2y + 3)(3y - 2)$ **b.** $(2y + 3)(3y + 2)$
36. **a.** $(x + y)(x + y)$ **b.** $(x - y)(x - y)$
37. **a.** $(x + y)(x - y)$ **b.** $(a + b)(a - b)$
38. **a.** $(5x - 4)(5x + 4)$ **b.** $(3y - 2)(3y + 2)$
39. **a.** $(x + 2)^2$ **b.** $(x - 2)^2$
40. **a.** $(x + 4)^2$ **b.** $(x - 3)^2$
41. **a.** $(a + b)^2$ **b.** $(a - b)^2$

Simplify the expressions in Problems 42–68.

42. $(x + y - z) + (2x - 3y + z)$
43. $(x - y - z) + (2x + y - 3z)$
44. $(x + 2y - 3z) + (x - 3y + 5z)$
45. $(x + 3y - 2z) + (3x - 5y + 3z)$
46. $(x + 2y) - (2x - y)$
47. $(2x - y) - (2x + y)$
48. $(x - 3y) - (5x + y)$
49. $(6x - 4y) - (4x - 6y)$
50. $(2x + y + 3) - (x - y + 4)$
51. $(x + y - 5) - (2x - 3y + 4)$
52. $(5x^2 + 3x - 5) - (3x^2 + 2x + 3)$
53. $(6x^2 - 3x + 2) - (2x^2 + 5x + 3)$
54. $(3x^2 + 2x + 6) - (2x^2 + 5x + 3)$
55. $(2x^2 - 3x - 5) - (5x^2 - 6x + 4)$
56. $(x^2 - 1) - (2 - x) + (x^2 - x)$
57. $(x^2 - x) + (x - 3) - (x - x^2)$
58. $(3x - x^2) - (5 - x) - (3x - 2)$
59. $(x^2 - 7) - (3x + 4) - (x - 2x^2)$
60. $(x^2 - 5) - (3x^2 + 2x + 5)$
61. $(x + 1)^3$
62. $(x - 1)^3$
63. $(3x^2 - 5x + 2) + (x^3 - 4x^2 + x - 4)$
64. $(5x + 1) + (x^3 - 4x^2 + x - 4)$
65. $(3x^2 - 5x + 2) - (5x + 1)$
66. $(x^3 - 4x^2 + x - 4) - (3x^2 - 5x + 2)$
67. $(5x + 1)(3x^2 - 5x + 2)$
68. $(3x - 1)(x^2 + 32x - 2)$

A.4
Factoring

A **factor** of a given algebraic expression is an algebraic expression that divides evenly into the given expression. The process of **factoring** involves resolving a given expression into its factors. The procedure we will use is to carry out a series of "tests" for different types of factors. Table A.1 lists these types in the order in which we should check them when factoring an expression.

TABLE A.1 Factoring Procedure

Type	Form	Comments
1. Common factors	$ax + ay + az = a(x + y + z)$	Use the distributive property. It can be applied with any number of terms.
2. Difference of squares	$x^2 - y^2 = (x - y)(x + y)$	Remember that the *sum* of two squares cannot be factored in the set of real numbers.
3. FOIL	$x^2 + (c + d)x + cd = (x + c)(x + d)$ $acx^2 + (ad + bc)xy + bdy^2 = (ax + by)(cx + dy)$	Use this trial-and-error procedure with trinomials, after checking for common factors and difference of squares. See the examples in this section.

Common Factoring

The distributive property leads to a very important type of factoring called **common factoring**. When simplifying expressions, we read the distributive property from left to right:

$$a(b + c) = ab + ac$$

When factoring common factors, we read it from right to left, as shown in the following box.

Common Factoring

$$ab + ac = a(b + c)$$
$$ab - ac = a(b - c)$$

EXAMPLE 1 $5x^2 - 25x = (5x)x - (5x)5$
$= 5x(x - 5)$

Common factoring extends to any number of terms.

EXAMPLE 2 $2x^3 - 20x^2 + 6x = 2x(x^2) - 2x(10x) + 2x(3)$
$= 2x(x^2 - 10x + 3)$

If we factor out the *greatest factor* common to each term, or if no other factor can be found, we say the polynomial is **completely factored**.

EXAMPLE 3 $10x^2y + 25xy^2 - 15x^2y^3 = 5xy(2x) + 5xy(5y) + 5xy(-3xy^2)$
$= 5xy(2x + 5y - 3xy^2)$

Common factors refer to the base numerals, not to the exponents, but notice that the smallest exponent (namely, one on the x and y in Example 3) on a common base number leads to the appropriate common factor. For example, $3x^3y^5 + 5x^4y^2$ has a common factor x^3y^2, since the smallest exponent on the common factor x is 3 and on the common factor y is 2.

We usually factor **over the set of integers**, which means that all the numerical coefficients are integers. If the original polynomial has fractional coefficients, however, we factor out the fractional part first.

EXAMPLE 4 $\frac{1}{36}x^2 - 5x + 1 = \frac{1}{36}x^2 - \frac{36}{36}5x + \frac{36}{36}(1)$
$= \frac{1}{36}(x^2 - 180x + 36)$

Common factors do not have to be monomials; they can be any algebraic expression.

EXAMPLE 5 $5x(3a - 2b) + y(3a - 2b) = (5x + y)(3a - 2b)$ The common factor is $(3a - 2b)$. ∎

The most important type of factoring used in calculus is finding common factors that are not monomials (as shown in Example 5), except in calculus they are often longer.

EXAMPLE 6 Factor $(x^2 + 5x - 8)(3)(5x + 2)^2(5) + (2x + 5)(5x + 2)^3$.

Solution You must first recognize the two terms in this expresssion:

$$(x^2 + 5x - 8)(3)(5x + 2)^2(5) \; + \; (2x + 5)(5x + 2)^3$$

Remember that the order of operations groups together the parentheses and the multiplications and the terms are separated by additions or subtractions. The common factor is $(5x + 2)^2$.

$$(5x + 2)^2 \, (x^2 + 5x - 8)(3)(5) + (5x + 2)^2 \, (2x + 5)(5x + 2)$$
$$= (5x^2 + 2)^2[15(x^2 + 5x - 8) + (2x + 5)(5x + 2)]$$
$$= (5x^2 + 2)^2(15x^2 + 75x - 120 + 10x^2 + 29x + 10)$$
$$= (5x^2 + 2)^2(25x^2 + 104x - 110)$$ ∎

Difference of Squares

The second type of factoring involves determining whether the expression is a difference of squares.

Difference of Squares

$$x^2 - y^2 = (x - y)(x + y)$$

EXAMPLE 7
a. $9x^2 - 25y^2 = (3x)^2 - (5y)^2 = (3x - 5y)(3x + 5y)$
b. $16x^4 - 1 = (4x^2)^2 - (1)^2$ This can be a mental step.
  $= (4x^2 - 1)(4x^2 + 1)$ Difference of squares
  $= (2x - 1)(2x + 1)(4x^2 + 1)$ Difference of squares again
c. $x^2 - 3$ is irreducible over the set of integers (3 is not a perfect square). Factoring over the set of integers rules out factoring this expression as

$$x^2 - 3 = (x - \sqrt{3})(x + \sqrt{3})$$

since the factors do not have integer coefficients. In this book, an expression is called completely factored if all fractions are eliminated by common factoring and if no further factoring is possible *over the set of integers*.
d. $x^2 + 4$ is irreducible over the set of integers (it is not a difference but a sum of two squares). ∎

FOIL—Factoring a Trinomial

Consider the product of two binomials:

$$(2x - 3)(x + 1) = 2x^2 - 3x + 2x - 3 = 2x^2 - x - 3$$

If you understand where the terms of the product came from, you will find it easier to reverse the process. In the following discussion we will concentrate on a particular product, a second-degree polynomial in a single variable, like the one above. You can then see how this case applies to similar products.

First, examine the second-degree (or leading) term and the last term (the constant):

$$2x^2 \qquad -3$$
$$(2x - 3)(x + 1)$$

These terms are the products of the variable terms and the constants of the binomial factors.

Now, recall the origin of the first-degree (or middle) term of the product:

$$(2x - 3)(x + 1)$$
$$-3x$$
$$2x$$

This term is the sum of the products of the variable and constant terms in the binomial factors.

Now consider a product and reverse the multiplication procedure to determine the factors:

$$3x^2 + 13x - 10$$

1. First, find two factors whose product is $3x^2$. These determine the variable terms of the factors and hence the form of the factors:

   $$(x \quad)(3x \quad)$$

2. Next, factor the constant term. These factors will yield all possible pairs of factors:

   $(x + 2)(3x - 5)$ $\qquad$ $(x - 2)(3x + 5)$ $\qquad$ $(x + 5)(3x - 2)$
   $(x - 5)(3x - 2)$ $\qquad$ $(x + 1)(3x - 10)$ $\qquad$ $(x - 1)(3x + 10)$
   $(x + 10)(3x - 1)$ $\qquad$ $(x - 10)(3x + 1)$

3. Then check each of the possibilities to see which gives the correct middle term:

   $$(x + 5)(3x - 2) = 3x^2 + 13x - 10$$

Thus we factor a polynomial by reversing our knowledge of multiplication. Not all examples are this easy, and, indeed, not all can be factored over the set of integers. If no possibility yields the correct middle term, the polynomial is not factorable over the set of integers. For example, $x^2 + x + 1$ must be of the form

$$(x \quad 1)(x \quad 1)$$

but the only possibilities for the constant terms are

$$(x - 1)(x - 1) \quad \text{and} \quad (x + 1)(x + 1)$$

and no possibility yields the correct middle term. Thus we say that $x^2 + x + 1$ is not factorable over the set of integers.

Procedure for Factoring a Trinomial

1. Find the factors of the leading term and set up the binomials.
2. Find the factors of the constant term, and consider all possible binomials.
3. Determine the factors that yield the correct middle term.
4. If no pair of factors produces the correct full product, then the trinomial is not factorable over the set of integers.

EXAMPLE 8 $x^2 - 8x + 15 = (x - 5)(x - 3)$

EXAMPLE 9 $6x^2 + x - 12 = (2x + 3)(3x - 4)$

EXAMPLE 10 $x^2 - 2xy + y^2 = (x - y)(x - y) = (x - y)^2$

If both the binomial factors are the same, then the expression is called a **perfect square**.

Factoring problems are generally not divided into categories as in Examples 1–10. We should always begin by looking for a common factor and then a difference of squares, and finally trying FOIL, as shown in Examples 11–13.

EXAMPLE 11 $3x^2 - 75 = 3(x^2 - 25)$ Common factor
$= 3(x - 5)(x + 5)$ Difference of squares

EXAMPLE 12 $(x + 3y)^2 - 1 = [(x + 3y) - 1][(x + 3y) + 1]$
$= (x + 3y - 1)(x + 3y + 1)$

EXAMPLE 13 $6ax^2 - 21ax - 12a = 3a(2x^2 - 7x - 4)$
$= 3a(2x + 1)(x - 4)$

In calculus, you must sometimes use common factoring with negative or fractional exponents, as illustrated by the next two examples.

EXAMPLE 14 Factor $-6x^{-4} - x^{-3} + x^{-2}$.

Solution The common factor is found by looking at the smallest power on the common base. It is x^{-4} for this example.

$$-6x^{-4} - x^{-3} + x^{-2} = x^{-4}(-6 - x + x^2)$$ Remember to add exponents when multiplying numbers with the same base.

$$= x^{-4}(x^2 - x - 6)$$ Factor trinomial, if possible.

$$= x^{-4}(x - 3)(x + 2)$$

EXAMPLE 15 Factor $2x^{1/2} + x^{-1/2} + x^{3/2}$.

Solution With a common base, x in this example, the common factor will be that common base with the smallest value of the exponents on that base. The smallest exponent in this example is $-\frac{1}{2}$, so the common factor is $x^{-1/2}$:

$$2x^{1/2} + x^{-1/2} + x^{3/2} = x^{-1/2}(2x + 1 + x^2)$$

$$= x^{-1/2}(x^2 + 2x + 1)$$ Factor trinomial, if possible.

$$= x^{-1/2}(x + 1)^2$$

A.4

Problem Set

Factor the expressions, if possible, in Problems 1–60.

1. $20xy - 12x$
2. $8xy - 6x$
3. $6x - 2$
4. $5y + 5$
5. $xy + xz^2 + 3x$
6. $a^2 - b^2$
7. $a^2 + b^2$
8. $s^2 + 2st + t^2$
9. $m^2 - 2mn + n^2$
10. $u^2 + 2uv + v^2$
11. $x^{-3} + x^{-1} + x^2$
12. $x^{-4} + x^{-1} + x$
13. $(4x - 1)x + (4x - 1)3$
14. $(a + b)x + (a + b)y$
15. $2x^2 + 7x - 15$
16. $x^2 - 2x - 35$
17. $3x^2 - 5x - 2$
18. $6y^2 - 7y + 2$
19. $2x^{-1} - 10x^{-2} - 48x^{-3}$
20. $4x + 1 - 21x^{-1}$
21. $2 - 5x^{-1} + 3x^{-2}$
22. $3x^{-1} + 8x^{-2} - 3x^{-3}$
23. $x^{1/3} + x^{4/3} + x^{7/3}$
24. $x^{1/6} + x^{7/6} + x^{13/6}$
25. $(x - y)^2 - 1$
26. $(2x + 3)^2 - 1$
27. $x^{3/2} + 6x^{1/2} + 9x^{-1/2}$
28. $2x^{3/2} - 7x^{1/2} - 4x^{-1/2}$
29. $(a + b)^2 - (x + y)^2$
30. $(m - 2)^2 - (m + 1)^2$
31. $2x^2 + x - 6$
32. $3x^2 - 11x - 4$
33. $6x^2 + 47x - 8$
34. $6x^2 - 47x - 8$
35. $6x^2 + 49x + 8$
36. $6x^2 - 49x + 8$
37. $4x^2 + 13x - 12$
38. $9x^2 - 43x - 10$
39. $9x^2 - 56x + 12$
40. $12x^2 + 12x - 25$
41. $10x^2 - 9 - x^4$
42. $5x^2 - 4 - x^4$
43. $(x^2 - \frac{1}{4})(x^2 - \frac{1}{9})$
44. $(x^2 - \frac{1}{4})(x^2 - \frac{1}{16})$
45. $(x^2 - 3x - 6)^2 - 4$
46. $(x + y + 2z)^2 - (x - y + 2z)^2$
47. $2(x + y)^2 - 5(x + y)(a + b) - 3(a + b)^2$
48. $2(s + t)^2 + 3(s + t)(s + 2t) - 2(s + 2t)^2$
49. $18x^2(x - 5)^3 + 6x^3(3)(x - 5)^2$
50. $12x^3(x - 8)^4 + 3x^4(4)(x - 8)^3$
51. $10x(3x + 1)^3 + 5x^2(3)(3x + 1)^2(3)$
52. $12x^3(2x - 1)^3 + 3x^4(3)(2x - 1)^2(2)$
53. $2x(4x + 3)^3 + 3x^2(4x + 3)^2(4)$
54. $3x^2(5x - 2)^2 + 2x^3(5x - 2)(5)$
55. $3(x + 1)^2(x - 2)^4 + 4(x + 1)^3(x - 2)^3$
56. $4(x - 5)^3(x + 3)^2 + 2(x - 5)^4(x + 3)$
57. $3(2x - 1)^2(2)(3x + 2)^2 + 2(2x - 1)^3(3x + 2)(3)$
58. $(2x - 3)^3(3)(1 - x)^2(-1) + 3(2x - 3)^2(1 - x)^3(2)$
59. $4(x + 5)^3(x^2 - 2)^3 + (x + 5)^4(3)(x^2 - 2)^2(2x)$
60. $5(x - 2)^4(x^2 + 1)^3 + (x - 2)^5(3)(x^2 + 1)^2(2x)$

A.5
Linear Equations and Inequalities

Linear Equations

A **linear equation in one variable** is an equation that can be written in the form

$$ax + b = 0 \qquad a \neq 0$$

where x is a variable and a and b are any real numbers. An **open** or **conditional equation** is an equation containing a variable that may be either true or false, depending on the replacement for the variable. A **root** or a **solution** is a replacement for the variable that makes the equation true. We also say that the root **satisfies** the equation. The **solution set** of an open equation is the set of all solutions of the equation. To **solve an equation** means to find its solution set. If there are no values for the variable that satisfy the equation, then the solution set is said to be **empty** and is denoted by $\emptyset$. If every replacement of the variable makes the equation true, then the equation is called an **identity**. Two equations with the same solution set are called **equivalent equations**. The process of solving an equation involves finding a sequence of equivalent equations; it ends when the solution or solutions are obvious. There are a few operations that produce equivalent equations, some of which are summarized in the following box.

Properties of Equations

If P and Q are algebraic expressions, and k is a real number, then each of the following is equivalent to $P = Q$:

ADDITION PROPERTY: $\quad P + k = Q + k$

SUBTRACTION PROPERTY: $\quad P - k = Q - k$

MULTIPLICATION PROPERTY: $\quad kP = kQ, \quad k \neq 0$

DIVISION PROPERTY: $\quad \dfrac{P}{k} = \dfrac{Q}{k} \quad k \neq 0$

EXAMPLE 1 Solve $3x + 5 = x - 3$.

Solution

$\quad 3x + 5 = x - 3$

$\quad 3x + 5 - x = x - x - 3 \qquad$ Subtract x from both sides. (Do this step in your head.)

$\quad 2x + 5 = -3 \qquad$ Simplify both sides.

$\quad 2x + 5 - 5 = -3 - 5 \qquad$ Subtract 5 from both sides (mentally).

$\quad 2x = -8$

$\quad x = -4 \qquad$ Divide both sides by 2 (mental step—not shown).

Check: $\quad 3(-4) + 5 = -4 - 3$

$\qquad\qquad -7 = -7 \qquad$ This is true, so the solution checks. ∎

EXAMPLE 2 Solve $2x - 5(x - 2) = 3(3 - x)$.

Solution
$$2x - 5(x - 2) = 3(3 - x)$$
$$2x - 5x + 10 = 9 - 3x \quad \text{Eliminate the parentheses, then simplify.}$$
$$-3x + 10 = 9 - 3x$$
$$10 = 9 \quad \text{Add 3x to both sides. This results in a false equation.}$$

This is a contradiction, and, since it is equivalent to the original equation, the solution set is empty. ■

EXAMPLE 3 Solve $2x - (7 - x) = x + 1 - 2(4 - x)$.

Solution
$$2x - (7 - x) = x + 1 - 2(4 - x)$$
$$2x - 7 + x = x + 1 - 8 + 2x \quad \text{Remove the parentheses first.}$$
$$3x - 7 = 3x - 7 \quad \text{Combine similar terms.}$$
$$-7 = -7 \quad \text{True equation}$$

This is an identity. Because it is equivalent to the original equation, the solution set is the set of all real numbers so that x can be replaced by any real number and the equation will be true. ■

It is frequently necessary to solve variable equations. The following example comes from Section 3.1.

EXAMPLE 4 Solve $2x + xy' + y + 2yy' = 0$ for y'. The variable y' is read "why-prime" and is not the same as y, nor is it y to the first power. Treat y' as a separate variable.

Solution Isolate the terms involving y' on one side.

$$xy' + 2yy' = -2x - y$$
$$(x + 2y)y' = -2x - y \quad \text{Isolate the } y' \text{ by factoring.}$$
$$y' = \frac{-2x - y}{x + 2y} \quad \text{Divide both sides by } x + 2y.$$
■

Linear Inequalities

Another type of statement is an **inequality**. Inequalities can be sentences that are always true (for example, $x - 1 < x$ or $5 < 7$), called **absolute inequalities**; always false (for example, $x > x$), called **contradictions**; or sometimes true and sometimes false (for example, $x > 2$), called **conditional inequalities**. The latter can be solved by using a set of properties similar to those for equations.

A notation called **interval notation** is useful in much of the following discussion. Interval notation uses the idea of an ordered pair listing the left and right endpoints of the interval as the first and second components, respectively. The ordered pair is enclosed in brackets or parentheses. Brackets are used when the endpoint is included in the interval, and parentheses are used when the endpoint is excluded from the interval. The smaller number of the pair of numbers a and b must be the first component. The idea is rather simple, as Table A.2 (page 420) shows.

TABLE A.2
Interval notation for line segments

	Interval notation	Inequality notation	Line graph
Closed interval	$[a, b]$	$a \leq x \leq b$	•——————• a b
	$[a, b)$	$a \leq x < b$	•——————∘ a b
	$(a, b]$	$a < x \leq b$	∘——————• a b
Open interval	(a, b)	$a < x < b$	∘——————∘ a b

The symbols ∞ and $-\infty$ are used to denote rays, as shown in Table A.3.

TABLE A.3
Interval notation for rays

Interval notation	Inequality notation	Line graph
$(-\infty, b]$	$x \leq b$	←——————• a b
$(-\infty, b)$	$x < b$	←——————∘ a b
$[a, \infty)$	$x \geq a$	•——————→ a b
(a, ∞)	$x > a$	∘——————→ a b

Note that the open interval notation ( ) is always used with ∞ and $-\infty$, and that [ ] is never used.

Several properties of inequalities are needed for solving linear inequalities.

Properties of Inequalities

If P and Q are algebraic expressions, and k is a real number, then each of the following is equivalent to $P < Q$:

ADDITION PROPERTY: $\quad P + k < Q + k$

SUBTRACTION PROPERTY: $\quad P - k < Q - k$

MULTIPLICATION PROPERTY:

$\quad$ Positive number k: $\quad kP < kQ$
$\quad$ Negative number k: $\quad kP > kQ \quad$ Note that the inequality is reversed.

DIVISION PROPERTY:

$\quad$ Positive number k: $\quad \dfrac{P}{k} < \dfrac{Q}{k}$

$\quad$ Negative number k: $\quad \dfrac{P}{k} > \dfrac{Q}{k} \quad$ Note that the inequality is reversed.

These properties also hold for $\leq$, $>$, and $\geq$.

Essentially, properties of inequalities allow any operation that is allowed for equations, *except that multiplying or dividing by a negative number reverses the sense of the inequality.*

EXAMPLE 5 Solve $5 - 2(3x - 4) < 4x - 7$.

Solution
$$5 - 2(3x - 4) < 4x - 7$$
$$5 - 6x + 8 < 4x - 7$$
$$-6x + 13 < 4x - 7$$
$$-10x + 13 < -7 \quad \text{Subtract 4x from both sides.}$$
$$-10x < -20 \quad \text{Subtract 13 from both sides.}$$
$$x > 2 \quad \text{Divide both sides by } -10; \text{ note that the order of the inequality has changed.}$$

We say the solution is the interval $(2, \infty)$. The graph of this solution is

Notice that the open point indicates that $x = 2$ is excluded. ∎

EXAMPLE 6
$$5(2x - 3) - 4(x - 2) \leq 3(x - 3)$$
$$10x - 15 - 4x + 8 \leq 3x - 9$$
$$6x - 7 \leq 3x - 9$$
$$3x - 7 \leq -9$$
$$3x \leq -2$$
$$x \leq -\frac{2}{3}$$

The solution is $(-\infty, -\frac{2}{3}]$. The graph of this solution is

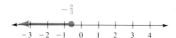

Notice that the closed point indicates that $x = -\frac{2}{3}$ is included. ∎

A "string of inequalities" may be used to show the order of three or more quantities. For example, $2 < x < 5$ states that x is a number *between* 2 and 5. The statement is a *compound inequality*, equivalent to $x > 2$ and $x < 5$ at the *same* time, and is graphed as the interval on the number line between 2 and 5:

Such inequalities may be solved in a way similar to that used for other inequalities. Note that what is done to one member of the string in each of the following inequalities is done to *all* three members. We try to isolate the x in the middle part of the string.

EXAMPLE 7

$$-3 \leq 2x - 5 \leq 7$$
$$-3 + 5 \leq 2x - 5 + 5 \leq 7 + 5 \quad \text{Add 5 to all three members.}$$
$$2 \leq 2x \leq 12$$
$$1 \leq x \leq 6 \quad \text{Divide all three members by 2 (the inequality keeps the same direction).}$$

This says the solution is $[1, 6]$.

EXAMPLE 8

$$-2 < 1 - 3x < 7$$
$$-3 < -3x < 6 \quad \text{Subtract 1.}$$
$$1 > x > -2 \quad \text{Divide by } -3; \text{ reverse the order of the inequalities.}$$
$$-2 < x < 1$$

The string of inequalities $1 < 2 < 3$ is equivalent to $3 > 2 > 1$. The first states the order of the three integers from smallest to largest, and the second from the largest to smallest. It is standard practice to use the ascending order from smallest to largest, so inequalities are stated with the less than ($<$) relation whenever possible. Notice that $1 > x > -2$ should be rewritten as $-2 < x < 1$ or $(-2, 1)$.

A.5
Problem Set

Solve each equation in Problems 1–16.

1. $3x = 5x - 4$
2. $2x = 9 - x$
3. $7x + 10 = 5x$
4. $9x - 8 = 5x$
5. $3x + 22 = 1 - 4x$
6. $5x - 4 = 3x + 8$
7. $2x - 13 = 7x + 2$
8. $3x + 32 = 18 - 4x$
9. $7x + 18 = -2x$
10. $6x - 1 = 13 - x$
11. $8x - 3 = 15 - x$
12. $-5x = 9 - 2x$
13. $2(x - 1) = 1 + 3(x - 2)$
14. $11 - x = 2 - 3(x - 1)$
15. $5 - 2x = 1 + 3(x - 2)$
16. $1 - 2(x - 3) = 2x - 1$

Solve each inequality in Problems 17–24.

17. $3x - 2 \leq 7$
18. $2x - 1 \geq 9$
19. $4 - 5x > 29$
20. $7 - 4x < 3$
21. $9x + 7 \geq 5x - 9$
22. $3x + 7 > 7x - 5$
23. $2(3 - 4x) < 30$
24. $3(2x - 5) \leq 9$

Solve the compound inequalities in Problems 25–30.

25. $7 < x + 2 < 11$
26. $12 < 5 + x < 14$
27. $-2 < x - 1 < 3$
28. $-5 \leq x - 2 \leq 4$
29. $9 < 1 - 2x < 15$
30. $-3 \leq 1 - 2x \leq 7$

Solve each equation in Problems 31–42 for y'.

31. $5x^4 - 5y^4 y' = 0$
32. $3x^2 y^2 y' + 2xy^3 = 0$
33. $4x^3 y^2 + 2x^4 yy' = 0$
34. $6x^5 y^3 + 3x^6 y^2 y' = 0$
35. $2x - (xy' + y) + 2yy' = 0$
36. $2y - (x + xy') + 2x = 0$
37. $3x^2 + 4x + (xy' + y) = 0$
38. $2x + (xy + y') + 2yy' = 0$
39. $\frac{1}{25}[2(x + 1)] + \frac{1}{4}[2(y - 1)y'] = 0$
40. $\frac{1}{9}[3(x - 4)] + \frac{1}{12}[3(y + 2)y'] = 0$
41. $3[x^2(3y^2 y') + 2xy^3] - 3[x(2yy') + y^2] + 5(xy' + y) = 0$
42. $5[x^2(5y^2 y') + 4xy^3] - 5[x(4yy') + y^2] + 6(xy' + y) = 0$

Solve each equation or inequality in Problems 43–60.

43. $2(x - 3) - 5x = 3(1 - 2x)$
44. $3(2x + 7) + 11 = 5(2 - x)$
45. $5(x - 1) + 3(2 - 4x) > 8$
46. $4(1 - x) - 7(2x - 5) < 3$
47. $4(x - 2) + 1 \leq 3(x + 1)$
48. $6(2x - 1) \geq 3(x + 4)$
49. $2(4 - 3x) > 4(3 - x)$
50. $3(2 - 5x) \geq 5(x + 2) + 36$
51. $2(3 - 7x) \geq -4 - (5 - x)$

52. $3(7 - x) < 2(x - 2)$
53. $3(1 - x) - 5(x - 2) = 5$
54. $5(5 - 3x) - (x - 8) = 1$
55. $6(2x + 5) = 4(3x + 1)$
56. $9(2 - 3x) = 6(5x + 3)$
57. $3(2x - 5) = 5(4x - 3)$
58. $2(4 - 3x) = 3(x - 2) - (9x - 14)$
59. $5(1 - 2x) = 3(x - 4) - (13x - 17)$
60. $6(1 - 4x) = 3(x - 1) - (5x + 13)$

A.6
Quadratic Equations and Inequalities

Quadratic Equations

A *quadratic equation in one variable* is an equation that can be written in the form

$$ax^2 + bx + c = 0 \quad a \neq 0$$

where x is a variable, and a, b, and c are real numbers. We will consider three different methods for solving quadratic equations—factoring, square root, and the quadratic formula.

The simplest method can be used if the quadratic expression $ax^2 + bx + c$ is factorable over the integers. The solution then depends on the following property of zero.

Property of Zero

$AB = 0$ if and only if $A = 0$ or $B = 0$ (or both)

Thus, if the product of two factors is zero, then at least one of the factors is zero. If a quadratic equation is factorable, this property provides a method of solution.

EXAMPLE 1 Solve $x^2 = 2x + 15$.

Solution $x^2 - 2x - 15 = 0$ Rewrite with a zero on one side (subtract 2x and 15 from both sides).

$(x + 3)(x - 5) = 0$ Factor.

$x + 3 = 0 \quad \text{or} \quad x - 5 = 0$ Since the product is zero, one of the factors must be zero.

$x = -3 \quad\quad x = 5$

Solution: $\{-3, 5\}$ ∎

You check a quadratic equation just as you check a linear equation, by substitution to see if you obtain a true or false equation. For Example 1,

Check $x = -3$: $(-3)^2 = 2(-3) + 15$ Check $x = 5$: $5^2 = 2(5) + 15$
$9 = -6 + 15$ $25 = 10 + 15$
$9 = 9 \checkmark$ $25 = 25 \checkmark$

Solution of Quadratic Equations by Factoring

To solve a quadratic equation that can be expressed as a product of linear factors:

1. Rewrite all nonzero terms on one side of the equation.
2. Factor the expression.
3. Set each of the factors equal to zero.
4. Solve each of the linear equations.
5. Write the solution set that is the union of the solution sets of the linear equations.

EXAMPLE 2   Solve $x^2 - 6x = 0$.

Solution   Factor, if possible: $x(x - 6) = 0$

Set each factor equal to zero: $x = 0$ $x - 6 = 0$

$$x = 6$$

Solution is $\{0, 6\}$. However, $x = 0, x = 6$ or $x = 0, 6$ is a sufficient answer. ∎

When the quadratic equation is not factorable, other methods must be employed. One such method depends on the **square root property**.

Square Root Property

$$P^2 = Q \quad \text{if and only if} \quad P = \pm\sqrt{Q}.$$

For example, the equation $x^2 = 4$ can be rewritten as $x^2 - 4 = 0$, factored, and solved. However, the square root property can also be used:

Square root property	*Factoring*
$x^2 = 4$	$x^2 - 4 = 0$
$x = \pm\sqrt{4}$	$(x - 2)(x + 2) = 0$
$x = \pm 2$	$x = 2, -2$

The square root property can be derived by using the following property of square roots:

Square Root of a Real Number

For all real numbers x,

$$\sqrt{x^2} = |x|$$

This means that, if $x \geq 0$, then $\sqrt{x^2} = x$ and if $x < 0$, then $\sqrt{x^2} = -x$. For example, $\sqrt{2^2} = |2| = 2$ and $\sqrt{(-2)^2} = |-2| = 2$.

The importance of the square root property is that it can be applied to *any* quadratic of the form $ax^2 + bx + c = 0$, $a \neq 0$. The result (which is derived by completing the square and is shown in most algebra textbooks) is called the **quadratic**

formula. This method for solving quadratic equations will solve any quadratic, so in practice we will usually try to solve a quadratic equation by factoring, and if that does not easily work we will go directly to the quadratic formula.

Quadratic Formula

If $ax^2 + bx + c = 0$ with $a \neq 0$, then
$$x = \frac{-b \pm \sqrt{b^2 - 4ac}}{2a}$$

EXAMPLE 3 Solve $5x^2 + 2x - 2 = 0$.

Solution   Note that $a = 5$, $b = 2$, and $c = -2$. Thus
$$x = \frac{-2 \pm \sqrt{4 - 4(5)(-2)}}{2(5)}$$
$$= \frac{-2 \pm 2\sqrt{1 + 10}}{2(5)}$$
$$= \frac{-1 \pm \sqrt{11}}{5}$$

EXAMPLE 4 Solve $5x^2 + 2x + 2 = 0$.

Solution $$x = \frac{-2 \pm \sqrt{4 - 4(5)(2)}}{2(5)}$$
$$= \frac{-2 \pm 2\sqrt{-9}}{10}$$

Since the square root of a negative number is not a real number, the solution set is *empty over the reals*.* That is, we say the solution set is ∅.

Since the quadratic formula contains a radical, the sign of the radicand (the number under the radical) will determine whether the roots will be real or nonreal. This radicand is called the **discriminant** of the quadratic, and its properties are summarized below:

Discriminant

If $ax^2 + bx + c = 0$ ($a \neq 0$), then $D = b^2 - 4ac$ is called the **discriminant**.

If $D < 0$, then there are *no real solutions*.
If $D = 0$, then there is *one real solution*.
If $D > 0$, then there are *two real solutions*.

---

* In algebra, you may have solved this equation by using complex numbers. However, since the domain for variables in this course is the set of real numbers, and since there is no real number whose square is negative, we see that there is no real number that satisfies this equation.

Quadratic Inequalities

A *quadratic inequality in one variable* is an inequality that can be written in the form

$$ax^2 + bx + c < 0 \qquad a \neq 0$$

where x is a variable, and a, b, and c are any real numbers. The symbol $<$ can be replaced by $\leq$, $>$, or $\geq$ and the inequality will still be a quadratic inequality in one variable.

The procedure for solving a quadratic inequality is similar to that for solving a quadratic equality. First, we use the properties of inequality to obtain a zero on one side of the inequality. Then we factor, if possible, the quadratic expression on the left. For example,

$$x^2 - x - 6 \geq 0$$
$$(x + 2)(x - 3) \geq 0$$

A value that causes one of the factors to be zero is called a **critical value** for the inequality; the critical values for the example above are $x = -2$ and $x = 3$. It follows, then, that the inequality must be either positive or negative for every real number that is not a critical value. We therefore examine the intervals defined by the critical values.

EXAMPLE 5 Solve $x^2 - x < 6$.

Solution First obtain a zero on one side, and then factor, if possible:

$$x^2 - x - 6 < 0$$
$$(x + 2)(x - 3) < 0$$

Next, find the critical values (one at a time), plot them on a number line, and label the parts of the number line to the left and right of the critical value "$+$" or "$-$" depending on whether the factor is positive or negative in that region.

Factor: $x + 2$ *Critical value:* $x + 2 = 0$
$$x = -2$$

Plot:

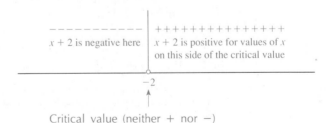

Critical value (neither $+$ nor $-$)

Factor: $x - 3$ *Critical value:* $x - 3 = 0$
$$x = 3$$

Plot (on the same number line):

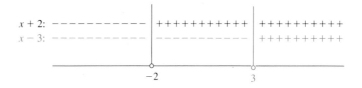

Finally, use the property of products to label the regions of the number line "positive" or "negative" as shown:

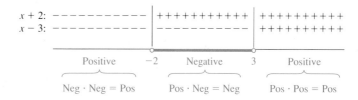

Shade in the appropriate portion of the number line indicated by the original inequality:

$x^2 - x - 6 < 0$ "<0" means you are looking for the portion of the number line labeled "negative."

Answer: $(-2, 3)$ That is, x is between -2 and 3. ∎

The previous example seems lengthy because all the steps are shown, but the process is quite easy and will be greatly simplified when you are doing the work. The process is summarized below, while Example 6 shows how your paper should look when doing this type of problem.

Solution of Polynomial Inequalities by Factoring

To solve an inequality that can be expressed as a product less than or greater than zero:

1. Rewrite all nonzero terms on one side of the inequality.
2. Factor the expression.
3. Determine the critical values.
4. Determine the signs of the factors on the intervals between critical values.
5. Select the interval or intervals on which the product has the desired sign.

EXAMPLE 6   Solve $5 + 4x - x^2 \leq 0$.

Solution $(5 - x)(1 + x) \leq 0$ The critical values are 5 and -1. Notice that in this example the critical values are included in the solution since equality is included.

```
              5 − x is positive for x < 5
5 − x:  ++++++++++ | +++++++++++ | ─────────
                                 1 + x is positive for x > −1
1 + x:  ──────────  | +++++++++++ | +++++++++++
(5 − x)(1 + x):    Negative   −1   Positive   5   Negative
                   Pos · Neg = Neg   Pos · Pos = Pos   Neg · Pos = Neg
```

Answer: Since the inequality we are solving is "≤", we look for those portions on the number line where the product is negative ("less than or equal to 0" means the values are negative with the endpoints included). We see that two of the three parts of the number line are labeled negative. This means that any x value in the left one *or* the right one makes the inequality true. The word "or" is written symbolically as "∪". The answer is

$$[-\infty, -1] \cup [5, \infty)$$

A.6

Problem Set

Solve each equation in Problems 1–20 over the set of real numbers.

1. $x^2 + 2x - 15 = 0$
2. $x^2 - 8x + 12 = 0$
3. $x^2 + 7x - 18 = 0$
4. $2x^2 + 5x - 12 = 0$
5. $10x^2 - 3x - 4 = 0$
6. $6x^2 + 7x - 10 = 0$
7. $x^2 + 5x - 6 = 0$
8. $x^2 + 5x + 6 = 0$
9. $x^2 - 10x + 25 = 0$
10. $x^2 + 6x + 9 = 0$
11. $12x^2 + 5x - 2 = 0$
12. $2x^2 - 6x + 5 = 0$
13. $5x^2 - 4x + 1 = 0$
14. $2x^2 + x - 15 = 0$
15. $4x^2 - 5 = 0$
16. $3x^2 - 1 = 0$
17. $3x^2 = 7x$
18. $7x^2 = 3$
19. $3x^2 = 5x + 2$
20. $3x^2 - 2 = -5x$

Solve the inequalities in Problems 21–40.

21. $(x - 6)(x - 2) \leq 0$
22. $(x + 2)(x - 8) \leq 0$
23. $x(x + 3) < 0$
24. $x(x - 3) \geq 0$
25. $(x + 2)(8 - x) \leq 0$
26. $(2 - x)(x + 8) \geq 0$
27. $(1 - 3x)(x - 4) < 0$
28. $(2x + 1)(3 - x) > 0$
29. $x^2 \geq 9$
30. $x^2 > 4$
31. $x^2 + 9 \leq 0$
32. $x^2 + 2x - 3 < 0$
33. $x^2 - x - 6 > 0$
34. $x^2 - 7x + 12 > 0$
35. $5x - 6 \geq x^2$
36. $4 \geq x^2 + 3x$
37. $5 - 4x \geq x^2$
38. $x^2 + 2x - 1 < 0$
39. $6x^2 - 10 > 59x$
40. $8x^2 < 2 + 15x$

A.7

Rational Expressions

Rational Expressions

If a variable is used in a denominator, then the expression is not called a polynomial. Instead, it is called a **rational expression**.

Rational Expressions

> A **rational expression** is an expression that can be written as a polynomial divided by a polynomial. Any values that cause division by zero are excluded from the domain.

APPENDIX A   REVIEW OF ALGEBRA   **429**

The fundamental property used to simplify rational expressions involves factoring both the numerator and denominator and then eliminating common factors according to the property

$$\frac{PK}{QK} = \frac{P}{Q} \quad Q, K \neq 0$$

Some rational expressions are simplified in Examples 1–5. Do not forget that all values for variables that cause division by zero are excluded from the domain.

EXAMPLE 1 $\dfrac{3xyz}{x^2 + 3x} = \dfrac{x(3yz)}{x(x + 3)} = \dfrac{3yz}{x + 3}$ ∎

EXAMPLE 2 $\dfrac{x - 2}{x^2 - 4} = \dfrac{1(x - 2)}{(x + 2)(x - 2)} = \dfrac{1}{x + 2}$ ∎

Sometimes the factors that are eliminated (as shown in color in Examples 1 and 2) are marked off in pairs as shown in Example 3. The slashes should be viewed as replacing the factor K by the number 1 since

$$\frac{PK}{QK} = \frac{P\cancel{K}}{Q\cancel{K}} = \frac{P \cdot 1}{Q \cdot 1} = \frac{P}{Q} \cdot 1 = \frac{P}{Q}$$

EXAMPLE 3 $\dfrac{(x - 5)(x + 2)(x + 1)}{(x + 1)(x - 2)(x - 5)} = \dfrac{\cancel{(x - 5)}(x + 2)\cancel{(x + 1)}}{\cancel{(x + 1)}(x - 2)\cancel{(x - 5)}}$

$\phantom{\dfrac{(x - 5)(x + 2)(x + 1)}{(x + 1)(x - 2)(x - 5)}} = \dfrac{x + 2}{x - 2}$ This is reduced since the x's are terms and *not* factors. You divide factors, not terms. ∎

EXAMPLE 4 $\dfrac{6x^2 + 2x - 20}{30x^2 - 68x + 30} = \dfrac{2(3x^2 + x - 10)}{2(15x^2 - 34x + 15)}$ Common factor first.

$\phantom{\dfrac{6x^2 + 2x - 20}{30x^2 - 68x + 30}} = \dfrac{2\cancel{(3x - 5)}(x + 2)}{2\cancel{(3x - 5)}(5x - 3)}$ Complete the factoring; then reduce.

$\phantom{\dfrac{6x^2 + 2x - 20}{30x^2 - 68x + 30}} = \dfrac{x + 2}{5x - 3}$ ∎

The laws of exponents can be extended to include rational expressions and negative exponents.

EXAMPLE 5 **a.** $\left(\dfrac{x}{y}\right)^3 = \left(\dfrac{x}{y}\right)\left(\dfrac{x}{y}\right)\left(\dfrac{x}{y}\right) = \dfrac{xxx}{yyy} = \dfrac{x^3}{y^3}$

b. $\dfrac{x^5}{x^3} = \dfrac{xxxxx}{xxx} = xx = x^2$

c. $\dfrac{x^3}{x^5} = \dfrac{xxx}{xxxxx} = \dfrac{1}{x^2} = x^{-2}$ ∎

Laws of Exponents

Let a and b be any real numbers, m and n be any integers, and assume that each expression is defined.*

FOURTH LAW: $\left(\dfrac{a}{b}\right)^m = \dfrac{a^m}{b^m}$

FIFTH LAW: $\dfrac{b^m}{b^n} = b^{m-n}$

The procedures for operation on rational expressions are identical to those for operations with fractions. However, with rational expressions, the numerators and denominators are any polynomial (division by zero is excluded) rather than constants. The procedures are summarized below.

Properties of Rational Expressions

Let P, Q, R, S, and K be any polynomials such that all values of the variable that cause division by zero are excluded from the domain.

EQUALITY: $\dfrac{P}{Q} = \dfrac{R}{S}$ if and only if $PS = QR$

FUNDAMENTAL PROPERTY: $\dfrac{PK}{QK} = \dfrac{P}{Q}$

ADDITION: $\dfrac{P}{Q} + \dfrac{R}{S} = \dfrac{PS + QR}{QS}$

SUBTRACTION: $\dfrac{P}{Q} - \dfrac{R}{S} = \dfrac{PS - QR}{QS}$

MULTIPLICATION: $\dfrac{P}{Q} \cdot \dfrac{R}{S} = \dfrac{PR}{QS}$

DIVISION: $\dfrac{P}{Q} \div \dfrac{R}{S} = \dfrac{PS}{QR}$

Some operations on rational expressions are performed in Examples 6–9. All values of variables that could cause division by zero are excluded.

EXAMPLE 6 $\quad \dfrac{13}{x-y} + \dfrac{2}{x-y} = \dfrac{15}{x-y} \quad$ Common denominator

* The first three laws of exponents are in Section A.3, page 410.

EXAMPLE 7
$$\frac{13}{x-y} - \frac{2}{y-x} = \frac{13}{x-y} + \frac{-2}{y-x}$$
It is helpful to write subtractions as additions.

$$= \frac{13}{x-y} + \frac{-2}{y-x} \cdot \frac{-1}{-1}$$
Multiply by 1 (written as $-1/-1$) in order to obtain a common denominator.

$$= \frac{13}{x-y} + \frac{2}{x-y}$$

$$= \frac{15}{x-y}$$

EXAMPLE 8
$$\frac{x+y}{x-y} + \frac{x-2y}{2x+y} = \frac{(x+y)(2x+y) + (x-y)(x-2y)}{(x-y)(2x+y)}$$

$$= \frac{2x^2 + 3xy + y^2 + x^2 - 3xy + 2y^2}{(x-y)(2x+y)}$$

$$= \frac{3x^2 + 3y^2}{(x-y)(2x+y)}$$

$$= \frac{3(x^2 + y^2)}{(x-y)(2x+y)}$$

EXAMPLE 9
$$\left(\frac{x^2 + 5x + 6}{2x^2 - x - 1} \cdot \frac{2x^2 - 9x - 5}{x^2 + 7x + 12}\right) \div \frac{2x^2 - 13x + 15}{x^2 + 3x - 4}$$

$$= \left[\frac{(x+2)(x+3)}{(x-1)(2x+1)} \cdot \frac{(2x+1)(x-5)}{(x+3)(x+4)}\right] \div \frac{(x-5)(2x-3)}{(x-1)(x+4)}$$

$$= \frac{(x+2)(x+3)(2x+1)(x-5)(x-1)(x+4)}{(x-1)(2x+1)(x+3)(x+4)(x-5)(2x-3)}$$

$$= \frac{x+2}{2x-3}$$

A.7
Problem Set

Simplify the expressions in Problems 1–30. Values that cause division by zero are excluded. All expressions should be reduced and negative exponents eliminated.

1. $\dfrac{2}{x+y} + 3$

2. $\dfrac{3}{x+y} - 2$

3. $\dfrac{x^2 - y^2}{2x + 2y}$

4. $\dfrac{x^2 - y^2}{3x - 3y}$

5. $\dfrac{3x^2 - 4x - 4}{x^2 - 4}$

6. $\dfrac{x^2 + 3x - 18}{x^2 - 9}$

7. $\dfrac{3}{x+y} + \dfrac{5}{2x+2y}$

8. $\dfrac{4}{x-y} - \dfrac{3}{2x-2y}$

9. $[7^3 + 2^5(3^3 + 4^4)]^0$

10. $[9^3 + 3^6(5^3 + 8^3)]^0$

11. $(x^2 - 36)\left(\dfrac{3x+1}{x+6}\right)$

12. $x^2 - 9 \div \dfrac{x+3}{x-3}$

13. $\dfrac{x+3}{x} + \dfrac{3-x}{x^2}$

14. $\dfrac{2}{x-y} + \dfrac{5}{y-x}$

15. $\dfrac{x}{x-1} + \dfrac{x-3}{1-x}$

16. $\dfrac{x+1}{x} + \dfrac{2-x}{x^2}$

17. $\dfrac{2x+3}{x^2} + \dfrac{3-x}{x}$

18. $\dfrac{1}{x^3} + 2xy + \dfrac{x^2}{y^2}$

19. $\dfrac{x}{y} + 2 + \dfrac{y}{x}$

20. $\dfrac{1}{x^3 y^2} + \dfrac{y}{x} + 2xy$

21. $\dfrac{x}{y} - 2 + \dfrac{y}{x}$

22. $\dfrac{1}{2y} + \dfrac{1}{x} + \dfrac{y}{2x^2}$

23. $\dfrac{1}{3xy^2} + xy + \dfrac{1}{x^3 y}$

24. $\dfrac{2x+y}{(x+y)^2} + \dfrac{x^2 - 2y^2}{(x+y)^3}$

25. $\dfrac{4x - 12}{x^2 - 49} \div \dfrac{18 - 2x^2}{x^2 - 4x - 21}$

26. $\dfrac{36 - 9x}{3x^2 - 48} \div \dfrac{15 + 13x + 2x^2}{12 + 11x + 2x^2}$

27. $\dfrac{1}{x^2 + 1} - \dfrac{x^2}{x^2 + 1}$

28. $(x^2 + 1) + \dfrac{x^2}{(x^2 + 1)^2}$

29. $\dfrac{4x^2}{x^4 - 2x^3} + \dfrac{8}{4x - x^3} - \dfrac{-4}{x + 2}$

30. $\dfrac{6x}{2x + 1} - \dfrac{2x}{x - 3} + \dfrac{4x^2}{2x^2 - 5x - 3}$

*A.8
Review

The material of this chapter is reviewed in the following list of objectives. After each objective there are some practice questions. For a sample test select the first question of each set and check your answers. The second question for each objective has no answer given. If you are having trouble with a particular type of problem, look back at the indicated section in the text. When you are finished reviewing these objectives, a sample examination is given at the end of this section.

[A.2]
Objective A.1: Classify numbers as natural, whole, integer, rational, irrational, or real.

1. Classify the following numbers by listing the set(s) into which they fall:

$$-8, \quad \dfrac{5}{6}, \quad .2\overline{3}, \quad \sqrt{169}, \quad \dfrac{3}{0}$$

2. Draw a diagram showing how the integers, rationals, and reals are related.

Objective A.2: Find the absolute value of a number or an expression.

3. $-|-5|$

4. $|\sqrt{1{,}000} - 35|$ (do not approximate)

5. $|x - 4|$ if $x < -2$

Objective A.3: Find the distance between pairs of points on a number line.

6. $A(-6)$ and $B(11)$

7. $C(2\pi)$ and $D(-4)$

8. $E(4)$ and $F(\sqrt{20})$

[A.3]
Objective A.4: Simplify numerical expressions.

9. a. -8^2 b. $(-8)^2$
   c. $-(-8)^2$ d. -1^0

10. $6 + 5 \cdot 2 - 8 + 4 + 2$

11. $\dfrac{6 + 4(-3)}{-2}$

12. $\dfrac{12 - 2(3)}{-4^2}$

Objective A.5: Evaluate algebraic expressions. Let $x = -1$, $y = 2$, and $z = -3$.

13. $x - (yz - 2z)$

14. $xy - yz + 4xz$

15. a. $-x^2$ b. $(-x)^2$

16. a. $(-y)^2$ b. $-y^2$

Objective A.6: Mentally multiply binomials using FOIL.

17. a. $(x - 2)(x + 7)$ b. $(x + 3)(x - 3)$

18. a. $(x + 3)^2$ b. $(2x - 1)^2$

19. a. $(x + 2y)(x - y)$ b. $(x - 3y)(x + 3y)$

20. a. $(2x - 3)(x + 5)$ b. $(3x - 1)(2x + 3)$

Objective A.7: Simplify algebraic expressions.

21. $(x - y - z) + (2x + y - 3z)$

22. $(5x^2 + 3x - 5) - (2x^2 + 2x + 3)$

23. $2(3 - x^2) - 3(5 - x) - (3x - 2)$

24. $(x + 2)(2x^2 + 3x + 2)$

[A.4]
Objective A.8: Factor polynomials.

25. $x^2 - 5x + 6$

26. $x^2 - 9x + 14$

27. $25 - x^2$

28. $4x^2 - 12x + 8$

29. $9x^4 - 40x^2 + 16$

30. $4x^4 - 13x^3 + 9x^2$

* Optional section.

[A.5]
Objective A.9: *Solve linear equations.*
31. $90 = 5x - 10$
32. $6 - 5x = 5 - 9x$
33. $3 - 4x = 3(5 - 2x)$
34. $5(x - 5) + 2 = 3(x - 3)$

Objective A.10: *Solve linear inequalities.*
35. $3x - 5 > 16$
36. $9 - x < -4$
37. $x < 3(2 + x)$
38. $2(x - 2) + 3 \le 5(x + 1)$

Objective A.11: *Solve compound inequalities.*
39. $-8 \le x + 5 \le -3$
40. $5 \le x + 1 < 15$
41. $6 < -x < 10$
42. $-1 < 5 - 2x \le 1$

[A.6]
Objective A.12: *Solve quadratic equations.*
43. $(3x - 1)(5x + 2) = 0$
44. $(2x + 1)(x - 4) = 11$
45. $6x^2 + 19x = -15$
46. $x^2 - 6x + 1 = 0$
47. $x^2 - 6x + 7 = 0$
48. $2x^2 + 1 = 2x$

Objective A.13: *Solve quadratic inequalities.*
49. $(x - 5)(x - 7) < 0$
50. $x^2 - 4x \ge 0$
51. $6x^2 > 11x + 10$
52. $x^2 - 2x + 3 < 0$

[A.7]
Objective A.14: *Simplify rational expressions.*
53. $\dfrac{2x}{15} - \dfrac{y}{12}$
54. $\dfrac{5x}{x - 1} + 1$
55. $\dfrac{3x + 2}{9x^2 - 6x + 1} + \dfrac{1}{3x - 1}$
56. $\dfrac{x^2 - 9}{x + 2} + \dfrac{x^2 + x - 6}{x^2 - 4}$

SAMPLE TEST

The following sample test (45 minutes) is intended to review the main ideas of this chapter.

1. Draw a diagram showing how the natural numbers, whole number, integers, rationals, and real numbers are related.
2. Write $|2\pi - 10|$ without absolute value symbols (do not approximate).
3. Find the distance between $A(\sqrt{10})$ and $B(5)$.
4. Simplify $\dfrac{1 + 3 \cdot 5}{-3^2}$.
5. Evaluate $-x^2 + 5x(3 - 2y)$ where $x = -6$ and $y = -3$.
6. Simplify $(3x + 1)(4x - 5)$.
7. Simplify $(x + y)^2$.
8. Simplify $(x + 3)(2x^2 - 5x - 4)$.
9. Factor $6x^2 - 29x - 5$.
10. Factor $1 - 9x^2$.
11. Solve $4 - 3x = 2(5 - 2x)$.
12. Solve $15 < -x$.
13. Solve $2(x + 1) + 5 < 3(1 - x)$.
14. Solve $-3 \le x - 5 < 0$.
15. Solve $x = 5x^2$.
16. Solve $x^2 + 2x - 2 = 0$.
17. Solve $(1 - x)(3x - 2) > 0$.
18. Solve $x^2 - 5 \ge 4x$.
19. Simplify $\dfrac{7x}{30} - \dfrac{y}{12}$.
20. Simplify $\dfrac{x^2 - 25}{x + 1} \div \dfrac{x^2 - 4x - 5}{x^2 + 2x + 1}$.

APPENDIX B
Computers

Software Accompanying Text

Microcomputers are changing the curriculum today as much as calculators changed the curriculum in the last decade. The personal computer is accessible to a great many students, and more and more schools and colleges have computer labs in which computer programs can be used.

The history of the marriage of software and textbooks is still in a state of transition. At first, software was written to accompany specific textbooks, but this proved less than desirable for at least two reasons. First, it is costly to develop *good* software, so many books claimed "tailor-made" software. However, it was not very general, and for the most part left both instructor and student wanting for better and more useful software. The second reason was that computer technology was changing so quickly that the software written for a textbook would either be released long after the textbook or else it became dated by the time the book came out. For these reasons I am recommending a stand-alone software program that I found *extremely* useful in understanding the material in this book. It is called The MATH LAB by Chris Avery and Charles B. Barker. I have included several COMPUTER APPLICATION boxes in this book to direct you to particular aspects of this program. MATH LAB comes in IBM format. Here is a quick summary of how you can use MATH LAB to help you with this course.

Calculus

A. Function plotter
B. Parametric equations
C. Implicit relations
D. Limits
E. Riemann sums
F. Numerical integration
G. Double integration
H. Sequences and series
I. Surface plotter
J. Differential equations

Spreadsheet Programs

SPREADSHEET PROGRAM	
A	**B**
1 x	y = 1/(3x^2-5x-2)
2 -3	@1/(3*A2^2-5*A2-2)
3 +A2+0.25	replicate
4 replicate	

x	y = 1/(3x^2-5x-2)
-3	0.025
-2.75	0.029038113
-2.5	0.034188034
-2.25	0.040920716
-2	0.05
-1.75	0.062745098
-1.5	0.081632653
-1.25	0.111888112
-1	0.166666667
-0.75	0.290909091
-0.5	0.8
-0.25	-1.77777778
0	-0.5
0.25	-0.32653061
0.5	-0.26666667
0.75	-0.24615385
1	-0.25
1.25	-0.28070175
1.5	-0.36363636
1.75	-0.64
2	ERR
2.25	0.516129032
2.5	0.235294118
2.75	0.144144144
3	0.1

Many of us now have access to a home computer as well as software for word processing, data bases, and spreadsheets. Spreadsheets provide a powerful, easy tool for processing data. The most commonly used spreadsheets today are versions of Excel and Lotus 1–2–3.* Although it is not appropriate to develop the techniques of spreadsheets in this book, suffice it to say that *it is very easy* to learn enough about spreadsheets to make them a useful tool in calculus.

The basic format for a spreadsheet is the arrangement of data into **cells** which are referenced by a row number (1, 2, 3,...) and a column number (A, B, C,...). The first spreadsheet shown in the text is reproduced in the margin. Look at the part labeled "SPREADSHEET PROGRAM." Cell A1 contains the letter x and cell A2 contains the numeral -3. What is contained in cell A3? (*Answer:* the formula "+A2+0.25.") A cell can contain a word or a letter (such as x), a number (such as -3), or a formula (such as +A2+0.25). This means that 0.25 is to be added to the contents of cell A2. The plus sign precedes a cell reference to distinguish the contents of a cell from a word. Almost any formula can be put into a cell. Notice the formula in cell B2:

Spreadsheet notation   *Algebraic notation*

$$1/(3*A2^2-5*A2-2) \qquad \frac{1}{3(x^2-5x-2)}$$

In the program (cell A4) you see the word **replicate**. As we will use it in this book, it means "copy the cell right above, and repeat for an entire column of entries." Thus, you see the first column of output; these entries were entered by using a COPY command on cell A3 and then replicated for the entire column. This means cell A4 contains "+A3+0.25", cell A5 contains "+A4+0.25", and cell A6 contains "+A5+0.25",.... The changing of the cell that is referenced is automatically carried out by the program when using the replicate command. Do you see how all of the entries in the first column are generated with a single formula and the replicate command? In turn, the second column is evaluating a function (in preparation to graphing it in Chapter 1).

Finally, notice in the program that the entry in cell C3 has a formula preceded by @. The way you communicate the type of entry you are putting into a cell is as follows:

First entry is a letter or "'" to indicate a word or text.
First entry is "+" or a numeral to indicate a number or to reference a cell.
First entry is "@" to indicate a formula.

From time to time in this book you will find inserts showing a spreadsheet program and its associated output.

---

* In preparing this book we use a Lotus 1–2–3-compatible spreadsheet called *Quattro-Pro*.

APPENDIX C
Answers to Odd-Numbered Problems

CHAPTER 1
1.1 What Is Calculus?

1. Answers vary. **3.** Answers vary. **5. a.** y is a function of x, since for each year, x, there will be one and only one closing price, y. **b.** y is not a function of x, since for each closing price, x, we may not be able to uniquely determine the year, y. **7. a.** \$.29 **b.** \$.53 **9. a.** 11 **b.** -11 **11. a.** 7 **b.** 47 **c.** -78 **d.** 497 **13. a.** 6 **b.** -26 **c.** 34 **d.** -394 **15. a.** 1 **b.** -1 **c.** -1 **d.** 1 **17. a.** All real numbers **b.** All real numbers **19. a.** All real numbers except 5 **b.** $[\frac{1}{2}, \infty)$. **21. a.** $5t-2$ **b.** $5w-2$ **23. a.** $5t+5h-2$ **b.** $5s+5t-2$ **25.** $5t+5h+38$ **27.** $2h^2+8h+1$ **29.** $10x^2-2$ **31.** $50x^2-60x+11$ **33.** $250x^2-100x-27$ **35.** $8x^4-32x^3-16x^2+96x+65$ **37.** 2 **39.** $4x+2h$ **41.** $4x+2h$ **43.** $2x+h-2$ **45. a.** \$.92 **b.** \$.45 **47.** \$1.14 **49. a.** \$.51 **b.** $e(1984)-e(1944)$ **51. a.** \$.03 **b.** The average change per year in the price of gasoline from 1944 to 1984. **53. a.** \$.02 **b.** \$.01 **c.** \$.02 **d.** \$.03 **e.** $\dfrac{s(1944+h)-s(1944)}{h}$ **55.** The population will be about 356,500,000. **57.** 208,900,000 **59. a.** $-14{,}800$ **b.** The average change in number of marriages from 1982 to year $1982+h$.

1.2 Functions and Graphs

1. A picture of the ordered pairs for the function. **3.** The x-intercepts are the points where the graph crosses (or touches) the x-axis. To find the x-intercept, set $y=0$. The y-intercept is the point where the graph crosses the y-axis. To find the y-intercept, set $x=0$. **5.** $(2, f(2))$ **7.** $(x_0, f(x_0))$ **9.** $(x_0, g(x_0))$ **11.** $(3, h(3))$ **13.** $(x_0+t, h(x_0+t))$ **15.** Domain is $x=-3$; range is $(-\infty, \infty)$; x-intercept $(-3, 0)$; not a function. **17.** Domain is $-2 \le x \le 5$; the range is $-5 \le y \le 3$; x-intercepts are $(-\frac{3}{2}, 0)$ and $(\frac{15}{4}, 0)$; y-intercept is $(0, 3)$; a function. **19.** The domain and the range are all real numbers; x-intercepts are $(-3, 0), (-1, 0)$ and $(2, 0)$; y-intercept is $(0, \frac{5}{2})$; the relation is a function.

21. 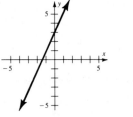 **23.** 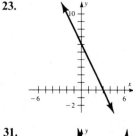 **25.** 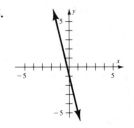 **27.**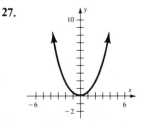

29. **31.** 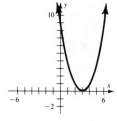 **33.** 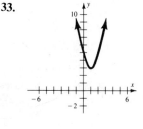 **35.**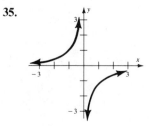

37. 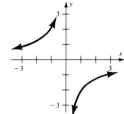 **39.** 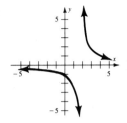 **41.** 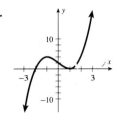 **43.**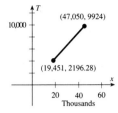

45. 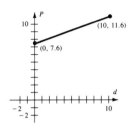 **47.**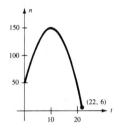

1.3 Linear Functions

1. The x-intercept is $(-2, 0)$; the y-intercept is $(0, 4)$. 3. The x-intercept is $(-1, 0)$; the y-intercept is $(0, -\frac{4}{3})$.
5. The x-intercept is $(-5, 0)$; the y-intercept is $(0, 2)$. 7. The y-intercept is $(0, -2)$. 9. $\frac{1}{3}$ 11. $\frac{5}{7}$ 13. 1
15. The slope is $m = 2$; the y-intercept is $b = 4$ or $(0, 4)$. 17. The slope is $m = 9$; the y-intercept is $b = 1$ or $(0, 1)$.
19. The slope is $m = \frac{2}{3}$; the y-intercept is $b = \frac{5}{3}$ or $(0, \frac{5}{3})$. 21. The slope is $m = 0$; the y-intercept is $b = 5$ or $(0, 5)$.
23. This is a vertical line; hence it has no slope and no y-intercept.
25. Slope $= 2$; y-intercept is -5 27. Slope $= -\frac{1}{4}$; y-intercept is 2   29. Slope $= \frac{3}{5}$; y-intercept is $\frac{2}{5}$

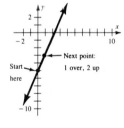 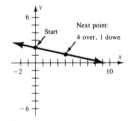 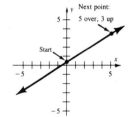

31. Slope is $-\frac{1}{3}$; y-intercept is 3   33. Slope $= \frac{2}{3}$; y-intercept is 0   35. (No slope and no y-intercept) $x = -\frac{5}{2}$

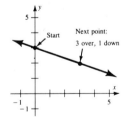 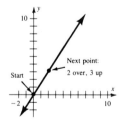 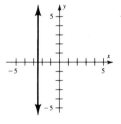

37. 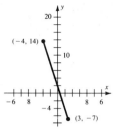 **39.** 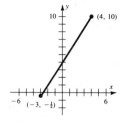 **41.**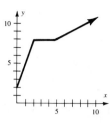

43. 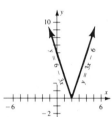 **45.** 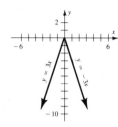 **47.**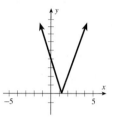

49. $2x + y + 3 = 0$ **51.** $y - 4 = 0$ **53.** $x + y - 2 = 0$ **55.** $3x - 5y - 27 = 0$ **57.** $2x - y - 4 = 0$
59. $y - 6 = 0$ **61.** $2x - y = 0$; four boxes could be supplied. **63.** $3x - 10y + 82 = 0$; 20.2 (million).
65. $y - 60 = 0$ **67.** The expected cost is \$535. **69.** The expected percentage is 63.0%.
71. Let x = amount on Form 1040, line 37; $T = .15x$ **73.**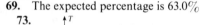

75. Let x = amount on Form 1040, line 37; $T = \begin{cases} .15x & \text{for } 0 < x \leq 32{,}450 \\ .28x + 4{,}867.50 & \text{for } 32{,}450 < x \leq 78{,}401 \\ .33x + 17{,}733.50 & \text{for } 78{,}401 < x \leq 185{,}730 \end{cases}$

1.4 Quadratic and Polynomial Functions

1. 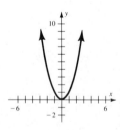 **3.** 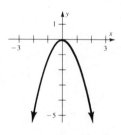 **5.** 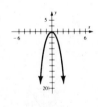 **7.**

9.   **11.**   **13.** Vertex is $(1, 0)$. 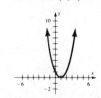 **15.** Vertex is $(-3, 0)$.

17. Vertex is $(1, 0)$.

19. Vertex is $(1, 2)$.

21. Vertex is $(1, 2)$.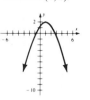

23. Vertex is $(-\frac{1}{3}, -\frac{2}{3})$.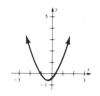

25. Vertex is $(\frac{3}{5}, -\frac{2}{5})$.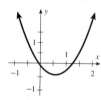

27. $y = (x - 2)^2$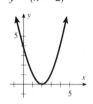

29. $y = -2(x + 1)^2$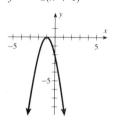

31. $y - 1 = 3(x + 2)^2$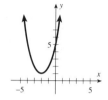

33. $y + 2 = 2(x + 1)^2$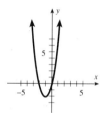

35. $y - 3 = -2(x + 1)^2$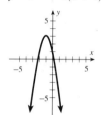

37. $y + 4 = -3(x - 2)^2$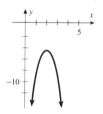

39. The maximum value of y is 3, occurring when $x = -1$. **41.** The maximum value of y is 1,250, occurring when $x = 450$. **43.** The minimum value of y is $-1{,}400$, occurring when $x = 560$. **45.** The maximum value of y is -14, occurring when $x = -3$. **47.** The minimum value for y is 13, occurring at $x = -3$. **49.** The maximum is $y = -4$, which occurs when $x = 1$. **51. a.** $-10x^2 + 1{,}040x$ **b.** The break-even points are $(19, 16150)$ and $(35, 24150)$. **53. a.** The maximum profit is \$1,156,250, which occurs when 375 boats are produced; that is, 375 should be produced. **b.** The loss is \$250,000. **c.** The maximum profit is \$1,156,150. **55.** The maximum profit is \$25, which occurs when $x = 15$. **57.** They intersect at $(20, 600)$ and $(10, 400)$. **59.** The maximum height was 3,456 units.

1.5 Rational Functions

1. A rational function is a function of the form $\frac{P(x)}{Q(x)}$ where $P(x)$ and $Q(x)$ are polynomials and $Q(x) \neq 0$.
3. A horizontal asymptote will be present if: (a) the degrees of the numerator and denominator are equal, in which case the ratio of leading coefficients gives the value of the asymptote or (b) the degree of the numerator is less than that of the denominator, in which case the x-axis will be the asymptote. **5.** Vertical asymptote is $x = \frac{3}{2}$; horizontal asymptote is $y = 0$ (the x-axis). **7.** Deleted point at $x = -3$. Vertical asymptote $x = 1$. The horizontal asymptote is $y = 0$ (the x-axis). **9.** The vertical asymptotes are $x = -\frac{3}{2}$ and $x = \frac{2}{3}$; the horizontal asymptote is $y = 0$ (the x-axis). **11.** There is a vertical asymptote at $x = \dfrac{-2 \pm \sqrt{2}}{2}$; the horizontal asymptote is the ratio of leading coefficients: $y = \frac{5}{2}$.

13. No vertical asymptote; the horizontal asymptote is $y = 0$. **15.** The vertical asymptotes are $x = -2$ and $x = 1$; the horizontal asymptote is $y = 0$ (the x-axis).

17.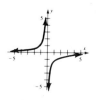

19.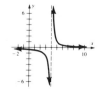

21.

23. 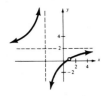 **25.** 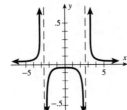 **27.**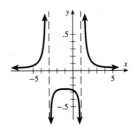

29. $106,667   **31.** 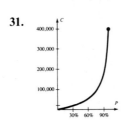   **33.** $72,000

35. 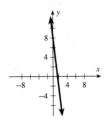   **37.** 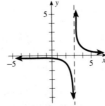   **39.**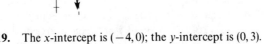

1.6 Review

Objective Questions
1. 47 **3.** 15 **5.** $3s^2 + 18s + 27$ **7.** $6xh + 3h^2$ **9.** The coordinates of A are $(2, 3)$; the domain is all real numbers; the range is $y \leq 4$. The x-intercepts are -1 and 3, and the y-intercept is 3. **11.** Coordinates of C are $(x_0 + h, f(x_0 + h))$. The domain is $s \leq x \leq t$; the range is $r \leq y \leq q$. The x-intercept is at d; the y-intercept is at q.

13. 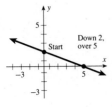   **15.**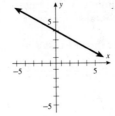

17. x-intercept $(\frac{4}{5}, 0)$; y-intercept $(0, -4)$ **19.** The x-intercept is $(-4, 0)$; the y-intercept is $(0, 3)$.   **21.** $\frac{3}{7}$ **23.** -5

25. 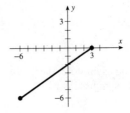   **27.**   **29.**

31. 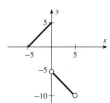   **33.** $5x - y = 0$ **35.** $x - 5y - 20 = 0$

37. 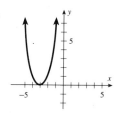   **39.**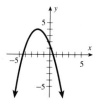

41. Maximum: $y = -250$; occurs when $x = 1,300$.   **43.** Minimum: $y = 250$; occurs when $x = 3$.
45. There is a vertical asymptote at $x = 5$; there is a horizontal asymptote at $y = 0$.
47. The vertical asymptotes are at $x = 3$ and $x = -2$; horizontal asymptote at $y = 0$.

49. 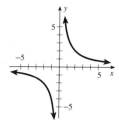   **51.**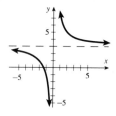

53. $\dfrac{Z(1985) - Z(1980)}{5}$ **55.** Call $S(x)$ the supply function: $S(x) = 30x - 4,000$
57. Break-even points occur at $(10, 35000)$ and $(50, 75000)$.   **59.** $380,000

Sample Test
1. 14   **3.** $-4t - 3$ **5.** -4 **7.**   **9.** Slope $= -8$

11. 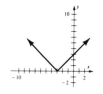   **13.** $5x + y - 170 = 0$ **15.** $x + 3 = 0$
17. The maximum value of y is 550, which occurs when $x = 40$.
19. $4,500

CHAPTER 2
2.1 Limits
1. 0   **3.** 8   **5.** 7   **7.** 6   **9.** 2   **11.** 2   **13.** 15   **15.** Limit does not exist.   **17.** 0   **19.** 5
21. $-\frac{1}{6}$ **23.** 1   **25.** 0   **27.** $\frac{1}{2}$ **29.** 4   **31.** 8   **33.** $\frac{1}{6}$ **35.** 0   **37.** Limit does not exist.
39. 0   **41.** $\frac{1}{4}$ **43.** 2   **45.** $\frac{3}{2}$ **47.** 5   **49.** Limit does not exist.   **51.** 1

53. The function is decreasing without bound. **55.** $\frac{4}{3}$ **57.** 3 **59.** 5 **61.** 10 **63.** $y = 3$ **65.** $y = \frac{2}{3}$
67. $y = 2$ **69. a.** $7.00 **b.** $12.00 **c.** $7.00 **d.** Limit does not exist. **71. a.** Limit does not exist. **b.** 80%
c. 70% **d.** 50% **73.** $\frac{1}{2}$ **75.** $-\frac{1}{9}$

2.2 Continuity

1. Suspicious point at $x = 4$; $x = 4$ is a point of discontinuity. **3.** Suspicious point at $x = 2$; $x = 2$ is a point of discontinuity.
5. Suspicious points at $x = 1$ and $x = 4$; $x = 4$ is a point of discontinuity. **7.** Suspicious points at $x = 1$, $x = 6$; $x = 1$ and $x = 6$ are points of discontinuity. **9.** Suspicious points at $x = 2$, $x = -2$; $x = -2$ and $x = 2$ are points of discontinuity.
11. There are no suspicious points; hence the function is continuous for $x > 0$. **13.** Yes **15.** Yes
17. Since there are no abrupt jumps or undefined values, it is continuous. The domain is $0 \le x < 24$, where x is the hour of the day.
19. Not continuous (unless the price remains constant). The domain is $0 \le x < 24$, where x is the hour of the day.
21. Not continuous; the domain is $t > 0$, where t is the number of minutes. **23.** Continuous on $[-5, 5]$
25. Continuous on $[5, 10]$ **27.** Discontinuities at $x = 1$ and $x = -1$ **29.** $x = -2$ is the only discontinuity on $[-5, 5]$.
31. Continuous on $[0, 2]$. **33.** Continuous on $[0, 5]$. **35.** $x = 3$ is a discontinuity. **37.** Discontinuous at $x = -2$.
39. Continuous on $[0, 5]$ **41.** $x = -2$ is the only suspicious point; $x = -2$ is a discontinuity.
43. Continuous on $[-5, 5]$ **45.** Continuous on $[-5, 5]$ **47. a.** Not continuous **b.** $.52, $.75, $.75
c. **49.** Answers vary.

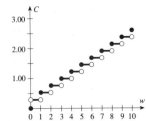

2.3 Rates of Change

1. $\frac{20}{3}$ ft/sec **3.** 40 ft/sec **5.** $\frac{10}{7}$ units/worker **7.** $\frac{5}{6}$ units/worker **9.** $-10 \frac{\text{points}}{\text{year}}$ **11.** $-9 \frac{\text{points}}{\text{year}}$
13. .1486 trillion/yr **15.** .2564 trillion/yr **17.** $59.3 \frac{\text{miles}}{\text{hour}}$ **19.** $36 \frac{\text{miles}}{\text{hour}}$ **21.** Slope $\to 0$

23. Slope $= 2$ for *each* value of Q **25.** Slope $\to 0$ **27.** -3 **29.** 0 **31.** 12 **33.** -9 **35.** $\frac{1}{6}$
37. -3 **39.** 0 **41.** 6 **43.** -3 **45.** 2 **47.** 0 **49.** 9 **51.** 8,900 **53.** 5,930
55. $60x + 30h - 100$ **57.** $60x - 100$ **59.** The instantaneous rate of change of cost at a given x.

2.4 Definition of Derivative

1. Derivative does not exist at $x = 1$. **3.** The derivative does not exist at $x = 3$. **5.** The derivative does not exist at $x = 4$.
7. $4x$ **9.** $-6x$ **11.** -5 **13.** $6x + 4$ **15.** $-6x - 50$ **17.** $6x^2$ **19.** $\frac{3}{x^2}$ **21.** $\frac{\sqrt{x}}{x}$
23. $30x + y + 45 = 0$ **25.** $5x + y - 4 = 0$ **27.** $8x - y + 6 = 0$ **29.** The marginal cost is $0.
31. The actual cost of producing an extra item is $1. **33.** $2x - 100$ **35.** The marginal cost is $20.
37. The actual cost is $21. **39.** -188 **41.** $\frac{31}{5}$ **43.** 20 miles per min

2.5 Differentiation Techniques, Part I

1. $7x^6$ **3.** $12x^{11}$ **5.** $-5x^{-6}$ **7.** 0 **9.** $32x^{-9}$ **11.** $\frac{-\sqrt{x}}{4x}$ **13.** $-40x^{-9}$ **15.** $15x^{1/4}$
17. $6x + 1$ **19.** $4x - 5$ **21.** $15x^2 - 10x + 4$ **23.** $\frac{2x^5 - x^2 - 3}{x^4}$ **25.** $8x^7 + 16x^3$ **27.** $\frac{-2x - 10}{x^3}$

APPENDIX C ANSWERS TO ODD-NUMBERED PROBLEMS 443

29. $-35x^6 + x^{-1/2} + 3x^{-2}$ **31.** $1 - 2x^{-3}$ **33.** $\frac{9}{2}x^{7/2} + \frac{15}{2}x^{3/2}$ **35.** $\frac{1}{3}x^{-2/3}$ **37.** π **39.** $4\pi r^2$
41. $\frac{C}{2\pi}$ **43.** $f'(x) = 5x^4$; $f'(1) = 5$ **45.** $f'(x) = -\frac{1}{2}x^{-3/2}$; $f'(1) = -\frac{1}{2}$ **47.** $f'(x) = 6x - 4$; $f(0) = -4$
49. $f'(x) = 2x + 2$; $f'(0) = 2$; $2x - y + 1 = 0$ **51.** $f'(x) = 4x^{-3}$; $f'(1) = 4$; $4x - y - 4 = 0$
53. $f'(x) = \frac{1}{3}x^{-2/3}$; $f'(1) = \frac{1}{3}$; $x - 3y + 2 = 0$ **55.** Marginal cost is $C'(x) = 40x + 500$ **57.** $\frac{dm}{dt} = \frac{-400}{t^2}$
59. The rate of change of earnings is 25.6 (thousand dollars per year).   **61.** The instantaneous rate of change of number of tasks (N) per hour (unit change in x) is $N' = \frac{25}{2\sqrt{x}}$. At the end of the fifth hour, the subjects are learning at the rate of 6 tasks per hour.
63. The rate of change of P with respect to x is about 20 foxes per rabbit.   **65. a.** The average rate of change in the CPI from 1990 to 1994 was 4.8.   **b.** 4.9   **67.** Answers vary.   **69.** Answers vary.

2.6 Differentiation Techniques, Part II

1. $20x^3 - 60x$ **3.** $2x - 1$ **5.** $24x^3 - 10x$ **7.** $60x^5 - 125x^4 + 20x^3$ **9.** $\frac{-3}{(x-3)^2}$ **11.** $\frac{-8}{(x-3)^2}$
13. $\frac{-16x}{(x^2-5)^2}$ **15.** $\frac{-1}{(x+2)^2}$ **17.** $10x^4 + 16x^3 - 3x^2 - 16x$ **19.** $20x^4 - 44x^3 - 9x^2 + 104x + 13$
21. $42x^6 - 20x^4 + 100x^3 - 81x^2 - 20x + 43$ **23.** $-\frac{25}{3}x^{-4/3} + 25x^{2/3}$ **25.** $30x^{3/2} - 15x^{-1/2}$
27. $\frac{-10x^2 + 6}{(5x^2 - 11x + 3)^2}$ **29.** $-\frac{35}{8}$ **31.** $\frac{83}{4}$ **33.** $-\frac{3}{2}$ **35.** $x + y = 0$ **37.** $83x - 4y - 180 = 0$
39. $x + 9y + 2 = 0$ **41.** $\frac{-200{,}000x - 1{,}500{,}000}{(x^2 + 15x + 25)^2}$ **43.** When the price is \$15, the demand is decreasing by about 20 items per unit change in price. The slope of the graph at $x = 15$ is -20.   **45.** $N'(x) = 575x^{-2}$ **47.** The slope of the graph at $x = 10$ is 5.75. The number of items is increasing at the point $x = 10$ since N' is positive.   **49.** $.01(.01 + .005t)^{-2}$
51. $\frac{-20t^2 + 10}{(2t^2 + 1)^2}$

2.7 The Chain Rule

1. $9(3x + 2)^2$ **3.** $20(5x - 1)^3$ **5.** $3(2x^2 + x)^2(4x + 1)$ or $3x^2(2x + 1)^2(4x + 1)$ **7.** $2(2x^2 - 3x + 2)(4x + 3)$
9. $4(x^3 + 5x)^3(3x^2 + 5)$ or $4x^3(x^2 + 5)^3(3x^2 + 5)$ **11.** $-2(2x^2 - 5x)^{-3}(4x - 5)$ or $-2x^{-3}(2x - 5)^{-3}(4x - 5)$
13. $-(x^4 + 3x^3)^{-2}(4x^3 + 9x^2)$ or $-x^{-4}(x + 3)^{-2}(4x + 9)$ **15.** $15(4x^3 + 3x^2)^2(12x^2 + 6x)$ or $90x^5(4x + 3)^2(2x + 1)$
17. $\frac{1}{4}(x^2 - 3x)^{-3/4}(2x - 3)$ or $\frac{2x - 3}{4(x^2 - 3x)^{3/4}}$ **19.** $x(x^2 + 16)^{-1/2}$ or $\frac{x\sqrt{x^2 + 16}}{x^2 + 16}$ **21.** $\frac{15}{2}x^2(x^3 + 8)^{-1/2}$
23. $\frac{1}{2}(3x + 1)^{-1/2}(9x + 2)$ **25.** $\frac{2}{3}(2x + 5)^{2/3}$ **27.** $-5(5x + 3)^{-2}$ **29.** $-4x(x^2 + 3)^{-3}$ **31.** $60x - y - 35 = 0$
33. $3x - 8y + 11 = 0$ **35.** $7x - 2y - 25 = 0$ **37.** $f'(x) = 0$ if $x = 0, 3, \frac{3}{2}$ **39.** $f'(x) = 0$ if $x = \frac{3}{2}$
41. $f'(x) = 0$ if $x = \frac{5}{2}$ **43. a.** $(2x + 1)^2[3(3x + 2)^2 3] + [2(2x + 1)(2)](3x + 2)^3$ **b.** $(2x + 1)(3x + 2)^2(30x + 17)$
45. a. $(5x + 1)^2[-1(4x + 3)^{-2}(4)] + [2(5x + 1)(5)](4x + 3)^{-1}$ **b.** $2(5x + 1)(4x + 3)^{-2}(10x + 13)$
47. a. $\frac{(5x + 3)[2(2x - 5)(2)] - 5(2x - 5)^2}{(5x + 3)^2}$ **b.** $\frac{(2x - 5)(10x + 37)}{(5x + 3)^2}$ **49. a.** $\frac{(2x - 5)^2[4(x + 5)^3] - (x + 5)^4[4(2x - 5)]}{(2x - 5)^4}$
b. $\frac{4(x + 5)^3(x - 10)}{(2x - 5)^3}$ **51. a.** $3(x^2 + 1)^{-1/2} - 3x^2(x^2 + 1)^{-3/2}$ **b.** $3(x^2 + 1)^{-3/2}$ **53.** The rate of change at any t is: $-5{,}400 \cdot 10^9$ bacteria per minute.   **55.** $\frac{dA}{dr} = 100{,}000(1 + r)^9$ **57.** $\frac{dP}{dr} = -100{,}000(1 + r)^{-11}$ **59. a.** The present enrollment is 4,000.   **b.** The enrollment in 10 years will be about 5,691.   **c.** $400(1 + .2t)^{-3/2}$ **d.** The annual enrollment is increasing by about 400 students per year.   **e.** The annual enrollment is increasing by about 77 students per year.

61. Answers vary; $\frac{f}{g} = fg^{-1}$. Using the product rule,

$$(fg^{-1})' = f'(g^{-1}) + f(g^{-1})'$$
$$= f'g^{-1} + f(-g^{-2}g')$$
$$= \frac{f'}{g} - \frac{fg'}{g^2}$$
$$= \frac{fg'}{g^2} - \frac{fg'}{g^2}$$
$$= \frac{gf' - fg'}{g^2}$$

2.8 Review

Objective Questions

1. 10 **3.** The limit does not exist. **5.** The limit does not exist. **7.** The limit is -5. **9.** 7 **11.** 0
13. Suspicious at $x = -3$ and $x = 4$; discontinuous at $x = 4$ **15.** Suspicious at $x = 7$; discontinuous at $x = 7$
17. Discontinuous **19.** Continuous **21.** $f(x)$ is continuous on $[-5, 5]$ **23.** Suspicious at $x = 8$; discontinuous at $x = 8$ **25.** 16 **27.** $\frac{\sqrt{2}}{2}$ **29.** 10 **31.** $\frac{1}{2}$ **33.** For a function $f(x)$, the derivative is $f'(x) = \lim_{h \to 0} \frac{f(x+h) - f(x)}{h}$ provided this limit exists. **35.** $\frac{-1}{(x-5)^2}$ **37.** $14x^{13}$ **39.** $-\frac{7}{9}x^{-16/9}$ **41.** 0
43. $6x^2 - 10x$ **45.** $3 - 54x$ **47.** $-\frac{1}{2}x^{-1/2}(x-1)^{-2}(x+1)$ **49.** $\frac{-6x}{(3x^2+1)^2}$ **51.** $\frac{-5x-3}{2x^{1/2}(5x-3)^2}$
53. $20(5x+9)^3$ **55.** $y' = \frac{1}{5}(5x^2 - 3x)^{-4/5}(10x - 3)$

Sample Test

1. 4 **3.** $\lim_{x \to 2} \frac{6x+1}{3-x} = 13$ **5.** Look for "jumps, holes, or poles" (vertical asymptotes). There is a hole at $x = 0$, a jump at $x = 3$, and a pole at $x = 10$. **7.** Let $f(x) = y = 5 + x - 2x^2$. The average rate of change is $\frac{f(3) - f(1)}{3 - 1} = \frac{f(3) - f(1)}{3 - 1} = \frac{(-10) - (4)}{3 - 1} = -7$. **9.** The derivative of $f(x)$ is $f'(x) = \lim_{h \to 0} \frac{f(x+h) - f(x)}{h}$ provided this limit exists. **11.** 0 **13.** $y' = 10x + 2x^{-2}$ **15.** $y' = \frac{-60}{(4x-5)^2}$ **17.** $y' = \frac{-(x-5)^2(4x+7)}{(1-2x)^2}$
19. The profit is decreasing by $.50 per unit.

CHAPTER 3
3.1 Implicit Differentiation

1. Explicitly: $\frac{dy}{dx} = -10x$; implicitly: $\frac{dy}{dx} = -10x$ **3.** Explicitly: $\frac{dy}{dx} = -5x^{-2}$; implicitly: $\frac{dy}{dx} = \frac{-y}{x}$
5. Explicitly: $\frac{dy}{dx} = \frac{1}{3} + \frac{50}{3}x^{-2}$; implicitly: $\frac{dy}{dx} = \frac{2}{3} - \frac{y}{x}$ **7.** Explicitly: $\frac{dy}{dx} = -x(4 - x^2)^{-1/2}$;
implicitly: $\frac{dy}{dx} = -\frac{x}{y}, y > 0$ **9.** Explicitly: $\frac{dy}{dx} = \frac{8}{15}x^3\left(\frac{2}{5}x^4 - \frac{7}{5}\right)^{-2/3}$; implicitly: $\frac{dy}{dx} = \frac{8x^3}{15y^2}$ **11.** $\frac{dy}{dx} = \frac{2x - y}{x - 2y}$
13. $\frac{dy}{dx} = \frac{-2(2x+1)}{3(3y-5)}$ **15.** $\frac{dy}{dx} = \frac{x - 10}{11y}$ **17.** $\frac{dy}{dx} = \frac{-6xy^3 + 3y^2 - 5y}{9x^2y^2 - 6xy + 5x}$ **19.** $\frac{dy}{dx} = \frac{-3x^2 - 4x - y}{x}$
21. $y - 2 = 0$ **23.** $8x + 15y + 23 = 0$ **25.** $x + 5y + 14 = 0$ **27.** $y - 4 = 0$ and $y + 2 = 0$ **29.** $x - y = 0$

APPENDIX C   ANSWERS TO ODD-NUMBERED PROBLEMS   445

31. $y - 3 = 0$ and $y + 1 = 0$ **33.** At $(1, -2)$, $\dfrac{dy}{dx} = -3$ **35.** At $(1, 2)$, $\dfrac{dy}{dx} = 2$ **37.** At $(3, 4)$, $\dfrac{dy}{dx} \approx -.2$

39. $\dfrac{dp}{dx} = \dfrac{1}{2p - 5}$

3.2 Differentials

1. $dy = 15x^2\, dx$ **3.** $dy = -10x^{-2}\, dx$ **5.** $dy = (300x^2 - 50)\, dx$ **7.** $dy = \frac{3}{2}(x - 1)^{-1/2}\, dx$ **9.** $dy = 5\, dx$
11. $dy = 2(5x - 3)(10x^2 - 3x - 15)\, dx$ **13.** $dy = -7(x - 2)^{-2}\, dx$ **15.** $dy = (2x - 1)\, dx$
17. $dy = (-5x^{-2} + 2x^{-3} + 3x^{-4})\, dx$ **19.** $dy = \dfrac{8x^2 - 6x + 7}{(x^2 + 3x - 2)^2}\, dx$ **21.** $dy = 1.8$, $\Delta y = 1.81$
23. $dy = .075$, $\Delta y \approx .0744457825$ **25.** $\Delta y \approx dy = \dfrac{57}{502^2} \approx .0002261869$ **27.** $\Delta y \approx \dfrac{40}{10^3}(.02) = .0008$
29. $\Delta y \approx dy \approx 46.105036$ **31.** $\Delta S \approx dS = 3,000$, so the sales will increase by about 3,000 units.
33. The alcohol will increase by about .01 percent.   **35.** The area of a circle is 1.256637 square miles.
37. The change in revenue is $\Delta R \approx dR = \$600$. The change in profit is $\Delta P \approx dP = \$300$.   **39.** 14,000 votes

3.3 Business Models Using Differentiation

1. 7   **3.** 46   **5.** 9   **7.** 8.9   **9.** -18 **11.** -14 **13.** $\bar{C}(x) = 200x^{-1} + 6 - x + x^2$;
$\bar{C}'(x) = -200x^{-2} - 1 + 2x$ **15.** $\bar{C}(x) = 5{,}000x^{-1} + .4x$; $\bar{C}'(x) = -5{,}000x^{-2} + .4$ **17.** $\bar{R}(x) = 50 - .5x$; $\bar{R}'(x) = -.5$
19. $\bar{P}(x) = x^2 - 8x + 2 + \dfrac{50}{x}$; $\bar{P}'(x) = 2x - 8 - \dfrac{50}{x^2}$ **21.** $\bar{P}(x) = x^2 - 50x + 5 + \dfrac{200}{x}$; $\bar{P}'(x) = 2x - 50 - \dfrac{200}{x^2}$
23. $R(x) = -.001x^2 + 30x$ **25.** $R'(x) = -.002x + 30$ This is approximately the change of revenue relative to a unit
change in x.   **27.** $P'(x) = -.002x + 21.5$; this is the change in profit relative to a unit change in x.   **29.** $\bar{C}'(x) = -50{,}000x^{-2}$
31. $R'(x) = 200 - .08x$ **33.** At a production level of 1,000 items, the revenue is increasing at a rate of \$120 per item. At a
production level of 2,500 items the revenue does not change per unit change in the number of items.   **35.** At a production
level of 1,000 items, the profit is increasing at \$70 per item. At 2,500 items the profit is decreasing at \$50 per item.
37.   **39.** $R'(x) - 5 - .002x$ **41.** Elastic demand   **43.** Inelastic demand

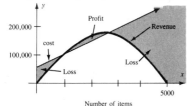

3.4 Related Rates

1. $\dfrac{dy}{dt} = -3$ **3.** $\dfrac{dy}{dt} = 1{,}000$ **5.** $\dfrac{dx}{dt} = 15$ **7.** $\dfrac{dy}{dt} = \dfrac{4}{5}$ **9.** $\dfrac{dx}{dt} = 0$ **11.** Profit is decreasing at a rate of \$123
per week.   **13.** The distance is changing at about 2.24 feet per second.   **15.** $\dfrac{dh}{dt} \approx .398$ **17.** $\dfrac{dV}{dt} = -.015$
19. $\dfrac{dV}{dt} = -.06$ **21.** The area is increasing at a rate of about 37.7 square miles per year.   **23.** The area is growing at
a rate of about 62.8 square feet per second.   **25.** The price is increasing at \$.11 per day.   **27.** $\dfrac{dV}{dt} \approx 60.32$ cm^3 per minute
29. $\dfrac{ds}{dt} = 8$ feet per second

3.5 Review

Objective Questions

1. $\dfrac{dy}{dx} = \dfrac{-5x^4}{4y + 1}$ **3.** $\dfrac{dy}{dx} = \dfrac{x - 3}{y + 1}$ **5.** $f'(1) = 6$ **7.** $f'(1) = 100$ **9.** $6x - y - 4 = 0$ **11.** $100x + y + 95 = 0$

13. $dy = 18x(3x^2 + 5)^{-1/2} dx$ 15. $dy = \dfrac{-x^2 - 10x - 3}{(3x^2 + x - 4)^2} dx$ 17. $dy = 10; \Delta y = 10.05$ 19. $dy = .1; \Delta y \approx .09918$

21. $\Delta y \approx .000045$ 23. $y = \dfrac{x^2 - 1}{x + 2}; \Delta y \approx dy \approx .9938$ 25. a. $1.18 b. $\dfrac{x^4 - 30x^2 + 400x}{(10 - x^2)^2}$

27. $2,500x^2(x - 20)(x - 12)$ 29. $-\frac{4}{5}\sqrt{5}$ 31. $\dfrac{dy}{dt} = -5$

Sample Test

1. $\dfrac{dy}{dx} = 20x - 6$ 3. $\dfrac{dy}{dx} = \dfrac{4x^3 - 6xy + 9y^2 + 5y}{3x^2 - 18xy - 5x}$ 5. $\dfrac{dy}{dt} = \pm\dfrac{27}{8}$ 7. $\dfrac{dx}{dt} = \dfrac{-2}{75}$

9. The marginal revenue is $4,000 - 200x$.

CHAPTER 4

4.1 First Derivatives and Graphs

1. 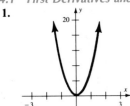 3. 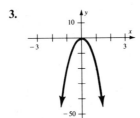 5.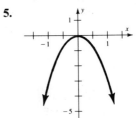

7. 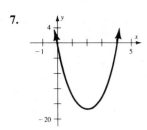 9. 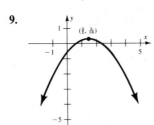 11.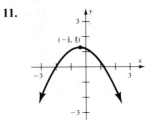

13. 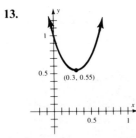 15.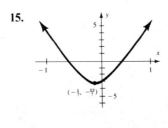

17. Increasing on $(-\infty, 2)$ and decreasing on $(2, \infty)$; horizontal tangent at $(2, 4)$. 19. Decreasing on $(-\infty, 0)$ and on $(5, \infty)$; increasing on $(0, 3)$; constant on $(3, 5)$; horizontal tangent at $(0, 3)$ and for all $3 < x < 5$ 21. Increasing on $(-\infty, 2)$ and $(2, \infty)$; no horizontal tangents 23. 25.

27. $f'(x) = 10 - 2x$ a. Critical value is 5. b. $f(x)$ is increasing on $(-\infty, 5)$ and decreasing on $(5, \infty)$.
29. $f'(x) = 2x - 12$ a. Critical value is 6. b. $f(x)$ is increasing on $(6, \infty)$ and decreasing on $(-\infty, 6)$.

31. $g' = 6x^2 - 6x - 36$ **a.** Critical values: 3, -2 **b.** $g(x)$ is increasing on $(-\infty, -2)$ and $(3, \infty)$ and decreasing on $(-2, 3)$.
33. $y' = 3x^2 + 10x + 8$ **a.** Critical values: $-\frac{4}{3}, -2$ **b.** y is increasing on $(-\infty, -2)$ and $(-\frac{4}{3}, \infty)$ and decreasing on $(-2, -\frac{4}{3})$. **35.** $y' = 12x^3 + 6x^2 - 18x$ **a.** Critical values: 0, $-\frac{3}{2}, 1$ **b.** y is increasing on $(-\frac{3}{2}, 0)$ and $(1, \infty)$ and decreasing on $(-\infty, -\frac{3}{2})$ and $(0, 1)$. **37.** $f'(x) = 15x^4 - 75x^2 - 540$ **a.** Critical values: 3, -3 **b.** $f(x)$ is increasing on $(-\infty, -3)$ and $(3, \infty)$ and decreasing on $(-3, 3)$. **39.** $y' = \dfrac{-3}{(x-2)^2}$ **a.** No critical values **b.** y is increasing on $(-\infty, 2)$ and $(2, \infty)$. **41.** $f'(x) = -x^{-2} - 2x^{-3}$ **a.** Critical values: 0, -2 **b.** $f(x)$ is increasing on $(-2, 0)$ and decreasing elsewhere. **43.** $S(x)$ is increasing when the amount spent on advertising is under \$33,333. **45.** Begin campaign April 1.

4.2 Second Derivatives and Graphs

1. $f'(x) = 10x^4 - 12x^3 + 3x^2 - 10x + 19$; $f''(x) = 40x^3 - 36x^2 + 6x - 10$; $f'''(x) = 120x^2 - 72x + 6$ **3.** $y = \sqrt{5}\, x^{1/2}$; $y' = \dfrac{\sqrt{5}}{2} x^{-1/2}$; $y'' = -\dfrac{\sqrt{5}}{4} x^{-3/2}$; $y''' = \dfrac{3\sqrt{5}}{8} x^{-5/2}$; $\dfrac{d^4 y}{dx^4} = \dfrac{-15\sqrt{5}}{16} x^{-7/2}$ **5.** $g'(x) = 12x^2 + 2x^{-2}$; $g''(x) = 24x - 4x^{-3}$; $g'''(x) = 24 + 12x^{-4}$; $g^{(4)}(x) = -48x^{-5}$ **7.** $y' = 5x^{2/3}$; $y'' = \dfrac{10}{3} x^{-1/3}$; $\dfrac{d^3 y}{dx^3} = y''' = \dfrac{-10}{9} x^{-4/3}$ **9.** $y = 3(x-1)^{-1}$; $y' = -3(x-1)^{-2}$; $y'' = 6(x-1)^{-3}$; $\dfrac{d^3 y}{dx^3} = y''' = -18(x-1)^{-4}$ **11.** $y' = \dfrac{x^2 + 8x + 1}{(x+4)^2}$; $\dfrac{d^2 y}{dx^2} = y'' = \dfrac{30}{(x+4)^3}$
13. Relative minimum 0 at $x = -\frac{1}{2}$ **15.** Relative minimum 6 at $x = 3$; relative maximum -6 at $x = -3$
17. Relative minimum 1 at $x = 0$ **19.** Relative minimum 0 at $x = 0$; relative maximum 16 at $x = -2$; relative maximum 16 at $x = 2$ **21.** Relative minimum $\frac{194}{27}$ at $x = \frac{2}{3}$; relative maximum 58 at $x = -4$ **23.** Relative maximum $\frac{7}{4}$ at $-\frac{1}{2}$; relative minimum -331 at $x = 5$ **25.** Relative minimum $2\sqrt{2}$ at $x = \sqrt{2}$; relative maximum $-2\sqrt{2}$ at $x = -\sqrt{2}$
27. Relative maximum 0 at $x = 0$; relative minimum 4 at $x = 2$ **29.** A relative minimum 0 occurs at $x = -1$.
31. a. $x = 6$ **b.** increasing for $x < 6$; decreasing for $x > 6$. **c.** concave down for all x **33. a.** $x = -\frac{1}{3}$ or 5 **b.** curve is increasing left of $x = -\frac{1}{3}$, decreasing for $-\frac{1}{3} < x < 5$, and increasing right of $x = 5$ **c.** curve is concave down for $x < \frac{7}{3}$ **35. a.** $x = \frac{5}{3}$ or -9 **b.** curve increasing left of -9, decreasing for $-9 < x < \frac{5}{3}$, and increasing right of $\frac{5}{3}$ **c.** curve is concave up for $x > -\frac{11}{3}$, and concave down for $x > -\frac{11}{3}$ **37. a.** $x = -\frac{2}{9}$ or $\frac{1}{2}$ **b.** curve is increasing left of $-\frac{2}{9}$, decreasing $-\frac{2}{9} < x < \frac{1}{2}$, and increasing right of $\frac{1}{2}$. **c.** curve is concave up for $x > \frac{5}{36}$ and is concave down for $x < \frac{5}{36}$
39. a. $x = 2$ **b.** graph decreasing on $(-\infty, 2)$ and $(2, \infty)$ **c.** concave up on $(-\infty, 2)$, concave down on $(2, \infty)$
41. The point of diminishing returns is at $x = 15$. **43.** $[0, 15)$ **45.** $x = 15$ is the point of diminishing returns with sales of 11,650 **47.** Relative maximum at $x = \dfrac{30 + \sqrt{980}}{2} \approx 30.65$; $x < 15$ concave up; $x > 15$ concave down

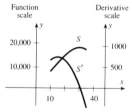

49. Relative maximum at $x = 68$; no relative minimums **51.** Revenue = (price)(number of items) **a.** $R(x) = 5x - \frac{1}{10,000} x^3$ **b.** $R'(x) = 5 - \frac{3}{10,000} x^2$ **c.** The marginal revenue is decreasing.

4.3 Curve Sketching—Relative Maximums and Minimums

1. Answers vary.   **3.** Answers vary.   **5.** Answers vary.

      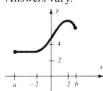

448 APPENDIXES

7.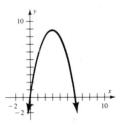

9.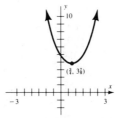

11.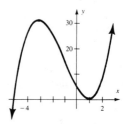

13.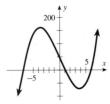

15.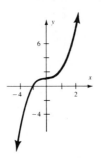

17.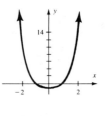

19.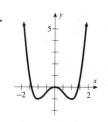

21.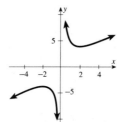

23.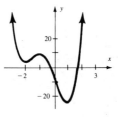

25.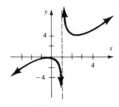

27.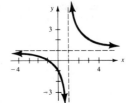

29.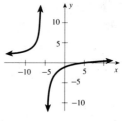

31.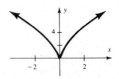

33.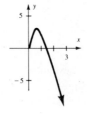

35.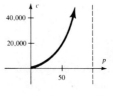

37.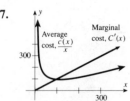

39.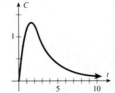

4.4 Absolute Maximum and Minimum

1. Maximum = 16 at $x = 4$ and minimum = -4 occurs at $x = 2$ and $x = -1$.   3. Maximum = 966 occurs at $x = -3$; minimum = $-3,050$ occurs at $x = -5$.   5. Maximum = 42 occurs at $x = 0$ and $x = 5$; minimum = 6 occurs at $x = 3$.
7. Maximum = 34 occurs at $x = 3$; minimum = -200 occurs at $x = 0$.   9. Maximum = 3 occurs at $x = 8$; minimum = 0 occurs at $x = -1$.   11. Maximum = 6 occurs at $x = 4$; minimum = 0 occurs at $x = 0$.   13. Maximum = 72 occurs at $x = 5$; minimum = $-\frac{256}{27}$ occurs at $x = \frac{5}{3}$.   15. No maximums or minimums   17. Maximum = 3 occurs at $x = 3$ and $x = -3$; minimum = 0 occurs at $x = 0$.   19. Maximum = $\frac{1}{8}$ occurs at $x = 3$; minimum = -1 occurs at $x = 0$.
21. $P(500) = 1,500$ 23. $P(24,000) = 57,520,000$ 25. $P(60) = 215$ 27. Absolute maximum at endpoint: $x = 50$; profit is 1,000 at a price of $100.   29. The minimum is 27, which occurs $6\frac{2}{3}$ miles from plant P_1.
31. The box has dimension 1 in. × 3 in. × 8 in. with a volume of 24 in.^{3}.   33. **a.** $25 + \dfrac{500}{x} - \dfrac{x}{20}$ **b.** Critical values 0, 100; decreasing on (0, 100) and increasing on (100, 200)   **c.** Minimal average cost is $25 at $x = \$100$
35. Maximum profit is $125 when $x = 15$. The price is $35.   37. The maximum profit is $69 when $x = 13$. The price is $37.
39. The minimum cost is $1,000 when $x = 100$.   41. The largest area occurs when the side parallel to the river is 500 ft.
43. Maximum is $62,500 which occurs when 50 sign up.   45. Plant 62 vines per acre.

4.5 Review

Objective Questions

1.
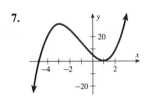

3.
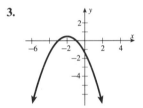

5.
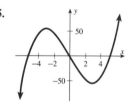

7.
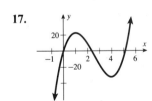

9. $y' = 12x^3 - 3x^2 + 10x$; $y'' = 36x^2 - 6x + 10$; $y''' = 72x - 6$; $y^{(4)} = 72$; $y^{(N)} = 0$ for $N > 4$
11. $f'(x) = -\frac{1}{2}x^{-3/2}$; $f''(x) = \frac{3}{4}x^{-5/2}$; $f'''(x) = -\frac{15}{8}x^{-7/2}$; $f^{(4)}(x) = \frac{105}{16}x^{-9/2}$
13. Relative maximum = 3, occurs at $x = -1$; relative minimum = $-\frac{175}{27}$, occurs at $x = \frac{5}{3}$
15. Relative maximum = 4, which occurs at $x = \pm\sqrt{2}$ and a relative minimum = 0, which occurs at $x = 0$

17.
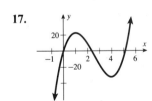

19.
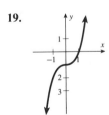

21. The maximum = 158 which occurs at $x = 1$; the minimum = $-1,408$ which occurs at $x = 4$.   23. Absolute minimum -27 at $x = 0$; absolute maximum 2,048 at $x = 5$ 25. 480 items   27. Produce 50 items   29. The marginal revenue is increasing at $x = 100$.   31. The rent should be $1,500 per month (60 will be rented with 40 vacancies).

Sample Test

1. **a.** $y' = -9x^2 + 5x^4$; $y'' = -18x + 20x^3$; $y''' = -18 + 60x^2$; $y^{(4)} = 120x$; $y^{(5)} = 120$; $y^{(6)} = y^{(7)} = \cdots = 0$
   b. $g' = 15x^4 + 3x^{-2}$; $g'' = 60x^3 - 6x^{-3}$; $g''' = 180x^2 + 18x^{-4}$; $g^{(4)} = 360x - 72x^{-5}$

3. 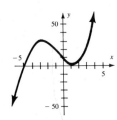   **5.**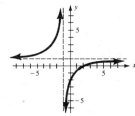

7. There is a relative maximum at $x = -2\sqrt{2}$ and a relative minimum at $x = 2\sqrt{2}$.
9. The marginal revenue is decreasing when $x = 100$.

Cumulative Review Chapters 2–4

1. The limit is 5.   **3.** 13   **5.** 15   **7.** Discontinuous at $x = 8$ **9.** At $x = 1$, $\dfrac{dy}{dx} = 1$ **11.** $y' = -24x - 15x^2$
13. $y' = -25(2 - 5x)^4$ **15.** $y' = \dfrac{-5y}{4x}$ **17.** $16x + y - 12 = 0$ **19.** $y' = 8x^3 - 3x^2 + 6x$; $y'' = 24x^2 - 6x + 6$; $y''' = 48x - 6$; $y^{(4)} = 48$; $y^{(5)} = y^{(6)} = \cdots = 0$ **21.** The maximum is $\frac{1}{4}$, which occurs at $x = \frac{1}{4}$; the minimum is -6, which occurs at $x = 9$.   **23.** Marginal revenue is decreasing at $x = 100$.   **25.** The frame should be 7 in. by 14 in.

CHAPTER 5

5.1 Exponential Functions

1. 5   **3.** Not a real number   **5.** -3 **7.** 7   **9.** $\frac{1}{10}$ **11.** $\frac{1}{1,000}$ **13.** 2^{-4} **15.** 3^{-3} **17.** $\frac{5}{6}$
19. 20.085537   **21.** 1.0512711   **23.** 1.0460279   **25.** $2x$ **27.** $1 + x$ **29.** $x + 2x^{1/2}y^{1/2} + y$

31.   **33.**    **35.**

37. 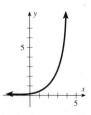   **39.**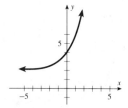

41. $54,598.15   **43.** $17,369.57   **45.** $10,081.44   **47.** $5,717.37   **49.** $5,627.05
51. If $p < x < q$ and $0 < b < 1$, then $b^q < b^x < b^p$.

5.2 Logarithmic Functions

1. $\log_2 64 = 6$ **3.** $\log_{1/3} 9 = -2$ **5.** $\log_b a = c$ **7.** $4^{1/2} = 2$ **9.** $10^{-2} = .01$ **11.** $e^2 = e^2$ **13.** 2
15. $\frac{1}{2}$ **17.** 3   **19.** 0   **21.** .033423755   **23.** 3.989449818   **25.** $-.493494967$ **27.** .81977983
29. .69314718

31.  **33.**  **35.**

37. 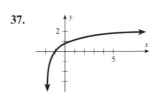 **39.** 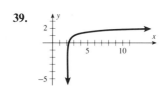 **41.**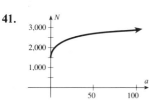

43. a. $t = -62.5\ln(1 - \frac{30}{80}) \approx 29$; it would take 1 month. **b.** No, since $N = 80$ gives $\ln 0$ which is not defined.

45. 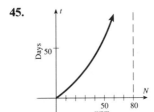   **47. a.** $M \approx 3.9$ **b.** $M = 8.8$

49. Because of the large numbers, it is difficult to graph (E, M); let $X = \log E$ so that
$M = \dfrac{\log 10^X - 11.8}{1.5} = \dfrac{2}{3}X - 7.867$ for $12 \leq X \leq 17$ as shown at the right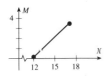

51. Since $b^x = b^x$, then $x = \log_b b^x$ by definition of logarithm.
53. Since $\log_b x = \log_b x$, then $b^{\log_b x} = x$ by definition of logarithm.

5.3 Logarithmic and Exponential Equations

1. 2   **3.** -1 **5.** $2\sqrt{7}$ **7.** $e^3 \approx 20.08553692$ **9.** 9.3   **11.** $\pm 2\sqrt{3}$ **13.** 3.5   **15.** No solution
17. 7   **19.** $\frac{2}{3}$ **21.** $\frac{1}{4}$ **23.** 5.6788736   **25.** e^4 **27.** 8.3571428   **29.** 2, 6   **31.** 25   **33.** 3   **35.** 2
37. 1.855903615   **39.** $-.143152$ **41.** $-.2772588$ **43.** $-.34161$ **45.** About 2004
47. To the nearest half-year, it would take 6 years.   **49.** 5 years, 283 days   **51.** 9.1551 years or 9 years 57 days.
53. a. 50%   **b.** About 13 seconds   **55.** $263.34   **57.** $5700\log_{1/2} P = t$
59. 15.3 years   **61.** Answers vary. Let $A = b^x$ and $B = b^y$, which means $\log_b A = x$ and $\log_b B = y$.
$\log_b \dfrac{A}{B} = \log_b \dfrac{b^x}{b^y} = \log_b b^{x-y} = x - y = \log_b A - \log_b B$

5.4 Derivatives of Logarithmic and Exponential Functions

1. $2e^{2x}$ **3.** $-e^{-x}$ **5.** $xe^x + e^x$ **7.** $2xe^{5x^2}[10x^3 + 3x + 5]$ **9.** $-2e^{5x}(6x + 15x^2 + 25)$ **11.** $\dfrac{1}{x-5}$
13. $4(2x^3 + e^{5x})^3(6x^2 + 5e^{5x})$ **15.** $\dfrac{4}{x}$ **17.** $\dfrac{1}{2x}$ **19.** $\dfrac{3 - 3x\ln x}{xe^x}$ **21.** $\dfrac{2}{x(x+2)}$ **23.** $e^{t^2-t}(1 + 2te^t - 2t)$
25. $2(e^{2x} - e^{-2x})$ **27.** $\dfrac{5x + 2 - 5x\ln|x|}{x(5x+2)^2}$ **29.** $\dfrac{e^x(x\ln|3x| - 1)}{x\ln^2|3x|}$ **31.** $\dfrac{-3{,}600e^{-3x}}{(1 + 6e^{-3x})}$ **33.** $\dfrac{100(e^{-2x} + 40 + 8x\ln|x|)}{xe^{-2x}(1 + 40e^{-.2x})^2}$

35. 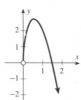 **37.** 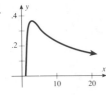 **39.**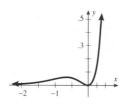

41. $R'(x) = 500\left[\dfrac{x - (x + 10)\ln(x + 10)}{x^2(x + 10)}\right]$ **43. a.** $N(0) = 1$ **b.** .09998 (people per day) **45.** After 1 year: $.1353P$; after 5 years: $.2187P$ **47.** $A'(t) = -.15e^{-.03t}$ **49.** $R'(x) = -50xe^{-.1x} + 500e^{-.1x} = 50e^{-.1x}(10 - x)$
51. Answers vary. **53.** Answers vary.

5.5 Review

Objective Questions
1. 25 **3.** $\sqrt{3}$ **5.** 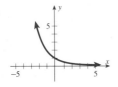 **7.**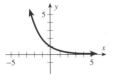

9. $e \approx 2.71828$ **11.** 2.8576511 **13.** $\log\sqrt{10} = .5$ **15.** $\log_9 729 = 3$ **17.** $10^0 = 1$ **19.** $2^6 = 64$
21. 1.0986123 **23.** -2.6777807 **25.**  **27.**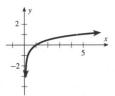

29. $x = 2$ **31.** $x = 3^5$ **33.** $x \approx .09691$ **35.** $x \approx .43065$ **37.** $\dfrac{dy}{dx} = -.49xe^{-.05x^2}$ **39.** $y' = x^2 e^{x^4}(4x^4 + 3)$
41. $y' = \dfrac{2}{x}$ **43.** $\dfrac{2x + 6 - x\ln x^2}{x(x + 3)^2}$ **45.** $102{,}637.89$ **47.** $t = 5{,}700\log_{1/2}\dfrac{A}{10}$ **49.** $P'(1) \approx -2.5023$ psi per mile
51. Equilibrium point at 5,756 items

Sample Test
1. $x = 2$ **3.** $x = 8$ **5.** $x = 3.39794$ **7.** $x \approx -.348707$ **9.** $y' = -5.5xe^{-.5x^2}$ **11.** $y' = \dfrac{5}{x}$
13. $y' = \dfrac{8 - 5x^3}{x(4 - x^3)}$ **15.** $y' = \dfrac{8 - 2x + x\ln x^2}{x(4 - x)^2}$ **17.** $t = \dfrac{1}{t}\ln\dfrac{A}{P}$ **19.** Equilibrium when $x = 7{,}675$ units

CHAPTER 6
6.1 The Antiderivative
1. $\dfrac{x^8}{8} + C$ **3.** $x^4 + C$ **5.** $3x + C$ **7.** $\dfrac{5x^2}{2} + 7x + C$ **9.** $x + C$ **11.** $6x^3 - 3x^2 + 5x + C$ **13.** $\dfrac{1}{9}x + C$
15. $-5x^{-1} + C$ **17.** $x - e^x + C$ **19.** $\dfrac{3}{4}x^{4/3} + \sqrt{2}x + C$ **21.** $2x^{3/2} - 2x^{1/2} + C$ **23.** $\dfrac{3}{5}x^{5/3} + C$
25. $\dfrac{x^2}{2} + x + \ln|x| + C$ **27.** $\dfrac{2}{9}y^{9/2} + C$ **29.** $z + \dfrac{2z^3}{3} + \dfrac{z^5}{5} + C$ **31.** $u^3 - \ln|u| + e^u + C$
33. $F(x) = x^3 + 2x^2 + x + 10$ **35.** $F(x) = 5x + \ln|x| + 4{,}995$ **37.** $C(x) = 2x^2 - 8x + 5{,}000$
39. $C(x) = .003x^3 + 18{,}500$ **41.** Since $P(t) = 450t + 400t^{3/2} + 420{,}000$, in 5 years the projected population is 426,722.

43. $P(x) = 100x^2 - 10{,}000x - 50{,}000$ 45. $N(t) = 120t^2 - t^3 - 69$, and on the tenth day $N(10) = 10{,}931$ (cases).
47. $v(t) = -32t - 72$ 49. $f(x) = x^3 + 5x - 4$

6.2 Integration by Substitution

1. $\dfrac{(5x+3)^4}{20} + C$ 3. $\dfrac{5}{3(5-x)^3} + C$ 5. $\dfrac{2(3x+5)^{3/2}}{9} + C$ 7. $\dfrac{(3x^2+1)^2}{2} + C$ 9. $2\sqrt{x^2+5x} + C$

11. $\tfrac{1}{5}\ln|6+5x| + C$ 13. $-\tfrac{1}{6}\ln|1-3x^2| + C$ 15. $\tfrac{1}{5}e^{5x} + C$ 17. $\tfrac{5}{12}e^{4x^3} + C$ 19. $-\dfrac{1}{4(4x^2-4x)} + C$

21. $\ln(x^4 - 2x^2 + 3) + C$ 23. $\dfrac{\ln^2|x|}{2} + C$ 25. $\dfrac{x^2}{2} + C$ 27. $\tfrac{3}{8}\ln(4x^2+1) + \tfrac{1}{3}e^{x^3} + C$ 29. $\tfrac{3}{10}(x^2+1)^{5/3} + C$

31. $\tfrac{2}{3}(x+1)^{3/2} - 2\sqrt{x+1} + C$ 33. a. $\dfrac{x^3}{3} + x + C$ b. $\dfrac{x^4}{4} + \dfrac{x^2}{2} + C$ c. $\dfrac{x^5}{5} + \dfrac{x^3}{3} + C$ d. $\dfrac{x^6}{6} + \dfrac{x^4}{4} + C$

35. a. $\dfrac{x^7}{7} + \dfrac{3x^5}{5} + x^3 + x + C$ b. $\tfrac{1}{8}(x^2+1)^4 + C$ c. $\dfrac{x^9}{9} + \dfrac{3x^7}{7} + \dfrac{3x^5}{5} + \dfrac{x^3}{3} + C$ d. $\dfrac{(x^2+1)^5}{10} - \dfrac{(x^2+1)^4}{8} + C$

37. a. $\dfrac{x^4}{4} + x + C$ b. $\dfrac{x^5}{5} + \dfrac{x^2}{2} + C$ c. $\dfrac{x^6}{6} + \dfrac{x^3}{3} + C$ d. $\dfrac{x^7}{7} + \dfrac{x^4}{4} + C$ 39. a. $\dfrac{x^{10}}{10} + \dfrac{3x^7}{7} + \dfrac{3x^4}{4} + x + C$

b. $\dfrac{x^{11}}{11} + \dfrac{3x^8}{8} + \dfrac{3x^5}{5} + \dfrac{x^2}{2} + C$ c. $\dfrac{(x^3+1)^4}{12} + C$ d. $\dfrac{x^{13}}{13} + \dfrac{3x^{10}}{10} + \dfrac{3x^7}{7} + \dfrac{x^4}{4} + C$

41. $P(t) = -\tfrac{20}{3}(5-t^2)^3 + \tfrac{2{,}530}{3} = \tfrac{20}{3}(t^2-5)^3 + \tfrac{2{,}530}{3}$ 43. $\$2{,}627.77$ 45. $\$19{,}511.88$ 47. 1,120,026 people
49. 714 million barrels

6.3 The Definite Integral

1. $\tfrac{19}{3}$ 3. $\tfrac{14}{3}$ 5. $9 - \sqrt{3}$ 7. $\tfrac{1}{2}(e^4 - 1)$ 9. $\ln 10$ 11. $\tfrac{35}{5}$ 13. $\ln 6$ 15. 12   17. 0   19. $\tfrac{1}{3}$
21. $\tfrac{17}{4} - \ln 4$ 23. 0   25. 0   27. $\tfrac{5}{33}$ 29. $\tfrac{3}{4}$ 31. a. 14   b. 5   c. 43   33. 6
35. The total consumption is 730 billion barrels.   37. The total consumption is 615 billion barrels.
39. About 14 years from 1985 to 1999.   41. The pollution level is 28,000 ppm.   43. The pollution level is 3,500 ppm.
45. $\$1{,}101{,}273.29$ 47. $\tfrac{1}{2}$ 49. $C(t) = \dfrac{t^2}{40} + 25{,}000$ and $R(t) = -5e^{-t^2} + 5$

6.4 Area Between Curves

1. Since $f(x) > g(x)$ on $[a, b]$, the area $= \displaystyle\int_a^b [f(x) - g(x)]\,dx$ 3. Since $g(x) > f(x)$ on $[a, b]$, the area $= \displaystyle\int_a^b [g(x) - f(x)]\,dx$

5. Since $f(x) > g(x)$ on $[a, b]$, the area $= \displaystyle\int_a^b [f(x) - g(x)]\,dx$ 7. 248   9. 21   11. 12   13. $e^2 - e^{-1}$ 15. 36

17. $\tfrac{86}{3}$ 19. $\tfrac{3}{2}$ 21. $\tfrac{14}{3}$ 23. $\tfrac{56}{3}$ 25. $\tfrac{3}{10}$ 27. $\tfrac{373}{6}$ 29. 4.6051702   31. $\tfrac{1}{6}$ 33. 243
35. 1.4459739   37. 4,192 tons   39. About 184,000 fish   41. 562.5; this area represents the total demand of 562.5 during the first 25 weeks.   43. 22.528776; the area represents the accumulated profit.

6.5 The Fundamental Theorem of Calculus

1. a. $\tfrac{28}{3}$ b. 10   3. a. 63   b. $\tfrac{3{,}006}{64} \approx 46.97$ 5. a. $\tfrac{16}{3}$ b. $\tfrac{25}{4}$ 7. .12   9. 3.43   11. .1080   13. 28
15. 5,700   17. $\tfrac{35}{3}$ 19. 280   21. $\tfrac{45}{28}$ 23. $-\tfrac{1}{5}(e^{-5} - 1) \approx .199$ 25. $x + 5$ 27. $(x+5)^2$ 29. $x^{1/2}$
31. $x^3 - 3x^2 + 5x - 7$ 33. $\sqrt[3]{x}$ 35. $x^2(x+1)^3$ 37. 3.9 million   39. About 68 million tons
41. 500 rabbits   43. The average temperature is about $63°F$.   45. 600 ft^2

6.6 Numerical Integration

1. $\Delta x = 5;\ A_1 = 10$ 3. $\Delta x = \tfrac{5}{3};\ A_3 = 11.81415$ 5. $\Delta x = 5;\ T_1 = 12.5$ 7. $\Delta x = \tfrac{5}{3};\ T_3 \approx 12.64748$
9. $\Delta x = \tfrac{5}{3};\ P_2 \approx 12.66503$ 11. $\Delta x = \tfrac{5}{6};\ P_6 \approx 12.66664$ 13. $\Delta x = 2;\ A_1 = .16$ 15. $\Delta x = \tfrac{2}{3};\ A_3 \approx .13546$
17. $\Delta x = 2;\ T_1 \approx .12938$ 19. $\Delta x = \tfrac{2}{3};\ T_3 \approx .12526$ 21. $\Delta x = 1;\ P_2 \approx .12476$ 23. $\Delta x = \tfrac{1}{3};\ P_6 \approx .12473$
25. $\Delta x = 1;\ A_4 \approx .14$ 27. $\Delta x = \tfrac{1}{2};\ A_4 \approx 8.36$ 29. $\Delta x = 1;\ T_4 \approx .11$ 31. $\Delta x = \tfrac{1}{2};\ T_4 \approx 14.86$
33. $\Delta x = 1;\ P_4 \approx .10$ 35. $\Delta x = \tfrac{1}{2};\ P_4 \approx 13.81$ 37. Area ≈ 890 square feet. About 100 yd^3 of fill will be needed.

6.7 Review

Objective Questions
1. $\frac{x^7}{7} + C$
3. $e^u + C$
5. $2x^3 - x^2 - 5x + C$
7. $\frac{1}{3}\ln|3x + 1| + C$
9. $F(x) = e^x + 9x + 9$
11. $F(x) = \ln|x|$
13. $C(x) = 3x^3 + x + 25,000$
15. $C(x) = 10x - 2\sqrt{x} + 1,500$
17. $-\frac{e^{-3x}}{3} - \frac{x}{e} + C$
19. $\frac{-2}{\sqrt{x^2 - 5x}} + C$
21. $\frac{5}{2}$
23. $15,657$
25. $\frac{74}{15}$
27. $\frac{20}{81}$
29. $\int_2^9 [g(x) - f(x)]\,dx$
31. $\int_{-20}^{-9} [g(x) - f(x)]\,dx + \int_{-9}^0 [f(x) - g(x)]\,dx$
33. 15
35. $\frac{2,197}{384} \approx 5.72$
37. 14.56
39. $\Delta x = 5.25; A \approx 24.33$
41. 658
43. About $2\frac{3}{4}$ million cars were sold in the last quarter.
45. 12
47. $\frac{1}{4}(e^9 - e) \approx 2,025.091411$
49. 1.2
51. $T \approx 1.10$
53. n must be *even* for Simpson's rule.
55. $P = 3.08$
57. $P(x) = 125x^2 - 5,000x - 1,500$
59. The concentration of waste in the tank is 5.8%.

Sample Test
1. $\frac{x^6}{6} + C$
3. $e^u + C$
5. $6x^3 - 3x^2 - 3x + C$
7. $\frac{1}{3}e^u + \frac{1}{e}x + C = \frac{1}{3}e^{3x+1} + \frac{1}{e}x + C$
9. $\frac{4}{3}$
11. $15,657.1875$
13. $\frac{74}{9}$
15. $\frac{1}{2}$
17. It will be economically feasible.
19. $t = 14.4$ years

CHAPTER 7
7.1 Business Models Using Integration
1. $512\frac{2}{3}$
3. $-.197$
5. 13
7. $\$2,500$
9. $\$1,229.56$
11. $\$663.60$
13. $\$61,607.44$
15. $\$2,563.55$
17. $\$38,913.42$
19. 116
21. 396
23. $\$4,028.35$
25. 6.32
27. $\$1,507.14$; additional monthly cost $\$31.40$
29. $\frac{4}{3}$
31. $\$126,241.19$
33. It will take $1\frac{1}{4}$ years.
35. The producers' surplus is $\$3,495.33$.
37. The producers' surplus is $\$2$.
39. $\$11,400$
41. 1.31 years
43. $\$7,453,560$

7.2 Integration by Parts
1. $4xe^{3x} - \frac{4}{3}e^{3x} + C$
3. $1 - 2e^{-1}$
5. $-\frac{2}{3}x(1-x)^{3/2} + \frac{2}{3}[-\frac{2}{5}(1-x)^{5/2}] + C$
7. $\frac{x(x+2)^4}{4} - \frac{(x+2)^5}{20} + C$
9. $(x+1)\ln(x+1) - \int dx = (x+1)\ln(x+1) - x + C$
11. $-2x(1-x)^{1/2} - \frac{4}{3}(1-x)^{3/2} + C$
13. $3x^2 e^{2x} - 3xe^{2x} + \frac{3}{2}e^{2x} + C$
15. $-\frac{x^2(1-2x)^{3/2}}{3} - \frac{2x(1-2x)^{5/2}}{15} - \frac{2(1-2x)^{7/2}}{105} + C$
17. $-\frac{3}{8}(1-x^2)^{4/3} + \frac{3}{14}(1-x^2)^{7/3} + C$ or $-\frac{3x^2(1-x^2)^{4/3}}{8} - \frac{9(1-x^2)^{7/3}}{56} + C$
19. $\frac{x^3 \ln x}{3} - \frac{x^3}{9} + C$
21. $\frac{3}{2}x^2 e^{4x} - \frac{3}{4}xe^{4x} + \frac{3}{16}e^{4x} + C$
23. $-(1-x^2)^{1/2} + \frac{1}{3}(1-x^2)^{3/2} + C$
25. $\frac{1}{2}x^2 e^{x^2} - \frac{1}{2}e^{x^2} + C$
27. $2\ln 6 - \ln 3 - 1 = \ln 12 - 1$
29. $5 - 4e$
31. $\ln 2$
33. $e - 2$
35. Answers vary.
37. **a.** $-\frac{1}{2}\ln|x^2 + 1| + \frac{1}{2}x^2 + C$ **b.** $\frac{x^2}{2} - \frac{1}{2}\ln|x^2 + 1| + C$
39. **a.** $N(t) = 401 - 20e^{-.05t}(t + 20)$ **b.** 20.5 million/liter
41. $\$32,559.20$

7.3 Using Tables of Integrals
1. Using Formula 9, $\int (1 + bx)^{-1}\,dx = \frac{1}{b}\ln|1 + bx| + C$
3. Using Formula 23, $\int \frac{x\,dx}{\sqrt{x^2 + a^2}} = \sqrt{x^2 + a^2} + C$
5. Using Formula 26, $\int \frac{dx}{x^2\sqrt{x^2 - a^2}} = \frac{\sqrt{x^2 - a^2}}{a^2 x} + C$
7. Using Formula 30, $\int x \ln x\,dx = \frac{x^2}{2}\ln x - \frac{x^2}{4} + C$
9. Using Formula 32 $(m = -1)$, $\int x^{-1} \ln x\,dx = \frac{1}{2}\ln^2 x + C$
11. Using Formula 35, $\int xe^{ax}\,dx = \frac{e^{ax}}{a^2}(ax - 1) + C$
13. Using Formula 29 $(a = 1)$, $\int \frac{x^2}{\sqrt{x^2 + 1}}\,dx = \frac{x}{2}\sqrt{x^2 + 1} - \frac{1}{2}\ln|x + \sqrt{x^2 + 1}| + C$
15. Using Formula 26 $(a = 4)$,

$\int \dfrac{dx}{x^2\sqrt{x^2+16}} = -\dfrac{\sqrt{x^2+16}}{16x} + C$ **17.** Let $u = 4x^2 + 1$, then $du = 8x\,dx$: $\int \dfrac{x\,dx}{\sqrt{4x^2+1}} = \tfrac{1}{4}\sqrt{4x^2+1} + C$

19. Using Formula 20 $\left(a = \tfrac{1}{3}\right)$, $\int \dfrac{dx}{x\sqrt{1-9x^2}} = -\ln\left|\dfrac{1+\sqrt{1-9x^2}}{3x}\right| + C$ **21.** Using Formula 23,

$\int \dfrac{x\,dx}{\sqrt{x^2+4}} = \sqrt{x^2+4} + C$ **23.** Using Formula 9, with $a = 1$, $b = 1$, and $n = 3$: $\int (1+x)^3\,dx = \dfrac{(1+x)^4}{4} + C$

25. Using Formula 10 with $a = 1$, $b = 1$, and $n = 3$: $\int x(1+x)^3\,dx = \tfrac{1}{5}(1+x)^5 - \tfrac{1}{4}(1+x)^4 + C$ **27.** Using Formula 13

with $m = 1$, $a = 1$, $b = 1$: $\int x\sqrt{1+x}\,dx = \dfrac{-2(2-3x)\sqrt{(1+x)^3}}{15} + C$ **29.** Using Formula 35 with $a = 4$:

$\int xe^{4x}\,dx = \dfrac{e^{4x}}{16}(4x - 1) + C$ **31.** Let $u = 2x$, then $x = \dfrac{u}{2}$ and $du = 2\,dx$. Using Formula 30, $\int x\ln 2x\,dx = \dfrac{x^2}{2}\ln 2x - \dfrac{x^2}{4} + C$

33. Let $u = 5x$, then $du = 5\,dx$. Then use Formula 31: $\int x^2 \ln 5x\,dx = \dfrac{x^3}{3}\ln 5x - \dfrac{x^3}{9} + C$ **35.** Using Formula 39 with

$a = 3$, $b = 5$, and $m = 1$: $\int \dfrac{dx}{3 + 5e^x} = \dfrac{x}{3} - \tfrac{1}{3}\ln|3 + 5e^x| + C$ **37.** Using Formula 39 with $a = 1$, $b = 1$, and $m = 2$:

$\int \dfrac{dx}{1 + e^{2x}} = x - \tfrac{1}{2}\ln|1 + e^{2x}| + C$ **39.** Let $u = 1 + x$, then $du = dx$. Using Formula 2: $\int \dfrac{dx}{\sqrt{1+x}} = 2(1+x)^{1/2} + C$

41. Using Formula 11 with $a = 1$, $b = 1$, and $n = 3$: $\int x^2(1+x)^3\,dx = \dfrac{(1+x)^6}{6} - \dfrac{2(1+x)^5}{5} + \dfrac{(1+x)^4}{4} + C$

43. Using Formula 11 with $a = 1$, $b = -1$, and $n = 3$: $\int 5x^2(1+x)^3\,dx = \dfrac{-5(1+x)^6}{6} + 2(1-x)^5 - \dfrac{5(1+x)^4}{4} + C$

45. Using Formula 13 with $m = 1$, $a = 1$, and $b = 2$: $\int 2x\sqrt{1-2x}\,dx = \dfrac{-2(1+3x)\sqrt{(1-2x)^3}}{15} + C$ **47.** Using Formula 11

with $a = 2$, $b = 3$, and $n = 3$: $\int x^2(2+3x)^3\,dx = \dfrac{1}{27}\left[\dfrac{(2+3x)^6}{6} - \dfrac{4(2+3x)^5}{5} + (2+3x)^4\right] + C$ **49.** Let $U = 4x^3 + 1$, then

$du = 12x^2\,dx$. Using Formula 2, $\int x^2\sqrt{4x^3+1}\,dx = \tfrac{1}{18}(4x^3+1)^{3/2} + C$ **51.** Using Formula 36 with $m = 2$ and $a = 3$:

$\int x^2 e^{3x}\,dx = \dfrac{x^2 e^{3x}}{3} - \dfrac{2}{3}\int xe^{3x}\,dx$. This last integral can be done using Formula 35 with $a = 3$: $\dfrac{e^{3x}}{27}(9x^2 - 6x + 2) + C$

53. Using Formula 14 (twice): $\dfrac{-\sqrt{(2+9x)^3}}{2x} + \dfrac{9}{4}\left[2\sqrt{2+9x} + \sqrt{2}\ln\left|\dfrac{\sqrt{2+9x}-\sqrt{2}}{\sqrt{2+9x}+\sqrt{2}}\right|\right] + C$ **55.** Let $u = 3x$, then

$du = 3\,dx$: $\int \dfrac{\sqrt{2+9x^2}}{x}\,dx = \int \dfrac{\sqrt{2+u^2}}{u}\,du$. Using Formula 21 with $a = \sqrt{2}$: $\sqrt{9x^2+2} - \sqrt{2}\ln\left|\dfrac{\sqrt{2}+\sqrt{9x^2+2}}{3x}\right| + C$

57. Let $u = 3x$, then $du = 3\,dx$: $\int \dfrac{1}{\sqrt{2+9x^2}}\,dx = \int \dfrac{1}{\sqrt{2+u^2}}\,\dfrac{du}{3}$. Using Formula 16 with $a = \sqrt{2}$: $\tfrac{1}{3}\ln|3x + \sqrt{9x^2+2}| + C$

59. Let $u = 9 - 16x^2$, then $du = -32x\,dx$: $\int 5x\sqrt{9-16x^2}\,dx = -\tfrac{5}{48}(9-16x^2)^{3/2} + C$ **61.** Using Formula 28 with $a = 1$:

$\int \dfrac{\sqrt{x^2-1}}{x^2}\,dx = -\dfrac{\sqrt{x^2-1}}{x} + \ln|x + \sqrt{x^2-1}| + C$ **63.** Using Formula 11 with $n = \tfrac{1}{2}$, $a = 3$, and $b = 10$:

$\int x^2(3+10x)^{1/2}\,dx = \dfrac{(3+10x)^{7/2}}{3{,}500} - \dfrac{3(3+10x)^{5/2}}{1{,}250} + \dfrac{3(3+10x)^{3/2}}{500} + C$. If instead of Formula 11 you use Formula 13, you will obtain the following (equivalent) form: $\dfrac{6 - 30x + 125x^2}{4{,}375}(3 + 10x)^{3/2} + C$ 65. The total amount will be $\displaystyle\int_1^{12} \dfrac{5{,}000}{t\sqrt{100 + t^2}}\,dt$.

Using Formula 19 with $a = 10$: $\displaystyle\int_1^{12} \dfrac{5{,}000}{t\sqrt{100 + t^2}}\,dt = -500\left[\left|\dfrac{10 + \sqrt{244}}{12}\right| - \ln\left|\dfrac{10 + \sqrt{101}}{1}\right|\right] \approx 1{,}120$ tons.

67. The accumulated sales are $\displaystyle\int_{12}^{24} te^{-.1t}\,dt$. Using Formula 35 (or integration by parts) with $a = .1$:

$\dfrac{e^{2.4}}{.01}(2.4 - 1) - \dfrac{e^{1.2}}{.01}(1.2 - 1) \approx 1{,}476.842$ thousands or about \$1,480,000.

7.4 Improper Integrals
1. 1   3. 2   5. $\tfrac{1}{2}$ 7. Diverges   9. 10   11. Diverges   13. Diverges   15. 0   17. 1   19. $2e^{-1}$
21. $\dfrac{\sqrt{5} - 1}{4}$ 23. \$336,666.67   25. Answers vary.   27. 4,000 millirems   29. $\tfrac{1}{4}$ 31. \$625,000
33. 4,000 (in thousands of barrels) or 4 million barrels

7.5 Probability Density Functions
1. Let X = number of people who responded "yes," which is a discrete random variable. The sample space is $\{0, 1, 2, \ldots, 10\}$.
3. Let X = number of words with an error, which is a discrete random variable. The sample space is $\{0, 1, 2, \ldots, 499, 500\}$.
5. Let x = number of boys (or the number of girls), which is a discrete random variable. The sample space is $\{bbb, bbg, bgb, bgg, gbb, gbg, ggb, ggg\}$.
7. Let X = height in inches.

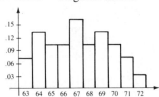

9. Let X = time (in minutes) spent on each transaction.

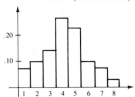

11. $f(x)$ is not a probability density function.
13. This is a probability density function.   15. $f(x)$ is not a probability density function.   17. $f(x)$ is not a probability density function.   19. No value of k will make $f(x)$ a probability density function.   21. $\tfrac{3}{125}$ 23. $\tfrac{2}{3}$ 25. $\tfrac{3}{14}$
27. $\tfrac{4}{3}$ 29. .8907   31. .0455   33. .3202   35. .40   37. .075   39. .1587   41. .0548   43. .5403
45. .6065

7.6 Review

Objective Questions
1. 135.13   3. 2.207   5. \$181.33   7. \$1.33   9. 5,000   11. \$809.01   13. \$488,629.85
15. \$184,027.30   17. \$208.33   19. \$83.33   21. $-\dfrac{e^{2x}}{4}(2x + 1) + C$ 23. $x\ln\sqrt{2x} - \tfrac{1}{2}x + C$
25. Use Formula 25 with $a^2 = 16$: $\tfrac{x}{4}\sqrt{(x - 16)^3} + 2x\sqrt{x^2 - 16} - 32\ln|x + \sqrt{x^2 - 16}| + C$ 27. Let $u = 3x$, then $du = 3\,dx$. Use Formula 34, then 33: $x\ln^3 3x - 3x\ln^2 3x + 6x\ln 3x - 6x + C$ 29. $-\tfrac{1}{2}x^2 e^{-2x} - \tfrac{1}{4}e^{-2x}(2x + 1) + C$
31. Let $u = 4x^2 + 5x - 3$, then $du = (8x + 5)\,dx$: $\ln|4x^2 + 5x - 3| + C$ 33. $\tfrac{1}{2}$ 35. 1   37. $\{2, 3, 5, \ldots, 12\}$; X = sum of the top faces; discrete.   39. $[0, \infty)$; X = number of minutes a rat takes to go through a maze; continuous.

41.

Roll	2	3	4	5	6	7	8	9	10	11	12
Frequency	1	3	3	2	7	7	4	6	3	3	1
Rel. Frequency	$\frac{1}{40}$	$\frac{3}{40}$	$\frac{3}{40}$	$\frac{2}{40}$	$\frac{7}{40}$	$\frac{7}{40}$	$\frac{4}{40}$	$\frac{6}{40}$	$\frac{3}{40}$	$\frac{3}{40}$	$\frac{1}{40}$

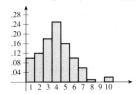

43.

Number of Minutes	1	2	3	4	5	6	7	8	9	10
Rel. Frequency	$\frac{10}{100}$	$\frac{12}{100}$	$\frac{18}{100}$	$\frac{25}{100}$	$\frac{16}{100}$	$\frac{10}{100}$	$\frac{6}{100}$	$\frac{1}{100}$	$\frac{0}{100}$	$\frac{2}{100}$

45. This is a probability density function.

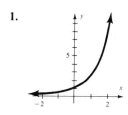

47. This is not a probability density function. **49.** $k = \frac{5}{128}$ **51.** $k = \frac{2}{495}$ **53.** .5 **55.** .7926 **57.** 276
59. $4,800,000 **61. a.** 31 **b.** 1 **c.** 60 **d.** 2.3%

Sample Test

1. $\frac{1}{8}(2x-1)^4 + C$ **3.** By parts: $\dfrac{x^2(2x-1)^4}{8} - \dfrac{x(2x-1)^5}{40} + \dfrac{(2x-1)^6}{480} + C$. If instead, you use Formula 11, you will obtain the following (equivalent) form: $\dfrac{(2x-1)^6}{48} + \dfrac{(2x-1)^5}{40} + \dfrac{(2x-1)^4}{32} + C$ **5.** $\ln|4x^2 + 5x - 3| + C$
7. $3x - \ln|1 - e^{3x}| + C$ **9.** $x\ln^3 x - 3x\ln^2 x + 6x\ln x - 6x + C$ **11.** Formula 14 **13.** $\dfrac{e}{3} \approx .9061$
15. $307,543.02 **17.** $50(1 - e^{-.1t})$

Cumulative Review for Chapters 5–7

1. **3.** $605 = x$ **5.** .301 **7. a.** 0 **b.** $b - a$ **c.** $-\int_b^a f(x)\,dx$ **d.** $\int_a^b f(x)\,dx$
e. $k\int_a^b f(x)\,dx$ **f.** $\int_a^b f(x)\,dx \pm \int_a^b g(x)\,dx$
g. $u(x)v(x)\big|_{x=a}^{x=b} - \int_{x=a}^{x=b} v\,du = u(b)v(b) - u(a)v(a) - \int_{x=a}^{x=b} v\,du$ where u and v are functions of x.

9. a. $x^5 + x^3 + 5x + C$ **b.** $\frac{1}{2}e^{2x} + C$ **11.** $x\ln^4 5x - 4x\ln^3 5x + 12x\ln^2 5x - 24x\ln 5x + 24x + C$
13. $\ln|4x^2 - 3x + 2| + C$ **15.** 0 **17.** $\frac{512}{3}$ **19.** $t = 4$ years **21.** 20.12 years **23.** 1,850 trees
25. .5488; about $\frac{1}{2}$ second response time

CHAPTER 8

8.1 Three-Dimensional Coordinate System

1. The area of a rectangle with dimensions 15 by 35 is $K(15, 35) = (15)(35) = 525$ square units.   **3.** The volume of a box 3 by 5 by 8 is $V(3, 5, 8) = (3)(5)(8) = 120$ cubic units.   **5.** The simple interest from $500 investment at 5% for 3 years is $I(500, .05, 3) = 500[1 + (.05)(3)] = 500(1.15) = 575.   **7.** The future value of a $2,500 investment at 12% compounded quarterly for 6 years is $A(2500, .12, 6, 4) = 2,500(1 + \frac{.12}{4})^{(4)(6)} = 2,500(1.03)^{24} \approx 2,500(2.0327941) = $5,081.99$
9. The future value of a $110,000 investment at 9% compounded monthly for 30 years is
$A(110,000, .09, 30, 12) = 110,000(1 + \frac{.09}{12})^{(30)(12)} = 110,000(1.0075)^{360} \approx 110,000(14.730577) = $1,620,363$
11. $f(2, 3) = 2^2 - 2(2)(3) + 3^2 = 1$ **13.** $f(-2, 5) = (-2)^2 - 2(-2)(5) + 5^2 = 49$ **15.** $g(2, 1) = \dfrac{2(2) - 4(1)}{2^2 + 1^2} = 0$

17. $g(5, -3) = \dfrac{2(5) - 4(-3)}{5^2 + (-3)^2} = \dfrac{11}{17}$ **19.** $h(0, 5) = \dfrac{e^{(0)(5)}}{\sqrt{0^2 + 5^2}} = \dfrac{1}{5}$ or .2

21. $h(-2, -3) = \dfrac{e^{(-2)(-3)}}{\sqrt{(-2)^2 + (-3)^2}} = \dfrac{e^6}{\sqrt{13}} \approx \dfrac{403.42879}{3.6055513} = 111.89$ **23.**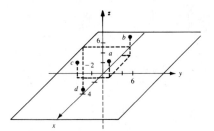

25. 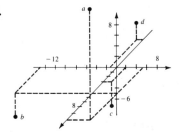   **27.** 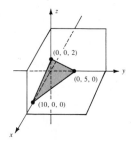   **29.**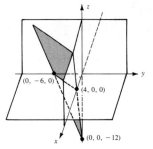

31. 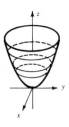   **33.**   **35.**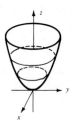

37. 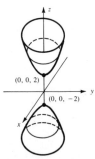   **39.** 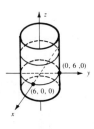   **41.**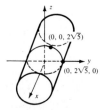

43. a. $19,950 **b.** $21,450 **c.** $21,200 **d.** $20,200   **45.** The total cost is $C(x, y, z) = 2xy + 1.25xy + 1.5xz + 1.5yz$
47. a. 3.875×10^8 mℓ **b.** 3.06×10^{10} mℓ **c.** 1.92×10^7 mℓ

8.2 Partial Derivatives
1. $10x - 9x^2y^4$ **3.** $-15(4x - 3y)^4$ **5.** -134 **7.** $-759,375$ **9.** $-36x^2y^3$ **11.** $-240(4x - 3y)^3$
13. 10 **15.** 0 **17.** $-69,120$ **19.** $15,360$ **21.** $3e^{3x+2y}$ **23.** $\dfrac{-3y}{\sqrt{x^2 - 3y^2}}$ **25.** $3e^7$
27. $g_y(3, -2)$ does not exist.   **29.** $6e^{3x+2y}$ **31.** $\dfrac{3xy}{(\sqrt{x^2 - 3y^2})^3}$ **33.** $9e^4$ **35.** $\dfrac{6}{e^3}$ **37.** 0 **39.** 0

41. $f_x = 1 + 2y + \lambda y;\ f_y = 2x + \lambda x;\ f_\lambda = xy - 10$ **43.** $f_x = 2x - 3\lambda;\ f_y = 2y - 2\lambda;\ f_\lambda = -3x - 2y + 6$

45. $\dfrac{\partial f}{\partial b} = 100m + 50b + 1;\ \dfrac{\partial f}{\partial m} = 200m + 100b + 2$

47. It might be easier to rewrite $f(b,m)$:
$f(b,m) = (m + b + 1)^2 + 4(m + b + 1)^2 + 9(m + b + 1)^2 = 14(m + b + 1)^2$

$\dfrac{\partial f}{\partial b} = 28(m + b + 1) \cdot \dfrac{\partial}{\partial b}(m + b + 1) = 28(m + b + 1)$

$\dfrac{\partial f}{\partial m} = 28(m + b + 1) \cdot \dfrac{\partial}{\partial m}(m + b + 1) = 28(m + b + 1)$

49. a. $\dfrac{\partial P}{\partial a} = 2p - .2ap$ **b.** $\dfrac{\partial P}{\partial p} = 2a + 50 - 20p - .1a^2$. At a fixed level of advertising the rate of change of profit per unit change in price is $2a + 50 - 20p - .1a^2$.   **51.** The rate of decrease in profit is \$1,000 per unit increase in x, assuming y stays constant, $P_y = 350$, $P_y(5, 10) = 350$. The rate of increase in profit is \$350 per unit increase in y, assuming x stays constant.

53. For most people answers will vary between 900 and 1,700 square feet.   **55.** $A_r = \dfrac{\partial A}{\partial r} = 500(1 + r)^4$

57. $P_r = 300\left[\dfrac{rt(1 + \frac{r}{12})^{-12t-1} - 1 + (1 + \frac{r}{12})^{-12t}}{r^2}\right]$. For $t = 5$: $P_r = -\dfrac{300}{r^2}[1 - 5r(1 + \frac{r}{12})^{-61} - (1 + \frac{r}{12})^{-60}]$

8.3 Maximum–Minimum Applications

1. Saddle point at $(0, 0, f(0,0)) = (0, 0, 0)$ **3.** Relative minimum at $(-1, 2, f(-1, 2)) = (-1, 2, -3)$
5. Saddle point at $(0, 0, f(0,0)) = (0, 0, 0)$; relative maximum at $(-1, 1, f(-1, 1)) = (-1, 1, 1)$
7. Relative maximum at $(-\frac{9}{11}, -\frac{41}{11}, \frac{45}{11})$ **9.** Saddle point at $(3, -2, 2)$ **11.** No relative maximums, minimums, or saddle points   **13.** No relative maximums, minimums, or saddle points   **15.** Saddle point at $(0, 0, 0)$; relative maximums at $(-1, -1, 2)$ and $(1, 1, 2)$ **17.** Saddle point at $(0, 0, 0)$; relative maximum at $(-1, -1, 1)$
19. The maximum is 141 (thousand dollars), which occurs at $(a, n) = (6, 1)$.   **21.** Profit will be maximized when about 50 standard and 75 deluxe rings are sold.   **23.** The box should therefore be 3.04 by 3.04 by approximately 2.32 feet.
25. Answers vary.

8.4 Lagrange Multipliers

1. 100   **3.** $-\dfrac{3,456}{11} \approx -314.18$ **5.** $\dfrac{5,488}{27} \approx 203.26$ **7.** $f(12, 12) = 288$ is a relative minimum.
9. $f(3, 3) = 5$ is a relative minimum.   **11.** $f(8, 8) = 128$ is a relative minimum.   **13.** Both numbers are 5.
15. 119 is a relative maximum; he should apply $\frac{250}{59}$ acre-feet of water and $\frac{75}{59}$ pounds of fertilizer.
17. The rancher should build his fence $166\frac{2}{3}$ by $166\frac{2}{3}$ for a maximum area of 27,778 square feet.
19. \$.95 is a maximum; the optimum mixture is $\sqrt{10} \approx 3.16$ ounces of meat and $5\sqrt{10} \approx 15.81$ ounces of vegetables.
21. $10\sqrt{3}$ is a maximum.   **23.** $T(\frac{10}{3}, \frac{10}{3}, \frac{10}{3}) = \frac{200}{3}$

8.5 Multiple Integrals

1. $2x^2 + \frac{8}{3}$ **3.** $\frac{15}{2}(y^2 - 1)$ **5.** $\dfrac{x}{3}(x^2 + 8)^{3/2} - \dfrac{x^4}{3}$ **7.** $3e^{2y+2} - 3e^{2y+1}$ **9.** $\dfrac{e^{3y+25}}{2} - \dfrac{e^{3y+1}}{2}$ **11.** x^2

13. 9   **15.** 10   **17.** $\frac{91}{3}$ **19.** $\displaystyle\int_0^2\int_0^{y^2} xy\,dx\,dy = \dfrac{16}{3}$ **21.** $\displaystyle\int_0^5\int_0^x (2x + y)\,dy\,dx = \dfrac{625}{6}$

23. $\displaystyle\int_0^1\int_{x^3}^{x^2} x^2 y\,dy\,dx = \dfrac{1}{63}$ **25.** $\displaystyle\int_1^3\int_0^4 dy\,dx = 9$ **27.** $\displaystyle\int_0^1\int_1^3 (x + 3y)\,dx\,dy + \displaystyle\int_1^3\int_x^3 (x + 3y)\,dy\,dx = \dfrac{91}{3}$

29. $\displaystyle\int_0^5\int_y^5 (2x + y)\,dy\,dx = \dfrac{625}{6}$ **31.** $\displaystyle\int_0^3\int_0^2 (x^2 + y)\,dx\,dy = 17$ **33.** $\displaystyle\int_0^3\int_0^{1-x} \dfrac{1}{x}\,dy\,dx = \ln 3 - 2$

35. $\displaystyle\int_1^2\int_0^{y^2} y^3 e^{xy}\,dx\,dy = \dfrac{e^8}{3} - \dfrac{e}{3} - \dfrac{7}{3}$ **37.** $\displaystyle\int_2^4\int_1^3 xy\sqrt{x^2 + y^2}\,dx\,dy \approx 92$ **39. A.** $\displaystyle\int_0^1\int_2^{\sqrt{x}} (x + y)\,dy\,dx = \dfrac{3}{10}$

B. $\displaystyle\int_0^1\int_{y^2}^{\sqrt{y}} (x + y)\,dx\,dy = \dfrac{3}{10}$ **41. A.** $\displaystyle\int_0^5\int_0^{5-x} 2xy\,dy\,dx = \dfrac{625}{12}$ **B.** $\displaystyle\int_0^5\int_0^{5-x} 2xy\,dx\,dy = \dfrac{625}{12}$

43. a. $\int_0^2 \int_{x^2}^{8-x^2} dy\,dx = \dfrac{32}{3}$ **b.** $\int_0^4 \int_0^{\sqrt{y}} dx\,dy + \int_4^8 \int_0^{\sqrt{8-y}} dx\,dy = \dfrac{32}{3}$ **45.** $\int_{-3}^2 \int_1^5 5\,dx\,dy = 100$

47. $\int_0^1 \int_0^1 xy\sqrt{x^2+y^2}\,dx\,dy = \dfrac{1}{15}(2^{5/2}-2) \approx .2437902833$ **49.** $\int_0^1 \int_{x^2}^{\sqrt{x}} 5xy\,dy\,dx = \dfrac{5}{12}$ **51.** $\int_0^3 \int_{x^2}^{3x} 6\,dy\,dx = 27$

53. $\int_0^1 \int_{x^2}^{x} xy^2\,dy\,dx = \dfrac{1}{40}$ **55.** -2 **57.** $\dfrac{16}{35}\sqrt[4]{250} \approx 1.81776166$

8.6 Correlation and Least Squares Applications

1. Yes **3.** No **5.** Yes **7.** No **9.** No **11.** No
13. .995 **15.** $-.9847$ **17.** .970

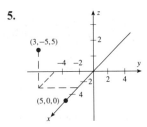

There is a linear correlation at 5%. There is a linear correlation at 1%. There is a linear correlation at 1%.

19. $y = 3.9x - 3$ **21.** $y = .56x + 41.58$ **23.** $y = -2.71x + 30.06$ **25.** .842, significant at 1%
27. $y = 2.45x + 6.09$ **29.** $r = .358$, no significant correlation **31.** $r = -.936$, significant at 1%
33. $y = .468x + 19.235$ **35.** $y = -6.186x + 87.179$

8.7 Review

Objective Questions

1. 339 **3.** -60 **5.** **7.**

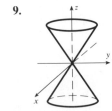

9. **11.** **13.** $24(2x-5y)^{11}$

15. $-1{,}000y(x^2+y^2)^{-3/2}$ **17.** $60x^3 - 60x^2y^3 + 4$ **19.** $-60x^3y^2$ **21.** No relative maximums or minimums
23. A saddle point at $(0,0,0)$ **25.** $f(5,5) = -38$ **27.** The solution is $x = 125, y = 125$ $(\lambda = -125)$; maximum is 15,625.
29. $\dfrac{26y^3}{3}$ **31.** $(6 - \tfrac{32}{9}\sqrt{2})y^2$ **33.** 54 **35.** $\int_1^2 \int_0^{1/y} y^3\,dx\,dy = \dfrac{7}{3}$ **37.** 16.2 **39.** $\tfrac{16}{3}$ **41.** $\dfrac{1{,}875}{4} = 468.75$

43. $\int_0^1 \int_0^2 x^2 y^3 \, dx \, dy = \frac{2}{3}$ **45.** 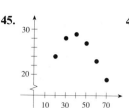   **47.**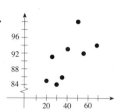

49. $r = -.914$, significant at both 1% and 5%   **51.** $y = -.12x + 30.4$ **53.** $y = .218x + 81.821$
55. At a capital investment level of 4 million dollars and a labor force of 2 (in hundreds), the rate of increase of productivity per unit change in labor (in hundreds) is 4 (units per employee per day).   **57.** The maximum productivity is .3511 when $x = \frac{16}{27}$ (≈ 59 workers) and $y = \frac{8}{9}$ ($\approx \$888{,}889$).

Sample Test
1. $P(10, 50) = 10^{.4} 50^{.2} \approx 5.49$ **3.** 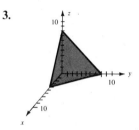   **5.** $f_x = 10(x - 3y)^9$

7. $f_x = 16x^3 - 9x^2y^2 + 4x$; $f_{xx} = 48x^2 - 18xy^2 + 4$ **9.** 18   **11.** Since $D > 0$, $(0, 0, 0)$ is a saddle point.
13. $\int_0^2 \int_1^{e^x} dy \, dx = e^2 - 3 \approx 4.39$ **15.** $R(5, 8) = 910$; $C(5, 8) = 1{,}286.63$

CHAPTER 9
9.1 First-Order Differential Equations
1. $y = \frac{x^3}{3} + C$ **3.** $y = 4x^2 - 10x + C$ **5.** $y = x^4 - x^3 - 5x + C$ **7.** $y = \frac{1}{5}e^x + C$ **9.** $y = \frac{1}{90}(5x + 1)^{3/2} + C$
11. $y^2 = x^2 + C$ **13.** $y^3 = \frac{3x^4}{4} - 9x + C$ **15.** $y = Ce^{x^2}$ **17.** $50 \ln|P| = t + C$ or $P = Me^{.02t}$
19. $1{,}000 \ln|N| = t + C$ **21.** $y^3 = x^3 + 125$ **23.** $P = e^3 e^{.02t}$ or $\ln P = 3 + .02t$ **25.** $\ln|y| = 2\sqrt{x} - 2$
27. $\ln|y| = \frac{1}{2} \ln(1 + x^2) + \frac{1}{2} \ln 2$ **29.** 4,725   **31.** 44.2 hours

9.2 Applications—Growth Models
1. 45,553   **3.** About 10,560 years old   **5.** 176 items per month   **7.** \$11,419 billion   **9.** 275,072,323
11. 17,145,076   **13.** 2,669,808   **15.** 277   **17.** 18 times per week   **19.** $P = 1 - e^{-kt}$

9.3 Special Second-Order Differential Equations
1. $y = -4x^3 + 28x + 5$ **3.** $y = 5x^2 - x^3 + 7x + 8$ **5.** $y = \frac{1}{6}x^4 - \frac{1}{2}x^3 + \frac{5}{2}x^2 + \frac{1}{6}x + 3$
7. $y = -\frac{1}{2}x^3 + 2x^2 - 415x + 4{,}500$ **9.** $y = \frac{4}{9}x^3 - 4$ **11.** $y = \left(\frac{e^2}{1 - e^6}\right)e^{-x/2} + \left(\frac{e^3}{e^6 - 1}\right)e^x$
13. $y = e^x + Ax + B$ **15.** $y = \frac{1}{2}x^2 - e^{-x} + Ax + B$ **17.** $y = \frac{5}{12}x^4 - x^2 + Ax + B$ **19.** $y = x^2 - \frac{1}{6}x^3 + Ax + B$
21. $y = Ae^{x/6} + Be^{-4x}$ **23.** $y = Ae^{3x} + Be^{-3x}$ **25.** $y = \frac{A}{x} + B$ **27.** $y = A(8x + x^4) + B$
29. $D(t) = Ae^{\sqrt{M}t} + Be^{-\sqrt{M}t}$ **31.** Answers vary.

9.4 Review
Objective Questions
1. $y = 2x^4 - \frac{2}{3}x^3 + x + C$ **3.** $y = \frac{1}{2}x^2 - \frac{5}{3}x + C$ **5.** $y^2 = \frac{1}{4}x^2 + \frac{5}{2}x + C$ **7.** $y = Ae^{5x/2}$ **9.** $y^3 = \frac{3}{4}x^4 - 125$
11. $y = e^{x^2 + 1}$ **13.** $y = -\frac{1}{4}x^4 + \frac{15}{2}x^2 - 10x + 3$ **15.** $y = \frac{1}{10}x^5 - 2x^2 + Ax + B$ **17.** $y = Ae^{4x} + Be^{-4x}$
19. $y = \frac{-2e^2}{(1 - e^2)^2} e^x + \frac{2e^2}{(1 - e^2)^2} e^{-x}$ **21.** 1,452   **23.** $k = .0283792013$; \$100

Sample Test

1. $y = 3x^4 - 2x^3 + 5x + C$
3. $y = \frac{7}{9}x^2 - 2x + C$
5. $8y^2 = x^2 + 6x + C$
7. $y = Ce^{5x^2}$
9. $3x^5 - 5y^3 - 5 = 0$
11. $y^8 = e^{x^2+y^2-2}$
13. $y = \frac{1}{20}x^5 - 4x^2 + C_1 x + C_2$
15. $y = Ae^{x/8} + Be^{2x}$
17. $y = \frac{4e^2}{e^5 - 1}e^{3x} - \frac{4e^2}{e^5 - 1}e^{-2x}$
19. $k \approx .047947121$; it will take about 29 months.

Cumulative Review

1. Answers vary. The limit of a function $f(x)$, as x approaches a, is the number L if the value of $f(x)$ approaches L as x gets closer to a. The values of x get closer and closer (but never reach) a, the values of $f(x)$ get closer and closer (and may or may not) reach L.
3. Answers vary. Integration is the inverse process of differentiation; it is antidifferentiation. $\int f(x)\,dx = F(x)$ means that $F'(x) = f(x)$. That is, when looking for the integral of $f(x)$ we are seeking a function, $F(x)$, whose derivative is $f(x)$. The definite integral is used to find areas, as well as to sum or total functional values. Finding areas under a curve can be used in calculating consumer and producer surplus, as well as certain probabilities if we are given a probability density function.
5. $-5x^{-2} + \frac{3}{2}x^{-1/2}$
7. $\frac{-11}{1 - 11x}$
9. $\int_0^2 4x\,dx = \frac{4x^2}{2}\Big|_0^2 = 8$
11. $\int_0^1 \int_{e^x}^e \frac{1}{\ln y}\,dy\,dx = \int_1^e \int_0^{\ln y} \frac{1}{\ln y}\,dx\,dy$

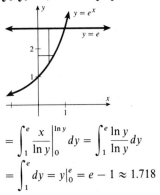

$$= \int_1^e \frac{x}{\ln y}\Big|_0^{\ln y}\,dy = \int_1^e \frac{\ln y}{\ln y}\,dy$$

$$= \int_1^e dy = y\Big|_1^e = e - 1 \approx 1.718$$

13. a. 64 units b. $8,320
15. The maximum is 36 (thousand) when the price is $2.00.
17. 9

19.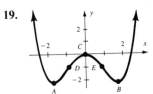

APPENDIXES

Appendix A: Review of Algebra

A.1 Algebra Pretest

Part I
1. E   3. A   5. C   7. D   9. D   11. C   13. B   15. D   17. F   19. F   21. F
23. F   25. T   27. T   29. T

Part II
1. B   3. C   5. C   7. A   9. B   11. D   13. C   15. A   17. D

A.2 Real Numbers

1. a. I, Q, R b. Q', R
3. a. N, W, I, Q, R b. N, W, I, Q, R
5. a. Q, R b. Q, R
7. a. Q, R b. Q', R
9. a. Undefined.   b. W, I, Q, R
11. Since all prime numbers are natural numbers P is drawn inside N.

13. All even numbers are whole numbers but all are not natural numbers (0 is even but not natural); so E is drawn inside W but not entirely inside N.   **15.** False   **17.** False   **19.** True   **21.** True   **23.** True   **25.** False
27. True   **29.** True   **31.** True   **33.** False   **35.** False   **37.** False   **39.** True   **41.** 9   **43.** 19
45. $-\pi$ **47.** $\pi - 2$ **49.** $10 - \pi$ **51.** $\sqrt{20} - 4$ **53.** $8 - \sqrt{50}$ **55.** $7 - 2\pi$ **57.** $|x + 3| = -(x + 3)$
59. $|y - 5| = y - 5$ **61.** Therefore, $|4 + 3t| = -(4 + 3t)$ **63.** 10   **65.** 11   **67.** $\pi - 2$ **69.** $2 - \sqrt{3}$

A.3 Algebraic Expressions
1. 13   **3.** 11   **5.** 36   **7.** -25 **9.** -49 **11.** 33   **13.** 6   **15.** 9   **17.** 4   **19.** 1   **21.** 0
23. 42   **25.** -71 **27.** -1 **29. a.** $x^2 + 4x - 12$ **b.** $x^2 + x - 20$ **31. a.** $x^2 - 8x + 15$ **b.** $x^2 - x - 12$
33. a. $2y^2 - y - 1$ **b.** $2y^2 - 5y + 3$ **35. a.** $6y^2 + 5y - 6$ **b.** $6y^2 + 13y + 6$ **37. a.** $x^2 - y^2$ **b.** $a^2 - b^2$
39. a. $x^2 + 4x + 4$ **b.** $x^2 - 4x + 4$ **41. a.** $a^2 + 2ab + b^2$ **b.** $a^2 - 2ab + b^2$ **43.** $3x - 4z$
45. $4x - 2y + z$ **47.** $-2y$ **49.** $2x + 2y$ **51.** $-x + 4y - 9$ **53.** $4x^2 - 8x - 1$ **55.** $-3x^2 + 3x - 9$
57. $2x^2 - x - 3$ **59.** $3x^2 - 4x - 11$ **61.** $x^3 + 3x^2 + 3x + 1$ **63.** $x^3 - x^2 - 4x - 2$ **65.** $3x^2 - 10x + 1$
67. $15x^3 - 22x^2 + 5x + 2$

A.4 Factoring
1. $4x(5y - 3)$ **3.** $2(3x - 1)$ **5.** $x(y + z^2 + 3)$ **7.** $a^2 + b^2$ does not factor.   **9.** $(m - n)^2$
11. $x^{-3}(1 + x^2 + x^5)$ **13.** $(4x - 1)(x + 3)$ **15.** $(2x - 3)(x + 5)$ **17.** $(3x + 1)(x - 2)$ **19.** $2x^{-3}(x - 8)(x + 3)$
21. $x^{-2}(2x - 3)(x - 1)$ **23.** $x^{1/3}(x^2 + x + 1)$ **25.** $(x - y - 1)(x - y + 1)$ **27.** $x^{-1/2}(x + 3)^2$
29. $(a + b - x - y)(a + b + x + y)$ **31.** $(2x - 3)(x + 2)$ **33.** $(6x - 1)(x + 8)$ **35.** $(6x + 1)(x + 8)$
37. $(4x - 3)(x + 4)$ **39.** $(9x - 2)(x - 6)$ **41.** $-(x - 1)(x + 1)(x - 3)(x + 3)$
43. $\frac{1}{36}(2x - 1)(2x + 1)(3x - 1)(3x + 1)$ **45.** $(x^2 - 3x - 8)(x - 4)(x + 1)$ **47.** $(2x + 2y + a + b)(x + y - 3a - 3b)$
49. $18x^2(x - 5)^2(2x - 5)$ **51.** $5x(3x + 1)^2(15x + 2)$ **53.** $2x(4x + 3)^2(10x + 3)$ **55.** $(x + 1)^2(x - 2)^3(7x - 2)$
57. $6(2x - 1)^2(3x + 2)(5x + 1)$ **59.** $2(x + 5)^2(x^2 - 2)^2(5x^2 + 15x - 4)$

A.5 Linear Equations and Inequalities
1. $x = 2$ **3.** $x = -5$ **5.** $x = -3$ **7.** $x = -3$ **9.** $x = -2$ **11.** $x = 2$ **13.** $x = 3$ **15.** $x = 2$
17. $x \le 3$ or $(-\infty, 3]$ **19.** $x < -5$ or $(-\infty, -5)$ **21.** $x \ge -4$ or $[-4, \infty)$ **23.** $x > -3$ or $(-3, \infty)$
25. $5 < x < 9$ or $(5, 9)$ **27.** $-1 < x < 4$ or $(-1, 4)$ **29.** $-7 < x < -4$ or $(-7, -4)$ **31.** $y' = \dfrac{x^4}{y^4}$
33. $y' = \dfrac{-2y}{x}$ **35.** $y' = \dfrac{2x - y}{x - 2y}$ **37.** $y' = \dfrac{-3x^2 - 4x - y}{x}$ **39.** $y' = \dfrac{-4(x + 1)}{25(y - 1)}$ **41.** $y' = \dfrac{3y^2 - 6xy^3 - 5y}{9x^2y^2 - 6xy + 5x}$
43. $x = 3$ **45.** $x < -1$ or $(-\infty, -1)$ **47.** $x \le 10$ or $(-\infty, 10]$ **49.** $x < -2$ **51.** $x \le 1$ or $(-\infty, 1]$
53. $x = 1$ **55.** No solution   **57.** $x = 0$ **59.** Any real number is a solution.

A.6 Quadratic Equations and Inequalities
1. $x = -5$ or $x = 3$ **3.** $x = -9$ or $x = 2$ **5.** $x = -\frac{1}{2}$ or $x = \frac{4}{5}$ **7.** $x = 1$ or $x = -6$ **9.** $x = 5$
11. $x = -\frac{2}{3}$ or $x = \frac{1}{4}$ **13.** No real solutions   **15.** $x = \pm\frac{\sqrt{5}}{2}$ **17.** $x = 0$ or $x = \frac{7}{3}$ **19.** $x = -\frac{1}{3}$ or $x = 2$
21. $[2, 6]$ **23.** $(-3, 0)$ **25.** $(-\infty, -2] \cup [8, \infty)$ **27.** $(-\infty, \frac{1}{3}) \cup (4, \infty)$ **29.** $[-\infty, -3] \cup [3, \infty)$
31. There is no real solution.   **33.** $(-\infty, -2) \cup (3, \infty)$ **35.** $[2, 3]$ **37.** $[-5, 1]$ **39.** $(-\infty, -\frac{1}{6}) \cup (10, \infty)$

A.7 Rational Expressions
1. $\dfrac{2 + 3x + 3y}{x + y}$ **3.** $\dfrac{x - y}{2}$ **5.** $\dfrac{3x + 2}{x + 2}$ **7.** $\dfrac{11}{2(x + y)}$ **9.** 1   **11.** $3x^2 - 17x - 6$ **13.** $\dfrac{x^2 + 2x + 3}{x^2}$
15. $\dfrac{3}{x - 1}$ **17.** $\dfrac{3 + 5x - x^2}{x^2}$ **19.** $\dfrac{x^2 + 2xy + y^2}{xy} = \dfrac{(x + y)^2}{xy}$ **21.** $\dfrac{x^2 - 2xy + y^2}{xy} = \dfrac{(x - y)^2}{xy}$
23. $\dfrac{x^2 + 3x^4y^3 + 3y}{3x^3y^2}$ **25.** $\dfrac{-2}{x + 7}$ **27.** $\dfrac{1 - x^2}{x^2 + 1} = \dfrac{(1 - x)(1 + x)}{x^2 + 1}$ **29.** $\dfrac{4(x - 1)}{(x + 2)(x - 2)}$

A.8 Review

Objective Questions

1.

Number	Natural	Whole	Integer	Set Rational	Irrational	Real
-8	No	No	Yes	Yes	No	Yes
$\frac{5}{6}$	No	No	No	Yes	No	Yes
$.2\overline{3}$	No	No	No	Yes	No	Yes
$\sqrt{169}$	Yes	Yes	Yes	Yes	Yes	Yes
$\frac{3}{0}$	No	No	No	No	No	No

3. -5 **5.** $4-x$ **7.** $2\pi + 4$ **9. a.** -64 **b.** 64 **c.** -64 **d.** -1 **11.** 3 **13.** -1 **15. a.** -1 **b.** 1 **17. a.** $x^2 + 5x - 14$ **b.** $x^2 - 9$ **19. a.** $x^2 + xy - 2y^2$ **b.** $x^2 - 9y^2$ **21.** $3x - 4z$ **23.** $-2x^2 - 7$ **25.** $(x-2)(x-3)$ **27.** $(5-x)(5+x)$ **29.** $(3x+2)(3x-2)(x+2)(x-2)$ **31.** $x=20$ **33.** $x=6$ **35.** $(7, \infty)$ **37.** $(-3, \infty)$ **39.** $[-13, -8]$ **41.** $(-10, -6)$ **43.** $x = \frac{1}{3}, x = -\frac{2}{5}$ **45.** $x = -\frac{5}{3}, -\frac{3}{2}$ **47.** $3 \pm \sqrt{2}$ **49.** $(5, 7)$ **51.** $(-\infty, -\frac{2}{3}) \cup (\frac{5}{2}, \infty)$ **53.** $\dfrac{8x - 5y}{60}$ **55.** $\dfrac{6x + 1}{(3x-1)^2}$

Sample Test

1.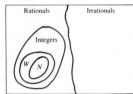

3. $5 - \sqrt{10}$ **5.** -306 **7.** $x^2 + 2xy + y^2$ **9.** $(6x+1)(x-5)$ **11.** $x=6$ **13.** $(-\infty, -\frac{4}{5})$ **15.** $x=0$ or $x=\frac{1}{5}$ **17.** $(\frac{2}{3}, 1)$ **19.** $\dfrac{14x - 5y}{60}$

Index

A posteriori model, 306
A priori model, 306
Abscissa, 16
Absolute inequality, 419
Absolute maximum, 153, 175
  procedure for finding, 176
  second-derivative test, 179
Absolute minimum, 153, 175
  procedure for finding, 176
  second-derivative test, 179
Absolute value, 403
Absolute value function, 25
Abstraction, 5
Acceleration, 53
Addition
  polynomials, 407
  property of equations, 418
  property of inequalities, 420
  property of rational expressions, 430
Algebra
  pretest, 400
  review, 399
Algebraic function, 192
American Paper Institute, 266
Amount of an annuity, 285
Annual compounding, 195
Annuity, 285
Answers, 436
Antiderivative, 224
Antidifferentiation, 225
Approximations for the definite integral, 267
Area
  between curve and *x*-axis, 247
  between curves, 244, 249
    definition, 250
    intersecting curves, 250
  function, 244
  probability distribution, 308
  rectangle, 244
  trapezoid, 245, 270
  triangle, 245
  under a curve, 245
Arkansas River, 40
Asymptote
  graphing, 169
  horizontal, 45–46, 65
  procedure for finding, 46
  slant, 172
  vertical, 42, 46
Auxiliary equation, 393
Average
  change, 12
  expense, 292
  marginal analysis, 135

Average (*continued*)
  rate, 77
  rate of change, 79, 82
  speed, 78
  total cost, 135
  total profit, 135
  total revenue, 135
  value of a function, 263
Axis of abscissas, 16
Axis of ordinates, 16

Bacterial growth, 383, 388
Base, 192, 406
Battle of Trafalgar, 375
Bernoulli, 125
Best-fitting line, 364
Bifolium, 124
Binomial, 407
Binomial multiplication, 410
Bonaparte, Napoleon, 375
Boundary condition, 377
Boyer, Carl, 69
Brace, 402
Break-even analysis, 39
Brief Integral Table, inside front and back covers
British Embassy, 321
Brown, Joseph, 191
Business growth, 388
Business models, 131, 282

C-curve, 388
Calculus
  definition, 4
  differential, 224
  fundamental theorem, 254, 262
  integral, 224
Cape Trafalgar, 375
Capital
  expansion, 381
  formation, 292
  value, 304
Carbon-14 dating, 211, 220, 383, 389
Cartesian coordinate system, 16
Cartesian plane, 16
Catalina Island, 389
Chain rule, 106
  composite functions, 108
Change, rate of, 77, 82
Change of base theorem, 207
Charlotte, NC 230
Chemical Rubber Company's *Standard Mathematical Tables*, 297
Chernobyl nuclear disaster, 389
Chicago, IL 278

465

INDEX

Circle, 124
Circular cylinder, 332
Classic Coke, 351
Cobb–Douglas production function, 325, 363
Coefficients, fractional, 413
Common factor, 412
Common logarithm, 201
Comparison, property of, 404
Comparison of compounding period, 196
  formula, 195
Comparison symbols, 403
Completely factored, 413
Completing the square, 35
Complex number, 402
Component, 16
Composite, 405
Composite function, 13, 108
Compound inequality, 421
Compound interest, 195
  formula, 195
Compounding period, 195–196
Computers in mathematics, 223
Concave downward, 34, 160–161, 169
  second-derivative test, 165
Concave upward, 34, 160–161, 169
  second-derivattive test, 165
Conditional equation, 418
Conditional inequality, 419
Constant, 406
  function, 21
  of integration, 224
  rate, 77
  rule, 96, 99
  rule, differential form, 129
  times a function, 96
  times a function rule, 99
Constraint, 346
Consumers' surplus, 287, 289
Continuity, 68
  at a point, 69–70
  polynomial, 72
  theorems, 72
Continuous compounding, 196
Continuous income stream, 284
Continuous mathematics, 54
Continuous random variable, 307
Continuous stream, 284
Contradiction, 419
Converge, integral, 303
Coordinate axes, 16
Coordinate of a point, 16, 402
Coordinate plane, 16, 328
Coordinate system, 16, 402
  one-dimensional, 402
  three-dimensional, 327
  two-dimensional, 16
Correlation, 363
  coefficient, definition, 365
  coefficient, table, 366
Cost, 131, 135
Cost–benefit model, 46
Cost function, 37
Counting number, 402

Critical point, functions of several variables, 340
Critical value, 153
  correlation, 366
  definition, 154
  derivative tests, 155
  inequalities, 426
Cumulative distribution, 314
Cumulative Review, 398
  Chapters 2–4, 189
  Chapters 5–7, 322
Curve
  area under, 245
  falling, 161
  rising, 161
  sketching, 169
Cylinder, right circular, 332

Daily compounding, 196
Death Valley, CA, 77
Debert site, 389
Debt, public, 395
Decay
  function, 194
  graph of, 388
Decreasing curve, 169
Decreasing function, 150
Dedekind, J. W. R., 69
Definite integral, 224, 262
  approximations, 267
  definition, 238
  evaluating, 238
  properties of, 239
  substitution method, 240
Degree, 407
Deleted point, 44
Δx, 22
Δy, 22
Demand curve, 388
Demand function, 37
Density function, 309
Department of Health and Human Services, 389
Dependent variable, 29
  function of two or more variables, 326
Depreciation, 291, 388
Derivative, 78
  definition, 88
  differential approximation, 128
  exponential, 214
  formulas, summary, 227
  higher-order, 159
  higher-order partial, 335
  logarithm, 212
  notation for, 94
  partial, 333
  second, 159
  test, first, 154
  test, second, 164–165
DERIVE, 223
Descartes, René, 16
Deterministic model, 5
Difference rule, 97, 99
Difference of squares, 412, 414

Differentiable functions, 88
  functions that are not, 90
Differential, 125
  approximation, 128
  calculus, 224
  equation
    first-order, 376
    second-order, 390
    theorem four, 393
    theorem one, 376
    theorem three, 380
    theorem two, 379
  formulas, 129
  of x, 125
  of y, 125
Differentiation (*See also* Derivative)
  business models, 131
  explicit, 120
  implicit, 120, 122
  techniques, 94, 100
Direct integration formulas, 227
Discontinuity, points of, 72
Discontinuous function, 72
Discrete mathematics, 54
Discrete random variable, 307
Discriminant, 425
Disease spread, 388
Distance, 404
Distribution
  exponential, 312
  frequency, 307
  normal, 313
  normal, cumulative table, 314
  probability, 307
  uniform, 310
Distributive multiplication, 410
Distributive property, 409
Diverge, integral, 303
Division
  property of equations, 418
  property of inequalities, 420
  rational expressions, 430
Divorce rate, 389
Domain, 8
  agreement in this book, 9
  function, 16
  function of two or more variables, 326
  variable, 406
Domar's capital expansion model, 381
Double integral, 353, 355
  definition, 354
  over a variable region, 356
Dummy variable, 238

e, 197
Edge, amusement ride, 84
Elasticity of demand, 137
Element, 402
Ellipsoid, 331
Elliptic cone, 331
Elliptic paraboloid, 331
Empirical probability, 306
Empty set, 402
Environmental Protection Agency (EPA), 47
Equal, rational expressions, 430

Equal-to symbol, 403
Equation
  conditional, 418
  contradiction, 419
  equivalent, 418
  exponential, 199
  linear, 418
  open, 418
  properties of, 418
  quadratic, 423
Equilibrium point, 48, 287
Equilibrium price, 287
Equivalent equations, 418
Erlanger, J., 281
Evaluate
  definite integral, 238
  expression, 408
  function, 10
  indefinite integral, 228
  logarithm, 200
Even number, 405
Event, 306
Everready Long-Life battery, 318
Exact interest, 196
Experiment, 306
Explicit differentiation, 120
Exponent, 192, 199, 406
  definition, 406
  laws of, 192, 410
  squeeze theorem, 193
Exponential
  derivative, 214
  distribution, 312
  equation, 199, 206
  growth, 388
  property of equality, 200
  rule, 227
Exponential function
  definition, 192
  graphing techniques, 216
  inverse of logarithmic, 202
Expression, 406
Extremum, 153 (*See also* Maximum; Minimum)

Factor, 406, 412
Factoring, 412
  over the set of integers, 413
  trinomial, 415
  types of, 412
Ferndale, CA, 319, 386
Finite mathematics, 54
Finite set, 402
First component, 16
First-derivative test, 154, 169
First octant, 329
First-order differential equation, 376
Fixed cost, 37
FOIL, 411
  factoring, 415
  multiplication, 411
Folium of Descartes, 124
Form 1040, 20
Fractional coefficients, 413
French Embassy, 321

French Revolution, 375
Frequency
  distribution, 307
  relative, 306
Function, 3
  algebraic, 192
  area, 244
  composite, 13, 108
  decay, 194
  decreasing, 150
  defined by an algebraic formula, 9
  defined by a graph, 9
  defined by a table, 8
  defined by a verbal rule, 8
  definition, 8
  domain (agreement in this book), 9
  equal derivatives, 226
  exponential, 192
  growth, 194
  increasing, 150
  linear, 21
  logarithmic, 199
  n independent variables, 326
  notation for, 11
  polynomial, 39
  quadratic, 32
  rational, 18, 41
  tabular, 258
  transcendental, 192
  two independent variables, 326
Function behavior principle, 154
Functional notation, 11
Fundamental property of rational expressions, 430
Fundamental Theorem of Calculus, 254, 262
Future value, 195
  formula, 195

Galileo, 69
Gasser, H. S., 281
Geitz, Robert, 325
General situation, related rates, 138, 141
General solution, differential equation, 377
Generalized power rule, 110
Gini index, 243–244
Gottlieb, Sheldon F., 281
Grade on a curve, 316
Graph
  first derivative, 150
  function, 16, 18
  number, 339, 402
Graphing a line, procedure, 24
Graphing strategy, 169
Great America Theme Park, 84
Greater than, 403
Greatest common factor, 413
Gross National Product (GNP), 86, 389
Growth models, 381
  function, 194
  human population, 208, 383
  inhibited growth, 382
  limited growth, 382
  summary, 388
  uninhibited growth, 382
Gutenberg, 203

Half-life, 220
Higher-order derivative, 159
Higher-order partial derivative, 335
Histogram, 307
Hoover Dam, 230
Horelick, Brindell 281
Horizontal asymptote, 45–46, 65, 169
Horizontal change, 22
Horizontal line, 21, 30
Houston, TX, 322
Hullian model, 390
Human population growth, 208, 383
Hyperbolic paraboloid, 331
Hyperboloid of one sheet, 331
Hyperboloid of two sheets, 331

IBM, 14, 76
Identity, 418
Implicit differentiation, 120, 122
Impossible event, 306
Improper integral, 303
Increasing curve, 169
Increasing function, 150
Indefinite integral, 225
Independent variable, 29
Indeterminate form, 62
Inequality, 419
  absolute, 419
  compound, 421
  conditional, 419
  polynomial, 427
  properties of, 420
  quadratic, 426
  symbols, 403
Infinite discrete random variable, 307
Infinite set, 402
Infinity, limit at, 63–64
Inflation, 388
Inflection point, 160, 176
  growth models, 387
Information diffusion, 388
Inhibited growth, 382, 384
  graph, 388
  model, 385
  summary, 388
Initial condition, 377
Initial value, 228
Instantaneous rate of change, 82
Instantaneous speed, 78
Integer, 402, 405
Integrable, 262
Integral
  calculus, 224
  constant, 227
  converge, 303
  definite, 224, 238
  difference, 227
  diverge, 303
  double, 354
  improper, 303
  indefinite, 225
  iterated, 354
  sum, 227
  symbol, 225
Integrand, 225, 238, 354

Integration
  constant of, 224
  double, 353
  exponential rule, 227
  formulas, inside front cover
  formulas, summary, 227
  limits of, 238
  logarithmic rule, 227
  numerical, 266
  by parts, 293
  power rule, 227
  procedural formulas, 227
  region of, 354
  reversing the order, 359
  substitution, 230
  substitution, procedure, 232
  by table, 297
  variable limits of, 355
Intercept, 19, 169
Interest, 195
  compound, 195
  compounded continuously, 196, 383, 388
  compounding periods, 195–196
  exact, 196
  ordinary, 196
  simple, 195
Interval notation, 419
Intravenous infusion, 388
Inverse functions, 202
Investment flow, 292
Irrational number, 402, 405
Irreducible polynomial, 414
Iterated integral, 354

J-shaped curve, 388

Knievel, Evel, 41
Koont, Sinan, 281

L-shaped curve, 388
Lagrange, Joseph, 346
Lagrange multiplier, 346
Lambda, 346
Lanchester, Frederick, 397
Law of cooling, 388
Learning curve, 388
Least squares line, 368
Least squares method, 364
Leibniz, Gottfried, 5, 69
Lemniscate of Bernoulli, 125
Less than, 403
Limit, 54
  constant, 58, 62
  difference, 62
  of f as x approaches c, 56
  infinity, 63–64
  intuitive notion, 54
  left-hand, 55
  notation, 56
  polynomial, 60, 62
  power, 62
  product, 62
  quotient, 61–62
  right-hand, 55
  root, 62

Limit (*continued*)
  sum, 62
Limited growth, 382, 386
  model, 387
  summary, 388
Limits of integration, 238
  variable, 355
Lindenmeier site, 211, 383
Lindstrom, Peter A., 53
Line, 17
  best-fitting, 364
  least squares, 368
  parallel, 26
  perpendicular, 26
Linear correlation coefficient, 364
Linear equation, 21, 418
  summary of forms, 30
Linear function, 21
Log of both sides theorem, 204
Logarithm
  change of base, 207
  common, 201
  definition, 199
  derivative, 212
  exponential, 206
  evaluate, 200
  laws of, 205
  natural, 201
Logarithmic equation, 204
Logarithmic function
  definition, 199
  graphing, 201
  inverse of exponential, 202
Logarithmic rule, 227
Logistic growth, 382
  graph, 388
Long-term population growth, 388
Loss, 39

Mahan, A. T., 397
MAPLE, 223
Marginal, 131
  analysis, 131
  analysis, average, 135
  cost, 92, 131
  cost, average, 135
  definition, 92
  profit, 83, 131
  profit, average, 135
  revenue, 131
  revenue, average, 135
Marginal willingness to spend, 287
Marriage rate, 389
Marshall, Alfred, 132
MATHEMATICA, 223
Mathematical model, 5
Maximum
  absolute, 153, 175
  method of Lagrange multipliers, 346
  profit, 180
  relative, 153, 169
  relative, for functions of several variables 339
  second-derivative test, 179
  value, 36

Mazda 626, 53
Mean, 313
Member, 402
Minimum
  absolute, 153, 175
  method of Lagrange multipliers, 346
  relative, 153, 169
  relative, for functions of several variables, 339
  second-derivative test, 179
  value, 36
Modeling, 5
  a posteriori, 306
  a priori, 306
  applications, 3, 53, 119, 191, 223, 325, 375, 381
  business, 282
  growth, 381
  mathematical, 5
  probability, 305
Modeling Applications
  Battle of Trafalgar, 375
  Cobb–Douglas production function, 325
  computers in mathematics, 223
  gaining a competitive edge, 3
  health care pricing, 149
  instantaneous acceleration, 53
  nervous system, 281
  publishing, 119
  world running records, 191
Money flow, 284–285
Monomial, 407
Monthly compounding, 196
Multiple integral, 352
Multiplication
  polynomials, 409
  property of equations, 418
  property of inequalities, 420
  rational expressions, 430
Multivariate function, 326
MUMATH, 223

Napoleon, 375
Nash, David H., 397
Natural logarithm, 201
Natural number, 402, 405
Natural resources, 388
Negative correlation, 365
Negative number, 402
Nervous system, model for, 281
Net excess profit, 286
Net investment flow, 292
New Haven, CT, 230
Newton, Isaac, 5
Newton's law of cooling, 388
Nievergelt, Yves, 129
Norm, 259
Normal curve, 312
  standard, 313
  z-score, 317
Normal distribution, 312–313
  cumulative table, 314
Normally distributed data, 313
No slope, 23, 29
Not-equal-to symbol, 403

INDEX

Null set, 402
Number
  comparison of types, 405
  complex, 402
  composite, 405
  counting, 402
  even, 405
  integer, 402
  irrational, 402
  line, 402
  negative, 402
  odd, 405
  positive, 402
  prime, 405
  rational, 402
  real, 402
  relationship among types of, 405
  whole, 402
Numerical coefficient, 406
Numerical integration, 266

Objective function, 346
Objectives
  Appendix A, 432
  Chapter 1, 49
  Chapter 2, 114
  Chapter 3, 144
  Chapter 4, 187
  Chapter 5, 219
  Chapter 6, 275
  Chapter 7, 319
  Chapter 8, 371
  Chapter 9, 395
Octant, 329
Odd number, 405
Oil consumption, 236, 243
  demand, 279
One-dimensional coordinate system, 402
Open equation, 418
Order of operations, 407
Ordered pair, 16
Ordered triplet, 327
Ordinary interest, 196
Ordinate, 16
Origin, 16, 402
Overhaul time, 291

Parabola, 17, 32, 34–35
Parabolic approximation, 272
Parallel lines, 26
Partial derivative, 333
  definition, 334
  geometric interpretation, 334
  higher-order, 335
Particular solution, differential equation, 377
Partition, regular, 261
Pearson correlation coefficient, 366
Perfect square, 35, 416
Perpendicular lines, 26
Petroleum
  demand, 279
  prices, 236, 243
  world use, 211
pH, 203
Piecewise linear function, 25

Plane, 329
  Cartesian, 16
  coordinate, 16
  xy, 16, 328
Plot a point, 16, 169
Point of inflection, 160, 176
  growth models, 387
Point–slope form, 27, 30
Poiseuille's law, 142, 338
Polynomial, 407
  addition, 407
  classified, 407
  function, 39
  general form, 407
  inequality, 427
  irreducible, 414
  multiplication, 409
  subtraction, 407
Population growth, 208
  long-term, 388
  short-term, 388
  U.S., 389
Positive correlation, 365
Positive number, 402
Power, 406
Power rule, 95, 99, 227
  differential form, 129
  generalized, 110
Present value, 194
Pretest, algebra, 400
Price, 388
  elasticity of demand, 137
Prime, 405
Principal, 194
Probabilistic model, 5, 305
Probability
  definition, 308
  density function, 309
  distribution, 307
  empirical, 306
  event, 306
  exponential distribution, 312
  normal cumulative table, 314
  normal distribution, 313
  relative frequency, 306
  subjective, 306
  theoretical, 306
  uniform distribution, 310
Procedural formulas, 227
Producers' surplus, 287, 289
Product rule, 100
  differential form, 129
Production function, 325
Profit, 39, 131, 135
  function, 37
  marginal, 83
  maximum, 180
Public debt, 395
Publishing, 119

Quadrant, 16
Quadratic equation, 423
  solution by factoring, 424
  solution by formula, 425
Quadratic formula, 425

Quadratic function, 32
Quadratic inequalities, 426
Quadric surface, 330
  summary, 331
Quarterly compounding, 195
Quotient rule, 103
  differential form, 129

Radioactive decay, 383, 388
Random variable, 307
Range
  function, 8, 16
  function of two or more variables, 326
Rate
  average, 77, 79
  of change, 77, 82
    directly proportional to population, 383
    directly proportional to quantity, 380
    proportional to both population and remaining room for growth, 385
    proportional to remaining room for growth, 387
  constant, 77
  of flow, 284
  related, 138
Rational expression
  definition, 428
  fundamental property, 430
  properties of, 430
  simplify, 429
Rational function, 18, 41
Rational number, 402, 405
Rationalize, 63
Real number, 402
  negative, 403
  positive, 402
  relationship of subsets, 405
Rectangular approximation, 257, 267
Region of integration, 354
Regression analysis, 364
Regression line, 369
Related rates, 138
  procedure for solving, 141
Relative frequency, 306
Relative maximum, 153, 169
  first-derivative test, 155
  functions of several variables, 339
    procedure for finding, 342
  second-derivative test, 165
Relative minimum, 153, 169
  first-derivative test, 155
  functions of several variables, 339
    procedure for finding, 342
  second-derivative test, 165
Revenue, 38, 131, 135
Revenue function, 37
Reversing the order of integration, 359
Reviews
  Appendix A, 432
  Chapter 1, 48
  Chapter 2, 114
  Chapter 3, 144
  Chapter 4, 186
  Chapter 5, 219
  Chapter 6, 275

Reviews (*continued*)
  Chapter 7, 319
  Chapter 8, 371
  Chapter 9, 395
  Cumulative, 398
    Chapters 2–4, 189
    Chapters 5–7, 322
Richter scale, 203
Riemann sum, 259
Right circular cylinder, 332
Rise, 22
Roller coaster, 80
Root, 418
Royal Gorge, 40
Run, 22

S-shaped curve, 388
Saddle point, 341
  procedure for finding, 343
Sales
  fads, 388
  new products, 388
Sample space, 306
San Antonio, TX, 209, 236
Santa Clara, CA, 84
Santa Rosa, CA, 229
SAT scores, 86
Satisfy an equation, 16, 418
Savick, Wayne, 282
Scatter diagram, 364
Schedule X, 20
Schedule Y-1, 32
Secant line, 78
Second component, 16
Second derivative, 159
Second-derivative test, 164–165, 169
  absolute maximum and minimum, 179
  functions of two variables, 342
Second-order differential equation, 390
Semiannual compounding, 195
Semicubical parabola, 124
Set, 402
  empty, 402
  null, 402
  numbers, 402
  solution, 418
Set-builder notation, 10
Short-term population growth, 388
Significance level, 366
Similar terms, 408
Simple event, 306
Simple interest, 195
Simpson's rule, 272
Slant asymptote, 172
Slope of a line, 22
Slope–intercept form, 23, 30
Snake River, 41
Solution set, 418
Solve an equation, 418
Specific situation, related rates, 138, 141
Speed, 78
Sphere, 331
Square, perfect, 35, 416

Square root
  definition, 424
  property, 424
Squeeze theorem for exponents, 193
Standard deviation, 313
Standard form, 28, 30
Standard normal cumulative distribution table, 314
Standard normal curve, 313
Standard position parabola, 32
Steepness of a line, 22
Stopping distance of an automobile, 41
Straight-line depreciation, 291
String of inequalities, 421
Subjective probability, 306
Substitution method for integration, 230
  definite integrals, 241
Subtraction
  polynomials, 407
  property of equations, 418
  property of inequalities, 420
  rational expressions, 430
Sum rule, 97, 99
  differential form, 129
Summation notation, 365
Supply curve, 388
Surface, 328
  quadric, 330–331
Suspicious points, 73
Symbolic manipulation, 223
Symmetric, 34
Symmetry, 169

Tabular function, 258
Tandy, 14
Tangent line, 81, 89
Tax rate schedule, 20, 32
Term, 406
Tests, Sample
  Appendix A, 433
  Chapter 1, 50
  Chapter 2, 116
  Chapter 3, 145
  Chapter 4, 188
  Chapter 5, 220
  Chapter 6, 278
  Chapter 7, 321
  Chapter 8, 373
  Chapter 9, 396
  Cumulative, 398
  Chapters 2–4, 189
  Chapters 5–7, 322
Theoretical probability, 306
Third derivative, 159
Three dimensions, 326
Three-dimensional coordinate system, 327
Total
  cost, 131
  cost function, 37
  income, 284
  money flow, 285
  profit, 131
  profit function, 37
  revenue, 131

Total (*continued*)
  revenue function, 37
  value, 283
Trace, 330
Transcendental function, 192
Trapezoid, 270
Trapezoidal approximation, 270
Trend line, 29
Trichotomy property, 404
Trinomial, 407
  factoring, 415
Triola, Mario F., 367
Two-dimensional coordinate system, 16
Two-leaved rose, 125

Ullmann, Steven, 149
Unbounded function, 42
Uniform distribution, 310
Uninhibited growth, 382
  graph, 388
  model, 383
  summary, 388
United Nations, 305
Unlimited growth, 382
U.S. Postal Service, 345
U.S. Public Health Service, 15–16

Value of a function, 8
Variable, 406
Variable limits of integration, 355
Velocity, 84
Vertex, 34
Vertical asymptote, 42, 46, 169
Vertical change, 22
Vertical line, 21, 30
  test, 18, 21
Volume, 360

Walton, W. U., 275
Washington, DC, 321
Weierstrass, Karl, 69
Whole number, 402, 405
World running records, 191

x-axis, 16
x-intercept, 169
  defined, 19
  procedure for finding, 21
Xerox, 49
xy-plane, 16, 328
xz-plane, 328

y-axis, 16
y-intercept, 169
  defined, 19
  procedure for finding, 21
yz-plane, 328

z-score, 317
Zero
  property of, 423
  slope, 23, 29
Zero of a function, 19

TO THE OWNER OF THIS BOOK:

We hope that you have found *Calculus with Applications*, 2nd Edition, useful. So that this book can be improved in a future edition, would you take the time to complete this sheet and return it? Thank you.

School and address: _____

_____

Department: _____

Instructor's name: _____

1. What I like most about this book is: _____

_____

2. What I like least about this book is: _____

_____

3. My general reaction to this book is: _____

_____

4. The name of the course in which I used this book is: _____

_____

5. Were all of the chapters of the book assigned for you to read?   Yes   No

   If not, which ones weren't? _____

6. What specific suggestions do you have for improving this book?

_____

_____

_____

_____

_____

_____

Optional:

Your name: _____ Date: _____

May Brooks/Cole quote you, either in promotion for *Calculus with Applications*, 2nd Edition, or in future publishing ventures?

Yes: _____ No: _____

Sincerely,
Karl J. Smith

FOLD HERE

NO POSTAGE
NECESSARY
IF MAILED
IN THE
UNITED STATES

BUSINESS REPLY MAIL
FIRST CLASS   PERMIT NO. 358   PACIFIC GROVE, CA

POSTAGE WILL BE PAID BY ADDRESSEE

ATT: *Karl J. Smith*

**Brooks/Cole Publishing Company
511 Forest Lodge Road
Pacific Grove, California 93950-9968**

FOLD HERE

991779

515
Smi

DATE DUE ON LINE 9/9	
JAN 0 3 2002	
AUG 1 6 2004	
GAYLORD	PRINTED IN U.S.A.

New Providence Memorial Library
377 Elkwood Avenue
New Providence, NJ 07974

Brief Integral Table (continued)

Forms Containing $\sqrt{x^2 \pm a^2}$, $\sqrt{a^2 - x^2}$

15. $\displaystyle\int \sqrt{x^2 \pm a^2}\, dx = \frac{x}{2}\sqrt{x^2 \pm a^2} \pm \frac{a^2}{2}\ln|x + \sqrt{x^2 \pm a^2}| + C$

16. $\displaystyle\int \frac{dx}{\sqrt{x^2 \pm a^2}} = \ln|x + \sqrt{x^2 \pm a^2}| + C$

17. $\displaystyle\int x\sqrt{x^2 \pm a^2}\, dx = \frac{1}{3}\sqrt{(x^2 \pm a^2)^3} + C$

18. $\displaystyle\int x\sqrt{a^2 - x^2}\, dx = -\frac{1}{3}\sqrt{(a^2 - x^2)^3} + C$

19. $\displaystyle\int \frac{dx}{x\sqrt{x^2 + a^2}} = -\frac{1}{a}\ln\left|\frac{a + \sqrt{x^2 + a^2}}{x}\right| + C$

20. $\displaystyle\int \frac{dx}{x\sqrt{a^2 - x^2}} = -\frac{1}{a}\ln\left|\frac{a + \sqrt{a^2 - x^2}}{x}\right| + C$

21. $\displaystyle\int \frac{\sqrt{x^2 + a^2}\, dx}{x} = \sqrt{x^2 + a^2} - a\ln\left|\frac{a + \sqrt{x^2 + a^2}}{x}\right| + C$

22. $\displaystyle\int \frac{\sqrt{a^2 - x^2}\, dx}{x} = \sqrt{a^2 - x^2} - a\ln\left|\frac{a + \sqrt{a^2 - x^2}}{x}\right| + C$

23. $\displaystyle\int \frac{x\, dx}{\sqrt{x^2 \pm a^2}} = \sqrt{x^2 \pm a^2} + C$

24. $\displaystyle\int \frac{x\, dx}{\sqrt{a^2 - x^2}} = -\sqrt{a^2 - x^2} + C$

25. $\displaystyle\int x^2\sqrt{x^2 \pm a^2}\, dx = \frac{x}{4}\sqrt{(x^2 \pm a^2)^3} \mp \frac{a^2 x}{8}\sqrt{x^2 \pm a^2} - \frac{a^4}{8}\ln|x + \sqrt{x^2 \pm a^2}| + C$

26. $\displaystyle\int \frac{dx}{x^2\sqrt{x^2 \pm a^2}} = \mp \frac{\sqrt{x^2 \pm a^2}}{a^2 x} + C$

27. $\displaystyle\int \frac{dx}{x^2\sqrt{a^2 - x^2}} = -\frac{\sqrt{a^2 - x^2}}{a^2 x} + C$

28. $\displaystyle\int \frac{\sqrt{x^2 \pm a^2}}{x^2}\, dx = -\frac{\sqrt{x^2 \pm a^2}}{x} + \ln|x + \sqrt{x^2 \pm a^2}| + C$

29. $\displaystyle\int \frac{x^2\, dx}{\sqrt{x^2 \pm a^2}} = \frac{x}{2}\sqrt{x^2 \pm a^2} \mp \frac{a^2}{2}\ln|x + \sqrt{x^2 \pm a^2}| + C$